Anthropological Genetics

Anthropological genetics is a field that has been in existence since the 1960s and has been growing within medical schools and academic departments, such as anthropology and human biology, ever since. With the recent developments in DNA and computer technologies, the field of anthropological genetics has been redefined. This volume deals with the molecular revolution and how DNA markers can provide insight into the processes of evolution, the mapping of genes for complex phenotypes and the reconstruction of the human diaspora. In addition to this, there are explanations of the technological developments and how they affect the fields of forensic anthropology and population studies, alongside the methods of field investigations and their contribution to anthropological genetics. This book brings together leading figures from the field to provide an up-to-date introduction to anthropological genetics, aimed at advanced undergraduates to professionals, in genetics, biology, medicine and anthropology.

Michael H. Crawford is Professor of Anthropology and Genetics at the University of Kansas, USA.

Anthropological Genetics
Theory, Methods and Applications[1]

Michael H. Crawford
University of Kansas

[1]Under the sponsorship of the American Association of Anthropological Genetics

CAMBRIDGE UNIVERSITY PRESS
Cambridge, New York, Melbourne, Madrid, Cape Town,
Singapore, São Paulo

Cambridge University Press
The Edinburgh Building, Cambridge CB2 2RU, UK

Published in the United States of America by
Cambridge University Press, New York

www.cambridge.org
Information on this title: www.cambridge.org/9780521838092

First published 2007

Printed in the United Kingdom at the University Press, Cambridge

A Catalog record for this publication is available from the British Library

Library of Congress Cataloging-in-Publication data

Anthropological genetics : theory, methods and applications / edited
by Michael H. Crawford.
     p. ; cm.
  Includes bibliographical references.
  ISBN-13: 978-0-521-54697-3 (pbk.)
  ISBN-10: 0-521-54697-4 (pbk.)
  ISBN-13: 978-0-521-83809-2 (hardback)
  ISBN-10: 0-521-83809-6 (hardback)
1. Human population genetics. 2. Human evolution. 3. Human genetics.
  I. Crawford, Michael H., 1939- II. Title.

  GN289.A68 2006

  576.5′8–dc22

                                        2006018837

ISBN-13 978-0-521-83809-2 hardback
ISBN-10 0-521-83809-6 hardback

ISBN-13 978-0-521-54697-3 paperback
ISBN-10 0-521-54697-4 paperback

Contents

Contributors

Laura Almasy, Department of Genetics, Southwest Foundation for Biomedical Research, San Antonio, TX 78245-0549

Barbara Arredi, Istituto di Medicina Legale, Universita Cattolica del Sacro Cuore di Roma, Rome, Italy

Guido Barbujani, Department of Genetics, University of Ferrara, Ferrara, Italy

John Blangero, Department of Genetics, Southwest Foundation for Biomedical Research, San Antonio, TX 78245-0549

Michael H. Crawford, Laboratory of Biological Anthropology, University of Kansas, Lawrence, KS 66045

Eric J. Devor, Molecular Genetics and Bioinformatics, Integrated DNA Technologies, 1710 Commercial Park, Coralville, Iowa 52241

Alan G. Fix, Department of Anthropology, University of California, Riverside, CA 92521

Mary Katherine Gonder, Department of Biology, University of Maryland, College Park, MD 20742

Henry C. Harpending, Department of Anthropology, University of Utah, Salt Lake City, UT 84112

Joseph H. Lee, Department of Epidemiology, Gertrude H. Sergievsky Centre, Taub Institute, Columbia University, New York, NY 10032

Lorena Madrigal, Department of Anthropology, University of South Florida, Tampa, FL 33620

Elizabeth Matisoo-Smith, Department of Anthropology, University of Auckland, Auckland, New Zealand

Phillip Melton, Laboratory of Biological Anthropology, University of Kansas, Lawrence, KS 66045

James H. Mielke, Department of Anthropology, 622 Fraser Hall, University of Kansas, Lawrence, KS 66045

Dennis H. O'Rourke, Department of Anthropology, University of Utah, Salt Lake City, UT 84112

Estella S. Poloni, Department of Anthropology, University of Geneva, Geneva, Switzerland

John H. Relethford, Department of Anthropology, State University of New York College at Oneonta, Oneonta, NY 13820

Rohina C. Rubicz, Laboratory of Biological Anthropology, University of Kansas, Lawrence, KS 66045

Francisco M. Salzano, Departamento de Genetica, Instituto de Biociencias, Universidade Federal do Rio Grande do Sul, Caixa Postal 15053, 91501-970 Porto Allegre, RS, Brazil

Moses Schanfield, Department of Forensic Sciences, 2036 H Street, NW, 103 Samson Hall, George Washington University, Washington DC 20052

Joseph D. Terwilliger, Department of Psychiatry, Columbia University, New York, NY 10032

Sarah A. Tishkoff, Department of Biology, University of Maryland, College Park, MD 20742

Chris Tyler-Smith, The Wellcome Trust Sanger Institute, Hinxton, United Kingdom

Jeff T. Williams, Department of Genetics, Southwest Foundation for Biomedical Research, San Antonio, TX 78245-0549

Sarah Williams-Blangero, Department of Genetics, Southwest Foundation for Biomedical Research, San Antonio, TX 78245-0549

Preface

The primary purposes of this volume are: (1) to define the field of anthropological genetics in lieu of the recent developments in molecular genetics; (2) to update the materials presented in earlier volumes on anthropological genetics. This update is essential because the recent technological developments have provided tools that can now be utilized to answer controversial evolutionary and historically based questions. The earlier four volumes dealing with anthropological genetics (Crawford and Workman, 1973; Mielke and Crawford, 1980; Crawford and Mielke, 1982; Crawford, 1984) preceded the molecular genetics revolution and focused primarily on population structure and the beginnings of genetic epidemiology. A glimpse of some of the technological and methodological innovations and their applications to anthropological genetics was contained in the special issue of *Human Biology*, devoted to the arrival of the new millennium and the future of the field (Crawford, 2000).

I was approached by the Executive Committee of the American Association of Anthropological Genetics (AAAG) and asked to develop a volume that would define the field of anthropological genetics and could be used as a textbook for advanced undergraduate students and graduate students. After several meetings with committees to discuss the contents of such a volume, I proposed that the book be divided into an Introduction, Conclusion plus four topical categories: Theory, Methods, General Applications, and The Human Diaspora. Initially, 19 sets of authors agreed to write chapters covering what I considered the most significant topics in anthropological genetics. The Introduction was to trace the historical roots of the field and set the stage for what was to follow. The Theory section was to include the following chapters: (1) Theory of Evolution and how its measurement in humans was altered by the molecular genetic revolution; (2) Decomposition of genetic variation and its implications about race and its classification; (3) The use of genetic isolates in the measurement of gametic disequilibrium and gene mapping. The section on Methods would include: (1) field research methodology; (2) use of historical demography; (3) molecular markers; (4) use of quantitative genetic traits in the reconstruction of population structure; and (5) ancient DNA methodology. The section on General Applications would include: (1) use of DNA for forensic purposes; (2) emerging technology and its uses; (3) application of linkage analysis and molecular markers in mapping QTLs; (4) use of DNA in behavioural genetics of primates. The fourth section would focus on the Human Diaspora and would include the following chapters: (1) out of Africa; (2) peopling of Europe; (3) peopling of Oceania; (4) colonization of the Americas; and (5) peopling of Asia.

This volume was to have been completed by a combination Overview and Concluding chapter.

A few months before the deadline for the completion of the volume, I ran into the author of the Evolution chapter at a professional meeting and he assured me that he was making great progress and that it would be completed on time. However, as the deadline approached and passed, the chapter failed to appear. Similarly, another author had promised the completion of a chapter on molecular markers, but informed me one month prior to the final deadline that he could not complete this critical chapter. Given the loss of these two essential and informative chapters (Forces of Evolution and Molecular Markers) plus a uniform lack of enthusiasm by my colleagues in writing these chapters at such short notice, I decided to 'bite the bullet' and do it myself. Fortunately, two of my advanced graduate students and research assistants (Rohina Rubicz and Phillip Melton) were willing to set aside most of their duties and help research and write a chapter within a two-month timeframe. We decided to combine the chapter on molecular markers with the forces of evolution, resulting in Chapter 6, which covers the availability of molecular markers, the analytical methods and the changes in the measurement of the forces of evolution based on the recent technological breakthroughs.

As the final deadline approached, two unexpected lacunae developed. I had contracted an overview on the molecular, archaeological and linguistic evidence for the peopling of Asia. Unfortunately, the author who had agreed to prepare this chapter apparently 'dropped off the face of the earth' and would not answer emails, telephone calls or letters. As a result, this volume does not document the human diaspora into Asia. However, there is some discussion of genetic variation in Siberia and theories concerning movements of humans out of Asia into the Americas and Oceania. I was informed one month after the final deadline that the chapter on primate molecular genetics and its implications to the understanding of behaviour would not be submitted. After the dust settled, the volume consisted of 16 chapters with Asia and Behavioural Genetics being omitted.

This is the last volume on the developments in anthropological genetics that I plan to edit. Hopefully additional updates on the field will periodically be generated by the next generation of anthropological geneticists.

Michael H. Crawford

Chapter 1

Foundations of Anthropological Genetics

Michael H. Crawford

Department of Anthropology University of Kansas, Lawrence, KS 66045

What is anthropological genetics?

Anthropological genetics is a synthetic discipline that applies the methods and theories of genetics to evolutionary questions posed by anthropologists. These anthropological questions concern the processes of human evolution, the human diaspora out of Africa, the resulting patterns of human variation, and bio-cultural involvement in complex diseases. How does anthropological genetics differ from its kin discipline, human genetics? Both fields examine various aspects of human genetics but from different perspectives. With the synthetic volume of 1973 (Methods and Theories of Anthropological Genetics), it became evident that the questions posed by the practitioners of anthropological genetics and human genetics tended to be somewhat different. I compared and contrasted these two fields in the introduction to the special issue of *Human Biology* (2000) on Anthropological Genetics in the twenty-first century (see Table 1.1). What distinguishes anthropological genetics from human genetics is its emphasis on smaller, reproductively isolated, non-Western populations, plus a broader, biocultural perspective on evolution and on complex disease etiology and transmission. Judging from the contents of the *American Journal of Human Genetics* (premiere journal in the field of human genetics) there is a greater emphasis on the causes and processes associated with disease, and the examination of these processes in affected phenotypes (probands) and their families. Anthropological geneticists tend to focus more on normal variation in non-Western reproductively isolated human populations (Crawford, 2000). Anthropological geneticists also attempt to measure environmental influences through co-variates of quantitative phenotypes, while human geneticists less often attempt to quantify the environment in order to assess the impact of environmental-genetic interactions.

Table 1.1. Differences between human genetics and anthropological genetics (Crawford, 2000).

Anthropological genetics	Human genetics
1. Broader biocultural perspective on genetic/environmental interactions	1. Mechanisms and processes — particularly in disease
2. Population focus, pedigrees utilized to measure familial resemblance	2. Families of proband, twins and twin families
3. Small, reproductively isolated populations — often, non-Western	3. Larger, urban, and clinical samples
4. Culturally homogeneous populations	4. Populations may be heterogeneous by race, socio-economic factors, occupation, and lifestyle
5. Sampling representative of normal variation in population	5. Sampling based on clinical ascertainment
6. Attempts made to characterize and measure the environment	6. Environmental variation rarely assessed. It is assumed that $e^2 = 1 - h^2$
7. Study of normal variation in complex traits	7. Dichotomy of disease vs. normality — usually observed

History of anthropological genetics

The ancestral roots of the field of anthropological genetics are intimately intertwined with the developments in evolutionary biology, population genetics, and biological anthropology. O'Rourke (2003) correctly noted that this modern amalgamated discipline was further cross-fertilized by molecular biology and bioinformatics. Through cross-fertilization this hybrid field has acquired the analytical and laboratory tools to dissect the molecular and genetic bases of human variation, a traditional focus within biological anthropology. The addition of genome scanning and linkage analyses have contributed to the fluorescence of genetic epidemiology and the mapping of genes involved in complex phenotypes, particularly those associated with chronic diseases.

Anthropological genetics of the late 1960s and early 1970s was preceded by almost a century of discovery and development in evolutionary theory and genetics. Many of the ideas associated with natural selection can be traced to the publication of Charles Darwin's *Origin of Species* in 1859 (see Table 1.2). Because Darwin was unaware of Gregor Mendel's experiments on the particulate nature of genes (using characteristics of pea plants) Darwin lacked specific mechanisms for generating new variation and had to settle for a blending form of inheritance. Darwin also used Lamarck's concept of the inheritance of acquired characteristics, a concept that persisted well into the twentieth century.

Table 1.2.	Time-line of significant developments in genetics and anthropological genetics.*
1859	Publication of Darwin's *Origin of Species*
1860	Meischer first isolated DNA
1880	Weismann demonstrated the separation of the germ plasm from somatic cells
1900	Rediscovery of Mendel's laws of inheritance
1901	Documentation of first polymorphism in humans, ABO blood group system, by K. Landsteiner
1902	Garrod demonstrated that the mode of inheritance of inborn errors of metabolism were Mendelian in nature
1908	Formulation of the principle of genetic equilibrium, generally attributed to Hardy and Weinberg, but preceded by Castle in 1903
1919	Population variability in the frequency of blood group genes demonstrated in World War I by Hirschfeld and Hirschfeld
	Fisher integrated Darwin's theory of natural selection with Mendel's formulations
1930-2	Fisher, Haldane and Wright publish the mathematical basis of modern population genetic theory
1937	Dobzhansky published *Genetics and the Origin of Species* and further fleshed-out the modern synthesis by reconciling the evidence of the naturalists with the geneticists
1944	DNA is shown to be heritable material
1949	Molecular basis for sickle cell disease demonstrated by J. V. Neel (1957), Pauling *et al.* (1949), and H. A. Itano and L. Pauling (1961)
1953	Watson and Crick break the genetic code
1954	Allison reveals relationship between sickle-cell trait and malaria
1956	Human chromosomal numbers correctly characterized by J. HinTjio and A. Levan (1956)
1955	Smithies (1955, 1959) develops starch-gel electrophoresis, a method for separating protein variation based on charge and size of molecules
	Y-chromosome shown to determine the sex of organisms
1972	Lewontin apportioned human genetic diversity and demonstrated the 85% is within populations
1973	Publication of the first major synthesis of anthropological genetics, Crawford and Workman
1977	DNA sequencing methods described
1978	Restriction Fragment Length Polymorphism (RFLP) first described
1981	Human mtDNA genome sequenced
1984	Methods of DNA fingerprinting first described by Jeffries
1985	Development of Polymerase Chain Reaction (PCR) methods
1987	Development of laser based fluorescent detection of DNA
1988	Beginning of the Human Genome Project
1991	Human Genome Diversity Project Proposed
1997	First Neandertal mtDNA sequence
1998	Completion of sequencing of the first human chromosome (Ch. 22)
2001	Draft of human genome sequence

*Time-line was based in part on Jobling, Hurles and Tyler-Smith (2004).

Despite the brilliant research of August Weismann, who demonstrated the separation of germ plasm from the soma, Lamarckian concepts were adopted in Stalinist Soviet Union because they better fitted the ideology. Taken to extremes, there was a belief that changes in the phenotype affect the genotype, which is then transmitted to the next generation. However, later geneticists demonstrated that the alternation of the phenotype does not get inherited by subsequent generations because of the separation of the sex cells from other somatic cells. The concept of mutation initially arose from Hugo DeVries' research on primroses. He concluded that most mutations had drastic effects and that speciation was driven by mutations. However, the creative research of Thomas Hunt Morgan on the fruit fly demonstrated that mutations introduced variation in populations at incremental levels but rarely resulted in the formation of new species.

Measures of human variation

Blood groups

At the turn of the twentieth century, Karl Landsteiner's (1901) immunological characterization of the ABO blood group system and its mode of inheritance provided a genetic marker for the measure of human variation. Ludwik Hirschfeld and Hanka Hirschfeld (1919), during World War I, demonstrated that military personnel of various so-called 'racial groups' or ethnicities differed in the frequencies of the ABO blood groups. In the few decades that followed, additional blood group systems, such as the Rhesus, MNS, Duffy and Diego, were shown to vary in human populations. These blood group systems were polymorphic and differed in allelic frequencies in human regional populations. Yet, the function of these complex gycolipid (sugar/fat) or glycoprotein (sugar/protein) molecules expressed on the surface of human erythrocytes were unknown until relatively recently. For example, the Rh (Rhesus blood group system), discovered by Landsteiner and Weiner in 1941, came to medical attention because of its importance in pregnancy and the risk of maternal/fetal incompatibility. It only became evident in the year 1997 that the evolutionary history of the Rhesus blood group system was of great antiquity and that the function of the RH glycoproteins is the transportation of ammonium ions ($NH4+$) across the cell membrane (Marini and Urrestarazu, 1997; Marini et al., 2000). My chapter, The Use of Genetic Markers of the Blood in the Study of the Evolution of Human Populations, contains a summary of the known genetic variation in human blood group systems (Crawford, 1973).

Protein variation

In the mid 1950s, Orville Smithies developed a method (starch gel electrophoresis) for separating protein mixtures on the basis of both the electric charge and the size of the molecules (Smithies, 1955).

Thus, the degree of genetic variation in the serum proteins in populations could be ascertained electrophoretically (Smithies and Connell, 1959; Smithies, 1959). This methodological innovation provided a glimpse into the genetic variation contained within the human genome, using primary gene products such as serum and red blood cell proteins and enzymes (extracted from the red blood cells into hemolysates). Refinements in electrophoretic methods, from filter paper electrophoresis (which separated proteins on the basis of molecular charge) to starch gel electrophoresis (separation based on both the charge and the size of the molecule), to isoelectric focusing (IEF, a method of electrophoresis performed on a gel containing a pH gradient), were suggestive of a human genome consisting of approximately 100,000 genes. With the sequencing of the human genome, this estimate was later down-sized to approximately 30,000 genes.

Modern evolutionary synthesis

Most fields of inquiry are fortunate to have one, or maximally two, highly innovative 'founders', such as a Charles Darwin or an Albert Einstein. However, in addition to Charles Darwin, evolutionary theory was developed by three contemporaneous major figures, namely: Sewall Wright, J.B.S. Haldane, and R.A. Fisher (Table 1.2 contains a time-line of the significant genetic breakthroughs). They set the mathematical foundations for the modern synthetic evolutionary theory and provided the formal underpinnings for the measurement of natural selection and statistical methods for estimating the effects of stochastic processes. Other scientists, such as Thomas Hunt Morgan and Ernest Muller, using animal (fruit fly) models provided an understanding of mutation, the source of new genetic variation – which had eluded Charles Darwin. In an essay celebrating his 100th birthdate, the last survivor from that period of the development of the evolutionary synthesis, the eminent German evolutionary biologist, Ernst Mayr, recently reminisced about that era of evolutionary theory development (Mayr, 2004).

The next generation of population geneticists included the distinguished Russian émigré geneticist, Theodosius Dobzhansky, the great Chinese agronomist, C. C. Li, and US-born human geneticist, James Crow. They collectively added further refinements and detail to the synthetic theory of evolution. Although Dobzhansky's 'animals of choice' were the beetle and the lowly fruit fly (*Drosophila melanogaster*), he applied the principles of evolution learned from these models to humans and synthesized the available information on human evolution in a readable form. Similarly, C. C. Li synthesized much of the theory of population genetics in his concisely written primers, which assisted in the training of subsequent generations of evolutionists. James Crow coalesced demographic characteristics with population genetics by developing a method for assessing the opportunity of natural selection in human populations, based on fertility and mortality components derived from church records and civil documents. Together with his former student, Arthur Mange, Crow also developed methods for estimating levels of inbreeding in human populations using marital records and the likelihood of individuals with the same surname marrying (isonymy). These methods were applied and further elaborated by anthropological geneticists, such as Gabriel W. Lasker (a Harvard Ph.D., trained, in part, by Ernst Hooton). James Spuhler, another student of Hooton's, was greatly influenced by Sewall Wright and applied some of the path methods for the computation of inbreeding coefficients to Ramah Navajo populations (Spuhler and Kluckhohn, 1953). Derek F. Roberts, an Oxford-trained biological anthropologist, applied some of Wright's formulations to an island population in the south Atlantic, Tristan da Cunha, and demonstrated the importance of unique historical events and founder effects on the population of this small, remote island (Roberts, 1969). He also described the high incidence of forms of congenital deafness and mental retardation in the Tristan population (1969) and more recently showed the reduction in genetic variation as assessed by mtDNA (Soodyall et al., 1997).

In the late 1950s and early 1960s, with the publication of Sewall Wright's insights into the actions of stochastic processes, physicians and medical geneticists discovered the usefulness of small, genetically isolated populations for the understanding of rare genetic diseases and anomalies. Recessive mutations (normally of low incidence in large populations) may appear at high frequencies in some of these small populations because of the founder effect and chance segregation. Victor McKusick, of Johns Hopkins University, spearheaded the study of rare genetic anomalies in Pennsylvania Amish populations. The value of this approach was further demonstrated by the discovery of rare, familial genetic conditions, such as Christmas hemophilia, forms of dwarfism, and adenylate kinase deficiency in Amish kindred. Physicians/geneticists such as Victor McKusick (1964), Arno G. Motulsky (1965) and James V. Neel (1957) integrated biochemical genetic methodologies with evolutionary theory to elucidate human adaptation to diseases such as malaria. L. L. Cavalli-Sforza, another medically trained geneticist, examined allelic frequency fluctuations due to stochastic processes in small villages of Parma, northern Italy. Recently, together with colleagues Moroni and Zei, Cavalli-Sforza expanded this research into a tome on consanguinity, inbreeding, and genetic drift in Italy (Cavalli-Sforza et al., 2004). During the 1960s, Motulsky followed up his biochemical genetic interests in metabolic diseases to conduct fieldwork in populations of the Congo (Motulsky 1960; Motulsky et al., 1966). Similarly, J. V. Neel, together with Brazilian geneticist, Francisco Salzano, mounted a highly successful research programme on the genetics of tribal populations of South America (1964). Thus, in the 1950s and 1960s, the margins between the fields of anthropological genetics and human genetics were somewhat blurred, with geneticists and physicians conducting anthropological research and anthropological geneticists working in the realm of human genetics.

Early anthropological genetics

Anthropologists with training in genetics were useful to the medical profession in studies of small, highly isolated, non-Western populations. Unfortunately, until the 1950s, there were few anthropologists with adequate training in human genetics. The reason behind this paucity was that most physical anthropologists were traditionally trained in morphology and racial classification based on typology. However, several of Albert Hooton's last group of doctoral students at Harvard, namely Gabriel Lasker, Frederick Hulse and James Spuhler, had some training and interest in genetics. Lasker was influenced by the writings of Sewall Wright and applied these ideas to his field investigations with Mexican and Peruvian populations (Lasker, 1954, 1960). Hulse examined linguistic barriers to gene flow and blood group variation in Native American populations of northwestern United States (Hulse, 1955, 1957). In addition, he measured the effects of heterosis and exogamy in small-sized, Alpine Swiss communities (Hulse, 1957). James Spuhler collaborated with the cultural anthropologist, Clyde Kluckhohn, in applying Sewall Wright's pathway methods and measured the level of inbreeding among the Ramah Navajo (Spuhler and Kluckhohn, 1953).

Frank Livingstone, a former student of Neel and Spuhler at Michigan, conducted a study on the effects of culture (i.e. the introduction of slash-and-burn agriculture into sub-Saharan Africa) on the distribution of falciparum malaria. He demonstrated in his classic dissertation and subsequent publications that the destruction of the tropical rain forest resulted in the creation of standing bodies of water, a prerequisite for the successful breeding conditions of the Anopheles mosquito (Livingstone, 1958). The increased parasitization caused a shift from epidemic to endemic malarial infection and the action of natural selection against various phases of the life cycle of *Plasmodium falciparum*. Livingstone and Neel also trained a number of anthropological geneticists at Michigan, e.g. Kenneth Weiss, Alan Fix and the late Richard Ward — all went on to distinguished careers in anthropological genetics.

Several graduate anthropology students of W. W. Howells and Albert Damon at Harvard applied population genetic principles to anthropological populations. Eugene Giles tested theory of genetic drift on field populations of New Guinea. He sought to document that gene frequency fluctuations were due to genetic drift in small, isolated villages (Giles *et al.*, 1966). Jonathan Friedlander, a graduate student of Damon's, conducted anthropological genetic investigations in the Solomon Islands (Friedlander, 1971).

Richard Lewontin (a population geneticist) statistically partitioned genetic variation within populations and between populations on the basis of 15 protein loci (Lewontin, 1967). He demonstrated that

85% of human genetic diversity is within populations. Thus, a much smaller percentage, 15%, is between populations. This research has been used to discourage genetic comparisons between so-called geographical 'races' because most of the variation is contained within the populations. Barbujani (1997) retested Lewontin's findings based on DNA markers and confirmed that 84.4% of the variation was within populations, 4.7% among samples, within groups, and 10.8% among groups (see Chapter 2, Madrigal and Barbujani). However, a controversial analysis of single nucleotide repeat (SNP) diversity (Seielstad *et al.*, 1998) indicated that while autosomal and mtDNA SNPs provide a pattern similar to that observed by Lewontin and Barbujani (within populations 85.5% and 81.4% of the variation is subsumed), Y-chromosomal SNPs apportion almost 53% of the variation between continental populations.

Foundations of anthropological genetics

In 1988, when I assumed the editorship of the journal *Human Biology*, I inherited few manuscripts of publishable quality. Kenneth Weiss (a member of the editorial board) suggested that in celebration of the 60th anniversary of the journal I should consider publishing an issue of the journal containing the 'best' anthropological genetics articles that had graced the pages of *Human Biology* during its history. This special issue would, on one hand, provide the needed manuscripts for publication plus, on the other hand, connect the past with the new focus of the journal. I titled this special issue 'Foundations of Anthropological Genetics'. Gabriel Lasker and I selected the 'top-ten' most significant articles and had most of the original authors update their thoughts on the topic (Crawford and Lasker, 1989). Two of these classic articles were written almost 50 years ago, thus necessitating the preparation of the updates by willing specialists, namely David Hay and Robert Sokal, rather than the original authors. This special issue does establish connections between the early research in human genetics and the developments in anthropological genetics. The ten articles selected included distinguished authors such as J. B. S. Haldane, James Spuhler, D. F. Roberts, James Crow, J. V. Neel, Frank B. Livingstone, A. G. Motulsky, Morris Goodman, and P. T. Wilson. Their research and publications established a solid base, or foundation, for the field of anthropological genetics. While only Spuhler, Livingstone and Roberts were considered biological anthropologists, the field of anthropological genetics was built on the research and formulations of many disciplines and theoretical approaches to human evolution.

The synthesis

In 1970, due in part to prompting by my colleague at the University of Pittsburgh, the late C. C. Li, I organized a symposium on methods and theories of anthropological genetics at the School of American Research in Santa Fe, New Mexico. This symposium had a blend

Fig 1.1 The participants in the School of American Research symposium on Anthropological Genetics, held in 1970, in Santa Fe, New Mexico. Back row (standing, from the left): Steven Vandenberg, Jean Benoist, Frank Livingstone, Gabriel W. Lasker, Peter L. Workman, Eugene Giles, Christy Turner III, Francis Johnston, and James Spuhler. Front row (seated): Michael H. Crawford, Derek F. Roberts, and William W. Howells.

of senior, established scholars, such as James Spuhler, William W. Howells, Gabriel W. Lasker, Frank Livingstone, Steven G. Vandenberg, and Derek F. Roberts. In addition, a number of younger researchers were invited: Eugene Giles, Peter L. Workman, Jean Benoist, Christy G. Turner, III, and Francis Johnston. Figure 1.1 identifies the original participants in the 1970 symposium in Santa Fe, New Mexico. During this symposium, the participants instructed Peter Workman and me to serve as the editors of the volume that was to be compiled and published in the series established by Douglas Schwartz with the University of New Mexico Press. Because we were limited to a small number of participants at the symposium, Workman and I decided to solicit a few additional chapters to fill the obvious lacunae. We added two contributions by human geneticists and six by anthropologists of a 'genetic persuasion'. The genetic additions included Newton E. Morton (who had elaborated on Malecott's bioassay approaches to the study of population structure and applied them to populations of Micronesia) and Jean W. McCluer (who at that time was applying computer simulation methods to the demographic structures of South American Native populations). The anthropological geneticists added to the mix included Henry Harpending (R-matrix analysis of South African populations), Russell Reid (synthesis of theory on inbreeding), and Solomon Katz (fearless prognosticator of the evolutionary future of humans), the late Richard H. Ward (genetic structure of Amazonian populations), Nancy Howell (the feasibility of characterizing the demographic structures of small populations), and Kenneth Morgan (historical demography of a Navajo community). Initially, Workman and I debated whether to name the volume *Methods and Theories of Anthropological Genetics* or *Methods and Theories of Genetic Anthropology*. We eventually agreed on the use of Anthropological Genetics because it connoted a commingling of anthropology and genetics, yet this title suggested a unique anthropological approach to the field of genetics. This volume was published in 1973 and it comprised the first attempt by multiple authors of synthesizing this field. I later discovered that D. F. Roberts had preceded us in referring to anthropological genetics in a lecture that he had given to the Royal Anthropological Society (Roberts, 1965).

From 1973 to the 1980s, there was considerable research activity in anthropological genetics and related fields. The most significant developments were in the applications of quantitative genetic methodologies to complex phenotypes, particularly in chronic diseases. Developments in computer technology and programming facilitated the use of linkage methods, path analytical approaches of Sewall Wright, and segregation analyses to complex phenotypes. These methodologies provided information as to the mode of transmission of a complex phenotype and the chromosomal mapping (through linkage analysis) of Mendelian traits. These new developments, punctuated by a pronouncement from Newton Morton that all of the major questions in population structure have been answered and we should instead refocus on genetic epidemiology, prompted me to consider an update of the 1973 volume. After I was awarded a National Institutes of Health Career Development Award in 1976 a portion of my university salary was released by the administration of the University of Kansas. This award freed funds for a lecture series by distinguished speakers, each coming to Lawrence for one week, providing a public lecture, interacting with faculty and graduate students, and presenting one seminar to the graduate students and faculty. It was at this time that James H. Mielke (a former student of Peter Workman) was added to the faculty at Kansas and he joined me in developing this lecture programme. This collaboration resulted in the first volume of a three-volume series, *Current Developments in Anthropological Genetics: Theory and Methods*, published in 1980 by Plenum Press. Volume 2 focused on the effects of ecology on population structure and was released in 1982 (Crawford and Mielke, 1982). That volume contained a number of innovative approaches to population structure, including Robert Sokal's initial application of spatial autocorrelation to human populations of the Solomon Islands. In 1984, the final volume in the series was published. It was based on my research in Belize, Guatemala and St Vincent Island and was used as a case study applying the theories of population genetics to a series of historically related populations of the Caribbean and Central America (Crawford, 1984). This volume documented an evolutionary 'success story' of the Garifuna (Black Caribs). Although no unadmixed Carib or Arawak Native Americans now remain on St Vincent Island, their genes have been dispersed over a wide geographic expanse on the coast of Central America. In 1800, with fewer than 2,000 Black Caribs transported to Honduras, the population rapidly grew to more than 100,000 descendants and colonized much of the coast of Belize, Guatemala, Honduras and Nicaragua.

Molecular revolution

Since the publication of the last of the three volumes of the series in 1984, the field of anthropological genetics has been swept along by the molecular revolution (see Crawford, 2000). With the breaking of the genetic code by Watson and Crick in 1954 and the sequencing of the human genome in 2001, new methods and molecular markers became available for evolutionary studies. Since 1984, there has been a shift in emphasis in anthropological genetics, primarily from population structure (based on blood group and protein markers) and genetic epidemiology, to the study of human origins and the human diaspora. In addition, genetic epidemiology evolved from the initial analysis of the nature of the genetic component (based on heritability studies and segregation analyses) to the actual mapping of genes and Quantitative Trait Loci (QTLs) for complex phenotypes. These shifts in emphasis were made possible by the unique characteristics of DNA, i.e. the absence of recombination of mitochondrial DNA and a non-recombining portion of the Y-chromosome marker (NRY). These markers were more informative than polymorphic blood groups and proteins, plus they enabled the reconstruction of migration patterns from either the male or female perspectives. Mitochondrial DNA and NRY chromosome markers, because of the absence of recombination, can also be used to build chronometers based on the accumulation of mutations over a time period (see Chapter 7). The plethora of Short Tandem Repeats (STRs) and Single Nucleotide Polymorphisms (SNPs) distributed throughout the genome, provides anonymous markers that can be linked to complex phenotypes on specific chromosomes. In this way, genes and QTLs could be mapped and their gene products identified.

New questions can now be posed and answered using molecular genetic markers. For example, the question concerning whether Neandertals evolved into modern *Homo sapiens* or whether they were replaced by modern humans has apparently been answered using mtDNA sequences (see Chapter 13 by Arredi *et al.*). Comparisons of mismatch distributions between humans and six Neandertal remains show discontinuity. In addition, the Neandertal specimen from Germany does not resemble more closely the mtDNA of contemporary Germans (as predicted under the multi-regional model) than do Africans or Asians (Krings *et al.*, 1997). The final bit of evidence in favour of the replacement hypothesis was that there is DNA discontinuity between the Neandertals and the three 24,000-year-old anatomically modern Europeans (Caramelli *et al.*, 2003). The multi-regionalists have countered these findings by noting that there is continuity in some morphological traits suggestive of common origins. Certainly, Neandertals shared common ancestry with humans at least 500,000 years ago. In addition, morphological traits are highly plastic and tend to be sculpted by environmental factors. The pioneering research by Alphonse Riesenfeld demonstrated that the morphology of rat crania and long bones can be modified by functional and endocrine changes (Riesenfeld, 1974). More recently, the research of Susan Herring has demonstrated that physical factors, such as weight, pressure and other mechanical forces, shape of the crania of experimental animals such as pigs. Bone growth is influenced by a large number of environmental factors moulding the skull around the expanding brain and is further influenced by the pressures of muscles associated with chewing. Measures of heritability of human cranial features, revealed, in Mennonite populations, that there is a large environmental component as measured by familial resemblance (Devor *et al.*, 1986).

This volume

This volume is divided into four sections: Theory, Methods, Applications and the Human Diaspora. In addition, these sections are introduced by this chapter, on the 'Foundations of Anthropological Genetics', and concluded by an overview, 'Anthropological Genetics: Present and Future', written by Henry Harpending. This volume has an international as well as interdisciplinary flavour with contributions coming from anthropological geneticists, human geneticists, and population geneticists from Brazil, Costa Rica, Italy, Korea, New Zealand, United Kingdom, and United States. In addition, these scholars come from a variety of institutions: the private sector, universities, research institutes, and medical centres. Two chapters, one on the effects of the molecular revolution on the theory associated with the forces of evolution and another on the peopling of Asia were not produced by the authors. In order to compensate for their absence, several chapters had to cover some of the topical omissions.

The section on Theory is introduced by a chapter authored by Lorena Madrigal and Guido Barbujani which partitions the genetic variation observed in contemporary human populations and considers the relationship of genetic variation to the concept of race. The second chapter of this section, by Terwilliger and Lee, provides an overview of the concept of genetic isolate and illustrates its application to studies of genetic epidemiology.

The section on Methods consists of five chapters, all of which were written by anthropological geneticists. The first chapter of this section (Chapter 4) is authored by the editor of this volume (Michael H. Crawford) and introduces the importance of field investigations to the discipline of anthropological genetics. This is followed by a collaborative chapter (Chapter 5) by James Mielke and Alan Fix,

integrating the demographic processes with anthropological genetics. Chapter 6, by Rohina Rubicz, Philip Melton and Michael Crawford,
provides a survey of the available genetic and molecular markers
and analytical methods for the study of human phylogeny and the
genetic structure of populations. The fourth chapter in this section
on methodology (Chapter 7), by John Relethford, introduces the
reader to the use of quantitative traits for the study of human
evolution. The fifth chapter of this section (Chapter 8) by Dennis
O'Rourke, focuses on the application of ancient DNA (extracted
from skeletal remains) to the reconstruction of human phylogeny
and history.

The third section of this volume, consisting of three chapters,
deals with the Applications of Anthropological Genetics to various
related fields and disciplines. The first chapter of the section
(Chapter 9), written by Moses Schanfield, focuses on the application
of anthropological genetics to forensic sciences. This discipline
has received wide publicity and notoriety with celebrated cases in
the United States, such as the OJ Simpson case, the Peterson case,
the identification of skeletal materials among the 'disappeared'
of Argentina, the victims of Kosovo, and the executed of Iraq.
Anthropological genetics and osteology helped scientists and legal
investigators disentangle the identities of thousands of victims
of brutality and genocide. The second chapter of this section
(Chapter 10), by Eric Devor, summarizes the state-of-the-art technologies coming from the private sector that are available to anthropological geneticists. In particular, this chapter focuses on the
creative use of fluorescent techniques in molecular genetics.
The final chapter of this section (Chapter 11), authored by John
Blangero, Jeff Williams, Laura Almasy and Sarah Williams-Blangero,
explores the application of anthropological genetics to the mapping of genes influencing QTLs involved complex diseases such as
diabetes, atherosclerosis, obesity, hypertension, depression, alcoholism, osteoporosis, and cancer.

The fourth section of this volume, titled the Human Diaspora,
traces the movement of humans out of Africa to Europe, Asia into the
Americas and Oceania. The first chapter of this section (Chapter 12),
written by Sarah Tishkoff and Mary Katherine Gonder, examines
the genetic variation observed in Africa and utilizes these molecular
data to reconstruct the migration dynamics of *Homo sapiens* within
the continent and to adjacent geographical regions. The second
chapter of the Human Diaspora section (Chapter 13), by Barbara
Arredi, Estella Poloni and Chris Tyler-Smith, examines the patterns
of genetic variation observed in Europe and their implications to
the peopling of that continent. Elizabeth Matisoo-Smith applies DNA
data, both from humans and rats, in reconstructing the patterns of
migration of the peoples of Oceania. This third chapter of the
section (Chapter 14) on the Human Diaspora employs highly creative
methodology of mitochondrial DNA variation in a commensal
species that accompanied humans in their extended oceanic voyages

to Polynesia. Francisco Salzano's reconstruction of the peopling of the Americas (Chapter 15), based on molecular genetic evidence, concludes this section.

The concluding chapter to this volume (Chapter 16), written by Henry Harpending, provides some fearless predictions about theories and methodologies of anthropological genetics in the future. Harpending was one of the original contributors to the first volume of Anthropological Genetics (Crawford and Workman, 1973). His chapter on R-matrix analyses of South African populations, co-authored with Trefor Jenkins, is the most widely cited chapter of that classic volume (Harpending and Jenkins, 1973).

References

Allison, A. C. (1954). Protection afforded by sickle-cell trait against subtertian malarial infection. *British Medical Journal* 1, 290–294.

Barbujani, G., Magagni, A., Minch, E., and Cavalli-Sforza, L. L. (1997). An apportionment of human DNA diversity. *Proceedings of the National Academy of Sciences USA* **94**(9), 4516–19.

Caramelli, D., Lalueza-Fox, C., Vernesi, C., Lari, M., Casoli, A., Mallegni, F., Chiarelli, B., Dupanloup, I., Bertranpetit, J., Barbujani, G., and Bertorelle, G. (2003). Evidence for a genetic discontinuity between Neandertals and 24,000-year-old anatomically modern Europeans. *Proceedings of the National Academy of Sciences USA* **100**(11), 6593–7.

Cavalli-Sforza, C. C., Moroni, A., and Zei, G. (2004). *Consanguinity, Inbreeding and Genetic Drift in Italy.* Princeton University Press.

Crawford, M. H. (1973). The use of genetic markers of the blood in the study of the evolution of human populations. In: *Methods and Theories of Anthropological Genetics* (MH Crawford, and PL Workman, eds.), University of New Mexico Press, Albuquerque, pp. 19–38.

Crawford, M. H. (2000). Anthropological Genetics in the 21st Century: Introduction. *Human Biology* **72**(1), 3–13.

Crawford, M. H. and Lasker, G. W. (eds). (1989). *Foundations of Anthropological Genetics.* In a special issue of *Human Biology* (**5–6**), 615–958.

Crawford, M. H. and Mielke, J. H. (eds). (1982). *Current Developments in Anthropological Genetics, Vol.* **2** *Ecology and Population Structure,* New York, Plenum Press.

Crawford, M. H., and Workman, P. L. (eds). (1973). *Methods and Theories of Anthropological Genetics.* School of American Research Books, Albuquerque, New Mexico, University of New Mexico Press.

Crawford, M. H. (ed.). (1984). *Current Developments in Anthropological Genetics. Vol. 3. Black Caribs, A Case Study in Biocultural Adaptation,* New York, Plenum Press.

Devor, E. J., McGue, M., Crawford, M. H., and Lin, P. M. (1986). Transmissible and non-transmissible components of anthropometric variation in the Alexanderwohl Mennonites. II. Resolution by path analysis. *American Journal of Physical Anthropology* **69**(1), 83–92.

Friedlander, J. (1971). The population structure of South-Central Bougainville. *American Journal of Physical Anthropology* **35**, 13–26.

Garrod, A. E. (1902). The incidence of alkaptonuria: A study of chemical individuality. *Lancet* **ii**, 1616–1620.

Giles, E., Ogan E., Walsh R. J., and Bradley M. A. (1966). Microevolution in New Guinea: The role of genetic drift. *Annals of the New York Academy of Science* **134**, 655–65.

Hardy, G. H. (1908). Mendelian proportions in a mixed population. *Science* **28**, 49–50.

Harpending, H., and Jenkins, T. (1973). Genetic distance among Southern African populations. In: *Methods and Theories of Anthropological Genetics* (M. H. Crawford, and P. L. Workman, eds.), University of New Mexico Press, Albuquerque, pp. 177–199.

Hin Tjio, J., and Levan, A. (1956). The chromosome number of man. *Hereditas* **42**, 1–6.

Hirschfeld, L., and Hirschfeld, H. (1919). Serological differences between the blood of different races. The results of researches on the Macedonian front. *Lancet* **197**, 675–9.

Hulse, F. S. (1955). Blood types and mating patterns among Northwest Coast Indians. *Southwest Journal of Anthropology* **11**, 93–104.

Hulse, F. S. (1957). Linguistic barriers to gene flow. The blood-groups of the Yakima, Okanagon and Swinomish Indians. *Amer. J. Phys. Anthrop.* **15**, 235–46.

Itano, H. A., and Pauling, L. (1961). Thalassemia and the abnormal human haemoglobins. *Nature* **191**, 398–9.

Jobling, M. A. Hurles, M. E. and Tyler-Smith, C. (2004) *Human Evolutionary Genetics. Origins, Peoples and Disease.* New York, Garland Science.

Krings, M., Stone, A., Schmitz, R. W., Krainitzki, H., Stoneking, M., and Paabo, S. (1997). Neandertal DNA sequences and the origin of modern humans. *Cell* **90**(1), 19–30.

Landsteiner, K. (1901). Uber Agglutinionscheinungen normalen menschlichen blutes. *Wiener klinische Wochenschrift* **14**, 1132–4.

Landsteiner, K., and Weiner, A. S. (1941). Studies on an agglutinogen (Rh) in human blood reacting with anti-Rhesus sera and with human isoantibodies. *Journal of Experimental Medicine* **74**, 309–20.

Lasker, G. W. (1954). Human evolution in contemporary communities. *Southwest Journal of Anthropology* **10**, 353–65.

Lasker, G. W. (1960). Migration, isolation and ongoing human evolution. *Human Biology* **32**, 80–88.

Lewontin, R. C. (1967). An estimate of average heterozygosity in man. *American Journal of Human Genetics* **19**(5), 681–5.

Livingstone, F. B. (1958). Anthropological implications of sickle-cell gene distribution in West Africa. *American Anthropologist* **60**, 533–562.

Mayr, E. (2004). 80 years of watching the evolutionary scenery. *Science* **305**, 46–7.

Marini, A. M., and Urrestarazu, A. (1997). The Rh (Rhesus) blood group polypeptides are related to NH 4+ transporters. *Trends in Biochemical Sciences* **22**, 460–1.

Marini, A. M., Matassi, G., Raynal, V. V., Andre, B., Cartron, J. P., and Cherif-Zahar, B. (2000). The human Rhesus-associated RhAG protein and a kidney homologue promote ammonium transport in yeast. *Nature Genetics* **26**(3), 341–4.

McKusick, V. A., Hostetler, J. A., Egeland, J. A., and Eldridge, R. (1964). The distribution of certain genes in the Old Order Amish. *Cold Spring Harbor Symposium on Quantitative Biology* **XXIX**, 137–45.

Mielke, J.H., and Crawford, M.H., (eds). (1980). *Current Developments in Anthropological Genetics. Vol. 1. Theory and Methods.* New York, Plenum Press.

Motulsky, A. G. (1960). Metabolic polymorphisms and the role of infectious diseases in human evolution. *Human Biology* **32**, 28–62.

Motulsky A.G. (1965). Theoretical and clinical problems of glucose-6-phosphate dehydrogenase deficiency. In: *Abnormal Haemoglobins in Africa* (JHP Jonxis, ed.), Blackwell Scientific Publishers, Oxford, pp. 143–96.

Motulsky, A.G., Vandepitte J, and Fraser, G.R. (1966). Population genetic studies in the Congo. I. Glucose-6-phosphate dehydrogenase deficiency, hemoglobin S, and malaria. *American Journal of Human Genetics* **18**(6), 514–37.

Neel, J.V. (1949). The inheritance of sickle cell anemia. *Science* **110**, 64–66.

Neel, J. (1957). Human hemoglobin types: their epidemiologic implications. *New England Journal of Medicine* **256**(4), 161–171.

Neel, J.V. (1957). Human hemoglobin types; their epidemiologic implications. *New England Journal of Medicne* **256**(4), 161–71.

Neel, J.V., and F.M. Salzano (1964). A prospectus for genetic studies of the American Indian. In: *Cold Spring Harbor Symposia on Quantitative Biology* **29**, 85–98.

O'Rourke, D.H. (2003). Anthropological genetics in the genomic era: A look back and ahead. *American Anthropologist* **105**(1), 101–9.

Pauling, L., Itano H.A., Singer S.J., and Wells I.C. (1949). Sickle cell anemia, a molecular disease. *Science* **110**, 543–548.

Riesenfeld, A. (1974). Endocrine and biomechanical control of craniofacial growth. An experimental study. *Human Biology* **46**(3), 531–72.

Roberts, D. F. (1965). Assumption and fact in anthropological genetics. *Journal of the Royal Anthropological Institute* **95**, 87–103.

Roberts, D.F. (1969). Consanguineous marriages and calculation of the genetic load. *Annals of Human Genetics (London)* **32**, 407–10.

Rosenberg, N, Pritchard J.K., Weber J.L., Cann H.M., Kidd K.K., Zhivotovsky L.A., and Feldman M.W. (2002). Genetic structure of human populations. *Science* **298**, 2381–2385.

Seielstad, M., Minch E., and Cavalli-Sforza, L. (1998). Genetic evidence for a higher female migration rate in humans. *Nature Generics* **20**, 278–80.

Smithies, O. (1955). Zone electrophoresis in starch gels: Group variations in the serum proteins of normal human adults. *Biochemistry Journal* **61**, 629.

Smithies, O. (1959). An improved procedure for starch-gel electrophoresis: Further variations in the serum proteins of normal individuals. *Biochemistry Journal* **71**, 585.

Smithies, O., and Connell, G.E. (1959). Biochemical aspects of the inherited variations in human serum haptoglobins and transferrins. *Ciba Foundation Symposium., Biochemistry of Human Genetics.* London, J. and A. Churchill, Ltd, 178–89.

Soodyall, H., Jenkins, T., Mukherjee, A., duToit, E., Roberts, D. F., and Stoneking, M. (1997). The founding mitochondrial DNA lineages of Tristan da Cunha Islanders. *American Journal of Physical Anthropology* **104**(2), 157–66.

Spuhler, J.N., and Kluckhohn, C. (1953). Inbreeding coefficients of the Ramah Navajo populations. *Human Biology* **25**, 295–317.

PART I

Theory

Michael H. Crawford

Chapter 2

Partitioning of Genetic Variation in Human Populations and the Concept of Race

Lorena Madrigal

University of South Florida

Guido Barbujani

University of Ferrara, Italy

In this chapter we have two major purposes in mind. Firstly, we review data on the distribution of human genetic variation, to address the question of the existence of races as valid biological entities for our species. Following this review, we focus on the concept of races as it is frequently used in biomedical and epidemiological work.

Human races

Introducing geographic variation in his classical text *Systematics and the Origin of Species*, Ernst Mayr (1947) makes a distinction between species in which biological changes from population to population are continuous, and species in which groups of populations with different character combinations are separated by borders. In the latter, the entities separated by borders are subspecies or geographic races. Similarly, in some human genetic textbooks (see for example Vogel and Motulsky, 1986) races are envisaged as large populations of individuals who have a significant fraction of their genes in common, and can be distinguished from other races by their common gene pool. Under both definitions, an increased level of genetic similarity among members of a race is the result of shared ancestry.

Human races have been the subject of intense scientific scrutiny long before genetics existed. A description of the rationale of these studies, and of the related debate on monogenism versus polygenism (in today's terms, mono- or poly-phyletic origin of humans) is in Cohen (1991) and a summary of modern studies quantifying genetic differences within and between major human groups is in Barbujani (2005).

A widespread misconception is that the analysis of morphological traits, such as skeletal measures or skin colour, demonstrates a clear racial subdivision of humankind; accordingly, the genetic studies that fail to confirm that subdivision had presumably focused on the wrong genes (Miller, 1994). However, a review of the relevant literature shows that that is not the case (see for example Brown and Armelagos, 2001). Indeed, the number of races proposed by serious scientists in the course of time ranges from three to hundreds, and no consensus seems to ever have been reached. One reason for that can depend on the fact that even distantly related populations are similar for certain traits if they evolved under the same selective pressures, which is called evolutionary convergence. Based on skin colour, possibly the most popular phenotype for racial classification, populations of sub-Saharan Africa, Southern India, Australia and Melanesia cluster together, yet these populations occur in distant branches of evolutionary trees based on genetic distances (Bowcock et al., 1994; Cavalli-Sforza et al., 1988). Thus, although skin colour is genetically transmitted, skin colour similarity is a poor indicator of shared ancestry (Jablonski, 2004; Jablonski and Chaplin, 2000; Relethford, 2002) as are other traits that evolved under selection. Similarly, studies on 'pygmy' populations from South East Asia (the so-called 'negritos') indicate that the phenotypic similarities of these groups with African pygmies are a result of convergent evolution: although similar, these populations do not share a recent ancestry (Thangaraj et al., 2003). Therefore, studies of race should focus on traits whose variation presumably reflects the consequences of genome-wide evolutionary pressures such as migration and drift, rather than on traits whose polymorphism is maintained by natural selection (Luikart et al., 2003).

Although only in 1962 were doubts explicitly raised that human diversity can be described in terms of racial differences (Livingstone, 1962) ever since Linnaeus no consensus has emerged on how many races exist, and which populations belong to which race. Lists compiled by scientists include up to 200 different races (Armelagos, 1994; Kittles and Weiss, 2003). An admittedly incomplete list of human races proposed by various authors is shown in Box 2.1, and is based on Armelagos (1994), Cohen (1991) and Risch et al. (2002). We should note that the lack of agreement on the number of human races is not taken to invalidate the concept itself by Levin (2002). On the contrary, Relethford (2003) argued that unless agreement is reached on the number of classification units, any classification system is useless. We are not in a position to tackle the philosophical aspects of this question here, but we note that it is impossible to test hypotheses on the degree of genetic differentiation among human groups if the definition of these groups is vague or controversial.

Apportioning human genetic diversity

Apportioning diversity means to start from putative races and test whether and to what extent they differ genetically. Under this approach three measures of diversity are often compared, namely the variances between: (i) members of the same population, (ii) populations of the same race, and (iii) different races (Box 2.2). In a seminal paper, Lewontin (1972) demonstrated that most human variation is found within, not among, human races. In the absence of an established list of races, seven-race (Latter, 1980; Lewontin, 1972) or three-race classification systems (Ryman *et al.*, 1983) were initially chosen by various authors, but in all of these studies individual differences between members of the same population accounted for 85% of the global human variance (Table 2.1). Belonging to different populations added between 3% and 8% to that value, and to different races between 6% and 11%. These proportions, inferred from protein variation, appeared at first counterintuitive, but analyses of autosomal DNA polymorphisms, whether microsatellite, restriction-site, or insertion/deletion polymorphisms, confirmed them with remarkable precision (Barbujani *et al.*, 1997; Jorde *et al.*, 2000; Romualdi *et al.*, 2002). Box 2.2 discusses the computation of these variances.

Box 2.1

Author	No. of races	Races proposed
Linnaeus (1735)	6	Europaeus, Asiaticus, Afer, Americanus, Ferus, Monstruosus
Buffon (1749)	6	Laplander, Tartar, South Asian, European, Ethiopian, American
Blumenbach (1795)	5	Caucasian, Mongolian, Ethiopian, American, Malay
Cuvier (1828)	3	Caucasoid, Negroid, Mongoloid
Deniker (1900)	29	
Weinert (1935)	17	
Von Eickstedt (1937)	38	
Biasutti (1959)	53	
Coon (1962)	5	Congoid, Capoid, Caucasoid, Mongoloid, Australoid
US Office of Management and Budget (1997)	5	African-American, White, American Indian or Alaska Native, Asian, Native Hawaiian or Pacific Islander
Risch et al. (2002) Fig. 2.1	5	African, Caucasian, Pacific islanders, East Asian, Native American
Risch et al. (2002) Table 2.3	5	African Americans, Caucasians, Hispanic Americans, East Asians, Native Americans

Box 2.2 | Analysis of genetic variance

In most studies of global human variation the overall genetic variance is broken down into three components, representing the differences between members of the same population, between populations of the same race, and between different races. Lewontin (1972) estimated for each the Shannon information measure, H, which is similar to the heterozygosity or gene diversity index. Three measures of dispersion were estimated, around the average allele frequencies of each population (Hpop), the average allele frequency across populations of a race (Hrace), and the average allele frequency for the entire species (Hspecies). The global genetic diversity was then partitioned at the three levels as follows:

Variance within populations = Hpop/H species

Variance between populations, within races = (Hrace − Hpop)/Hspecies

(the effect of belonging to different populations of the same race)

Variance between races = (Hspecies − Hrace)/Hspecies.

(the effect of belonging to different races)

The actual values of these variances do not contain much useful information, since they also depend on the level of genetic polymorphism at the loci considered, and on the population samples available. That is why the proportion of the total variance observed at each level is generally reported, rather than the variance itself.

AMOVA is a non-parametric method for the analysis of variance suitable for molecular data (Excoffier et al., 1992). Genetic variances are estimated from allele-frequency differences between populations, or from measures of molecular difference between alleles, namely number of substitutions for sequence polymorphisms, or allele-length differences for microsatellites. Much like in Lewontin's approach, the overall genetic variance is then subdivided into three hierarchical components: between individuals within populations, between populations of the same group, and between groups. The significance of each component is then tested by a randomization approach in which each individual or population is reassigned to a random location, according to three resampling schemes. The molecular variances are recalculated, and the procedure is repeated many times, so as to obtain empirical null distributions for the three variances, against which the observed values are finally compared.

As a rule, higher diversity between continents was observed for the Y chromosome (with one exception; Table 2.1). Possible explanations included reduced male mobility (Jorde et al., 1997), polygyny (Dupanloup et al., 2003), selection, or combinations of these factors; in other words, processes affecting the social structure of populations. At any rate, the effective population size for uniparentally transmitted markers such as the Y chromosome and the mitochondrial DNA, is one-fourth the effective population size for autosomal markers, and hence a greater diversity between populations is to be expected for the former because of the greater impact on them of genetic drift. Thus, what is unexpected and calls for an explanation is not so much the higher variances between continents observed for the Y chromosome, but the lower values observed for mtDNA. A recent study (Wilder et al., 2004) suggests that these differences would be less marked, or can even be zero, if random regions of the mitochondrial genome and of the Y chromosome are sequenced, thus eliminating the effect of ascertainment bias on the polymorphisms. On the other hand, the lowest variances between continents were estimated in the broadest study

Table 2.1. Estimated fractions of the global human diversity at three hierarchical levels of population subdivision.

Polymorphism	Reference	N of loci	Within population	Between populations, within race or continent	Between races or continents
Proteins[a]	Lewontin, 1972	17	85.4	8.3	6.3
Proteins[a]	Latter, 1980	18	85.5	5.5	9.0
Proteins[b]	Ryman et al., 1983	25	86.0	2.8	11.2
mtDNA	Excoffier et al., 1992	(HV-I)[c]	75.4	3.5	21.1
Autosomal DNA	Barbujani et al., 1997	109	84.4	4.7	10.8
mtDNA	Seielstad et al., 1998	(HV-I and -II)[c]	81.4	6.1	12.5
Y chromosome	Seielstad et al., 1998	10	35.5	11.8	52.7
Autosomal DNA	Jorde et al., 2000	90	84.8	1.6	13.6
mtDNA	Jorde et al., 2000	(HV-I)[c]	71.5	6.1	23.4
Y chromosome	Jorde et al., 2000	10	83.3	18.5	−1.8[f]
Alu insertions	Romualdi et al., 2002	21	82.9	8.2	8.9
Y chromosome	Romualdi et al., 2002	14	42.6	17.3	40.1
Beta-globin	Romualdi et al., 2002	1	79.4	2.8	17.8
Autosomal DNA	Rosenberg et al., 2002	377	94.1	2.4	3.6
Median, all loci[d]			82.5	5.8	11.7
Median, autosomal[e]			86.2	3.8	10.0

[a]Seven races considered.

[b]Three races considered.

[c]HV-I and HV-II are the hypervariable regions I and II, respectively, of the mitochondrial genome control region.

[d]This value was obtained by considering all studies equally informative, calculating the median among their results, and normalizing it by dividing by 1.007 Because Ref. 19 contains roughly as many loci as all other combined studies, the median estimated giving equal weight to each locus would be very close to the upper bound of the range of within-population variance, and to the lower bounds of the between-population and between-continent variances in that study, respectively 93.8, 2.5, 3.9.

[e]Normalization obtained by dividing by 0.987.

[f]Because two variances are obtained by subtraction, it may occasionally happen that one of them takes a negative value. This value (which, however, is not significantly different from 0) means that, on average, members of different continents do not differ for those markers more than members of the same continent.

of autosomal polymorphisms (Rosenberg *et al.*, 2002), although in a later reanalysis based on a more appropriate mutation model these values appeared very close to the values estimated in all other studies of autosomal loci (Excoffier and Hamilton, 2003).

Also, of the 4,199 alleles observed at 377 microsatellite loci, 7.4% are continent-specific and 66% are shared by Africa, Asia and Europe (Rosenberg *et al.*, 2002). This is in agreement with the distribution of haplotype blocks, i.e. 20 to 50 kb regions of the genome that appear mildly affected, if at all, by historical recombination (Gabriel *et al.*, 2002). Less than 4% of haplotype blocks appear restricted to either Asia or Europe, 25% are Africa-specific, and the rest are shared among continents, with more than 50% occurring worldwide. In brief, extensive allele and haplotype sharing across continents is the rule, not the exception. Such distribution of human genetic variation argues against the possibility to identify races, each characterized by a specific set of diagnostic alleles.

A similar conclusion is reached by several recent reviews on human genetic diversity (Long and Kittles, 2003). Based on analyses of Alu insertion, beta-globin and Y-chromosome polymorphisms, Romualdi *et al.* (2002) found that there is little if any evidence of a clear subdivision of humans into biologically defined groups. Serre and Pääbo (2004) reanalysed Rosenberg *et al.*,'s (2002) dataset, concluding that at the worldwide scale human diversity forms geographical clines, and thus major genetic discontinuities between different continents or races are not documented. Tishkoff and Kidd (2004) show that racial classifications are inadequate descriptors of the distribution of human genetic variation. Kittles and Weiss (2003) note that the pattern of genetic variation in humans does not correspond to races but is best described as isolation by distance. Jorde and Wooding (2004) note that although clusters of individuals based on genetic data are correlated to some extent with traditional concepts of race, these correlations are imperfect because variation is distributed in a continuous and overlapping manner among populations.

The small variances between populations and continents do not imply that all populations are equal of course. Using eight different methods of assignment on five datasets, genotypes whose origin was temporarily disregarded could be allocated to the right continent with an error ≥30% using autosomal markers, and ≥27% using the Y-chromosome (Romualdi *et al.*, 2002). Predictably, the precision of the assignment decreased sharply at the subcontinental level. These results confirm that substantial geographic structure exists in humans, as already demonstrated by classical protein studies in which, with sufficient data, significant differences emerged between even adjacent pairs of populations (Ward *et al.*, 1975). Such a geographic population structure could be legitimately regarded as racial structure only if analyses of different sets of loci define the same clusters of populations.

Identifying human population structure

In this context, identifying structure means to test whether analyses of different loci lead to identification of the same clusters of populations. In a sense, this approach is reversed with respect to those based on the analysis of molecular variance, where the definition of races is preliminary to the genetic analysis. A likelihood-based approach, implemented in the software package structure, considers all populations as resulting from the admixture of k founder populations, each in Hardy-Weinberg equilibrium. The software estimates the most likely number of clusters in a dataset, k, and assigns probabilistically individuals to them (Box 2.1; Pritchard et al., 2002). Analysing by structure two datasets of Alu insertions, Romualdi et al., (2002) found evidence for three (two worldwide distributed, one Eurasian) and four (African-Oceanian, Asian-American, European, Eurasian) clusters, respectively, which overlapped very poorly. The distribution of X-chromosome microsatellites showed four clusters (Africa, Western Eurasia, China, New Guinea, with most Ethiopians falling in the second cluster) which are uncorrelated with variation at loci of pharmacogenetic relevance (Wilson et al., 2001). In the studies by Rosenberg et al. (2002) and Bamshad et al. (2003) the most likely k was not estimated from the data, but rather given arbitrary values between 2 and 6. With $k = 6$ Rosenberg et al. (2002) found genetic clusters corresponding to: (1) Africa; (2) Europe, Western Asia and part of Central Asia; (3) the Kalash of Pakistan; (4) East Asia and part of Central Asia; (5) Oceania; and (6) South America. Bamshad et al. (2003) observed a separation between Africa and Eurasia with $k = 2$, a split between Asia and Europe with $k = 3$, and two African clusters with $k = 4$, confirming that variation within Africa exceeds that among other continents (Kittles and Weiss, 2003; Yu et al., 2002). To summarize, these studies do not appear to consistently suggest an unequivocal clustering of populations. The fact that different sets of populations were considered in the analyses of different loci may have acted as a confounding factor, but it is impossible to claim that a discontinuous population structure with well-identified clusters has emerged so far, and even less so that the clustering based on a set of loci is a reliable predictor of the clustering that would be observed using a different set of loci (Bamshad et al., 2004).

Despite these results, some authors maintain that human population structure can be regarded as essentially discontinuous (Risch et al., 2002, Burchardt et al., 2003). The results of Rosenberg et al. (2002) specifically have been interpreted as showing that there are six major genetic groups in humankind corresponding to common notions of races (Bamshad et al., 2004), although one such group, the Kalash of Pakistan, does not correspond to any historically recognized race. This interpretation can be critiqued from at least three perspectives. First, that number six was not estimated, but assumed to be true;

reanalysis of subsets of the same data by Serre and Pääbo (2004) shows that the assignment of individuals to founder populations changes dramatically depending on the choice of k. Second, the six clusters do not correspond to any previously proposed racial classification (Box 2.1). Third, as Kittles and Weiss (2003) remarked, Risch *et al.* (2002) and Rosenberg *et al.* (2002) do not consider that the clusters they obtained are due at least in part to the sampling of their studies, sampling which emphasized diverse populations which fit the US folk concept of race. Rosenberg *et al.* (2002) were aware of the third problem, because they prudently concluded that the general agreement they found of genetic data and what they called 'predefined populations' suggests that self-reported ancestry can facilitate assessments of epidemiological risks. However, this statement may create more problems than it solves, since obviously the subjects of the study were not asked to report their ancestry. Had they been asked to do so, there is no guarantee that their answer would have been the same given by a US observer; as we all know, the same individual may easily be regarded as white in, say, Bombay, and black in London (Foster and Sharp, 2002). The importance of the sampling procedure as a causal factor of Rosenberg *et al.*,'s conclusions is also stressed by Serre and Pääbo (2004).

In their search for human global population structure, other authors looked for diagnostic genes by which individuals could be assigned to groups. There are two problems with this approach. First, pathologic population-specific alleles are diagnostic in a sense, but they are present in just a few individuals, and so this exercise defines a set of subjects at risk (perhaps a sub-population) but certainly not a race (Kittles and Weiss, 2003). Second, an allele that is diagnostic in the United States, where groups of very different origins came into contact in the last few centuries, may not be diagnostic at all at a world scale, where there was potential for population admixture through many millennia. In other words, it is easy to find a number of alleles that, in the US context, are carried only by one, sometimes small, group of immigrants or Native Americans (Shriver *et al.*, 1997), because these groups were subjected to large drift effects, respectively in the phase of immigration, and during the population bottlenecks they underwent. But the only way to see whether these alleles are diagnostic of major subdivisions of humankind that could be equated with races is to collect samples of populations in the areas where these populations evolved and see if the distribution of those alleles is really discontinuous and group-specific. Thus, although some studies have shown a significant difference in allele frequencies among main ethnic groups in the United States, and among world populations selected to represent those groups, these studies have excluded large geographic areas and their populations, groups such as Oceanians, West and Central Asians, and Native Americans (Bamshad *et al.*, 2003; Shriver *et al.*, 1997). Kittles and Weiss (2003) rightly note that a racial view of human diversity excludes a few 'pesky' billion people from India, whose classification into racial boxes is problematic. Studies that emphasize population discontinuity may be useful for forensic identification in the United States, but, despite claims to the contrary (Bamshad *et al.*, 2003), contain no information on the global structure of the human species and can actually be misleading. For instance, measures of skin colour appeared correlated with several diagnostic markers in US whites and blacks (Shriver *et al.*, 2003) but not in Brazil (Parra *et al.*, 2003) showing that that correlation was an artefact of population stratification. Often considered in these comparisons were Hispanics (for example, Risch *et al.*, 2002) a category that does not fit any definition of race, comprising US immigrants of New and Old-world origin with varying degrees of European, Amerindian and African ancestry, who speak Spanish and or Portuguese (depending on the definition) and who would never define themselves as Hispanics in their country.

To summarize, no analysis at the world scale provided the same description of human population structure, and different sets of genes led to different clusterings of the individuals in the same study (Romualdi *et al.*, 2002; Wilson *et al.*, 2001). This result is hardly surprising in evolutionary terms. Our species is comparatively young, and its current diversity can be traced back to one or a few founder populations that expanded from Africa not long ago (Tishkoff and Williams, 2002). Under that scenario genetic differences between continents are expected to be small, and geographic variation to be continuous (Cavalli-Sforza *et al.*, 1994), so that populations that appear to form a cluster when studied for certain markers do not cluster together when analysed for other markers. Therefore, groups inferred from the genes or from the physical aspect of people are not reliable predictors of variation at other independently transmitted loci (see especially Wilson *et al.*, 2001). By studying genotypes we can identify with good approximation the geographic origin of most individuals, but humans do not come in neat racial packages. If races are subspecies, there is no such thing in humans, and if races are not subspecies somebody should make it clear what they are instead.

If races don't exist, why are forensic scientists so good at finding them?, asked Sauer (1992), this being a frequently advanced argument in favour of a racial classification. By considering separate databases for different ethnic groups, forensic scientists minimize the probability of unfair decisions against members of minorities (Jobling *et al.*, 2004). However, this seems a case in which people say race but mean something else. If races are biological realities, they must be the same everywhere, whereas forensic race catalogues differ across countries. In the United Kingdom we find light-skinned European, Afro-Caribbean, Indian Subcontinent, South East Asian and Middle Eastern: (www.forensic.gov.uk/forensic/foi/foi_docs/43L_Commonplace_characteristics.pdf), only two of which (the first and the fourth) correspond to races in the United States (Box 2.1). Which list is right? Our answer is that neither gives a sensible, all-purpose description of human diversity, but both can help categorize people in specific urban areas, where boundaries between groups are sharper than at the world scale. A concept of population diversity seems all that is needed for gene hunting and forensic identification.

'Race' in epidemiological studies

Perhaps nowhere else is the debate on the usefulness of races as lively as it is in the epidemiological and medical literature. For example, it has been proposed that races are important for gene hunting (Phimister, 2003) because alleles that cause monogenic disorders are enriched in some populations; examples include Mexicans in Texas, Ashkenazi Jews and the inhabitants of Tristan da Cunha. However, none of these populations can possibly enjoy the status of a race, according to any of the definitions listed at the beginning of this chapter. We do not need races to study population-specific pathologic alleles, and the presence of these alleles cannot dictate medical treatment for an entire continent, whose inhabitants will mostly carry the normal alleles. Indeed, studies on human genetic diversity do not provide support for the proposal that most health differences among groups usually labelled as 'races' are a result of genetic differences between them. Kittles and Weiss (2003) note that whereas rare alleles are geographically localized, they vary greatly among different groups classified by geography or race. They also note that such high frequencies of disease-causing alleles in localized groups have mistakenly reinforced folk notions of race. Jorde and Wooding (2004) state that an individual's population affiliation may be a faulty indicator of the presence or absence of an allele used for diagnosis or important for drug responses. Tishkoff and Kidd (2004) simply state that identifiers based on race will often be insufficient for biomedical studies.

Few multigenic diseases are well understood at the genetic level, but some populations can conceivably carry certain combinations of predisposing alleles in linkage disequilibrium (Arcos-Burgos and Muenke, 2002). Ioannidis et al. (2004) performed a large meta-analysis of 83 meta-analyses genetic-association studies for various complex diseases across groups which they labelled 'races' between inverted commas (of note, none of these included the ubiquitous 'Hispanics' of US-based studies). Of these 83, only 43 showed either overall significant results or significant results for at least one racial group. However, the genetic effects of such markers on actual disease frequency were usually not significant. Specifically, the odds ratios between 'races' were heterogeneous in only 14% of the studies. Ioannidis et al. (2004) interpret their results as further support for the presence of more variation within than among groups, and for a lack of significant modifications to a common genetic background in our species. They also indicate that risk factors such as lifestyle and environment, just like genetic make-up, are likely to show more variation within than among groups. They suggest that literature claims for 'racial' differences in genetic risks should be scrutinized cautiously.

Although many aspects of human population structure are still to be clarified, the available evidence strongly suggests that to efficiently predict disease risk, classifying individuals by racial group is scientifically arbitrary, and almost useless from the practical standpoint. What matters, instead, is the individual genotype, which can only be described by studying the individual. Indeed, there is a virtually overwhelming agreement in recent medical and epidemiological literature with the proposition that human races are not valid biological entities for epidemiological studies. For example, Lin and Kelsey (2000) state that the groups traditionally used to classify humans in the United States are a result of the history of the country, and have no biological foundation. In an editorial for the journal *Epidemiology*, Kaufman (1999) refers to groups such as 'Blacks' and 'Whites' as absurdly heterogeneous, and to race as an indicator of innate risk, as appearing in the literature with embarrassing regularity. In an editorial in the journal *Archives of Pediatric and Adolescent Medicine*, Rivara and Finberg (2001) ask authors not to use the terms race/ethnicity, variables that lack any scientific basis, when variables such as educational level, household income, etc., can be measured directly. Anderson *et al.* (2001) note that in the United States before 1989, a baby born to a 'White' and a 'Non-White' parent would have been assigned the race of the 'Non-White' parent, but that after 1989, s/he would have been assigned the race of the mother. Clearly, this change in racial assignment indicates that race is not a biological entity, but a social construct (Anderson *et al.*, 2001). The same point is made by Bowman (2000) who notes that in the United States a black mother cannot have a white baby, but a white mother can have a black baby. Kaplan and Bennett (2003) argue that because race is socially constructed, it should not be viewed as an inherent attribute of an individual in epidemiological studies. Perhaps what is the strongest statement about epidemiology and medicine's rejection of races as biological entities is the Institute of Medicine's (IOM) recommendation to the National Institutes of Health (NIH)'s that the latter re-evaluate its use of the term because it lacks scientific validity (Oppenheimer, 2001).

Just as biomedical and epidemiological workers appear to be ready to discard race as a biological entity, several researchers emphasize that just the same, ethnicity is a powerful social category which affects people's health (Brondolo *et al.*, 2003; Jones, 2001; Kaufman and Cooper, 2001; Krieger, 2000a). In the United States in particular, groups labelled by terms such as Hispanics, African-American, Euro-American, Amerindian, etc., have different epidemiological profiles. Indeed, the data overwhelmingly show that African-Americans have a higher probability of suffering from hypertension (East *et al.*, 2003; Price and Fisher, 2003), Amerindians from diabetes (Acton *et al.*, 2002), etc. But writers such as Jones (2001), who argue for the continued research of 'race'-associated health disparities, stress that 'race' is a social, not a biological construct, and that there is more diversity within than among 'racial' groups. Even when epidemiologists have controlled for variables such as education and socioeconomic status, they should not assume that different health outcomes in different groups are best explained by genetic differences. On the contrary, the genetic data show that there is more variation within than among groups, whereas socioeconomic variables do show more variation among than within these groups (Root, 2003). Therefore, it appears that in societies in which health outcomes are tied to ethnicity, it is perhaps a good idea to collect the data by these groups. But this should be done with the full acknowledgement that the disparities are due to non-genetic differences among these groups.

In the same manner, the available data suggest that a good way to predict whether certain individuals will benefit from pharmaceutical treatment is to study their genes, not to assign them to races. For example, Shimizu *et al.*, (2003) have shown in two large samples of Asian and Europeans that for CYP2D6, a locus coding for one of the main drug-metabolizing enzymes of the cytochrome, depending on the genotype, people benefit from standard treatment (normal metabolizers), have side-effects (slow metabolizers), or eliminate the drug before it has exerted its action (fast metabolizers). Genotype frequencies are almost identical in Asian and European samples. The point here is that both groups contain the whole spectrum of genotypes, and hence in both populations there will be fast, normal and slow metabolizers. Therefore, there is no hope to develop different drugs, or drug dosages, specific for the Asian or the European market. Much more informative, and now technically feasible at least in principle, is the genetic typing of individuals, which in turn may lead to tailoring specific pharmacological treatment for fast, normal and slow metabolizers.

Why is race such a relentless term?

Given that the fact that there is more variation within than among groups was established since 1972, it is worth asking why race as a valid *biological* (not cultural/environmental) entity continues to be used. The only reason we can fathom is that the authors of such studies have not questioned their own culturally learned human taxonomy scheme. That human taxonomies are culture-specific has been shown repeatedly by anthropologists. That is, humans learn to classify other humans from their culture, just as they learn their linguistic, cooking and religious traditions (Marks, 1996).

For example, Ember and Ember (2002) mention that whereas in the United States, Japanese and Chinese people would be considered to be 'Asian', in Apartheid South Africa the former were considered to be 'White', the latter 'Asian', this a reflection of their different socioeconomic standing. Kottak (2004) notes that in Japan, the majority of the population considers the ethnic group *burakumin* as different from them biologically, whereas in the United States both the Japanese majority and minority groups would be considered part of the same race. Another example of a culture-specific racial classification of which we are particularly fond is that of the Bri-Bri Indians of Southern Costa Rica. The Bri-Bri classify humans as follows: The Bri-Bri and the Cabecar (their neighbour) Indians are people, everybody else is *ña* (excrement) (Murillo, 1982). Many individuals who have spent considerable time in more than one culture become adept at recognizing how a particular person would be classified differently in different cultures. Indeed, one of us belongs to two different races, depending on the culture she is. Which culture is right?

The answer of course is that neither is right. As members of their culture, geneticists, medical and epidemiological practitioners have learned to classify humans in their own culture-specific manner. However, they should not assume that their racial taxonomy is supported by genetic data. This is particularly crucial for a New-World, admixed population such as that of the United States. A large number of papers indicates that the US groups traditionally identified in studies such as those of Risch *et al.* (2002) (Whites, African American, Hispanics and Amerindians) can only be characterized as admixed (Bonilla *et al.*, 2000; Bowman, 2000; Parra *et al.*, 2001; Scozzari *et al.*, 1997; Tseng *et al.*, 1998; Williams, 1986). Therefore, representing US Whites, African Americans, Hispanics and Amerindians as homogeneous, clearly distinct genetic populations, fails to acknowledge the large amount of gene flow that produced them, plus the high amount of variation found within their parental populations. 'Blacks' and 'Whites' of the United States are not the native groups from Africa and Europe studied by geneticists interested in the distribution of human variation worldwide (Root, 2001). Jackson (2000) aptly notes that the mentality that separates patients by a line of colour has frequently undermined the practice of good science, by interfering with the accurate measurement, analysis and valid interpretations of human variation. She goes on to note that the history of science is replete with instances of 'virtual science' used to camouflage a political and social agenda. Krieger (2000b) states that to study racial/ethnic disparities in health without reference to racial discrimination is itself an ideological statement, which undermines rigorous science.

The importance of culture as a determinant of how an individual scientist decides to classify humans is relevant to this review of the distribution of human variation in yet another point. As noted previously, Rosenberg *et al.* (2002) produced six clusters which corresponded largely to major geographic regions, and proposed that self-reported population ancestry (SRPA) is a likely proxy for genetic ancestry. However, the subjects analysed by Rosenberg *et al.* were not asked to self-report their population ancestry. Instead, Rosenberg *et al.* assumed that their subjects would have characterized themselves in the same manner as the authors would have. Rosenberg *et al.* assumed that their own culturally learned taxonomy is universal, although their ideas of SRPA have no more scientific base than do the Bri-Bri's. As Marks (1996) plainly puts it, classification of people into races involves cultural, not biological knowledge.

In a direct contradiction to Rosenberg *et al.*'s conclusion that SRPA is a valid genetic proxy, Kaplan and Bennett (2003), in a set of guidelines on how biomedical publications should deal with race and ethnicity, state that 'Race/ethnicity should not be used as a proxy for genetic variation'. They argue that given the overwhelming evidence for more variation within than among races, it is unlikely that differences among groups are genetic in nature but are more likely due to other socially derived factors such as poverty and racism.

Conclusion

Species that are subdivided in essentially isolated reproductive units, such as some bats (Miller-Butterworth *et al.*, 2003) tend to form races, species where gene flow prevails such as the coyote (Vila *et al.*, 1999) tend to show continuous variation. Apparently, humans are closer to the latter than to the former. Still, race is a social reality, and as such, it will continue affecting our life. Racial categorization has a long history, and may be related to a deep-rooted psychological need to quickly identify potential enemies and allies (Kurzban *et al.*, 2001). Having said this, the biological reality is different and, for humans, it is one of continuous variation (Cavalli-Sforza *et al.*, 1994), clines, and genetic boundaries that cross the geographic space without surrounding and thus defining specific isolated groups of populations (Barbujani and Sokal, 1990). If we are to understand human diversity, and if we are to exploit the potential represented by the ever-increasing genomic data, race is neither an accurate nor a useful concept, unless it is used in such a loose sense as to mean population, in which case a rigorous usage of words seems advisable (Keita *et al.*, 2004). Because the effects of single-locus genotypes are influenced by variation at other loci (Weatherall, 2001), predicting the phenotype is difficult for monogenic diseases, and extremely difficult for complex diseases. However, if we want to understand human evolution, and if our aim is to treat multifactorial pathologies, there is no alternative to coming to terms with the continuous nature of human genetic diversity.

References

Acton, K. J., Burrows, N. R., Querec, L., Geiss, L. S. and Engelgau, M. M. (2002). Trends in diabetes prevalence among American Indian and Alaska Native children, adolescents, and young adults. *American Journal of Public Health* **92**(9), 1485–1490.

Anderson, M. R., Moscou, S., Fulchon, C. and Neuspiel, D. R. (2001). The role of race in the clinical presentation. *Family Medicine* **33**(6), 430–4.

Arcos-Burgos, M. and Muenke, M. (2002). Genetics of population isolates. *Clinical Genetics* **61**(4), 233–47.

Armelagos, G. J. (1994) Racism and physical anthropology: Brues' review of Barkan's The Retreat of Scientific Racism. *Am. J. Phys. Anthropol.* **93**, 381–3.

Bamshad, M., Wooding, S., Salisbury, B. A. and Stephens, J. C. (2004). Deconstructing the relationship between genetics and race. *Nature Reviews Genetics* **5**(8), 598–U2.

Bamshad, M. J., Wooding, S., Watkins, W. S., Ostler, C. T., Batzer, M. A. and Jorde, L. B. (2003). Human population genetic structure and inference of group membership. *American Journal of Human Genetics* **72**(3), 578–89.

Barbujani, G. (2004). Human races: Classifying people vs. understanding diversity. *Curr. Genomics* **6**(4), 215–26.

Barbujani, G. (2005). Human races: Classifying people vs. understanding diversity. *Current Genomics* **6**(4), 215–26.

Barbujani, G. and Sokal, R. R. (1990). Zones of sharp genetic change in Europe are also linguistic boundaries. *Proceedings of The National Academy of Sciences of the United States Of America* **87**(5), 1816–19.

Barbujani, G., Magagni, A., Minch, E. and Cavalli-Sforza, L. L. (1997). An apportionment of human DNA diversity. *Proceedings of the National Academy of Sciences of the United States of America* **94**(9), 4516–19.

Bonilla, C., Parra, E. J., Pfaff, C. L., Hiester, K. G., Sosnoski, D. M., Dios, S., Gulden, F. O., Ferrell, R. E., Hamman, R. F. and Shriver, M. D. (2000). Native American admixture and its relationship to diabetes in a Hispanic population of the Southwest. *American Journal of Human Genetics* **67**(4), 48.

Bowcock, A. M., Ruiz Linares, A., Tomfohrde, J., Minch, E, Kidd, Jr and Cavalli-Sforza, L. L. (1994). High-resolution of human evolutionary trees with polymorphic microsatellites. *Nature* **368**(6470), 455–7.

Bowman, J. E. (2000). Anthropology: From bones to the human genome. *Annals of the American Academy of Political and Social Science* **568**, 140–53.

Brondolo, E., Rieppi, R., Kelly, K. P. and Gerin, W. (2003). Perceived racism and blood pressure: A review of the literature and conceptual and methodological critique. *Annals of Behavioral Medicine* **25**(1), 55–65.

Brown, R. A. and Armelagos, G. J. (2001). Apportionment of racial diversity: A review. *Evol. Anthropol.* **10**, 24–40.

Burchard, E. G., Ziv, E., Coyle, N., Gomez, S. L., Tang, H., Karter, A. J., Mountain, J. L., Perez-Stable, E. J., Sheppard, D. and Risch, N. (2003). The importance of race and ethnic background in biomedical research and clinical practice. *New England Journal of Medicine* **348**(12), 1170–5.

Cavalli-Sforza, L. L., Menozzi, P. and Piazza, A. (1994) *The History and Geography of Human Genes*. Princeton University Press.

Cavalli-Sforza, L. L, Piazza, A., Menozzi, P. and Mountain, J. (1988). Reconstruction of human-evolution – bringing together genetic, archaeological, and linguistic data. *Proceedings of the National Academy of Sciences of the United States of America* **85**(16), 6002–6.

Cohen, C. (1991). Les races humaines en histoire des sciences. In *Aux Origines d'Homo sapiens*, Hublin, J.J. and Tillier, A.M. (eds), pp. 9–56, Presses Universitaires de France.

Dupanloup, I., Pereira, L. and Bertorelle, G., *et al.* (2003). A recent shift from polygyny to monogamy in humans is suggested by the analysis of world-wide Y-chromosome diversity. *Journal of Molecular Evolution* **57**(1), 85–97.

East, M.A., Jollis, J.G., Nelson, C.L., Marks, D. and Peterson, E.D. (2003). The influence of left ventricular hypertrophy on survival in patients with coronary artery disease: Do race and gender matter? *Journal of the American College of Cardiology* **41**(6), 949–54.

Ember, C.R. and Ember, M. (2002). *Cultural Anthropology*, 10th edn. Prentice Hall.

Excoffier, L. and Hamilton, G. (2003). Comment on 'Genetic structure of human populations'. *Science* **300**, 1877

Excoffier, L., Smouse, P.E. and Quattro, J.M. (1992). Analysis of molecular variance inferred from metric distances among DNA Haplotypes – Application to human mitochondrial-DNA restriction data. *Genetics* **131**(2), 479–91.

Foster, M.W. and Sharp, R.R. (2002). Race, ethnicity, and genomics: Social classifications as proxies of biological heterogeneities. *Genome Res.* **12**, 844–50.

Gabriel, S.B., Schaffner, S.F., Nguyen, H., Moore, J.M., Roy, J., Blumenstiel, B., Higgins, J., DeFelice, M., Lochner, A., Faggart, M., Liu-Cordero, S.N., Rotimi, C., Adeyemo, A., Cooper, R., Ward, R., Lander, E.S., Daly, M.J. and Altshuler, D. (2002). The structure of haplotype blocks in the human genome. *Science* **296**(5576), 2225–9.

Ioannidis, J.P.A., Ntzani, E.E. and Trikalinos, T.A. (2004). 'Racial' differences in genetic effects for complex diseases. *Nature Genetics* **36**(12), 1312–18.

Jablonski, N.G. (2004). The evolution of human skin and skin color. *Annual Review of Anthropology* **33**, 587–623.

Jablonski, N.G. and Chaplin, G. (2000). The evolution of human skin coloration. *Journal of Human Evolution* **39**, 57–106.

Jackson, F.L.C. (2000). Anthropological measurement: The mismeasure of African Americans. *Annals of the American Academy of Political and Social Science* **568**, 154–71.

Jobling, M.A., Hurles, M.E. and Tyler-Smith, C. (2004). *Human Evolutionary Genetics: Origins, Peoples and Disease*. New York: Garland Science.

Jones, C.P. (2001). Invited commentary: 'Race', racism, and the practice of epidemiology. *American Journal of Epidemiology* **154**(4), 299–304.

Jorde, L.B. and Wooding, S.P. (2004). Genetic variation, classification and 'race'. *Nature Genetics* **36**(11), S28–S33.

Jorde, L.B., Watkins, W.S., Bamshad, M.J., Dixon, M.E., Ricker, C.E., Seielstad, M.T. and Batzer, M.A. (2000). The distribution of human genetic diversity: A comparison of mitochondrial, autosomal, and Y-chromosome data. *American Journal of Human Genetics* **66**(3), 979–88.

Jorde, L.B., Rogers, A.R., Bamshad, M., Watkins, W.S., Krakowiak, P., Sung, S., Kere, J. and Harpending, H.C. (1997). Microsatellite diversity and the demographic history of modern humans. *Proceedings of the National Academy of Sciences of the United States Of America* **94**(7), 3100–3.

Kaplan, J.B. and Bennett, T. (2003). Use of race and ethnicity in bio-medical publication. *JAMA-Journal of the American Medical Association* **289**(20), 2709–16.

Kaufman, J. S. (1999). How inconsistencies in racial classification demystify the race construct in public health statistics. *Epidemiology* **10**(2), 101–3.

Kaufman, J. S. and Cooper, R. S. (2001). Commentary: Considerations for use of racial/ethnic classification in etiologic research. *American Journal of Epidemiology* **154**(4), 291–8.

Keita, S. O. Y., Kittles, R. A., Royal, C. D. M., Bonney, G. E., Furbert-Harris, P., Dunston, G. M. and Rotimi, C. N. (2004). Conceptualizing human variation. *Nature Genet.* **36**, 517–20.

Kittles, R. A. and Weiss, K. M. (2003). Race, ancestry, and genes: Implications for defining disease risk. *Annual Review of Genomics and Human Genetics* **4**, 33–67.

Kottak, C. P. (2004) Anthropology: *The Exploration of Human Diversity*, 10th edn. New York: McGraw-Hill.

Krieger, N. (2000a). Refiguring 'race': Epidemiology, racialized biology, and biological expressions of race relations. *International Journal of Health Services* **30**(1), 211–16.

Krieger, N. (2000b). Epidemiology, racism, and health: The case of low birth weight. *Epidemiology* **11**(3), 237–9.

Kurzban, R., Tooby, J. and Cosmides, L. (2001). Can race be erased? Coalitional computation and social categorization. *Proceedings of the National Academy of Sciences of the United States of America* **98**(26), 15387–92.

Latter, B. D. H. (1980). Genetic differences within and between populations of the major human subgroups. *Am. Nat.* **116**(2), 220–37.

Levin, M. (2002). Race unreal? *Mankind Quarterly* **42**(4), 413–17.

Lewontin, R. C. (1972). The apportionment of human diversity. *Evol. Biol.* **6**, 381–98.

Lin, S. S., Kelsey, J. L. (2000). Use of race and ethnicity in epidemiologic research: Concepts, methodological issues, and suggestions for research. *Epidemiologic Reviews* **22**(2), 187–202.

Livingstone, F. B. (1962). On the nonexistence of human races. *Curr. Anthropol.* **3**, 279–81.

Long, J. C. and Kittles, R. A. (2003). Human genetic diversity and the nonexistence of biological races. *Human Biology* **75**(4), 449–71.

Luikart, G., England, P. R., Tallmon, D., Jordan, S. and Taberlet, P. (2003). The power and promise of population genomics: From genotyping to genome typing. *Nature Reviews Genetics* **4**(12), 981–94.

Mayr, E. (1947). *Systematics and the Origin of Species*, 3rd edn. New York: Columbia University Press.

Marks, J. (1996). Science and race. *American Behavioral Scientist* **40**, 12–133.

Miller, E. M. (1994). Review of history and geography of human genes, by Cavalli-Sforza, L. L. *et al. Mankind Quart.* **35**, 71–108.

Miller-Butterworth, C. M., Jacobs, D. S. and Harley, E. H. (2003). Strong population substructure is correlated with morphology and ecology in a migratory bat. *Nature* **424** (6945), 187–91.

Murillo, C. (1982). Cultural Anthropology Lecture. University of Costa Rica.

Oppenheimer, G. M. (2001). Paradigm lost: Race, ethnicity, and the search for a new population taxonomy. *American Journal of Public Health* **91**(7), 1049–55.

Parra, F. C., Amado, R. C., Lambertucci, J. R., Rocha, J., Antunes, C. M. and Pena, S. D. J. (2003). Color and genomic ancestry in Brazilians. *Proceedings of the National Academy of Sciences of the United States Of America* **100**(1), 177–82.

Parra, E. J., Kittles, R. A., Argyropoulos, G., Pfaff, C. L., Hiester, K., Bonilla, C., Sylvester, N., Parrish-Gause, D., Garvey, W. T., Jin, L., McKeigue, P. M., Kamboh, M. I., Ferrell, R. E., Pollitzer, W. S. and Shriver, M. D. (2001).

Ancestral proportions and admixture dynamics in geographically defined African Americans living in South Carolina. *American Journal of Physical Anthropology* **114**(1), 18–29.

Pritchard, J., Falush, D. and Stephens, M. (2002). Inference of population structure in recently admixed populations. *American Journal of Human Genetics* **71**(4), 177.

Phimister, E. G. (2003). Medicine and the racial divide. *New England Journal of Medicine* **348**(12), 1081–82.

Price, D. A. and Fisher, N. D. L. (2003). The renin-angiotensin system in blacks: Active, passive, or what? *Current Hypertension Reports* **5**(3), 225–30.

Relethford, J. H. (2002). Apportionment of global human genetic diversity based on craniometrics and skin color *Am. J. Phys. Anthropol.* **118**, 393–8.

Relethford, J. H. (2003). *The Human Species. An Introduction to Biological Anthropology*, 5th edn. New York: McGraw-Hill.

Risch, N., Burchard, E., Ziv, E. and Tang, H. (2002). Categorization of humans in biomedical research: Genes, race and disease. *Genome Biol.* **3**, 2007.1–11.

Rivara, F. P., Finberg, L. (2001). Use of the terms race and ethnicity. *Archives of Pediatrics and Adolescent Medicine* **155**(2), 119.

Romualdi, C., Balding, D., Nasidze, I. S., Risch, G., Robichaux, M., Sherry, S. T., Stoneking, M., Batzer, M. A. and Barbujani, G. (2002). Patterns of human diversity, within and among continents, inferred from biallelic DNA polymorphisms. *Genome Research* **12**(4), 602–12.

Root, M. (2003). The use of race in medicine as a proxy for genetic differences. *Philosophy of Science* **70**(5), 1173–83.

Root, M. (2001). The problem of race in medicine. *Philosophy of the Social Sciences* **31**(1), 20–39.

Rosenberg, N. A., Pritchard, J. K., Weber, J. L., Cann, H. M., Kidd, K. K., Zhivotovsky, L. A. and Feldman, M. W. (2002). Genetic structure of human populations. *Science* **298**(5602), 2381–85.

Ryman, N., Chakraborty, R. and Nei, M. (1983). Differences in the Relative Distribution of Human-Gene Diversity between Electrophoretic and Red and White Cell Antigen Loci. *Human Heredity* **33**(2), 93–102.

Sauer, N. J. (1992). Forensic Anthropology and the Concept of Race – If Races Don't Exist, Why Are Forensic Anthropologists So Good At Identifying Them? *Social Science and Medicine* **34**(2), 107–11.

Scozzari, R., Cruciani, F., Santolamazza, P., Sellitto, D., Cole, D. E. C., Rubin, L. A., Labuda, D., Marini, E., Succa, V., Vona, G. and Torroni, A. (1997). mtDNA and Y chromosome-specific polymorphisms in modern Ojibwa: Implications about the origin of their gene pool. *American Journal of Human Genetics* **60**(1), 241–4.

Seielstad, M. T., Minc, E. and Cavalli-Sforza, L. L. (1998). Genetic evidence for a higher female migration rate in humans. *Nature Genetics* **20**(3), 278–80.

Serre, D. and Pääbo, S. P. (2004). Evidence for gradients of human genetic diversity within and among continents. *Genome Research* **14**(9), 1679–85.

Shimizu, T., Ochiai, H., Asell, F., Yokono, Y., Kikuchi, Y., Nitta, M., Hama, Y., Yamaguchi, S., Hashimoto, M., Taki, K., Nakata, K., Aida, Y., Ohashi, A. and Ozawa, N. (2003). Bioinformatics research on inter-racial differences in drug metabolism. I. Analysis on frequencies of mutant alleles and poor metabolizers on CYP2D6 and CYP2C19. *Drug Metab. Pharmacokin.* **18**, 48–70.

Shriver, M. D., Parra, E. J., McKeigue, P. and Kittles, R. (2003). Skin pigmentation, biogeographical ancestry and admixture mapping. *American Journal of Physical Anthropology*, **191**, Suppl. 36

Shriver, M. D., Smith, M. W., Jin, L., Marcini, A., Akey, J. M., Deka, R. and Ferrell, R. E. (1997). Ethnic-affiliation estimation by use of population-specific DNA markers. *American Journal of Human Genetics* **60**(4), 957−64.

Templeton, A. (1999) Human races: A genetic and evolutionary perspective. *Am. Anthropol.* **100**, 632−650.

Thangaraj, K., Singh, L., Reddy, A. G., Rao, V. R., Sehgal, S. C., Underhill, P. A., Pierson, M., Frame, I. G. and Hagelberg, E. (2003). Genetic affinities of the Andaman Islanders, a vanishing human population. *Current Biology* **13**(2), 86−93.

Tishkoff, S. A. and Kidd, K. K. (2004). Implications of biogeography of human populations for 'race' and medicine. *Nature Genetics* **36**(11), S21−S27, Suppl. S.

Tishkoff, S. A. and Williams, S. M. (2002). Genetic analysis of African populations: Human evolution and complex disease. *Nature Reviews Genetics* **3**(8), 611−21.

Tseng, M., Williams, R. C., Maurer, K. R., Schanfield, M. S., Knowler, W. C. and Everhart, J. E. (1998). Genetic admixture and gallbladder disease in Mexican Americans. *American Journal of Physical Anthropology* **106**(3), 361−71.

Vila, C., Amorim, I. R., Leonard, J. A., Posada, D., Castroviejo, J., Petrucci-Fonseca, F., Crandall, K. A., Ellegren, H. and Wayne, R. K. (1999). Mitochondrial DNA phylogeography and population history of the grey wolf Canis lupus. *Molecular Ecology* **8**(12), 2089−2103.

Vogel, F. and Motulsky, A. G. (1986). *Human Genetics: Problems and Approaches*, 2nd edn. Berlin: Springer-Verlag.

Ward, R. H., Gershowitz, H., Layrisse, M. and Neel, J. V. (1975). Genetic structure of a tribal population, Yanomama Indians, 11. Gene-frequencies for 10 blood-groups and ABH-Le Secretor traits in Yanomama and their neighbors − uniqueness of tribe. *American Journal of Human Genetics* **27**(1), 1−30.

Weatherall, D. J. (2001). Phenotype-genotype relationships in monogenic disease: Lessons from the thalassaemias. *Nat. Rev. Genet.* **2**, 245−55.

Wilder, J. A., Kingan, S. B., Mobasher, Z., Pilkington, M. M. and Hammer, M. F. (2004). Global patterns of human mitochondrial DNA and Y-chromosome structure are not influenced by higher migration rates of females versus males. *Nature Genetics* **36**(10), 1122−1125.

Williams, R. C., Steinberg, A. G., Knowler, W. C. and Pettitt, D. J. (1986). GM 3,5,13,14 and stated-admixture − independent estimates of admixture in American-Indians. *American Journal of Human Genetics* **39**(3), 409−13.

Wilson, J. F., Weale, M. E., Smith, A. C., Gratrix, F., Fletcher, B., Thomas, M. G., Bradman, N. and Goldstein, D. B. (2001). Population genetic structure of variable drug response. *Nature Genetics* **29**(3), 265−9.

Yu, N., Chen, F. C., Ota, S., Jorde, L. B., Pamilo, P., Patthy, L., Ramsay, M., Jenkins, T., Shyue, S. K. and Li, W. H. (2002). Larger genetic differences within Africans than between Africans and Eurasians. *Genetics* **161**(1), 269−74.

Chapter 3

Natural Experiments in Human Gene Mapping: The Intersection of Anthropological Genetics and Genetic Epidemiology

Joseph D. Terwilliger

Department of Psychiatry, Department of Genetics and Development, Columbia Genome Center, Columbia University, New York, NY, USA; Finnish Genome Center, University of Helsinki, Helsinki, Finland; Division of Medical Genetics, New York State Psychiatric Institute, New York, NY, USA

Joseph H. Lee

Department of Epidemiology, Gertrude H. Sergievsky Center, Taub Institute, Columbia University, New York, NY, USA

Researchers in anthropology, epidemiology, and human genetics have been investigating the sources of human variation and the etiology of human diseases over the past century, largely independently of one another, using very different methodologies to study environmental factors, infectious agents, genetic variation, and their respective effects on human phenotypic variation. Human geneticists have had success in identifying DNA sequence variations which 'cause' a variety of clearly inherited 'Mendelian' conditions (Terwilliger and Goring, 2000), similar to the successes of epidemiologists in identifying 'causal' risk factors which by themselves dominated the trait's etiology. However, once many of the major 'sledgehammer' risk factors were identified for each of these flavours of risk factor (genetic, environmental, and infectious), researchers began to focus in earnest on the more complex family of common traits related to normal variation, which are of predominant public health relevance, and which are hypothesized to be influenced in a small way by each of numerous risk factors, potentially of each type (Weiss and Terwilliger, 2000) — that is to say epidemiologists and human geneticists have been moving into areas where

Corresponding author: Joseph D. Terwilliger, Columbia University, 1150 St Nicholas Avenue, Room 504, New York, NY 10032 (joseph.terwilliger@helsinki.fi).

anthropologists have been for decades, applying a new set of tools to old problems. However, the standard toolkit of epidemiologists and medical geneticists are not the most appropriate for this problem, as anthropologists have been aware for a long time, and this application of unmodified tools from classical health science has been very difficult to translate to studies of normal variation in humans, since the etiology is much more complex. Such traits have been amenable to investigation in animals and plants, but these results cannot be induced to therefore work in humans, as there is no way to experimentally control or modify exposures to putative risk factors to see what happens, the way we can and do in experimental and agricultural organisms (Terwilliger and Goring, 2000; Terwilliger *et al.*, 2002).

There have been numerous attempts to develop *post-hoc* statistical methods to approximate what one would do experimentally in a model organism, but generally these have proven unsatisfactory for the same reasons one does not apply random sampling with *post-hoc* statistical correction in model organisms – it simply does not have much power (Weiss, 2002; Weiss, 1993; Weiss, 1996). Mouse geneticists do not choose to work with inbred strains because they are easy to maintain and are a good model for naturally occurring variation in the species. They work with inbred strains because they can simplify the problem dramatically by controlling the amount of variation in exposures to genetic and environmental factors artificially, so they can figure out, at least in some specific case, what variants *can* affect what phenotypes. Of course, nobody seriously believes that variants discovered related to obesity in inbred mice will be major predictors of obesity in natural populations of mice. But working with mice collected from the subway would not be a powerful way to identify genes or environmental risk factors because of their enormous genetic and environmental exposure variation. And we, as a species, are not inbred strains, but rather more like subway mice. To this end, while human geneticists cannot design breeding experiments, and epidemiologists cannot expose subjects to noxious risk factors or infectious agents to perform this task, we do have the ability to use anthropological methods to identify natural experiments – populations in which the exposures we wish to study or the mating experiments we would like to design in an experimental system have happened naturally. Thus if we are to become successful in understanding the etiology of common disease or normal variation, the work of anthropologists and anthropological geneticists will be critically important (Terwilliger and Goring, 2000). There are a wide variety of human populations in the world, and if one searches hard enough, it is quite often possible to find natural experiments that would approximate what a given researcher would like to do if he had a more workable species to study (Terwilliger, 2001b; Terwilliger and Goring, 2000; Terwilliger *et al.*, 2002; Terwilliger, 2003). The purpose of this chapter will be to discuss some of the approaches epidemiologists, human geneticists, and experimental biologists use to address questions of etiology, and some ways anthropological geneticists can play an integral role in identifying natural experiments in human populations that approximate the experimental ideal to some degree (Terwilliger and Goring, 2000; Terwilliger, 2003).

Etiology of complex traits

Virtually all human traits are etiologically influenced by multiple environmental, cultural, and genetic factors (Weiss, 1993; Weiss and Terwilliger, 2000). Figure 3.1 illustrates a schematic of the sorts of factors that may influence a given phenotype. As geneticists, we hope there will be 'major genes' (i.e. genes with sufficient impact on the trait to be identifiable) influencing the trait of interest. Furthermore, we assume there are usually numerous genetic factors with minor impact that we do not expect to identify—so-called 'polygenes'. Polygenic variation influences may be further subdivided based on its influences on inter-family variation within a population, and variation between populations. Thus in the figure the polygenic component of trait variation is separated into familially correlated effects, and those which are relevant more for the correlations among individuals within populations but which do not significantly vary across families within a population.

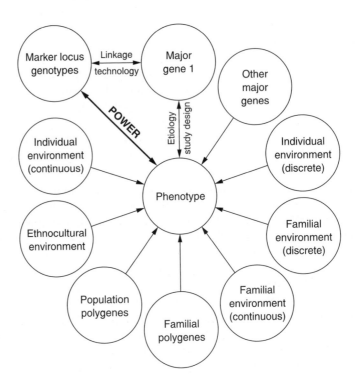

Fig 3.1 Schematic overview of the various types of etiological factors likely to influence a complex phenotype. Note that in reality exposures to each sort of factor are likely confounded in real populations, but these exposure correlations are not indicated in the figure for purposes of clarity.

Environmental risk can be subdivided in an analogous manner, separating exposures which vary among individuals within families (individual environment), those which do not vary within families but vary across families within a population (familial environment) and those which vary across populations but not within (ethnocultural environment). And these correlations can be subdivided into two types — discrete and continuous (loosely analogous to the differences between major genes and polygenes). For example, individual discrete factors, like whether or not someone smokes can have an enormous individual effect on traits distributions, while continuous factors (like how much an individual smokes) can have more quantitative relationships to risk. And smoking as a discrete trait is correlated among relatives, as is how much someone smokes. Likewise, there is cultural variation as well, which influence the prevalence and quantity of exposure. Similarly, there is a quantitative correlation among siblings in terms of how much they exercise or how much fat they eat, and this can be confounded with the polygenic background effect. On the population level similar factors have a role, such as increased cancer incidence around sites of nuclear disasters or effects of the population distribution of fat intake on serum cholesterol or triglycerides. In practice, the effects of discrete environmental factors mimic the discrete effects of major genes, while the continuous environmental factors mimic the continuous effects of polygenes, though continuous environmental factors, unlike polygenes, may be detectable. The correlation of environmental exposures in families and populations can falsely lead a naive investigator to overestimate the genetic component of a trait, as such estimates are based on familial correlations in phenotype, often assuming the only sources of such correlation are genetic, as they are in experimental systems in which we control and normalize the environment of all individuals. Of course genetic factors can influence the choice of environment, and vice versa, and all the factors shown in the schematic are correlated, with their effects on the trait almost never actually being independent.

All of this complexity is typically ignored in traditional medical genetics studies, whether genetic epidemiological or gene mapping studies, rather simple marginal models are often employed in such work, based on what is schematically illustrated in Figure 3.2 as the effect of 'Gene 1' independent of all other factors in the model. In this model, the effects of a single genetic variant on the phenotype is assumed, labelled 'Etiology' in the figure, typically parameterized in terms of the penetrance, or P(Phenotype | Genotype). In the real world, however, the power of a study is related to the inverse of this probability, P(Genotype | Phenotype), or 'detectance', a parameter which can be influenced by the study design (Weiss and Terwilliger, 2000). While the etiological model is a biological constant (at least conditional on exposures to the other genetic and environmental risk factors), the study designs we use can alter the detectance significantly and thus affect the power. Marginal

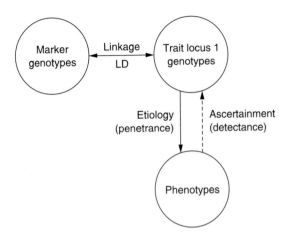

Fig 3.2 The gene mapper's view of the etiology of disease: Markers are typically genotyped throughout the genome, hoping that one will be arbitrarily close to some putative trait locus, and therefore correlated with it because of linkage and linkage disequilibrium. It is hoped that ascertainment of individuals conditional on their phenotype will predict these trait locus genotypes, such that correlations between phenotype and marker genotypes may be observed in the study — and therefore the gene can be localized in the genome.

models allow one to quickly and easily perform power calculations, but they make the typically erroneous assumption that all individuals in the study have risks that are independent and identically distributed (i.i.d), which always makes things appear more tractable than they really are. This assumption is most often patently violated and especially so when considering pedigree data, in which both genetic and environmental exposures are correlated among subjects. There is no such thing as i.i.d. replicate sampling in human genetics (excepting the rare case of MZ twins), as we are all genetically unique and different at millions of sites and have been exposed to a different history of environmental and cultural exposures. This violates every tenet of classical statistical theory, however, and leads to gross overestimation of power to detect specific genetic risk factors when power is inferred from such marginal models, and even leads to naive embracing of dangerously misguided study designs (Weiss, 2002; Weiss, 1996; Weiss and Buchanan, 2003; Weiss et al., 1999; Weiss and Fullerton, 2000).

Although much of the public health research investment in the industrialized world is focused on disease, genetic and environmental risk factors do not typically impact clinical disease directly, but rather act via proximal quantitative phenotypes, as illustrated in Figure 3.3. Genetic variation itself can only directly impact the structure or regulation of a protein or RNA molecule. It cannot give you schizophrenia or diabetes directly. The same is generally true for environmental risk factors, as even a major risk factor like smoking does not cause lung cancer per se, but rather increases mutagenesis rates in certain cells, which leads to increased cancer

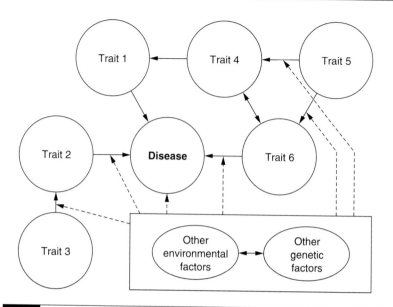

Fig 3.3 Of course, each disease phenotype is usually influenced by numerous quantitative traits which are often closer to the action of the genetic or environmental factors we hope to identify. For example, the disease 'heart attack' is influenced by various quantitative traits like cholesterol level, blood pressure, body mass index, and so on. Note that each of these quantitative traits has its own etiological complexity of the sort indicated in Figure 3.1.

risk. To this end, it is important to evaluate the detectance functions for quantitative traits as well as disease outcomes, in order to predict what phenotypes and study designs might be optimal for detection of genotypic variants which impact on phenotypic variation (Mackay, 2001; Perola *et al.*, 2000; Terwilliger, 2003; Todd, 1999; Weiss *et al.*, 1999; Wright *et al.*, 1999). Note that each 'trait' described in Figure 3.3 would have an etiological architecture of its own resembling that shown in Figure 3.1.

No matter how one conducts a study, in either humans or in experimental organisms, the goal is to maximize the detectance, that is to say we ascertain samples from a population (or design a specific experiment) in order to enrich for some subset of the total etiological universe. We attempt to enrich for one or few risk factors artificially, such that we might identify them as factors which *can* influence the trait in some situations, whether or not they have high attributable fraction (i.e. whether or not they are important in the general population). Of course such studies are not useful for estimation of the penetrance – different study designs more related to those used in classical epidemiology can and should be used to estimate their population-level effects after they have been identified and shown to be relevant in experimental contexts (Terwilliger, 2003). For purposes of this chapter, we focus primarily on the objective of identification of genetic risk factors, as this must be done prior to estimating what exactly their effect would be in uncontrived situations. In the terminology of Figure 3.2, the detectance is what we seek to artificially maximize such that phenotypic variation in an experimental population will predict the risk factor (genetic or environmental) exposure variation with maximal probability.

Epidemiology of complex traits in non-human organisms

There are numerous techniques applied in experimental systems to determine whether genetic and/or environmental factors influence a given phenotype, to estimate how big a role they play, and to identify the specific factors involved (Terwilliger and Goring, 2000; Terwilliger, 2003; Weiss, 1993). In fact, the ability to do such experimental manipulations is the primary reason for the enormous investment in studies of model systems, since one cannot do this sort of research directly in humans for obvious reasons. Furthermore, plant and animal geneticists would be reluctant to study outbred natural populations for purposes of identifying etiological factors, since this is not a powerful approach. There is nothing magical about humans which makes this any less of a problem. However, the medical system in human societies does allow for simpler identification of rare phenotypes, which would not be readily noticed in natural populations of rats or grass. Thus we do have a greater opportunity to retrospectively identify populations with unusual traits, unusual histories, unusual environmental exposure patterns, and unusual mating structures, since these have been recorded in many instances, and certainly diseased individuals are readily noticed in modern societies. This is the one advantage we do have working with humans – it is potentially feasible to look for models of the experiments we might do in experimental systems. To this end, it is important to review the major genetic approaches used to study biological variation in other species.

Natural experiments in genetic epidemiology: Is my trait influenced by genes?

One can work with a variety of inbred strains, and look for phenotypic differences among them, controlling the environment; one can take the same strain and expose it to a variety of environmental factors to see if they have an influence on the distribution of the trait; and of course one can combine these and expose different strains to a set of putative environmental risk factors to see if they have different effects on different genetic backgrounds. In non-inbred organisms, one might do selective breeding experiments in which extreme phenotypic outliers in a given population are mated to each other to see if the offspring have a more extreme phenotypic distribution. Continually doing this breeding of extreme members of a population in subsequent generations until the phenotypic distribution no longer changes allows breeders to estimate the number of genetic factors involved, and to estimate the mode of inheritance. Also, as more and more loci become fixed through this process, identification of specific genes becomes easier as the genetic complexity is gradually reduced though gradual reduction of genetic variation from the population in genes related to the trait of interest. Of course simply looking at large pedigrees of organisms itself can shed light on the mode of inheritance of a trait as well (Wright, 1968).

When moving to human studies, however, it becomes much more complicated, as one cannot homogenize environmental exposures the way one can in mice, one cannot design crosses of the type described by breeding experiments (though one can look for where they may have happened naturally). Even when some phenotype has a different distribution across different ethnic populations, it is not only the genes but also the cultural and environmental exposures which vary between populations making inference much more complex. In humans many types of natural experiments have been used by genetic epidemiologists to try to determine whether or not given traits have genetic components, but these do tend to systematically overestimate the contribution of genes because of confounding of genes and environment and cultural exposures, as alluded to above. Nevertheless, it is important to look at some of the types of natural experiment we use in the absence of a chance to design the studies which would be scientifically optimal.

Table 3.1 gives a detailed overview of a few natural experiments which have been exploited in human research to try to estimate the relevance of various genetic and environmental contributions to the observed phenotypic variation in a population, as outlined in Figure 3.1. Twin studies, adoption studies, family studies, migrant studies, and various combinations thereof are the traditional approaches used for this purpose in humans. Monozygotic twins are the only clonally identical genetic humans, while DZ twins and full siblings share half their genetic variation identical by descent from their parents, and half-siblings share only 25% of their genetic variation – i.e. half the variants they received from the parent they share. More distantly related individuals share a proportion of their genetic variation which decreases by 50% for every meiosis separating them from their nearest common ancestor. Of course the environmental sharing also decays with how closely people are related, in most cases, but the rules for this are far less quantitative and regular. Basically, in genetic epidemiology studies, one attributes phenotypic correlation among relatives to genetic correlation among relatives – according to the rules outlined above. Exceptions to this, in which one directly attempts to assess environmental sharing and its phenotypic impact include adoption studies, in which sibs or twins

Table 3.1. Overview of various natural experiments used in genetic epidemiology: Different data structures provide different information about the contributions of genetic and environmental factors to the trait etiology, as shown in this table – we indicate the amount of correlation in each sort of exposure which is present in each data structure, such that in a genetic epidemiological investigation, the correlation among individuals at the trait can be parsed into these components based on the assumptions in this table. The last column indicates whether samples from each data structure can provide information in a gene mapping study using linkage, linkage disequilibrium (LD), or both.

Data structure	Familial polygenes	Familial environment	Individual environment	Population polygenes	Cultural environment	Major genes	In-utero environment	Gene mapping
MZ twins	Same	'Same' + MZ twins	Twin effect	Same	Same	Same	1 or 2 placentas	LD
DZ twins	50% shared	'Same' + DZ twins	Twin effect	Same	Same	50% shared	2 placentas	Linkage and LD
Sibs reared together	50% shared	Same		Same	Same	50% shared		Linkage and LD
Sibs reared apart	50% shared	Different	Adoption effect	Same	Same, except crosscultural adoption	50% shared		Linkage and LD
Non-sibs reared together	Independent	Same	Adoption effect	Same, except 'trans-racial' adoption	Same	Independent	Independent	LD
Half-sibs	25% shared	'Same' if raised in same household	Half-sib effect	Same, except if non-shared parent from different population	Same if raised in household from same culture	25% shared	Independent if father shared and mother not shared	Linkage and LD
Migrant studies			'Self-selection for migration' effect possible	'Same' within genetic population, different between	'Same' within cultural population, different between			LD
Family studies	Correlated by Mendelian laws	Correlated in less quantifiable patterns		Same	Same	Correlated by Mendelian laws		Linkage and LD
Migrant-family-adoption study	Varies	Varies	Varies	Varies	Varies	Varies	Varies	Linkage and LD
Inbred animal backcross	75% shared	Same	Same	Same	Same	75% shared	Same	Linkage equals LD

are raised in different families from their biological relatives, and migrant studies, in which groups from the same genetic population have been exposed to differences in environmental and cultural exposures as a result of migration. Furthermore, differences in phenotypic correlation between dizygotic twins and full siblings may be attributed to the greater environmental sharing among twins than normal siblings. These study designs are far less informative about relative importance of genetic and environmental factors than studies in experimental species – the extreme case of a backcross between two inbred strains is shown at the bottom of Table 3.1, in which everything is held constant across organisms except genetic variation, for which all animals share 75% of their genomic variation. Differences among such organisms can readily be attributed to genetic variants, whereas in human studies, genetic and environmental similarity of individuals are almost always confounded. To this end, it is critical to think about the hypotheses to be tested before collecting data for a human genetic investigation, such that one might hope to identify appropriate natural experiments. Clearly, despite the best wishes of medical scientists, one cannot accomplish every task in the context of a single data structure or study design!

Experimental approaches to identification of genetic and environmental risk factors

In model systems and agricultural species, there are numerous ways that scientists manipulate populations of organisms to identify specific genetic and/or environmental risk factors for a given phenotype (Mackay, 2001). For environmental exposures, one can simply expose organisms from each of several strains, breeds, or populations to some set of environmental factors, and see what happens. For example, different plant strains may be exposed to different pesticides, fertilizers or soil conditions to see how they respond, or a population of baboons might be fed a high cholesterol diet to estimate the effects of diet on serum cholesterol or blood pressure. In this way, it is possible to look not only at the environmental effects themselves, but also at potential etiological interactions between genetic and environmental variation by comparing the response across families, strains, and populations.

For identification of specific genetic variants related to phenotypic variation, however, much more sophisticated experimental manipulations are required, as we typically have no specific a priori hypotheses in gene mapping studies about which genes and which variants of those genes are responsible for the observed phenotypic variation. We simply try to identify which chromosomal regions or genetic variants are shared by organisms which are phenotypically similar, and iterate this process until we can identify unique sequence variants which can be concluded to influence

Table 3.2. Comparison of study designs used in experimental genetics, and examples of natural experiments we use to approximate the same experimental conditions in human studies: Note that none of the human studies are nearly as powerful as the corresponding experimental systems, but they represent attempts to get as close as possible within the constraints imposed by the need to identify natural experiments in studies of humans.

Experimental genetics	Observational genetics
Saturation mutagenesis	Mutation screening near nuclear disasters
Gene knockouts	Autosomal recessive diseases
Inbred organisms	Populations with low N_e and/or high F_{is}
Clonal population	Twin studies
Environmental manipulation	Migrant studies/Adoption/Epi studies
Random mating experiments	Family studies
Enormous pedigrees by design	Searching for populations with large family size
Selection for extreme traits	Populations with extreme bottlenecks due to historical selection events
Designed breeding experiments	Selective ascertainment on phenotypes
Inter-strain variation studies	Population/anthropological genetic studies
Highly inbred species	Cultures with inbreeding by design

the phenotypic distribution (Terwilliger, 2001a). To this end, various mutagenesis and breeding experiments are utilized. In humans it is far more difficult to approximate these processes than it is to look for environmental exposures that have naturally occurred, and the natural breeding and mutagenesis experiments we identify will always be less optimal than the experimental techniques they hope to approximate, though they have served us well for understanding simple problems in human genetics. It is hoped that through more integrated collaboration with the physical and cultural anthropologists who have expertise in identifying such natural experiments, there may be hope for dissection of some genetic variants which *can* influence variation in complex multifactorial traits in the coming century (Doerge, 2002; Kohler, 1994; Kreutz and Hubner, 2002; Kwitek-Black and Jacob, 2001; Lupu *et al.*, 2001; Mackay, 2001; Montagutelli, 2000; Nordborg *et al.*, 2002; Silver, 1995). Table 3.2 provides a brief outline of some of the basic techniques in experimental genetics, and analogous natural experiments in human genetics which have been used to approximate them.

Inbred strains and crosses

Numerous inbred strains of mice have been designed and maintained over many generations for application in research (Silver, 1995). Inbred mice are homozygous over their entire genome, and all mice from a given strain are assumed to be genetically identical, ignoring the consequences of new mutations now and then. Since they are all maintained under controlled environmental conditions, all differences between strains are assumed to be due to genetic factors (though phenotypic differences have been reported for some traits as a function of how the animals are housed (Mazelis *et al.*, 1987)). It is important to point out that most inbred mouse strains are closely related historically as well, meaning that the variation which exists in these strains is a very small subset of all naturally occurring variation in natural mouse populations, setting aside the obvious selection which has occurred in creating inbred strains, through which any variants which are extremely bad for the mouse in homozygous form will have been eliminated. One simple way in which such inbred strains are useful, as mentioned earlier, is in determining how a given phenotype varies in its distribution across strains. If two strains have a remarkably different distribution for some trait, it can be deduced that the differences have to do with the limited amount of genetic variation which distinguishes them, which is the logical basis of subsequent breeding experiments between those strains to identify which specific genes harbour the phenotypically relevant variants. Strains which are more similar are presumed to be genetically more similar at phenotypically relevant sites, and those which are more different are presumed to be less similar, though the latter is more readily deducible, as for traits influenced by many genes, there may be many ways to end up similar, but few ways to be very different (another reason geneticists tend to be more interested in unusual phenotypes).

When one wishes to study a specific phenotype in the mouse, one first selects inbred strains in which the phenotype of interest is distributed in a markedly different manner. To identify which factors are responsible for inter-strain differences, one might cross mice from the two strains, and then cross the resulting mice (which would be heterozygous at all loci where the two original strains differ, and homozygous at the remainder) with each other, generating a set of mice which have varying portions of their DNA from the two original strains, such that over the whole genome, roughly 25% of loci in a given mouse would be homozygous identical to strain 1, 25% would be homozygous identical to strain 2, and half would be heterozygous, with one allele from strain 1 and one from strain 2. Since each meiosis generates an independent assortment of the parental strains, there is genetic variation among the offspring of this second generation, and if they were again crossed with their siblings, over time greater and greater genetic variation between mice would result across the genome. Nevertheless, there would still be at most two segregating variants at any given site in the genome, and their frequencies in the offspring generations would all be 50–50 in expectation.

One can correlate phenotypic variation in these animals with genotypic variation across the genome, such that if a given chromosomal region harboured etiologically relevant variants, then animals sharing DNA from the same ancestral strain in that region should be phenotypically more similar than animals which received their DNA in that genomic region from different strains. This is the principal of linkage mapping in inbreds (which is equivalent to linkage disequilibrium mapping in this pedigree, because all variants which segregate in this pedigree identify ancestrally to one of the two original strains). This approach has been very useful in identification of DNA sequence variation related to quantitative traits in mouse (and rat) models. It should be noted that even in populations with effective population size (N_e) of two chromosomes (equivalent to one diploid founder individual!), it often requires tens or hundreds of meioses to identify QTLs (quantitative trait loci), and this is in a model system where environmental exposure variation and infection burden are kept artificially homogeneous.

In experimental crosses of inbred strains, when more than two strains are crossed, it becomes much more difficult to map the same genes, as a genetic variant may have variable effect conditional on the genotypes of other loci, and there may be multiple segregating variants at the same or different loci in different strains which influence the same trait. As the number of strains increases, the equivalence of linkage and linkage disequilibrium breaks down, as the same marker alleles may be found in multiple ancestral strains, especially for less-informative markers like SNPs (single nucleotide polymorphisms). For these reasons, it is very important for gene identification to experimentally minimize the amount of genetic variation in a study. To put it simply, experimental geneticists work with inbred strains because it is so much more powerful than working with natural populations that they see no other practical alternative, even though research on subway mice and rats might be easier and less expensive. Mouse geneticists do not randomly collect mice and use a mouse HapMap to find genes related to quantitative variation, not because it is not technically feasible, nor because mice are genetically more complicated than humans – of course they are not, but rather because it is not powerful compared to designed breeding experiments.

If we were to refer back to Figure 3.1, the essence of experimental crosses between inbred strains is to simplify dramatically the schematic in the figure, eliminating all the environmental sources of variation by experimental design, reducing 'population-level genetics' and 'familial level genetics' to a single factor. Furthermore there is no effect of variation in population allele frequencies for different variant sites, as all segregating variants in a cross between two inbred strains will have 50% allele frequency by design. The only distinction here between 'polygenic' and 'major genes' would be related to the actual penetrance function, whereas in humans the importance of a given variant to the distribution of a phenotype in the population is dominated by the allele frequency, such that rare alleles with huge effect in a given individual may have no effect on the distribution of the trait in the population, while common alleles of minor effect may have more consequence on phenotypic means and variances.. In this sense one gets as close as possible to the reductionist model shown in Figure 3.2. Of course, even in experimental crosses, we expect there to often be more than one segregating genetic risk factor, but at least such variation is artificially kept to a minimum.

Randomly mating populations in experimental genetics

In many experimental systems, inbred strains are not the only tool of choice. In studying *Drosophila*, for example, enormously large sample sizes can be obtained at minimal cost, because generation time is extremely short, many offspring can be generated per generation, and there are many genetic marker loci which are assessable based on their phenotypic effects, such that no expensive genotyping is needed for linkage analysis. This allows researchers to start with small selected populations of flies, or even pairs of flies, and let them breed for a few generations to generate a large set of descendants, whose phenotypes and genotypes can be ascertained at low cost with high speed, often using just a microscope, and no molecular markers whatsoever (at least for gross-scale mapping studies). Because one can manipulate the genomes of these species to greater degree, focused hypothesis-driven research can be done in mapping studies, rather than the sort of blind hypothesis-free fishing-expedition paradigm human geneticists are relegated to most commonly. Most *Drosophila* crosses require many more meioses for successful gene mapping than do rat or mouse inbred crosses, because there is much more genetic and environmental variation in the populations used for research (though they are still orders of magnitude simpler than the simplest of human populations).

Because it is so cheap to maintain flies, and because the generation time is so short, one can obtain the very large sample sizes needed to work with outbred populations at low cost, and thus learn many interesting lessons about quantitative genetics in less contrived populations. Note, however, that in studies of bristle number in *Drosophila*, one of the classical quantitative traits studied in any system, there are many chromosomal regions in which clear evidence of linkage has been found repeatedly, and yet identification of the specific DNA sequence variants in those regions responsible for this phenotypic variation remain elusive, despite application of modern molecular genotyping methods.

Similarly, large pedigrees with small effective population sizes are characteristic of dairy cattle and other species of agricultural interest. In dairy cattle, most males are sacrificed at birth, as they do not give milk and thus have little economic value. The few that are maintained must then be used to sire the next generation of cows, leading to (and maintaining) populations with very small N_e. Furthermore, selective breeding for economically desirable traits (like milk yield and quality, among others) has helped to further minimize the amount of residual genetic variation in many breeds of cattle, such that enormous populations can be studied in which N_e is remarkably small. Such studies have been successful in helping to unravel the genetic portion of the etiology of many quantitative traits in these species, despite the absence of large litter sizes or short generation times. In these species, economic considerations have produced experiments which can be used for real scientific advancement.

Returning to Figure 3.1, these large pedigree studies retain greater variation among segregating sites in their allele frequencies, as the parental organisms are no longer genetic clones. Environmental and cultural factors are still largely homogenized and controlled by the experimental design, and unless one conducts independent breeding experiments with different breeds or selected populations, there will not be issues of inter-population genetic and familial variation contributing to the trait variance. Nevertheless, relaxing of the assumptions of inbred crosses leads here to a substantial increase in requisite sample size, without introducing the complex environmental exposure variation seen in humans.

Mutagenesis

Inbred strains and manipulated populations of organisms do not have a lot of naturally occurring genetic variation. This is a very good thing, in terms of making it easier to identify those segregating variants which affect a given trait, but in natural populations, essentially every single nucleotide which can be varied and be compatible with life is varied in some organism somewhere (Terwilliger, 2001a). And especially in inbred systems, the variants which are found are those which do not significantly reduce viability or fecundity in homozygous state. What this means is that only a small subset of all possible variants will be found to vary in these populations. So how can one learn about the rest of the universe of genetic variation? In experimental organisms, numerous methods have been developed to artificially induce mutations, thus increasing the amount of variation present in a population. Two of the most significant approaches are targeted mutagenesis in the form of directed gene knockouts, and random 'saturation' mutagenesis experiments.

In knockout experiments, specific genes are targeted for mutagenesis, with the goal of essentially silencing the gene, so no gene product would be produced. This has been a useful tool in rodent studies, where one can engineer an animal identical at all loci to other animals in its strain, but missing one single gene product. This allows the investigator to determine what the consequences would be of lacking said gene, in a given background, to elucidate some aspects of the function of the missing gene.

Perhaps the most interesting observation to come from years of such study is that there are many instances where a given gene can be knocked out in one strain with no noticeable phenotypic consequences, and yet in another strain, knocking out the same gene can be disastrous. This should probably not be surprising given the complexity and redundancy characteristic of most genetic pathways and networks, but it is a critical point to bear in mind in thinking about complexity even in inbred systems, where no simple genotype→phenotype relationship exists even for sequence variation of this destructive a nature! (Lupu *et al.*, 2001; Montagutelli, 2000; Silver, 1995). While not a mutagenesis experiment, mouse and rat breeders also work with so-called congenic strains which are bred experimentally to be genetically identical to a given inbred strain with the single exception of one single gene or gene-region, which is derived from a second strain — allowing a researcher to look at the context-dependent effects of variation in a gene, when said variation is of a less dramatic nature than that generated by a targeted gene knockout.

In contrast to this targeted genetic manipulation, *Drosophila* researchers often use saturation mutagenesis experiments in which a population of organisms is exposed to some mutagenic agent to randomly generate mutations at random nucleotide positions. The resulting organisms would be bred and the resulting offspring (which carry any induced germ-line mutations from the mutagenesis experiment) would be screened phenotypically for traits of interest. Organisms with unusual phenotypes, or which are outliers from the distribution of some quantitative trait of interest in the original population, might then be used in subsequent breeding experiments, using the techniques outlined above to identify the new variant mutation by mapping as described above. In this way, one can identify how additional variants not existing in the original experimental population influence phenotypic variation. Whether or not those mutations exist in nature, however, is something one cannot determine without screening natural populations for variation in the same gene systematically, such that what is identified are variants which *can* influence trait variation whether or not they actually *do* in real populations.

Of course, by direct analogy, when working with experimental organisms it is also straightforward to test the effects of a variety of environmental factors as well (not just mutagenic agents used in these experiments) simply by taking genetically identical organisms and exposing them to different factors to see what happens to them, as described above. And gene-environment interaction studies can similarly be conducted by exposing genetically different organisms to the same sets of environmental agents to see if there is variation in the consequences. This is equivalent to looking back to Figure 3.1, and systematically altering the exposures — both genetic and environmental — in all possible ways to directly quantify the etiological effects. The beauty of such experiments is that one need not rely on indirect measures of etiology like detectance, but can directly assess penetrance itself.

Natural experiments for human gene mapping

When trying to understand the phenotypic consequences of variation in genetic and environmental exposures in human, it is naturally impossible to perform the sorts of experiment outlined above for experimental species, as would be scientifically desirable. With the exception of controlled clinical trials, it is very difficult to alter the kinds of specific environmental exposures we wish to study in a satisfactorily directed way, and of course, we cannot do mutagenesis experiments or designed breeding (especially not inbreeding) studies to assist in our mapping efforts. The science of epidemiology has focused for generations on how to detect and estimate quantitative relationships between specific environmental exposures and disease risk from observational data, where — for example — cases of a given disease and controls might be ascertained from a population and subjected to questioning about their earlier environmental exposures. This practice has certainly led to some important findings, but mostly of the 'sledgehammer' variety, for instance demonstrating that most lung cancer victims smoked at some point in their lives, while a much smaller portion of the healthy population did. Of course for more subtle relationships, the results have been less than clear. And as one would predict from basic statistical theory, the estimates of effect size tend to be upwardly biased, especially in the first studies that proved the existence of an etiological relationship (Goring et al., 2001). Indeed, even factors that have been initially thought to be well-understood 'sledgehammers' regularly turn out not to be so clear after all, as one clearly sees in the daily epidemiological study reports in major media outlets (Ghosh and Collins, 1996; Ha and Rees, 2000; Holtzman, 2001; Horrobin, 2001; Jorde et al., 2001; Kaprio, 2000; Le Fanu, 1999; Nevanlinna, 1974; Patterson, 2002; Proctor, 1995; Rees, 2002; Weiss and Buchanan, 2003; Weiss and Terwilliger, 2000; Zak et al., 2001).

In the absence of controlled experiments, human geneticists are forced to become detectives, and look for natural experiments, in which the sorts of matings and population histories we would like to observe have been generated naturally, without regard to our biomedical interests, or where environmental exposures have been intrinsically altered in such a way as to make for a useful resource for epidemiological investigation, again without wilful design by scientists. Thus, rather than being a formally experimental science, as would be the ideal for hypothesis testing and unravelling of the complexities of life and illness, human genetics must rely on statistical inference from observational data.

These are the primary contrasts between the more pure scientific culture of hypothesis-driven research in lower organisms and the more anthropological detective work we must rely on in human studies. This necessity to play detective and search for natural experiments has motivated whole industries of sophisticated technological advances, spawning the young sciences of genomics and bioinformatics to help us perform our detective work, or to make the most out of the undersized, underpowered, and suboptimally designed natural experiments with which we are able to work, even in the best cases. However, since the relationships between genetic variation and phenotypic variation are not engineering problems, technological advances are not solutions, but rather tools which are only useful with optimal data and good scientific hypotheses. Thus, geneticists must search the world for optimal data to apply these technological advances to. The real challenge in the genomic era is to remember that fundamentally our questions are scientific, largely requiring selection of appropriate study designs, and not matters of technology per se (Terwilliger, 2004).

Study designs for gene mapping

In gene mapping, there have been a number of changes over time in the philosophical approaches to study design. Initially most gene mapping studies were based on collecting large families of individuals, many of whom shared some unusual disease phenotype. The underlying assumption was that these individuals shared the same phenotype because they shared some underlying genotypic risk factor — that is to say detectance was complete, with phenotype determining genotype. Within a given family, affected individuals were reasonably assumed to share the risk alleles from some common source, and thus the gene involved could easily be localized to some chromosomal region by linkage analysis (Terwilliger, 2001a). For more complex multifactorial traits, however, there is no deterministic detectance function — in the best case scenarios, one hopes for slight increases in risk allele frequencies in affected individuals, not guarantees of a specific disease genotype. For this reason, in order to increase the power of such human mapping studies, one has to find a way to increase the predictive value of the ascertained phenotype on the underlying genotype.

It is easy to say that one needs ascertainment bias to improve the detectance functions, but how can this be done in practice? Many approaches have been used by geneticists to date, though it must be admitted that the track record of mapping studies has not been good, even with these approaches. Many investigators have identified those unusual large families with lots of individuals sharing their target disease phenotype (Enattah *et al.*, 2002; Hovatta *et al.*, 1999; Kuokkanen *et al.*, 1997; Liu *et al.*, 1998; Pajukanta *et al.*, 1999). The empirical evidence suggests that this approach is much more powerful than studies of larger numbers of small pedigrees with few affecteds, and there are theoretical arguments to support this observation (Terwilliger *et al.*, 2002; Terwilliger, 2001a). Another way to increase the predictive value would be to ascertain individuals who are as similar as possible environmentally and culturally, to minimize variability in risk due to such factors. One could also ascertain families of affected individuals who lack the known environmental and genetic risk factors for disease, and unaffected individuals who have many of the known risk factors for the same disease, to increase the likely differences between groups for the as-yet-unidentified risk factors (both genetic and environmental). Sampling from populations with as small a degree of genetic variability as possible can also be of use, because in such populations there should be a reduced number of unique genetic variants segregating (both rare and common) which could potentially influence disease risk. Of course, a combination of all such approaches must work better than any one of them individually.

There is something of a false dichotomy between linkage and linkage disequilibrium (LD) analysis — in animal crosses they are essentially identical, and even in outbred populations like human, we are ultimately all distantly related, and LD is nothing but linkage, in the enormous pedigree that connects us historically (Terwilliger, 2001a). The primary difference in human genetics has historically been related less to scientific issues, and more to study design used for such studies, with 'unrelated' individuals being used for LD analysis, and families being ascertained for linkage analysis. However, there is no reason one cannot perform a linkage disequilibrium analysis in large families, since this is essentially what was done for identifying pathogenic variants in monogenic diseases (cf. Enattah *et al.*, 2002; Nikali *et al.*, 1995; Tienari *et al.*, 1994; Norio, 2000). Moreover, there is no reason not to combine the data from all such studies for a joint linkage and LD analysis (Goring and Terwilliger, 2000a; Goring and Terwilliger, 2000b; Goring and Terwilliger, 2000c; Goring and Terwilliger, 2000d). Clearly tests of either hypothesis are made more powerful by joint consideration of both phenomena, especially since linkage disequilibrium is the same thing as linkage — just that in LD analysis, the affected individuals are assumed to be the bottom generation of some really large pedigree (i.e. they are assumed to have received the risk alleles from some common ancestor deep in history), on which linkage analysis is essentially being performed ... It is therefore illogical and counterproductive to speak of LD as something different from linkage, because the LD that is relevant to mapping exists only because of linkage. Despite all the rhetoric to the contrary

(Barton *et al.*, 2001; Brookes, 2001; Cardon and Bell, 2001; Carlson *et al.*, 2002; Chakravarti, 2001; Chanock, 2001; Collins, 1995; Collins, 1999b; Collins, 1999a; Collins and McKusick, 2001; Couzin, 2002; Gabriel *et al.*, 2002; Gottesman and Collins, 1994; Guyer and Collins, 1993; Johnson *et al.*, 2001; Jorde, 2000; Judson *et al.*, 2002; Kruglyak, 1997; Patterson, 2000; Patterson, 2002; Pennisi, 1998; Reich *et al.*, 2001; Reich and Lander, 2001; Risch and Merikangas, 1996), LD as a tool for mapping, even in genome scanning situations, is not a new idea – it was used for mapping many of the genes for rare monogenic diseases of the Finnish disease heritage, for example (Nikali *et al.*, 1995), as will be illustrated below – nevertheless, the reason it worked when it did was precisely the same reason linkage analysis works when it does – detectance was high and disease phenotypes accurately predicted risk genotypes. When detectance is low, LD mapping is just as problematic as linkage analysis, and may be worse because every individual in a random sample would have independent chromosomes drawn from the population, increasing heterogeneity dramatically as the number of independent chromosomes per phenotyped and genotyped individual is maximal in that sampling scheme.

Finnish population as natural experiment for rare monogenic diseases

Finland is a country in Northeastern Europe with a current population of roughly 5,000,000 inhabitants, with 99.5% of all people in the country as recently as 1991 having been themselves born in Finland, making it one of the countries in Europe with the lowest immigration rates (Eurostat, 2000; Jasinskaja-Lahti *et al.*, 2002; Pohjanpää *et al.*, 2003). While immigration has increased with the fall of the Soviet Union, Finland still has the lowest percentage of foreign-born residents in the European Union. While within Finland, there are small indigenous populations of Saami, and 5% of Finns speak Swedish as a first language, Finland remains one of the most genetically homogeneous and isolated countries in Europe (Peltonen, 1997; Peltonen *et al.*, 1999; Peltonen *et al.*, 1995). Despite its having been a colony of Sweden from the twelfth to the nineteenth century, and subsequently a Grand Duchy in the Russian Empire until the Russian Revolution in 1917, there was remarkably little immigration from these countries. Furthermore because the Finnish language is quite distinct from the dominant languages of its larger Indo-European speaking neighbours, there were additional barriers to mating and integration of these populations. For all of these historical reasons, the effective population size has been estimated to be only on the order of 10^3 due to founder effect in the original thousands of settlers, followed by rapid expansion with the advent of the agricultural economy in earnest (Kere, 2001; Norio *et al.*, 1973;

Peltonen *et al.*, 1999; Peltonen *et al.*, 2000; Perheentupa, 1995; Wiik, 2004). While today Finland is one of the most desirable places to live in the world, and could easily be a receiving country for immigrants, with its fantastically cosmopolitan cultural and intellectual life, advanced social, educational, and health care systems, and extremely high standard of living, few people migrate there to this day, making it one of the few 'large' populations in Europe with significant advantages for genetic study, largely because of its small N_e, combined with a sophisticated modern health care infrastructure.

Because of this rather unique history of genetic isolation followed by rapid expansion, today's 5,000,000 people have genomes that are mostly composed of segments inherited identical by descent from those original founder chromosomes, with each original chromosome being present in, on average, thousands of modern Finns. Because thousands of copies of each original allele are present in the population, alleles which are very rare elsewhere (and which may have been found in only a single founder of the modern Finnish population) have become quite common by chance in this population (Norio *et al.*, 1973). Among the alleles which increased in frequency because of this founder effect are many 'knock-out' alleles which are loss-of-function mutations in different genes which can be tolerated in single copy, but which produce disease when present on both chromosomes. The effect of this has been that many recessive diseases which are almost never seen elsewhere in the world have risen to appreciable frequency in Finland – the so-called 'Finnish Disease Heritage' (Norio *et al.*, 1973; Perheentupa, 1995). Finnish researchers noticed this high incidence of several unusual diseases and noticed that patients were often regionally isolated, and if they used the Finnish church books to trace the families of affected individuals back in history, very often they could find a common ancestor of a large proportion of the affected individuals (Varilo, 1999). Of course this also leads to regional clusters of disease in Finland as well, where the population is essentially one large family going back many generations.

In a population of $N_e = O(10^3)$, it is reasonable to assume that if a disease allele has a frequency of the order of 10^{-3}, that probably the plurality of disease alleles will be clonally identical from a common ancestor. If there were multiple ancestral alleles in the population, or new mutations occurred later in time, there would still be one major allele because of the small population size followed by rapid exponential growth. In a population with many more founders, one would expect much more heterogeneity in the molecular nature of the various ancestral knockout mutations responsible for disease, as is typically seen in recessive diseases in much larger Central European populations (there are many ways to knock out a gene by random mutagenesis). What this means, vis-à-vis the experiments we would like to do in animals, is that mapping in our disease population is analogous to mapping a gene using congenic animals, when all disease chromosomes come from a single ancestral background.

When studying monogenic recessive diseases in populations characterized by substantial homogeneity in the ancestral source of the disease alleles, the detectance, or P(Genotype | Phenotype) is enormously high – with all affected persons assumed to carry copies of the same disease allele, identical-by-descent from a common ancestor, along with all other variants contained on this ancestral chromosome carrying the disease allele in a neighbourhood sur-rounding the disease locus because of linkage disequilibrium (see (Terwilliger, 2001a) for mathematical details). By analogy to the mating system used to generate congenic animals (Kreutz and Hubner, 2002; Kwitek-Black and Jacob, 2001) consider the variants found on the founder chromosome on which the disease allele entered the population to be from 'strain 1', and all other chromo-somes in the population are lumped together as 'strain 2'. Over time, random matings occur in the population history, by which process, the disease allele and small segments of DNA surrounding it are transferred by recombination into people whose remaining DNA comes from 'strain 2' in this cross. After many generations, these 'congenic humans' eventually are identical to 'strain 2' with the exception of ever-smaller neighbourhoods around the disease-allele, on which DNA from 'strain 1' is maintained. Affected individuals must be homozygous for this allele from 'strain 1' and small neighbourhood around it, while everywhere else, they carry alleles from 'strain 2' on both chromosomes – essentially affecteds in this population are equivalent to congenics. By searching for these regions of the genome in which both chromosomes carry identical chromosomal segments in homozygous state over a region, the location of this 'congenic' region can be identified, and thus genes are mapped using a natural experiment in humans very similar to what we might engineer in rat genetics. This procedure was used with great success and elegance for mapping the genes for these 'Finnish Disease Heritage' diseases, as well as many other more common monogenic diseases which are likewise less heterogeneous in Finland than elsewhere in the world (Norio, 2000; Peltonen, 1997; Peltonen et al., 1999).

Natural experiments for gene mapping in complex multifactorial traits

Of course, the 'congenic' analogy of the Finnish population does not extend to the case of complex multifactorial traits, for which the detectance is much less deterministic. Furthermore, even if the etiology were not extremely complex, one expects allele frequencies on the order of 10^{-1} rather than 10^{-3} for the Finnish disease heritage alleles. If 10% of founder chromosomes harboured risk alleles at a given locus, then there would still be $O(10^2)$ founder chromosomes with this allele on average. This is not a situation for which the desirable properties of Finland as a natural experiment apply (Kere, 2001; Peltonen et al., 2000). When searching for more genes with higher frequency variants which alter the phenotypic distribution of some trait, one might wish to place greater focus on simplifying the etiology artificially by reducing the number of chromosomes segregating in a population, or at least in a sample therefrom. One approach is to look for populations with much more extreme reductions in N_e, in the hope that overall reductions in the background genetic variance will make mapping more straightforward. Of course, when alleles have frequencies on the order of 10^{-1}, it is probably not reasonable to expect clonal identity of all functional alleles in a sample, as even the smallest human populations harbour significantly more variation than inbred mouse crosses. However, as seen in the successful mapping studies in agriculture and more outbred species like *Drosophila* and *Xenopus*, it is possible to compensate somewhat for the added heterogeneity through increases in sample size, and homogenization of environmental exposures.

Use of large pedigrees for linkage studies can further simplify the problem, since the operative parameter for population variability would be the effective number of independent chromosomes from the population divided by the number of individuals in the overall sample. To this end, it is clear that, for example, unrelated individuals would have two chromosomes per sample, nuclear families with N offspring would have $4/(N+2)$ chromosomes per sample from the population. Furthermore, populations with smaller N_e would have even smaller effective numbers of chromosomes, due to the greater chance of sampling the same chromosome by chance from small populations compared to larger ones, such that the combination of large pedigrees and small populations with great cultural and environmental homogeneity would be another sort of natural experiment that would be useful in gene mapping studies.

Populations which practise consanguinity by design are likewise a useful means for reducing the effective number of segregating chromosomes in a pedigree sample, since spouses are increasingly likely to share chromosomal segments identical by descent, and become closer to the congenic or inbred cross models that we rely on in experimental species. Populations of this sort have been used in many cases for mapping of simple monogenic recessive diseases in Pakistan, for example, where there are populations that practise first-cousin marriage by design (Ahmad et al., 1993; Leutelt et al., 1995). Such pedigrees show increased incidence of rare recessive disorders because of the increased homozygosity resulting from inbreeding – another means of generating the sort of 'congenic' models that characterized the Finnish disease heritage. Of course, inbred populations should be much more valuable for complex multifactorial trait mapping than the Finnish population, because N_e

is generally much smaller (at least within a given breeding population), the cultural variation is minimal among individuals that practise this sort of mating, and pedigrees tend to be quite large in those populations as well, since ultimately all persons, including spouses, are related (Leutelt *et al.*, 1995; Denic *et al.*, 2005; Yaqoob *et al.*, 1998). Unfortunately, to date these advantages have not been exploited to a great extent for complex trait mapping, as researchers are still focused on the remaining hundreds of recessive monogenic traits remaining to be mapped in those populations.

Complex disease research, however, is still much simpler and less complex to do in Finland than in larger more cosmopolitan Western populations because of the availability of extensive medical diagnostic and population family registries, and top quality medical facilities at which sophisticated phenotyping can be performed. Further, the cooperative nature of the population, and their intrinsic interest in medical research make it a great laboratory for such research, and gives perhaps the best chance in the industrialized Western world for genetic epidemiology and mapping of genetic risk factors for such traits — of course everything would work better in more extreme isolates, and inbred populations, but most of those lack the sophisticated medical infrastructure close at hand. Finland is far from ideal in terms of population structure and history for mapping genes for common complex diseases, of course, with its relatively large effective population size, but anything that simplifies the problem is helpful, and one might be able to compensate with increased sample size for what one loses in population genetic advantages as one moves to more complex problems, as is the case when attempting to map genes in *Drosophila* or cattle, compared to inbred experimental crosses in mouse and rat. Nevertheless, it must be admitted that, to date, there has been little success in the early days of attempting to study more complex traits there (Hovatta *et al.*, 1999; Kuokkanen *et al.*, 1997; Pajukanta *et al.*, 1999; Peltonen *et al.*, 2000; Perola *et al.*, 2000; Terwilliger and Goring, 2000).

One example which we have worked on for a decade and more is the very complex disorder, schizophrenia, which has been a major target for gene mappers throughout the world, not because of any real specific genetic hypothesis, but rather because we know so little about the biology of this disease, and it is a major public health burden across the globe — not the best reasons for applying genetic methods in general. By linking together the Free Medicine and Disability Pension Registers from the Social Insurance Institution, the Hospital Discharge Register and the Population Register, a description of the prevalence and familiality of schizophrenia in Finland was obtained, as shown in Table 3.3 (Hovatta *et al.*, 1997). In this table are the nuclear family structures of all diagnosed schizophrenia patients born between 1940 and 1969, and having a reported diagnosis of schizophrenia in at least one of the registers between 1974 and 1991, presented as numbers of sibships in the population with N schizophrenics and M non-schizophrenics. Furthermore,

Table 3.3. Number of families with a fixed number of schizophrenic and non-schizophrenic individuals per sibship in all of Finland who were born between 1940 and 1969, who were diagnosed or hospitalized with schizophrenia between 1974 and 1991 (see text for details). For example, this table indicates that there were 89 sibships with 2 schizophrenics and 5 non-schizophrenics. Data from (Hovatta *et al.*, 1997).

Number of unaffected sibs in the sibship	Number of schizophrenic sibs in the sibship						
	1	2	3	4	5	6	7
0	3246	236	31	8	3	0	0
1	5345	374	42	7	1	1	0
2	4728	351	43	11	0	0	1
3	3250	238	38	4	2	0	0
4	2013	222	15	4	1	0	0
5	1381	89	13	5	0	1	0
6	463	56	21	1	0	1	0
7	312	52	8	1	0	0	0
8	181	24	2	1	1	0	0
9	91	18	1	1	0	0	0
10	41	4	3	0	0	0	0
11	28	2	1	0	0	0	0
12	14	2	0	0	0	0	0
13	10	1	0	0	0	0	0

within specific regions of the country, it was possible to build extremely large pedigrees connecting individuals together historically (Hovatta *et al.*, 1998; Varilo, 1999), in the hope that this technique would help simplify the problem, as had been the case for the 'congenic' models in the Finnish disease heritage. However, because of the massive complexity of schizophrenia, so far nothing conclusive has been found despite more than a decade of intense effort, extensive high-quality phenotyping of thousands of individuals from multiplex families, high quality genotyping of thousands of SNP and microsatellite markers, and the most sophisticated statistical analyses available. Not only is this sample selected because of Finland's relatively advantageous population history and cultural homogeneity, it is also by far the largest sample in the world of carefully phenotyped and well-characterized families that has ever been studied for schizophrenia. If nothing else, it sends a clear message to investigators working in other parts of the world, typically with *much* smaller sample sizes from *much* more cosmopolitan populations. Unless they have some very good scientific justification for why their study design should yield significantly more power than these studies in Finland, there is no point in pursuing them at this time without radically different methodologies, a sobering thought for people invested in this difficult and challenging disorder. But therein lies the rub – the reasons people do genetic studies of schizophrenia (and many other common diseases) is that they do not have any better idea how to approach the disease, and they know there is some familiality (sibling recurrence risks of about 2–3 (Hovatta *et al.*, 1997)). But one should use genetics as a tool when one has a real genetic hypothesis, and a reason to believe that genetic methods will work – as we had for the Finnish Disease Heritage, not just because the trait appears to be heritable or familial and we want to find the causes of the familiality. Better natural experiments may be necessary.

There are small indigenous populations in Finland and the other Nordic countries, each of which have the benefit of sophisticated modern medical facilities and population registers. Among these are the Greenlanders in the Kingdom of Denmark, and the Saami population inhabiting Northern Sweden, Norway, Finland and Russia. Each of these populations have significantly lower N_e than the Finnish population, as a result of centuries of isolation and absence of exponential growth spurts, as they are nomadic hunter-gatherer and reindeer-herding populations, not sedentary agriculturalists like the Finns were traditionally. Genetic epidemiology research in the Saami is in its early stages (Johansson *et al.*, 2005; Hassler *et al.*, 2001) but the hope is that due to the substantially reduced genetic variation and correspondingly long regions of high LD, they might be a more amenable population for such QTL mapping in complex traits (cf. Terwilliger *et al.*, 1998; Laan and Pääbo, 1997). Greenlanders offer the further advantage of being admixed with Danes in a very unique natural experiment, wherein two very distinct populations – one of which (Inuit) has extremely small N_e – have been mixing for centuries in Greenland, without additional heterogeneous populations entering the mix. Thus, one has the opportunity to look at some natural experiment analogous to interstrain crosses in agriculture or mouse, in which the initial populations are selected for maximally different phenotypes and genetic composition to look at the effects of admixture (Terwilliger and Goring, 2000; Bjerregaard and Young, 1998; Chakraborty and Weiss, 1988)). These populations offer the added benefit of access to population-based family structure and medical diagnostic registers characteristic of the Nordic countries. Hopefully, in the future further collaboration between epidemiologists, anthropologists and geneticists will allow for identification and detailed study of other such natural experiments throughout the world before the inevitable loss of genetic isolation and decline in family size that comes with Westernization.

Natural experiments for genetic epidemiology

It is the case, nevertheless, that in Central Asian republics of the former Soviet Union, like the Republic of Kazakhstan, that family sizes are actually increasing as the government has promoted population growth, given the vastness of the land mass and low population density in that part of the world. This will be a potentially fruitful resource for genetic epidemiology in the future given the highly educated populace, and present day population growth. Furthermore, Kazakhstan is an ethnic mosaic with numerous ethnic populations residing there and remaining largely endogamous over the last half century since many of these ethnic minorities were deported to Kazakhstan in the Stalin era, when Kazakhstan was one of the republics of the Soviet Union. During Soviet times, there was substantial pressure for the various ethnic groups to acculturate (i.e. to become good 'Soviets', which basically meant to adopt a Russian way of life in practice), leading to the sort of reduction in environmental and cultural variance which is useful for gene mapping studies. Thus in Kazakhstan, there exist many different genetic populations, with minimal inter-population cultural variation, allowing for the potential to examine the genetic component of different traits, by looking at inter-population phenotypic variation, and comparing this to the known inter-population levels of genetic variation. Nevertheless, with the decline and fall of the Soviet Union, there is a trend towards greater exogamy, which represents the human analogue of the inter-strain crosses and breeding experiments one might do in other species, allowing for interesting studies in the future, if we take advantage of the present opportunity for anthropologists and medical researchers to characterize the populations as extensively as possible before their inevitable mixing makes this impossible in the future. This will be a potentially valuable resource, assuming the trend toward larger families continues and applies to these inter-ethnic couples as well as to the titular majority population.

Korean diaspora project as natural experiment to examine gene-environment interaction

One of the most common study designs used in plant genetics to investigate the relative contributions of genetic and environmental factors to trait variation is to take identical plant strains (or populations) and grow them in different fields, under different environmental conditions (vis-à-vis soil composition, light exposure, water availability, exposure to fertilizers or pesticides, etc.) and look for variation in the mean and variance of some set of quantitative phenotypes. If one is looking at clonally identical plants, then all differences would be attributable to the environment, yet if different clonal populations responded differently to the environmental differences, this would be taken as evidence for gene-environment interaction (or differential response to the environment conditional on genotype). In non-clonal populations with some genetic variability, one would estimate the heritability of each trait for the same population exposed to different sets of environmental exposures – differences in heritability would be evidence of interaction between genetic and environmental factors, while differences in the mean of the phenotype in different environmental conditions would merely demonstrate an influence of the environmental factor on the phenotype, and would say nothing about genetic contributors. Such experiments have yielded some interesting results in agriculture, but it should be noted that proving gene-environment interaction to exist, even in such extreme experimental designs, has been notoriously difficult. This does not mean anyone doubts it likely exists, but rather that it is tough to prove, even in the best of situations.

In human studies, this is vastly more complicated still. Epidemiologists and geneticists these days speak about gene-environment interaction studies all the time, but really they are not engaged in experiments that disprove the null hypothesis of no interactions between the genetic and environmental risk factors — rather they look at contributions of both sorts of factor to the same trait in many cases. In other cases, they adopt a very extreme definition of the null hypothesis, related to additive action of all risk factors (or multiplicative independent relative risks in the case of discrete traits — a similarly restrictive definition), and they take evidence against this hypothesis to constitute proof of gene-environment interaction, despite the well-known fact that there are numerous ways these simple null hypotheses can be rejected without requiring that environmental effects on the phenotype act differently on individuals with different genotypes in a meaningful biological manner. This disconnect has led to numerous claims of gene-environment interaction having been proven to be widespread and epidemiologically relevant, yet virtually none of the existing data supports this conclusion in a way that would be equivalent to the meaning of 'gene-environment interaction' in the sort of experimentally designed studies described above, which are used all the time in agriculture. Essentially people believe G × E interaction to exist ubiquitously, and never bother to adequately define the null hypothesis in a manner amenable to testing it.

We have recently begun an investigation of G × E interactions in human quantitative traits which is based on the more pure definitions used in controlled agricultural experimental designs, by identifying a natural experiment that would be as close as possible to the 'same plants in different fields' design outlined above. For this purpose we take advantage of Korean emigration history as a natural experiment, which is probably as close as we can get in human populations to colonies from the same genetic population being exposed to radical differences in environmental and cultural exposures, largely as a result of the rather unique history of Korean migrations in the past 150 years (Lee, 2000).

The Korean peninsula had been relatively isolated culturally and genetically for centuries, prior to the political turmoil which began in the late nineteenth century. The isolation was largely aided by geographic factors, with a very steep mountain range, and a series of wide rivers on its Northern border with Manchuria, and seas on the other three sides, keeping them largely independent of its much larger neighbours, China and Japan. Starting in the 1860s, when a drought and famine hit the northern part of the peninsula, many Koreans began to cross one of the rivers and migrate to the Russian Far East. Over the next few decades, tens of thousands of Koreans migrated to the region following the agricultural success, with the blessing of the Russian government, as the region was largely uninhabited, and the Koreans made good economic contributions to the Russian Empire. After the fall of the last Chinese dynasty, Koreans were permitted also to cross the border and settle in Northeastern China, where even more Koreans went in the late nineteenth and early twentieth centuries. When the Japanese invaded and colonized Korea at the beginning of the twentieth century, these migrations accelerated, with many Koreans escaping to China and Russia, both as refugees and partisan fighters against the Japanese occupation. During this same period, the Imperial Japanese involuntarily sent hundreds of thousands of Koreans to Japan as forced labourers, many to the northern island of Karafuto.

In 1937, when Russia was fearing attacks from the Japanese army in the Far East, Stalin issued an order to deport all Koreans living in the Far East Maritime Province to Kazakhstan in Central Asia, as he feared they would spy for the Japanese, as strange as that may seem. In the course of one month in October 1937, the entire population of roughly 100,000 people were sent to Central Asia, with only the bare essential possessions with them. The Korean cultural institutions were eliminated, the libraries burned, and the entire population was forcibly acculturated, and moved to the steppes, mostly with the task of setting up collective farms in radically different conditions than they had been used to. Labelled as an 'enemy minority' until their rehabilitation with the fall of the Soviet Union in the early 1990s, this population was stripped of their cultural traditions, Russified in language and lifestyle, and were characterized by very low rates of exogamy, keeping the genetic variation in this group essentially unchanged, with radical alterations in environment and culture, in an involuntary manner. Furthermore, family size tends to be quite large in this population, with 11% of Korean women having more than four children even today, with even larger families among rural farming families (Pak, 2002).

The Koreans who crossed the border into China were also separated from their homeland by political barriers after the end of World War II, and the Chinese communist revolution. While physically located adjacent to the northern part of the Korean peninsula, the roughly 2,000,000 Koreans in China had no direct contact with South Korea for half a century, and had only limited interaction with the insular North Korean state. However, the Chinese minorities policy was quite different from that of the Soviet Union, and the Koreans were allowed to maintain their traditional culture, and set up an autonomous region in Northeastern

China, the Yanbian autonomous prefecture (Suh and Shultz, 1990). Because of the isolation there, however, this population of Koreans has a much more traditional lifestyle and environment than South Koreans today, and unlike Han Chinese, they are not restricted to one child per family, making for the possibility here as well of looking at large families of ethnic Koreans, living in an environment and lifestyle much like that of South Korea 30 years ago. Again, much like the population living in Central Asia, they were a population of involuntary migrants for the most part, and were forcibly isolated from their homeland by Cold War politics as well.

The Koreans in Japan as well represent a population of involuntary migrants, largely sent there during the Japanese occupation of the Korean peninsula in the first half of the twentieth century. Even second and third generation Koreans are often not allowed Japanese citizenship, and they suffer discrimination in employment, education, and other social situations through to the present day, again making for an interesting population of forced migrants with radically different environment and culture. As most Japanese Koreans speak Japanese on a daily basis, because of the discrimination, they often hide their ethnic background to avoid discrimination, and may avoid eating traditional Korean diets to avoid being discovered as Korean by the smell of the foods (Hicks, 1997).

The final population of forced migrants which are interesting for examination of relative contributions of genetic and environmental factors are the result of international adoption. After the end of the Korean War, there was enormous economic hardship in South Korea, and there were many infant children without fathers. Because there is not a strong tradition in Korea of adoption, excepting within the same family, orphanages were overloaded, and there was little chance of finding homes for so many displaced orphans in the chaos of post-war Korea. Many agencies then began to look for homes for these infants in Northern Europe and North America, and this was so successful that international adoption of Korean children remains a common occurrence to this day, with more than 100,000 such children having been sent to the United States, and slightly less than 10,000 to Sweden (and similar numbers to other Northern European countries) over the past 30 years. These children were then raised as if they were American or Swedish, with typically no exposure whatsoever to traditional Korean culture food or lifestyles. As such they are the most extreme example of 'same genes, different fields' in this natural experiment. The down-side is that they are typically without biological relatives who can be studied to look at trait heritabilities, with the exception of those children who have searched for and found their birth parents in Korea. As many of these Korean adoptees are now adults, they do make for a very interesting comparison group, nevertheless, especially the instances where full sibs and even twins have been sent internationally, and can be examined to look at G × E effects.

Of course, there are also many economic migrants from Korea, to North and South America, Australia, and Western Europe, but those populations tend to have more contact to their 'home culture', making for less extreme inter-field environmental exposure differences. Nevertheless, since there are millions of such economic migrants, many of whom have families of their own, they can provide some additional information for this natural experimental design.

We have only recently begun to collect data about these overseas Koreans, with a goal of eventually having data from sufficiently many of these populations that we can look at the effects of gene-environment interaction. For various scientific as well as economic reasons, we commenced with a study of the Koreans living in Kazakhstan, one of the newly independent republics of the former Soviet Union, while setting up collaborative plans to look at the other populations in the future.

Today, there are approximately 100,000 Kazakh Koreans, representing about 0.7% of the total population, in Kazakhstan (Pak, 2002). The number of Kazakh Koreans has declined by 10,000 in the past ten years following the collapse of Soviet Union, largely due to emigration to the Russian Federation. The ethnic Korean population is heavily populated in five areas, namely Almaty city (n = 19,090), Yuzhno-Kazakhstanskya oblast (n = 17,488), Qyzylordinskaya oblast (n = 11,020), Karagandinskaya oblast (n = 14,097), and Zhambylskaya oblast (n = 14,000). Similar to Kazakhs, the age distribution of Kazakh Koreans resembles that of most developing countries, where the number of elders is smaller than that in the younger generations. This is different from Korea where the age distribution is now heavy in the middle age groups. The distribution of sex is even: 49% male vs. 51% female.

In an initial pilot study, we randomly ascertained a set of ethnic Korean families, where the study eligibility criteria are that: (1) the proband was an ethnic Korean, with the requirement that all four grandparents were ethnic Koreans; and (2) they have family members willing to participate in the study. In this initial study, we obtained useful phenotypic and genotypic data from 107 individuals from 15 multiple-generation families residing in Almaty (urban centre) and 77 individuals from Ushtobe (rural area where most participants live on collective farms). The mean age of the sample was 40 in Almaty and 52 in Ushtobe, with 43% of the sampled individuals in Almaty under 40 years of age, whereas only 25.3% of the samples in Ushtobe were under 40, reflecting the tendency for young people to migrate from rural areas to urban centres for jobs.

The majority of the participants were born in Kazakhstan with the mean length of residence in Kazakhstan being 50 and 35 for Ushtobe and Almaty, respectively. Further, because of the political climate during the Stalin era, Kazakh Koreans readily assimilated to Russian culture. As a result, the majority of the subjects from both Almaty and Ushtobe spoke primarily Russian (98.6 in Almaty vs. 81.8% in Ushtobe). The majority of the participants reported that they sometimes eat Korean foods. However, it should also be noted that the Korean foods in Kazakhstan are quite different from those in the Korean peninsula. For example, the most 'famous' Korean dish in the former Soviet Union was a salad made from sliced carrots — something unheard of in Korea.

The prevalence of smoking is somewhat higher in Kazakh Koreans compared with Koreans in Korea, but lower than that among Korean Americans. Approximately 38.7% of the participants in Almaty reported smoking, compared with 29.6% in Ushtobe. In comparison, 16% of Korean Americans smoke (Kim and Chan, 2004), compared withy 82.6% of Korean men in Korea smoked and 9.2% of Korean women smoked (Kim, 2001). Drinking alcohol, as one might expect, is quite common in Kazakhstan: 57.5% in Almaty reported drinking regularly, compared with 43.7% in Ushtobe.

The potential problems of performing this sort of cross-cultural study were illustrated by the responses to the drinking and physical activity questions. Although the participants in Ushtobe were largely involved in physical labour, as farmers, while the participants in Almaty were more sedentary, working in white collar jobs, a greater proportion of individuals in Almaty (10.4%) reported regular physical activity than in Ushtobe (7.2%). It is clear that the respondents were largely reporting leisure activities, not physical daily activities. In the case of drinking alcoholic beverages, many participants responded that they do not drink, yet elsewhere indicated consumption of significant amounts of beer and wine. It turned out that they considered only vodka to be 'alcohol', thereby grossly underestimating the prevalence of drinking.

Among the phenotypes we examined in this initial pilot study, we compared BMI in the two populations of Kazakh Koreans. As expected, the proportion of obese (defined as BMI > 30) individuals was higher in the Almaty (urban) samples than that in the Ushtobe (rural) samples (13.2% vs. 9.5%, respectively), and BMI was, on average, considerably higher than reported statistics from South Korea (Kim, 2001). Although our study was rather small, we estimated the heritabilities for a large number of quantitative traits we collected in these individuals. Broadly speaking, we investigated numerous traits related to anthropometry, clinical chemistry, bone density, intraocular pressure, blood pressure, bioimpedance, and so on. We found results that were comparable to other reported findings in other populations, at least demonstrating that our phenotyping and family structure data was of reasonable quality. For example, we found heritabilities between 0.64 and 0.88 for quantitative traits related to liver functions, which was interesting, given the high prevalence of actual alcohol consumption, and high rates of hepatitis in this population. Other clinical chemistry measures, such as hemoglobin concentration and monocytes, were substantially heritable, whereas hemoglobin itself was not. Some of the morphologic features as measured by anthropometry, such as arm length ($h^2 = 1.0$) and sitting height ($h^2 = 0.55$), had a strong genetic influence in this population, while others (biiliocristal diameter and transverse chest) did not. Similar to reports in other populations, bone stiffness, measured by quantitative ultrasound measured at the heel, had a heritability of 0.86 in this population. It should be emphasized that this analysis is based on a very small sample size; thus, further study with a larger sample size will be needed, with ultimately samples of large families being measured from each of these populations, if we are to learn anything about relationships between genetic and environmental etiological agents through this natural experiment.

Summary

Humans are a very difficult species to study, largely because, rather than designing experiments to develop and test hypotheses, we are forced to search for populations and families that have the appropriate structures and phenotypic distributions to mimic the sort of experiments we would like to do ourselves. In a population as large as ours, characterized by numerous local isolates, and larger more cosmopolitan populations with a variety of histories and structures, living in a wide range of environments and cultures, it is often possible to identify natural experiments on which to perform our research. The incredible advancements in biotechnology, combined with clever identification of appropriate natural experiments, may prove to be a useful combination in elucidating what role various environmental and genetic risk factors play in the etiology of complex human disease. However, the trend at the moment, largely motivated by the desire to apply these technologies as soon as possible, is to take great technology and apply it to recycled epidemiological studies, from which blood samples can be found in someone's freezer somewhere. This sort of convenience sampling is not a powerful way to look for genetic risk factors, largely because these samples most often are the side-effects of large-scale epidemiological studies, aiming to identify environmental factors related to disease, often with sampling to intentionally avoid confounding from genetic risk factors. Studies designed in such a way are not appropriate natural experiments for gene identification studies, though they may one day be useful for estimation of the population-level effects of genetic factors after they are identified in other better designed studies. It is of the utmost importance to combine the expertise of physical and cultural anthropologists with that of clinicians, biotech engineers, human geneticists, epidemiologists and statisticians to make the unravelling of some of the major risk factors for complex traits less fantasy and more into the realm of possibility – but one must first override vested interests and technophilia which are rampant in this field, and get people back to thinking about human genetics as a scientific rather than an engineering discipline (Terwilliger, 2004)!

Acknowledgments

Support from the American taxpayers, via grant MH63749 (to JDT) from the National Institutes of Mental Health, grant AG20351 (to JL) from the National Institute of Aging is gratefully acknowledged, along with support from the Sigrid Juselius Foundation and from the Finnish Academy (to JT). We further acknowledge the input of several colleagues with whom we have taught courses based on this material over the years, most significantly Harald Göring, Markus Perola, and Ken Weiss. Our collaborators on the Korean diaspora project are likewise acknowledged for their contributions to the fieldwork, most significantly Dr Dong-Hwan Kim, who was the physician who was primarily responsible for phenotyping the individuals in Kazakhstan, and organizing the group that identified and ascertained the pedigrees used in this study.

References

Ahmad, M., H. Abbas and S. Haque, Alopecia universalis as a single abnormality in an inbred Pakistani kindred. *American Journal of Medical Genetics*, 1993, **46**(4), 369–71.

Barton, A. *et al.*, The single-nucleotide polymorphism lottery: How useful are a few common SNPs in identifying disease-associated alleles? *Genetic Epidemiology*, 2001, **21**, S384–S389.

Bjerregaard, P., and T. K. Young, *The Circumpolar Inuit: Health of a Population in Transition*, 1998, Copenhagen: Munksgaard.

Brookes, A. J., Rethinking genetic strategies to study complex diseases. *Trends in Molecular Medicine*, 2001, **7**(11), 512–16.

Cardon, L. R. and J. I. Bell, Association study designs for complex diseases. *Nature Reviews Genetics*, 2001, **2**(2), 91–9.

Carlson, C. S., T. L. Newman and D. A. Nickerson, SNPing in the human genome (Reprinted from *Current Opinion Chemical Biology*, Vol. **5**, pp. 78–85, 2001). *Trends in Genetics*, 2002, S2–S9.

Chakraborty, R. and K. M. Weiss, *Admixture as a tool for finding linked genes and detecting that difference from allelic association between loci.* Proceedings of the National Academy of Sciences of the United States of America, 1988, **85**, 9119–123.

Chakravarti, A., Single nucleotide polymorphisms, to a future of genetic medicine. *Nature*, 2001, **409**(6822), 822–3.

Chanock, S., Candidate genes and single nucleotide polymorphisms (SNPs) in the study of human disease. *Disease Markers*, 2001, **17**(2), 89–98.

Collins, F. S., Evolution of a vision: Genome project origins, present and future challenges, and far-reaching benefits. *Human Genome News*, 1995, **7**(3–4), 2–7.

Collins, F. S., Genetics: An explosion of knowledge is transforming clinical practice. *Geriatrics*, 1999, **54**(1), 41–7.

Collins, F. S. *The Human Genome Project and the Future of Medicine, in Great Issues for Medicine in the Twenty-First Century*, 1999, New York: New York Acad Sciences: pp. 42–55.

Collins, F. S. and V. A. McKusick, Implications of the human genome project for medical science. *Jama-Journal of the American Medical Association*, 2001, **285**(5), 540–4.

Couzin, J., New mapping project splits the community. *Science*, 2002, **296**(5572), 1391–2.

European social statistics: Migration, Eurostat (ed.), 2000, Office for Official Publications of the European Communities: Luxembourg.

Denic, S. *et al.*, Cancer by negative heterosis; breast and ovarian cancer excess in hybrids of inbred ethnic groups. *Medical Hypotheses*, 2005, **64**(5), 1002–6.

Doerge, R. W., Mapping and analysis of quantitative trait loci in experimental populations. *Nature Reviews Genetics*, 2002, **3**(1), 43–52.

Enattah, N. S. *et al.*, Identification of a variant associated with adult-type hypolactasia. *Nature Genetics*, 2002, **30**(2), 233–7.

Gabriel, S. B. *et al.*, The structure of haplotype blocks in the human genome. *Science*, 2002, **296**(5576), 2225–9.

Ghosh, S. and F. S. Collins, The geneticist's approach to complex disease. *Annual Review of Medicine*, 1996, **47**, 333–53.

Goring, H. H. H. and J. D. Terwilliger, Linkage analysis in the presence of errors I: Complex-valued recombination fractions and complex phenotypes. *American Journal of Human Genetics*, 2000, **66**(3), 1095–1106.

Goring, H. H. H. and J. D. Terwilliger, Linkage analysis in the presence of errors II: Marker-locus genotyping errors modeled with hypercomplex recombination fractions. *American Journal of Human Genetics*, 2000, **66**(3), 1107–18.

Goring, H. H. H. and J. D. Terwilliger, Linkage analysis in the presence of errors III: Marker loci and their map as nuisance parameters. *American Journal of Human Genetics*, 2000, **66**(4), 1298–1309.

Goring, H. H. H. and J. D. Terwilliger, Linkage analysis in the presence of errors IV: Joint pseudomarker analysis of linkage and/or linkage disequilibrium on a mixture of pedigrees and singletons when the mode of inheritance cannot be accurately specified. *American Journal of Human Genetics*, 2000, **66**(4), 1310–27.

Goring, H. H. H., J. D. Terwilliger and J. Blangero, Large upward bias in estimation of locus-specific effects from genomewide scans. *American Journal of Human Genetics*, 2001, **69**(6), 1357–69.

Gottesman, M. M. and F. S. Collins, The role of the human genome project in disease prevention. *Preventive Medicine*, 1994, **23**(5), 591–4.

Guyer, M. S. and F. S. Collins, The Human Genome Project and the Future of Medicine. *American Journal of Diseases of Children*, 1993, **147**(11), 1145–52.

Ha, T. and J. L. Rees, Keeping genetics simple. *Acta Dermato-Venereologica*, 2000, **80**(6), 401–3.

Hassler, S. *et al.*, Cancer risk in the reindeer-breeding Saami population of Sweden, 1961–1997. *European Journal of Epidemiology*, 2001, **17**(10), 969–76.

Hicks, G., *Japan's Hidden Apartheid: The Korean Minority and the Japanese*, 1997, Aldershot (UK): Ashgate Publishing Ltd.

Holtzman, N. A., Putting the search for genes in perspective. *International Journal of Health Services*, 2001, **31**(2), 445–61.

Horrobin, D. E., Realism in drug discovery – could Cassandra he right? *Nature Biotechnology*, 2001, **19**(12), 1099–1100.

Hovatta, I. *et al.*, Schizophrenia in the genetic isolate of Finland. *American Journal of Medical Genetics*, 1997, **74**(4), 353–60.

Hovatta, I. *et al.*, A genome-wide search for schizophrenia genes in an internal isolate of Finland suggesting multiple susceptibility loci. *American Journal of Medical Genetics*, 1998, **81**(6), 453–4.

Hovatta, I. *et al.*, A genomewide screen for schizophrenia genes in an isolated Finnish subpopulation, suggesting multiple susceptibility loci. *American Journal of Human Genetics*, 1999, **65**(4), 1114–24.

Jasinskaja-Lahti, I., L. K. and T. Vesala, *Rasismi ja Syrjintä Suomessa: Maahanmuuttajien kokemuksia*, 2002, Helsinki: Gaudeamus, 214.

Johansson, A. *et al.*, Linkage disequilibrium between microsatellite markers in the Swedish Sami relative to a worldwide selection of populations. *Human Genetics*, 2005, **116**, 105–13.

Johnson, G. C. L. *et al.*, Haplotype tagging for the identification of common disease genes. *Nature Genetics*, 2001, **29**(2), 233–7.

Jorde, L. B., Linkage disequilibrium and the search for complex disease genes. *Genome Research*, 2000, **10**(10), 1435–44.

Jorde, L. B., W. S. Watkins and M. J. Bamshad, Population genomics: a bridge from evolutionary history to genetic medicine. *Human Molecular Genetics*, 2001, **10**(20), 2199–2207.

Judson, R. *et al.*, How many SNPs does a genome-wide haplotype map require? *Pharmacogenomics*, 2002, **3**(3), 379–91.

Kaprio, J., Science, medicine, and the future – Genetic epidemiology. *British Medical Journal*, 2000, **320**(7244), 1257–9.

Kere, J., *Human population genetics: Lessons from Finland. Annual Review of Genomics and Human Genetics*, 2001, **2**, 103–28.

Kim, J. and M. M. Chan, Acculturation and dietary habits of Korean Americans. *Br J Nutr*, 2004, **91**(3), pp. 469–78.

Kim, J.-S., *Health and Disease of Koreans*, Vol. I. – III, 2001, Seoul: Singuang Publishing Company.

Kohler, R. E., *Lords of the Fly: Drosophila Genetics and the Experimental Life*, 1994, Chicago: University of Chicago Press, pp. xv, 321.

Kreutz, R. and N. Hubner, Congenic rat strains are important tools for the genetic dissection of essential hypertension. *Seminars in Nephrology*, 2002, **22**(2), 135–47.

Kruglyak, L., The use of a genetic map of biallelic markers in linkage studies. *Nature Genetics*, 1997, **17**(1), 21–4.

Kuokkanen, S. *et al.*, Genomewide scan of multiple sclerosis in Finnish multiplex families. *American Journal of Human Genetics*, 1997, **61**(6), 1379–87.

Kwitek-Black, A. E. and H. J. Jacob, The use of designer rats in the genetic dissection of hypertension. *Current Hypertension Reports*, 2001, **3**(1), 12–18.

Laan, M. and S. Pääbo, Demographic history and linkage disequilibrium in human populations. *Nature Genetics*, 1997, **17**, 435–8.

Le Fanu, J., *The Rise and Fall of Modern Medicine*, 1999, New York: Carroll and Graf Publishers.

Lee, K.-K., *Overseas Koreans*. Reports of the Overseas Koreans Foundation, Vol. **4**, 2000, Seoul: Jipmoondang Publishing Company, p. 260.

Leutelt, J. *et al.*, Autosomal recessive retinitis pigmentosa locus maps on chromosome 1q in a large consanguineous family from Pakistan. *Clinical Genetics*, 1995, **47**(3), 122–4.

Liu, J. *et al.*, A genome-wide scan for bipolar affective disorder. *American Journal of Medical Genetics*, 1998, **81**(6), 462.

Lupu, F. *et al.*, Roles of growth hormone and insulin-like growth factor 1 in mouse postnatal growth. *Developmental Biology*, 2001, **229**(1), 141–62.

Mackay, T. F. C., The genetic architecture of quantitative traits. *Annual Review of Genetics*, 2001, **35**, 303–39.

Mazelis, A. G. *et al.*, Relationship of stressful housing conditions to the onset of diabetes mellitus induced by multiple sub-diabetogenic doses of streptozotocin in mice. *Diabetes Research*, 1987, **6**(4), 195–200.

Montagutelli, X., Effect of the genetic background on the phenotype of mouse mutations. *Journal of the American Society of Nephrology*, 2000, **11**(Suppl 16), S101–15.

Nevanlinna, H. R., Finnish population as an object of epidemiological studies. *Duodecim.*, 1974, **90**(22), 1548–53.

Nikali, K. *et al.*, Random search for shared chromosomal regions in 4 affected individuals – the assignment of a new hereditary ataxia locus. *American Journal of Human Genetics*, 1995, **56**(5), 1088–95.

Nordborg, M. *et al.*, The extent of linkage disequilibrium in Arabidopsis thaliana. *Nature Genetics*, 2002, **30**(2), 190–3.

Norio, R., *Suomi Neidon Geenit. Tautiperinnön takana juurillemme johtamassa*, 2000.

Norio, R., H. R. Nevanlinna, and J. Perheentupa, Hereditary diseases in Finland; rare flora in rare soul. *Annals of Clinical Research*, 1973, **5**(3), 109–41.

Pajukanta, P. *et al.*, Genomewide scan for familial combined hyperlipidemia genes in Finnish families, suggesting multiple susceptibility loci influencing triglyceride, cholesterol, and apolipoprotein B levels. *American Journal of Human Genetics*, 1999, **64**(5), 1453–63.

Pak, A. P., *Demographic Characteristics of the Koreans of Kazakhstan*, 2002, Almaty: Almaty City Statistical Office.

Patterson, M., That damned elusive polygene. *Nature Reviews Genetics*, 2000, **1**(2), 86.

Patterson, M., Wake-up call for genome scanners. *Nature Reviews Genetics*, 2002, **3**(1), 9.

Peltonen, L., Molecular background of the Finnish disease heritage. *Annals of Medicine*, 1997, **29**(6), 553–6.

Peltonen, L., A. Jalanko and T. Varilo, Molecular genetics of the Finnish disease heritage. *Human Molecular Genetics*, 1999, **8**(10), 1913–23.

Peltonen, L., A. Palotie and K. Lange, Use of population isolates for mapping complex traits. *Nature Reviews Genetics*, 2000, **1**(3), 182–90.

Peltonen, L., P. Pekkarinen and J. Aaltonen, Messages from an isolate: lessons from the Finnish gene pool. *Biological Chemistry Hoppe-Seyler.*, 1995, **376**(12), 697–704.

Pennisi, E., A closer look at SNPs suggests difficulties. *Science*, 1998, **281**(5384), 1787–9.

Perheentupa, J., The Finnish disease heritage: a personal look. *Acta Paediatrica.*, 1995, **84**(10), 1094–9.

Perola, M. *et al.*, Genome-wide scan of predisposing loci for increased diastolic blood pressure in Finnish siblings. *Journal of Hypertension*, 2000, **18**(11), 1579–85.

Pohjanpää, K., S. Paananen and M. Nieminen, *Maahanmuuttajien elinolot: Venäläisten, virolaisten, somalialaisten ja vietnamilaisten elämöö Suomessa 2002. Elinolot*, Vol. **2003**:1, 2003, Helsinki: Tilastokeskus, 240.

Proctor, R. N., *Cancer Wars*, 1995, New York: Basic Books.

Rees, J., Complex disease and the new clinical sciences. *Science*, 2002, **296**(5568), 698–701.

Reich, D. E. *et al.*, Linkage disequilibrium in the human genome. *Nature*, 2001, **411**(6834), 199–204.

Reich, D. E. and E. St Lander, On the allelic spectrum of human disease. *Trends in Genetics*, 2001, **17**(9), 502–10.

Risch, N. and K. Merikangas, The future of genetic studies of complex human diseases. *Science*, 1996, **273**(5281), 1516–17.

Silver, L. M., *Mouse Genetics: Concepts and Applications*, 1995, New York: Oxford University Press, pp. xiii, 362.

Suh, D.-S., and E. J. Shultz, *Koreans in China*. Papers of the Center for Korean Studies, Vol. **16**, 1990, Honolulu: Center for Korean Studies, University of Hawaii.

Terwilliger, J. D., On the resolution and feasiblity of genome scanning approaches. *Adv Genet.*, 2001, **42**, 351–91.

Terwilliger, J. D., Population genetic epidemiology of complex disease. *American Journal of Human Biology*, 2001, **13**(1), 90.

Terwilliger, J. D., Science and engineering and their different roles in investigations of the genetic portion of the etiology of complex human traits. *Genomics and Informatics*, 2004, **2**(1), 1–6.

Terwilliger, J. D. and H. H. H. Goring, Gene mapping in the 20th and 21st centuries: Statistical methods, data analysis, and experimental design. *Human Biology*, 2000, **72**(1), 63–132.

Terwilliger, J. D. *et al.*, Study design for genetic epidemiology and gene mapping: The Korean diaspora project. *Shengming Kexue Yanjiu (Life Science Research)*, 2002, **6**(2), 95–115.

Terwilliger, J. D. *et al.*, Mapping genes through the use of linkage disequilibrium generated by genetic drift: 'Drift mapping' in small populations with no demographic expansion. *Human Heredity*, 1998, **48**(3), 138–54.

Terwilliger, J. D., Confounding, ascertainment bias, and the blind quest for a genetic fountain of youth. *Annals of Medicine*, 2003, **35**, 532–44.

Tienari, P. J. *et al.*, 2-locus linkage analysis in multiple-sclerosis (MS). *Genomics*, 1994, **19**(2), 320–5.

Todd, J. A., From genome to aetiology in a multifactorial disease, type 1 diabetes. *Bioessays*, 1999, **21**(2), 164–74.

Varilo, T., *The Age of the Mutations in the Finnish Disease Heritage: A Genealogical and Linkage Disequilibrium Study, in Department of Medical Genetics*, 1999, University of Helsinki: Helsinki, p. 99.

Weiss, K., Goings on in Mendel's Garden. *Evolutionary Anthropology*, 2002, **11**, 40–4.

Weiss, K. M., *Genetic Variation and Human Disease: Principles and evolutionary approaches*, 1st edn. Cambridge Studies in Biological Anthropology 11, 1993, Cambridge: Cambridge University Press p. 354.

Weiss, K. M., Is there a paradigm shift in genetics? Lessons from the study of human diseases. *Molecular Phylogenetics and Evolution*, 1996, **5**(1), 259–65.

Weiss, K. M. and A. V. Buchanan, Evolution by phenotype: a biomedical perspective. *Perspect Biol Med.*, 2003, **46**(2), 159–82.

Weiss, K. M. *et al.*, Evaluating the phenotypic effects of SNP variation: sampling issues. *American Journal of Human Genetics*, 1999, **65**(4), 8.

Weiss, K. M. and S. M. Fullerton, Phenogenetic drift and the evolution of genotype-phenotype relationships. *Theoretical Population Biology*, 2000, **57**(3), 187–95.

Weiss, K. M. and J. D. Terwilliger, How many diseases does it take to map a gene with SNPs? *Nature Genetics*, 2000, **26**(2), 151–7.

Wiik, K., *Suomalaisten Juuret*, 2004, Jyväskylä: Atena Kustannus Oy. 368.

Wright, A. F., A. D. Carothers, and M. Pirastu, Population choice in mapping genes for complex diseases. *Nature Genetics*, 1999, **23**(4), 397–404.

Wright, S., *Evolution and the Genetics of Populations*, Vol. **1–4**, 1968, Chicago: University of Chicago Press.

Yaqoob, M. *et al.*, Risk factors for mortality in young children living under various socio-economic conditions in Lahore, Pakistan: with particular reference to inbreeding. *Clinical Genetics*, 1998, **54**(5), 426–34.

Zak, N. B. *et al.*, Population-based gene discovery in the post-genomic era. *Drug Discovery Today*, 2001, **6**(21), 1111–15.

PART 2

Methods

Michael H. Crawford

Chapter 4

The Importance of Field Research in Anthropological Genetics: Methods, Experiences and Results

Michael H. Crawford

Laboratory of Biological Anthropology, University of Kansas, Lawrence, KS
Crawford@ku.edu

Introduction

Field research helps us answer some basic, almost universal, questions: 'Who are we? Where did we come from? 'How did we get here?'

As discussed in Chapter 1 of this volume, anthropological genetics was formalized as a field of investigation during the 1970s and 1980s, with the twin foci of genetic structure of human populations and genetic-environmental interactions in the dissection of the genetic architecture of complex phenotypes (Crawford, 2000b). What distinguishes anthropological genetics from human genetics is its emphasis on smaller, reproductively isolated, non-Western populations, plus a broader, biocultural perspective on complex disease etiology and transmission. Similarly, the reconstruction of the human diaspora and the phylogeny of our species, based on molecular genetic markers, required comparisons among human populations widely dispersed geographically. As a result of these foci, field research became an integral part of anthropological genetics and provided considerable insight into the understanding of human variation and disease processes.

This chapter summarizes the following aspects of field investigations: (1) The reasons for conducting field research in anthropological genetics. For most of us the air-conditioned laboratory is a much

Supported by grants from National Science Foundation (BNS576-11859; BSR910-1571; OPP-990590), PHS Research Career Development Award K04DE028-05, National Institutes of Health DE04115-02, Wenner-Gren Foundation for Anthropological Research, National Geographic Society, and General Research Fund from the University of Kansas 3507, 4932; Biomedical Sciences Research Grants 4349, 4309, 4932.

more comfortable place to conduct research than in some hot, tropical jungle, where you become part of the insect's food chain, or the frigid North with cold stress and summer mosquitoes. Why expose yourself to the risks of the hot sun and voracious insects if you can answer basic anthropological genetics questions in the relative comfort of your laboratory? (2) Describes the preparation necessary for developing a field programme. What kinds of permissions are required from governments and local authorities to be able to successfully conduct field research? What are the political pitfalls and problems that you face when dealing with human experimentation committees (IRBs), some with their unique political agenda? (3) Organization of field investigations. Are you planning to conduct research within a team framework or are you going to follow anthropological 'tradition' as a solitary researcher relishing the exposure to culture shock while languishing in a community for a prolonged period of time? (4) Data collection and choice of methodology for answering the questions that you pose. (5) Three examples of fieldwork in vastly different geographical regions of the world (Tlaxcala, Mexico; Evenki reindeer herders of Siberia; and the marine hunters and gatherers of the Aleutian Islands) on an assortment of evolutionary questions will be contrasted in the penultimate section of this chapter. (6) I conclude this chapter with a discussion of future field research in anthropological genetics and potential problems and political risks faced by the researchers.

Why field investigations?

In 1980 Derek F. Roberts, in an overview of the volume *Current Developments in Anthropological Genetics: Theory and Methods*, lamented:

> As one who has been fortunate enough to enjoy the stimulation and discomforts of field studies, and has learned to appreciate the extent of his own ignorance and blessings from the personal contact with others of different culture, I should be very reluctant to see anthropological genetics become an armchair discipline. From the contributions presented here there is indeed a suggestion that this is happening.
>
> (Roberts, 1980: 429).

He explained that political and financial problems made it progressively more difficult to get into the field during the 1980s when compared to earlier periods. During this millennium one can safely say that some of political baggage acquired by anthropologists during colonial administrations and the mistreatment of native peoples has further weighted down potential field researchers. Following Roberts' lament, in the ensuing 25 years did anthropological genetics become an armchair discipline? Although some field studies have continued into the new millennium, fewer researchers are venturing into the political morass found in some regions of the world. Strong nationalistic and ethnic movements have taken hostile stances towards scientists working in the communities. So, why did Roberts warn us against becoming an armchair discipline?

Field research is an essential tool and methodology of anthropological genetics because it provides:

(1) A *comparative dimension* to a study or analysis. Research on a single, Western population may yield results that may be misleading when dealing with complex phenotypes exposed to vastly different environmental conditions. For example, the purported association between APOE*4 allele and elevated plasma cholesterol levels, established in European American populations, was not verified in the Central Siberian reindeer-herding populations, such as the Evenki, who live under vastly different nutritional conditions (Kamboh *et al.*, 1996). A number of physicians from the United States had advocated a routine screening for mutations of APOE (a lipid transport molecule) as an indicator of risk from cardiovascular disease among patients. This association of APOE*4 allele with elevated total cholesterol levels had been described in several Western populations. Through fieldwork, theories of specific gene-environment interaction can be tested and either verified or rejected through the examination of the relationship of specific genes existing under different or unique environmental conditions.

Since the time of Franz Boas (1911), studies of plasticity of complex traits (such as body measurements) among migrants living under diverse environmental conditions has been a traditional methodology used by anthropologists in the field. These studies are based on the principles: 'different genes, same environment' or 'different environments, same genes'. Thus, the children of migrants (assumed to be similar genetically to the sedentary populations) are compared to the children of the population of origin. The underlying assumption is that the migrants are a genetically representative subset of the original population, relocated to regions with different environmental conditions – such as nutrition, temperature, humidity, or degree of hypoxic stress. The plastic traits (i.e. those affected by genetic/environmental interaction during maturation) are consequently altered through growth and development by different environmental conditions.

Cooper *et al.* (1997), in a comparison of blood pressure variation in populations of African ancestry, demonstrated the existence of a gradient for the incidence of hypertension, from a low in rural Nigerian populations (7%) to higher frequencies in the Caribbean (25–27%) to highest incidence of hypertension observed in urban African Americans of the United States (35%). This disparity in the incidence of hypertension in rural Africans and African Americans prompted Curtin (1992) to posit an explanation based on the differential survivorship of those slaves during the transatlantic voyage from Africa to the New World through the differential conservation of their salt levels. The slave survivors of the brutal transatlantic crossing had genes that once protected them from salt-wasting diseases (such as diarrhoeas and dehydration) but now these same genes may confer greater risk to hypertension. This research indicates the possible involvement of the RAAS (renin-angio-tensin-aldosterone system) through the action of selection, possibly operating on ACE (angiotensin converting enzyme) and elevated levels of aldosterone. The field research by Cooper's group contributed to our understanding of the physiology and molecular genetics involved in the regulation of blood pressure and the risks associated with hypertension.

(2) A *time dimension* can be added to the measurement of human evolution through the careful application of historical documents. Unlike fruit flies, bacteria or other experimental organisms that have been used to study evolution, humans have an extended generation time of 20 to 25 years, thus, it is difficult for researchers to patiently wait several genera-tions to test specific, time-related hypotheses. However, some societies maintain meticulous and precise records that can be used to assess multi-generation changes in the frequencies of genes in populations. In order to document human evolutionary change, within a temporal framework, popula-tions with reliable historic documentation of their time of divergence must be identified. In my study of the genetic, morphologic, and demographic differentiation of transplanted Tlaxcaltecan populations of Mexico, a time of divergence of 430 years (approximately 17 to 21 generations – depending on assumed length of generations) separated the village of Cuanalan from the populations currently residing within the Valley of Tlaxcala. During the pacification of the Aztecs, a garrison of Tlaxcaltecans was sent by their Spanish allies to the Valley of Mexico to protect irrigation dykes that could be destroyed and the farm lands flooded as a means of warfare. In 1591, 400 families from Tlaxcala were also relocated to San Esteban (currently the city of Saltillo) in northern Mexico, to relieve the Spanish garrison besieged by Chichimec tribes. These historical events provided a temporal separation of 378 years between the Tlaxcaltecans of Saltillo and those who remained in the Valley of Tlaxcala (Crawford, 1976). My field research programme (summers of 1969 to 1975) attempted to measure genetic, morphological and demographic divergence between the descendants of the transplanted populations and the inhabitants from the Valley of Tlaxcala that remained *in situ*. Although the transplanted Tlaxcaltecans were initially considered traitors by the other Mexicans, resulting in their reproductive isolation, in the last two generations reproductive

barriers deteriorated and gene flow complicated the measurement of population differentiation.

(3) *Unique social and demographic structures of populations* are essential for the understanding of the etiology and mechanisms of genetic transmission involving complex phenotypes, such as chronic degenerative diseases. For example, the majority of Black Carib (Garifuna) families, of Central America, practised a form of consecutive monogamy that produced a high percentage of half-sibs being raised within the matrilineal, consanguineal families (Gonzalez, 1984). This high incidence of half-sibs and their families in the Garifuna communities gives considerable statistical power for the partitioning of genetic variances of complex phenotypes into dominance deviation and the additive effect (Crawford, 1980a). This insight into the transmission of complex traits of biomedical significance would not be possible using standard nuclear family methods, commonly employed by European and American scientists.

In most urban populations (often the study population in human and medical genetics), it is difficult to identify groups that are genetically and culturally homogeneous. However, field research gives access to genetic isolates, such as the Mennonites of central Kansas, and the analyses of complex phenotypes, e.g. associated with biological aging (Crawford, 2000b; Duggirala et al., 2002). In addition to the well-documented population history of the Mennonites, the use of such a genetic isolate minimizes variation in the covariates of aging, such as smoking, differential activity patterns, different nutritional patterns, and alcohol consumption.

Methods of field research

Experimental design

After proposing a question, you have to ascertain if the hypotheses can in fact be tested successfully in the selected populations. Are the communities of adequate sizes to detect the changes in allelic frequencies that you hope to detect? If you are planning on studying genetic/environmental interactions in complex phenotypes, are the families used to measure family resemblance of sufficient sizes and numbers in the population? Is there sufficient genetic variability in the population from which the families are drawn? Before you begin a major project that requires funding from National Institutes of Health or the National Science Foundation, a pilot study may be useful for providing the documentation of the *power analysis* (i.e. determining the statistical power of a specific test). For Mendelian traits, power analysis is based on relatively simple algebraic equations consisting of the effective size of the population and the frequencies of the alleles (see Weiner and Lourie, 1969, for a discussion of the selection of sample size). However, for quantitative traits and the likelihood of demonstrating linkage with a particular LOD (logarithm of the odds) score requires computer simulation based on the sizes and numbers of families (Almasy and Blangero, 1998). In any proposal to a granting agency, you must be able to demonstrate that the hypotheses are in fact testable with the specified sample sizes and families.

Sampling procedures

Often claims are made in grant proposals and/or publications that the populations have been either *randomly* sampled or that the sample consisted of *unrelated* genomes. In a random sample all individuals within the population have equal probability of being chosen. It is extremely difficult to select truly random samples under field conditions. However, there are a number of statistical advantages to the use of a random sample of a population, namely: (1) it provides a probabilistic foundation for much of statistical theory; (2) it provides a baseline to which other methods can be compared. The achievement of a random sample minimally requires that a census be taken of the community, followed by the assignment of unique numbers to each community member. These numbers can be randomly drawn for each person of the sample, with the size of the sample defined by an appropriate power analysis. Members of the sample would then be asked to participate. Usually some individuals selected for the sample are not interested in volunteering for the research and subsequently refuse to participate. Those unwilling to participate must be replaced through a random process. Such a statistically based sampling strategy requires considerable investment of time by the researchers, the trust of the members of the community, an enumeration of the community, followed by the generation of a sample. However, what often happens is that researchers 'get who they can samples'. Occasionally newcomers to the village or curious volunteers from other communities are accidentally included in the study. If the 'get-what-you-can' sample is sufficiently large and representative of the community it is possible to select an adequate sub-sample. This sub-sample may be selected on the basis of the population structure – i.e. appropriate sex ratio, age stratification and representation of all the clans in the village or all the barrios of the municipio are included. One physician/ geneticist preceded by a few days our research team to an island population in the Bering Sea. He paid a local pediatrician to roundup all 'pure' natives from the village and to line them up the next morning at the clinic. He then paid each participant, obtained a blood sample and departed the following morning. Since the pediatrician knew the identity of the 'so-called' volunteers, we were able to demonstrate that the small sample consisted primarily of related individuals. Thus, the frequencies of mtDNA haplogroups based on this particular sample fail to represent their actual incidence in the total population.

Numerous publications and grant proposals purport to have *samples of unrelated individuals*. This strategy of sampling is time-consuming and requires the reconstruction of genealogies for the village and/or population. Few studies detail how the researchers managed to obtain the samples of unrelated individuals or genomes. In small, genetic isolates, most members of the community are at least distantly related – through the maternal, paternal or both sides of the family. Thus, if you require a sample of unrelated or 'minimally related' individuals for your study, the genealogies (pedigrees) must first be reconstructed prior to the collection of biological data. One recent study by a graduate student from a South African university explained that she collected unrelated genomes by asking all participants if any of their relatives had also come to the clinic. This approach has two complications: (1) those participating in the early phases of the project may not know which relatives would come later; (2) unless you reveal the complete list of participants who had come to the clinic at an earlier time (possibly violating the privacy of the participants) you may not be able to generate a list of unrelated individuals from a given community.

Permissions

IRBs

Before any research can begin, permission must be obtained from the institutional review board (IRB) of your university or institute. The purpose of this board is to protect research subjects from risk of physical, social or psychological harm resulting from the research. The IRB conducts a cost/benefit analysis of the risk of the research versus potential benefits (positive value related to health or welfare) to the subject or society. This board normally consists of scientists, physicians, ethicists, and the lay public and usually meets once a month in order to approve or disapprove potential research involving human subjects. These IRBs were constituted in 1974 as a result of the National Research Act (Public Law 93–348) signed into law and creating the National Commission for the Protection of Human Subjects of Biomedical and Behavioural Research. This commission was charged with the identification of the basic ethical principals that underlie research with human subjects and the development of guidelines for conducting research.

Normally, these IRBs are apolitical with a mandate to protect research subjects. However, there have been some rare cases of IRBs exerting their specific agenda and either blocking or delaying research from being done. Among the most egregious cases that I have found, an IRB prevented research from being done even though there was no risk involved and community approval had already been given. The research subjects did not have an opportunity to decide for themselves whether to participate in the study or not. This IRB made unilateral and paternalistic decisions on behalf of the subjects, while possibly violating federal regulations.

Government Local/Tribal

In order to conduct field research in another country, it is essential to learn about the appropriate governmental and local agencies from which permissions must be obtained. These sources of permission differ from country to country and may involve the Ministry of Health, officials from a counterpart university, centres for indigenous studies, local governmental and health agencies, and tribal councils. It is best to develop a working relationship with local academics or physicians from the country in which you hope to conduct research. Such persons could serve as guides to the political network in the country and possibly as co-principal investigators for the project.

There should be some technology transfer, particularly if the research is being done in a Third World country. In the grant application, the budget for the support of such a research programme should include funds for salaries, supplies or equipment for the foreign collaborating entities. These collaborative undertakings may require formal agreements and the establishment of possible exchange programmes between respective universities. On a local level, the researchers should learn what the members of the community desire in exchange for their participation in the study. In some cases, research subjects may prefer payment for their time, while in other cases, free medical and dental services may be needed and preferred. There is a fine line between paying research subjects (at local levels of compensation) for the time spent on the project versus payments that are considered exorbitant by the community. Such high payments could be perceived as 'coercive' in that the research subjects cannot afford not to participate and this could violate the principle of informed consent. In one impoverished barrio in Mexico, through the intercession of a local physician, we were able to assist in the establishment of a permanent medical clinic in the community. What's in it for the community must be discussed and agreed upon prior to any field investigations.

Data collection

Once permissions have been granted and field research is imminent, decisions must be made about the methods to be used in data collection. Is the researcher planning to follow cultural anthropological 'tradition' to conduct solitary, long-term research or to organize a team that is in the field for a shorter timeframe? The advantages of research by a solitary investigator include: (1) less social and economic disruption in the community. The behavioural dynamics may be altered by a 'herd' of strangers moving in and disrupting the community; (2) inter-observer error can be minimized. By having a single researcher collecting all of the data, this person would have a better understanding of how the data were collected, and the precision and quality of the data. However, in anthropological genetics, the disadvantages of solo field research tend to outweigh the advantages because more information can be collected over a shorter time period by a team of skilled and complementary researchers. Biological samples, such as DNA or blood, must be processed quickly so as to minimize potential degradation. In addition, some forms of data collection require specialized training and skill, e.g. phlebotomy in some countries requires certification, therefore the principal investigator (PI) should be able to bring such a specialist into the field. Medical and dental care can be provided by field members if the team includes physicians and/or dentists. Disadvantages of a large team project include: (1) team dynamics can be disruptive under stressful conditions with the teams fragmenting into warring factions that can sometimes jeopardize the research project; (2) The quality of data collection may vary depending on the background and training of the field members. Despite training and the standardization of methodologies, problems can arise in the field. For example, in the study of populations of Central America one so-called researcher on the team wasted precious research time and costs by measuring heights of women while ignoring the fact that the research subjects wore shoes with elevated heels. She also measured head circumferences, even though hair-pieces of the subjects were still in place, thus exaggerating the head sizes. These gaffs resulted in data that could not stand up to scientific scrutiny and consequently were discarded.

Informed consent

Federal regulations, administered by IRBs, require the collection of informed consent. The exact wording of the informed consent statement must be approved by the local institutional review board. The consent statement is normally administered to the participant in statement form. The research participant reads the statement, asks questions about the research, its implications and makes a decision, without prejudice, whether to participate in the study, part of the study, or not participate at all. If the subject decides to participate he/she signs two copies of the informed consent statement and keeps one copy. In some cases it may be inappropriate to use a written informed consent statement that requires a signature by the participant. However, with the approval of the IRB an oral informed consent statement can be read and explained to the research volunteer. This procedure may be required if the community is non-literate or if there is a belief that signing documents may place the volunteer at 'political risk' – e.g. a belief that signing documents constitutes a 'confession' of some sort. Under these conditions, it is best to have a witness sign a statement indicating that verbal informed consent was provided.

To obtain informed consent, the following information must be explained to the research subject:

(1) Purpose of the study – i.e. what is to be done.
(2) Any discomfort or risk must be explained.
(3) Benefit of the project to the participant or society.
(4) There should be an offer to answer questions.
(5) The subject should have the freedom to participate or to quit the study, without prejudice.
(6) Alternative procedures, if applicable, should be described. For example, if sufficient quantities of DNA can be obtained through buccal swabs, instead of blood samples withdrawn through venipuncture, the participant should have the option to select that method.

What's in it for the community

Communities, tribes, barrios are more likely to permit research to be carried out if there is something tangible in it for them. In some communities there may be sufficient interest in the science or the research outcome. For example, the Mennonite communities that I worked with for more than 25 years were interested in the results of the longevity/biological aging research, and individual information on the levels of serum lipids and possible risk of CHD research. They wanted to know why they live longer than the surrounding farming communities. They were also interested in the results of a series of blood chemistry tests that were done on all volunteers. We provided individual feedback of the blood chemistries, such as total cholesterol, HDL, LDL and triglyceride levels with an interpretation of the results by a physician. In addition, summary statistics were placed on a web site that members of the community could access. Lectures summarizing the results were provided to the congregations by the researchers involved in the study. In addition to peer reviewed publications, a volume, entitled *Different Seasons*, which summarized the biological aging study, was published with copies provided to the churches, libraries and the colleges in the region (Crawford, 2000b).

Communities that are not as well off economically as the Mennonites or the Aleuts of Alaska, often prefer that health care or dental care be provided in return for participation in the study. The exact preference of the community has to be discussed prior to the arrival of the researchers, announcing: 'We are here to study you'. It is understandable why some populations are reticent to allow any research in the community because of past abuses by researchers – providing little or no feedback of the results. Some communities prefer payment either to the tribe or to individual participants.

Applications of field methods

During the past 37 years, while focusing on numerous evolutionary and biomedical questions, I have conducted field investigations in

Table 4.1. Field research programmes conducted from 1968 to the present.

Populations	Chronology	Sites	Local sponsors	Funding agencies	Selected publications
Italian Alpine	1968	Valle Maira	University of Torino	U. of Pittsburgh	#9,45
Tlaxcaltecans	1969–75	Tlaxcala	Ministry of Health	Wenner-Gren Foundation	#6,7,21,30,49,56
		San Pablo del Monte Cuanalan Saltillo		University of Kansas NIH	
Irish Itinerants	1970	Dublin, Wexford	Medical-Social Board of Ireland	Wenner-Gren	#5,17,39
Black Caribs	1975–82	Belize Guatemala St Vincent	Ministry of Health	NIH Biomedical Research University of Kansas	#8,10,11,32
Indian Castes	1990–1	Andhra Pradesh	Andhra Pradesh Univ.	Fulbright (CIES) National Science Foundation (NSF)	#1,51
Eskimos	1976–7	St Lawrence Is. Wales, King Isl.	Norton Sound Health Corporation		#16,23 24
Fishing Villages	1981–3	Newfoundland	Canadian Red Cross	Wenner-Gren Foundation	#18,38
Mennonites	1979–2004	Kansas & Nebraska	General Conference Churches Old Order Mennonites	NIA Attorney General's Settlement Fund	#14,27
Farming Villages	1986–8	Tiszahat	Debrecen University	Earthwatch National Academy Sciences	#19,34
Siberian Populations	1989–2001	Central Siberia Kamchatka	Soviet Academy of Sciences Russian Acad. Med. Sciences	NSF NSF	#15,20,22 33,40,41,48,55,57
Aleuts	1999–05	Aleutian Is.	Aleut Corporation Aleut./Pribil of Isls. Assoc Tribal Councils	NSF	#46,47

Numbers listed in the Selected publications column of this table refer to the references that are followed by a bold number in the References section of this chapter.

many different geographical locales. Table 4.1 summarizes 11 of the most extensive field research programmes that I organized and implemented during those years. Some projects lasted a single summer, while others continued for as many as 27 years. The focus of the field research varied from questions concerning the peopling of the Americas to the measurement of the genetic components involved in differential biological aging (Duggirala *et al.*, 2002). The sizes of the research teams also varied, from as few as three researchers, e.g. George Gmelch (a cultural anthropologist), Ted Rebich, a dentist, and me, working with the Irish itinerants (Tinkers) to as many as 25 researchers involved with various facets of the Mennonite field investigations (Crawford and Gmelch, 1974; Crawford, 2000b).

My initial field research was conducted in Valle Maira, a breathtaking Alpine Valley in northern Italy, bordering France. Figure 4.1 is a schematic of Valle Maira, showing its location in northern Italy and the study community Acceglio. Brunetto Chiarelli, professor at Torino, enticed me to study genetic isolates, in this valley carved by the Maira River. Several years earlier, Gabriel Lasker and his colleagues (1972) had done similar genetic demographic research in an adjoining valley (Valle Variata). I hired four graduate students from the University of Torino to assist me with the demographic interviews and the transcription of the records. What became evident from the church and municipal records of Acceglio, Tetti and Elva was that these villages were experiencing increasing selective emigration of the young adults for economic reasons and that reproductive isolation was rapidly breaking down (Crawford, 1980b). In addition, the inhabitants of this valley were far from enthusiastic in volunteering for any genetic analyses. As a result, I terminated the programme after one summer's research in Valle Maira, convinced that these populations were no longer genetically isolated and that genetic studies could not be successfully completed.

In the remainder of this chapter I will: (1) provide three examples of my field experiences in Europe, the Americas and Siberia; and (2) these unique ecological settings, political milieu, and different evolutionary questions should provide the reader with some appreciation of the complexities and even 'minefields' associated with field

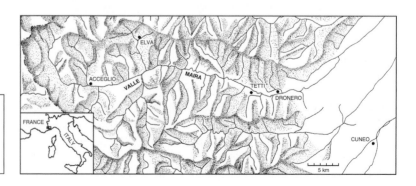

Fig 4.1 Map of Valle Maira, an Alpine Valley of Northwestern Italy. This valley was carved out by the River Maira, that drains the French/Italian Alps.

investigations; (3) discuss the lessons learned from 38 years of field investigations during 11 major research programmes located in four different continents.

Chronologically, my first field research with Native Americans was in the Valley of Tlaxcala, Mexico. This was followed by an extensive research programme with the Garifuna (Black Caribs) of Central America and the Caribbean, then the Yupik- and Inupik-speaking Eskimos of Alaska. Because of the so-called Cold War, I was unable to begin research with Siberian populations until 1989. My most recent field investigations are currently being conducted in the Aleutian Islands. In this chapter, I will focus first on the earliest research with Native American populations of Tlaxcala and for contrast discuss the more recent fieldwork in Siberia and the Aleutian Islands. The research conducted in Central America and the Caribbean with Garifuna has been fully described by my edited volume on Black Caribs and a summary article in the *Yearbook of Physical Anthropology* (Crawford, 1983; Crawford, 1984). In addition, there is an extensive discussion of the Garifuna research in my book on the *Origins of Native Americans* (Crawford, 1998).

Field research in the Americas

Tlaxcaltecan project

San Pablo del Monte and the City of Tlaxcala

While at the University of Pittsburgh, 1968, I learned that the Department of Anthropology administered a yearly summer field school on methods of cultural anthropology in Mexico. This field school, directed by Hugo Nutini and funded by National Science Foundation, provided first-hand field experience to cohorts of graduate students, selected through national competition. Nutini explained to me about the unique historical and demographic features of the Valley of Tlaxcala. Figure 4.2 locates the city of Tlaxcala and the village of San Pablo del Monte within the valley of Tlaxcala. Because of a military alliance (known as the *Segura de la Frontera*) with the Spanish Crown, the Valley of Tlaxcala did not experience the tragedy of colonization that befell most of Mexico (Halberstein *et al.*, 1973). To this day, villages continue to exist in the Valley of Tlaxcala with little or no Spanish admixture. In contrast, the colonial administrative centres within the State were built for the Spanish and these centres underwent considerable hybridization. Thus, it became evident that the Valley of Tlaxcala was an ideal site for the study of the processes of gene flow and admixture. I selected a village on the slopes of the volcano (LaMalinche) that because of its isolation from the administrative centres of Tlaxcala, had experienced minimal if any gene flow from the Spaniards. In contrast, the City of Tlaxcala, built as an administrative centre, was reputed to have experienced considerable gene flow.

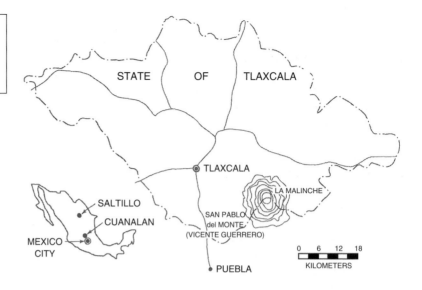

Fig 4.2 Map of the State of Tlaxcala, Mexico. This map indicates the location of the City of Tlaxcala and the municipio of San Pablo del Monte.

Fig 4.3 A market scene from the City of Tlaxcala, showing a Mestizo community of Central Mexico.

Figure 4.3 is a photograph of the weekly produce market held in the city of Tlaxcala. Thus, a reproductively isolated population (San Pablo del Monte) within the Valley could be used to represent parental gene frequencies of the Native component in the Mestizos (mixed) of the City of Tlaxcala. The Spanish parental frequencies were estimates based on mean frequencies of Spanish populations, derived from the literature.

A field team was organized, consisting of a physician (Thomas Aidan Cockburn), a dentist (Seishi W. Oka), two cultural anthropologists (Hugo Nutini and Thomas Schorr), and three graduate students from the University of Pittsburgh (Paul Scuilli, Robert Halberstein and Patricia Potzrebowski). The physician and dentist offered free medical and dental care to all who visited the clinic, set up in the Public Health Facilities (*Salubridads*) of San Pablo del Monte and the City of Tlaxcala. The nature of the study was explained to the participants by the cultural anthropologists and by other Spanish-speaking members of the research team. Adult volunteers were interviewed for demographic and genealogical information. The blood samples were drawn by the physician, while dental impressions were made by the accompanying dentist. A total of 395 Tlaxcaltecans participated in the initial year of this research project.

Fieldwork in Cuanalan

From the cultural anthropologists in the field and from Tlaxcaltecan historians, I discovered the unique role played by the Tlaxcaltecans in the history of Mexico. To this day, because of their role in the Conquest of Mexico, the Tlaxcaltecans are still regarded by some Mexicans as 'traitors'. During a professional meeting, I had discussed the fieldwork in Tlaxcala with a colleague (John McCullough), who had surveyed the Valley of Mexico for possible excavation sites. He recounted how he literally stumbled across a Tlaxcaltecan enclave in the barrio of Cuanalan, within the municipio of Acolman – almost within sight of the pyramids of Teotihuacan. He had learned that the residents of Cuanalan were descendants of the original Tlaxcaltecan garrison brought to the site to build and maintain a critical flood control dike for Lake Texcoco. While the exact size of the garrison brought to this site in the sixteenth century was not recorded, a total of 212 Indian and 10 Mestizo families were enumerated by the earliest census of Cuanalan from the eighteenth century (Crawford, 1976).

In summer of 1972, through the support of Wenner-Gren Foundation, an expedition was organized from the University of Kansas to study the genetic micro-differentiation of the Tlaxcaltecan enclave in Cuanalan. This research, in collaboration with the Department of Genetics, National Institute of Nutrition, was sponsored by the Ministry of Health of Mexico. The Mexican co-investigator, Ruben Lisker, analyzed the genetic markers of the blood used in the analytical portion of the study (Crawford *et al.*, 1976). The field team consisted of four graduate students from the University of Kansas: Robert Halberstein (who had also worked in Tlaxcala), John Hall (a cultural anthropologist), and Frank Lees and David Busija (graduate students in biological anthropology). In addition, a Mexican geneticist and physician, Ivanhoe Gamboa (Puebla University), explained the nature of the project and provided medical care to all volunteers. A total of 536 volunteers out of

2,040 residents of Cuanalan participated in this phase of the research project.

Fieldwork in Saltillo

The final phase of the Tlaxcala research programme, also sponsored by the Ministry of Health, was put in the field in 1974. Robert Halberstein and Francis Lees (both from the field team of Cuanalan) were joined by Renan del Barco and Eileen Mulhare (cultural anthropologists) and a local physician, who was hired to obtain informed consent and to provide medical care. A total of 280 volunteers from two barrios of Saltillo, plus a combined sample from other barrios, participated in the study. The specific barrios were selected because of the historical descriptions of the locations settled by the transplanted Tlaxcaltecans in the sixteenth century (Nava, 1969). However, sampling was complicated because: (1) the expanding commercial zones had impinged on the original Tlaxcaltecan sectors; (2) there was some stigma attached to being a Native American 'Indio' in a largely Mestizo city. As a result, many informants were reluctant to discuss their Tlaxcaltecan origins. This reluctance, in its extreme form, was illustrated by an incident in which an old, moustache-adorned native man, a veteran of the Pancho Villa campaign, was asked if he was a descendant of the Tlaxcaltecan founders of Saltillo. He replied with grave seriousness:

'No señor. I am not Tlaxcaltecan — but my brother is!'

(Crawford, 1976, p. 5).

Results

Measurement of admixture

Since this research occurred well before the molecular revolution, blood group and protein marker allelic frequencies were used to calculate genetic distances between the admixed Tlaxcaltecan population (City of Tlaxcala), San Pablo del Monte, and the Spanish parental population. The absence of European genetic markers in San Pablo suggested that this village received little, if any, Spanish gene flow. By contrast, the City of Tlaxcala exhibited numerous European genetic markers of the blood, indicative of considerable gene flow from the Spanish administrators and soldiers (Crawford *et al.*, 1974). The detection of a family with an apparent deletion of two of the Rhesus blood group loci led to the localization of the RH system to the distal arm of chromosome 1 (Turner *et al.*, 1975).

Starting with a historically derived biracial (bi-ethnic) model of gene flow from Spanish conquistadores and administrators to the Tlaxcaltecans living in the City of Tlaxcala, we estimated admixture using a maximum of 39 different alleles and four different analytical methods (see Table 4.2). The Spanish component in the Mestizo gene pool of the City of Tlaxcala varied from 23% to 32%. Since African admixture had been reported for populations along the east coast of Mexico (Vera Cruz region) and because of the presence of the

Table 4.2. Estimates of bi-ethnic hybridization in the City of Tlaxcala based upon five different methods (Crawford *et al.*, 1976).

Methods	Number of alleles[1]	Percentage of contribution	
		Spanish	Indian
Multiple regression analysis	39	32	68
$\sum$ Bernstein's m, weighted by σ^2	12	27	73
Roberts and Hiorn's	39	32	68
True least squares	39	31	69
Maximum likelihood	39	23	77

[1]With the exception of the 'Σ' method, hybridization estimates are based upon 15 loci: ABO, Duffy, Kidd, Diego, MNS-CHRM, MN, Rhesus, Kell, Haptoglobin, Group Specific Component, Phosphoglucomutase, Acid Phosphatase, Lewis, PTC Tasting, and Cerumen.

ce/D RH haplotype (an African marker), a tri-ethnic model was used to measure admixture in the City of Tlaxcala. The results of the tri-ethnic analyses indicated that 22% to 24% of the gene pool was of Spanish origin, 66% to 70% was of Tlaxcaltecan origin, and 6% to 9% was of African origin (see Table 4.2).

Genetic differentiation
The measurement of genetic differentiation of the transplanted populations, Saltillo and Cuanalan, was complicated by the recent gene flow from surrounding of Native American populations. Reconstruction of the genealogies revealed that two generations ago, migrants had started to move into the Tlaxcaltecan enclaves of Saltillo and Cuanalan. Their initial reproductive isolation, because they were considered traitors, was breaking down. There was evidence of both Spanish and African admixture, particularly in the barrio of LaMinita, a community derived from migrants who worked the mines of the region (Crawford, 1979). Therefore, it was not possible to partition out gene frequency changes during the 300+ years of divergence due to a combination of genetic drift, selection and gene flow. We were able to examine the specific loci that deviated significantly from expectation, if only selection was assumed (Crawford, 1978). The loci that deviated significantly from expectation included Duffy, Lewis, Diego blood group systems and Group Specific Component (GC). Based on immunoglobulin haplotypes, two of the Saltillo barrios display a large Spanish component (36% in LaMinita) and 45% in Chamizal (Table 4.3). This large Spanish admixture is not surprising given that the Tlaxcaltecans

Table 4.3. Admixture estimates for Cuanalan and Saltillo based on immunoglobulin allotype frequencies and Schanfield *et al.*, 1978.

Population and subdivision	Percentage European	Percentage African	Native American
Cuanalan			
Combined	15%	0	85%
Residents	4	0	96
Mixed	14	0	86
Immigrants	29	0	71
Saltillo			
Combined	42	5	53
Chamizal	45	3	52
LaMinita	36	6	58
Saltillo	53	6	41

were transplanted to an area containing a large Spanish garrison. Approximately 6% of the LaMinita genes are of African origin and 3% of Chamizal had an African parental contribution (Schanfield *et al.*, 1978).

Siberian field investigations

First attempts

In 1976 a grant proposal for the comparison of Siberian and Alaskan populations in the reconstruction of the peopling of the Americas was funded by the National Science Foundation. The Alaskan portion of this research was successfully carried out in 1978, and compared Inupik and Yupik-speaking Eskimo communities of Wales, King Island, Savoonga and Gambell, Alaska (Crawford *et al.*, 1981). The purpose of this project was to determine if languages served as reproductive barriers and created genetic discontinuity. The region where two language groups (Yupik and Inupik) abut was selected to determine if there was genetic discontinuity. Figure 4.4 is a map that locates St. Lawrence Island, Wales, Alaska and King Island and shows their geographic proximity to Siberia, Russia. However, the second part of the study, a comparison of Native Americans with the indigenous populations of Chukotka, Siberia, was not possible for political reasons. During my trip to Siberia, my stay was restricted to Rem Sukernik's laboratory in *Akademgorodok* (Academic town) of Novosibirsk. I provided computer programmes, enzyme substrates, and biochemical genetic laboratory procedures, but I was not permitted to conduct any field investigations with native peoples of Siberia. The Cold War was at its 'coldest', and joint field research with human populations was not possible. As a result, research concerning

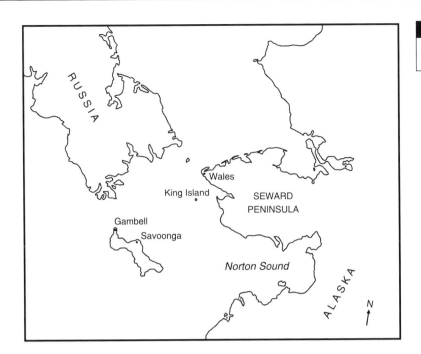

Fig 4.4 Map of St Lawrence Island, locating the two villages, Savoonga and Gambell.

the peopling of the Americas could not be initiated in Siberia at that time.

In the late 1980s, with the coming of '*perestroika*' (rebuilding) and '*glastnost*' (openness), opportunities for research with Siberian indigenous populations improved considerably. In 1988, under a Soros Foundation programme I was able to bring to the United States my long-time research collaborator, Rem Sukernik. The following year, after conducting a research programme with a Baboon colony (in Sukhumi, Abkhasia) that was experiencing an 'epidemic' of lymphoma, Dennis O'Rourke (a former student) and I had the opportunity to visit Novosibirsk. From there we were flown to Krasnoyarsk (a strategic nuclear weapons centre that had been off-limits to Western visitors), Baikit and then by helicopter to a number of Evenki reindeer herding villages of central Siberia. Figure 4.5 is a map that locates the Evenki villages of Surinda and Poligus along the Lower Tunguska River of central Siberia. This map also shows the location of Mendur-Sokhon in Gorno Altai. O'Rourke and I were the first Westerners to be seen in that part of Siberia and our presence stirred a commotion.

Evenki and Kets

During the summers of 1991 and 1992, under the sponsorship of the Man-in-the Biosphere Program (MAB), and funded by National Science Foundation, an international research team sampled two adjacent Evenki reindeer herding villages, Surinda and Poligus (both located in the Stony Tunguska region of the Evenky Okrug). We were also able to sample one small Ket village (Sulamai) located along

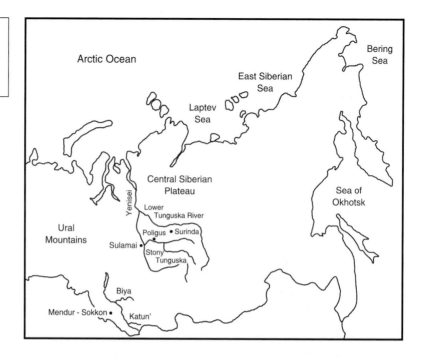

Fig 4.5 Map of Central Siberia, locating the Evenki reindeer herding communities, Sulamai, Poligus and the Altai village of Mendur-Sokhon.

Fig 4.6 A photograph of an Evenki reindeer herder preparing an evening meal.

the Yenisey River. The two Evenki villages and their herding units (brigades) each contained approximately 600 residents.

The Evenki are one of the largest of the indigenous populations of Siberia, numbering approximately 30,000 and distributed from Central Siberia to the Sea of Okhost. They speak a Tungusic language, a branch of the Altaic language family. The Evenki were socially organized into patrilineal clans that held group ownership of the reindeer herds for subsistence. See Figure 4.6, which is a photograph of an Evenki reindeer herder preparing meat for roasting. They were collectivized by the Soviets in the 1930s and the ownership of the herds was given to the collectives. In the late 1930s, many of the wealthier Evenki (*kulaks*) and the local shamans were executed by the Soviet authorities as they attempted to control these pastoral herders. However, with the breakup of the Soviet Union in the 1990s, the patrilineages had retaken control of the herds. Scarcity of fuel and absence of roads have once again isolated the Evenki from the surrounding Siberian populations.

The Kets are numerically one of the smallest indigenous groups of Siberia, with a population of fewer than 1,000 persons living along the Yenisey River. Their primary form of subsistence is fishing, supplemented by some hunting. The Kets have been of great interest to the scientific world because their language is unique and unintelligible to their neighbours and could not be included with the three major linguistic families of Siberia: Altaic, Uralic, and Paleoasiatic. As a result, there has been preoccupation by linguists and geneticists in regards to the origins of the Kets.

The research teams were made up of three United States and four Russian scientists from the Institute of Cytology and Genetics, Siberian Branch of the Russian Academy of Sciences in Novosibirsk. The researchers with equipment and instrumentation were flown in by helicopter and deposited in the reindeer herding brigades — widely dispersed throughout the Siberian boreal forest, or *taiga*. The research team camped in tents, set up outdoor laboratories, and conducted 'open-air' investigations. In addition to the small brigades, the Soviets had built collective villages (Surinda and Poligus), that contained the administration, the boarding schools, and huts for the elderly and those Evenks who did not follow traditional subsistence. Blood samples, demographic and genealogical information, anthropometric measurements, and various physiological measures of basal metabolic processes were collected by the research team. The Russian physician, Rem Sukernik and I, administered verbal informed consent in the Russian language. All of the Evenki and Kets had been educated as children in the boarding schools and spoke Russian fluently. The purposes of the study and the methods to be used were explained to each participant by the principal investigator of the project. Although all of the Evenki spoke Russian, they refused to sign any informed consent statements because of the negative connotations during the Soviet era associated with signing papers. In each community, only a proportion of residents (approximately 50%) agreed to volunteer for the study. Each participant was paid for their time, based on local rates.

Altai field research

The early analyses of Siberian mt-DNA haplogroups failed to detect the so-called Asian specific, 9 base-pair deletions that are characteristic of the B haplogroup, present in many of the indigenous populations of the Americas. Shields *et al.* (1993) reported the presence of a 9-bp deletion of the B haplogroup in the Altai portion of southern Siberia. The heterogeneous sample from the Altai (n = 17) was collected from three different villages – Ust Kan, Ulagan, and Chibit. Given the small sample size and the heterogeneous nature of the sample, I wanted to determine as to the frequency of the B haplogroup in the Altai. A small cattle herding village, Mendur-Sokhon, in the Republic of Gorno Altai, was selected by Ludmila Osipova (Institute of Cytology and Genetics, Siberian Branch of the Russian Academy of Sciences) in 1994. See Figure 4.5 for the location of Mendur-Sokhon. A research team from this Institute of Cytology and Genetics made the arrangements and accompanied me to Mendur-Sokhon on a truck that lacked shock absorbers and the numerous pot-holes along the way appear to have relocated my kidneys to another region of the abdomen.

Blood specimens were shipped from Siberia for analysis to the Laboratory of Biological Anthropology, University of Kansas, Lawrence, KS. DNA was extracted and analysed for DNA fingerprints (VNTRs – variable numbers tandem repeats) and RAPDs (random amplified polymorphisms of DNA) (McComb et al., 1995, 1996). Additional DNA was sent to Douglas Wallace's laboratory, where mt-DNA was typed and sequenced by Antonio Torroni and his staff. Y-chromosome markers were typed by Tyler-Smith's group in Oxford and by R. J. Mitchell at LaTrobe University. Collaborative analyses on several coding regions were also done:

(1) Apolipoprotein genes were analysed in M. Ilyas Kamboh's laboratory at the University of Pittsburgh (Kamboh et al., 1996).
(2) Alcohol and aldehyde dehydrogenase genes were genotyped by Holly Thomasson in T-K Li's laboratory at the University of Indiana Medical Center-Indianapolis (Thomasson et al., unpublished manuscript).
(3) Collagen gene (COL1A2) and dopamine transporter gene (DAT1) were characterized by R. J. Mitchell at LaTrobe University (Mitchell et al., 1999; Mitchell et al., 2000).
(4) Blood groups, serum proteins and red cell proteins were analysed at the Minneapolis War Memorial Blood Center and at Integrated DNA Technology of Denver.

Results

A comparison of mtDNA variation of the Evenki and nine other Siberian indigenous populations with Native Americans revealed the presence of haplogroups A, C, and D (all present in Native Americans), but the absence of haplogroup B (Torroni et al., 1993). The presence of B haplogroup (defined by a 9-bp deletion in region V), in addition to A, C, and D in the Altai, is suggestive that the Altai population shared common ancestry with Native Americans. Therefore, this molecular genetic study demonstrated what was suspected since the time of Blumenbach, that Native Americans came from Siberia (Crawford, 1998). This research also brought into question the chronology of the peopling of the Americas – previously based only on archaeological evidence and dating. Using a mtDNA evolution rate of 2% to 4% for haplogroups C and D, the time of divergence separating Siberia and the Americas was estimated, based on coalescence, between 17,000 and 34,000 years. Whether the mtDNA chronometer ticks at a steady rate has been questioned. However, without this fieldwork on both continents a comparison between Siberian and Native American populations would not have been possible.

The presence of the X-haplogroup in Native American and European populations (defined by RFLP markers -1715 DdeI, -10394 DdeI, +14465 AccI, and +16517 HaeIII), both contemporary and extant, has generated much controversy. This haplogroup was apparently absent in Siberians, therefore, Brown *et al.* (1998) proposed that the X haplogroup was brought to the Americas by ancestral European populations. This absence of X among Siberians and its presence among Europeans and Native Americans became a significant factor in the litigation associated with the repatriation of ancient Native American skeletal remains, such as Kennewick Man, reconstructed by biological anthropologists as a European. However, a publication by Derenko *et al.* (2001) indicated the presence of X at a frequency of 3.5% in a southern Altai population. The presence of this haplogroup in the Altai cannot be explained by Russian gene flow (estimated as 5%) because of its extremely low frequencies in Russian populations. Derenko *et al.* have argued that the presence of X in the Altai population suggests that this haplogroup is part of the ancestral gene pool of the Altai. Our investigations of the Kizhi-Altai support their conclusion (Krawczak *et al.*, 2006).

Field research in the Aleutian Islands

Although my interest in the peopling of the Americas and specifically the origin of the Aleuts goes back to the early 1970s, a conversation with an archaeologist (Dixie West) who was excavating the western Aleutian Islands, prompted the development of a grant proposal to NSF. I contacted Dennis O'Rourke (University of Utah) and we agreed to submit a collaborative grant proposal to NSF. The Utah group would focus on the analyses of skeletal materials and ancient DNA, while the Kansas collaborators would sample the contemporary populations of the Aleutian Islands. This experimental design would provide a rare opportunity to follow continuous genetic changes in human populations of known provenance for almost 9,000 years. The University of Kansas group would obtain DNA from those Aleutian Islands that are still inhabited, while the Utah group would sample skeletal remains from collections representing the same islands. These collections include the Smithsonian Institution, the

University of Alaska-Fairbanks Museum, and the Museum of the Aleutians.

Prior to the commencement of the project, Dennis O'Rourke and I visited the Aleut Corporation in Anchorage, met with the president and staff of the Corporation and reached an agreement for research cooperation. We also met with the Aleutian/Priblof Islands Association, responsible for cultural heritage and health services of the islands. We were instructed to obtain permission from each tribal council before any research could be commenced.

A field team, consisting of an Aleut elder and historian (Alice Petrivelli), a graduate student of anthropological genetics (Rohina Rubicz) and I, conducted fieldwork in 1999 and 2000 in Atka, Unalaska, Umnak, St Paul, St George and Anchorage with Aleuts originally from a number of different islands. See Figure 4.7, which is a map of the islands making up the Aleutian archipelago. Each tribe approved the project but the majority of the participants preferred to give buccal swabs and refused to allow blood specimens to be drawn. The Aleuts were interested primarily in their history. Where in Siberia did they come from? They voiced a lack of enthusiasm about any health-related research, primarily because they felt that they were depicted incorrectly in previous publications.

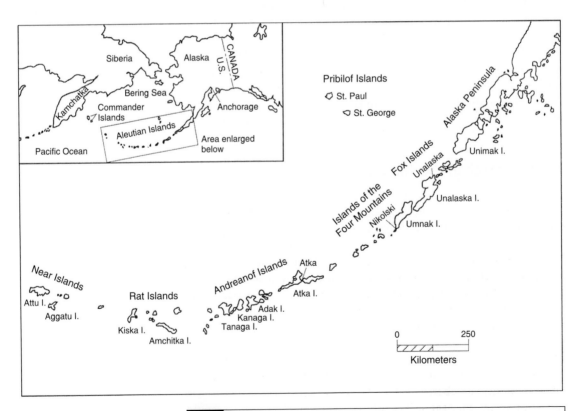

Fig 4.7 Map of the Aleutian Islands, stretching 1,500 kilometres from the Commander Islands off the coast of Siberia to the Alaskan Peninsula in the east.

The volunteers asked many penetrating and probing questions concerning the use of the DNA. They wanted to know if there is social risk associated with DNA analyses. If health insurance companies were able to obtain the results of this study could they refuse coverage to any Aleut who may be at risk for a chronic disease? I assured them of individual confidentiality and the fact that most of the systems or DNA regions that we analyse do not code for protein, therefore, disease risk cannot be assessed. A total of 177 Aleuts participated during the first two seasons.

In 2001, a Russian–United States expedition was organized to Bering Island of the Commander Islands – geographically the most westerly island currently inhabited by Aleuts. The two Commander Islands, Bering and Mednii, were unpopulated until 1825, when officials from the Russian-American Company transplanted Aleut hunters and their families from Atka, Attu, Unalaska, plus additional Natives from Kamchatka, and Russian and Gypsy workers have been added to the population since that time. These islands came to the attention of the Russian government when Admiral Vitus Bering and his party were stranded and Bering died on one of the islands in 1741. The Commander Islands, uninhabited by Aleuts at the time of Bering's expedition, had great faunal wealth with rookeries of fur seals, sea lions, sea otters, Arctic fox, and wide variety of birds, and the great northern sea cow or manatea (*Hyrodamalis gigas*), which became extinct in the nineteenth century. Laughlin (1963) argued that Commander Islands were uninhabited by humans prior to Russian contact because the sea cow would not have survived the habitation of either the Aleuts or Eskimos. In addition, Ales Hrdlička (1945) found no archaeological evidence of Aleut occupation on either Bering or Mednii Island prior to the eighteenth century. Steady contact was maintained between the families of the Commander Islands with their kin in Alaska until 1867, when Russia sold Alaska to the United States. Contacts further deteriorated until the Cold War, when all contact between the Aleuts of Commander Islands and their western kin was lost.

The primary questions and hypotheses generated for this collaborative project included: (1) Were the prehistoric Aleutian peoples members of a single, continuous population or were the Paleo-Aleuts replaced by a later migration of Neo-Aleuts from the Alaskan mainland or from Siberia? If there was replacement, then, significant statistical differences should be observed between the contemporary and ancient populations as measured in mt-DNA haplogroup frequencies and in unique DNA sequences. In addition, the Neo-Aleut remains should exhibit closer genetic affinity to the contemporary Aleut populations than do the Paleo-Aleuts. (2) What are the evolutionary relationships of the Aleuts to Eskimos, Northwest Coast Native Americans, Siberians? Are the Aleuts more closely related to the Eskimos or the Athabascan-speaking interior Native Americans? (3) Where in Siberia did the Aleuts come from? Do the Aleuts show closer genetic affinities to the populations of Kamchatka (Itelmens,

Koryaks and Evens) or do they genetically resemble the populations of Chukotka (Asian Eskimos and Chukchi)? (4) Do contemporary Aleut Island populations reflect genetic structure, such as geographically based subdivision? Or, did the migrations and relocations of the Aleuts during historic and prehistoric time periods obliterate the vestiges of isolation and its evolutionary consequences.

Results

A number of findings were of particular interest to the Aleut participants in the study. These include:

(1) Aleuts exhibit unique frequencies of mtDNA haplogroups. They have only two of the five mt-DNA haplogroups (A, B, C, D and X) shared by Native Americans and Siberians (Rubicz *et al.*, 2003). MtDNA haplogroup A was present in 29% while haplogroup D occurred in 71% of the Aleuts sampled. No non-native mtDNA haplogroups were detected among participants claiming maternal Aleut ancestry. The Aleuts are unique from all other Native American and Siberian populations in that they exhibit high frequencies of haplogroup D. Figure 4.7, a PCA plot based on mtDNA frequencies, compares Aleut populations to other Native American and Siberian groups.

(2) Even though the Aleuts underwent massive population reduction and genetic bottlenecks, considerable variation exists at the mitochondrial DNA locus. The sequencing results for the HVS-I region of the mtDNA genome reveals a total of 26 different control region haplotypes, characterized by 23 variable sites.

(3) Only Native American and Asian haplogroups were observed in all individuals claiming Aleut ancestry along the maternal line and were tested for mtDNA haplogroups. No European haplogroups were observed, nor any evidence for admixture with other Native Americans. These results indicate that the

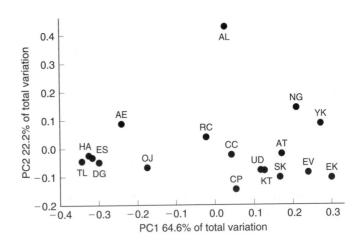

Fig 4.8 A principal components plot of Siberian and Native American populations based on mt-DNA haplogroup frequencies. AL abbreviation stands for Aleuts; AE = Asian Eskimos; NG = Nganasan; RC = Reindeer Chukchi; YK = Yukaghirs.

Table 4.4. | Selected frequencies of mtDNA haplogroups based on RFLPs (after Rubicz, 2001).

Population	N	mtDNA*A	mtDNA*B	mtDNA*C	mtDNA*D	Other
Aleut	179	28.5%	0%	0%	71.5%	0%
Altai	28	3.6	3.6	37.5	14.3	42.8
Asian Eskimo	50	80	0	0	20	0
Coastal Chukchi	46	24	0	21.7	8.7	45.7
Savoonga Eskimo	49	94	0	0	2	4
Dogrib	154	91	0	2	0	7
Evenki	51	4	0	84.3	9.8	2.0
Koryaks	155	5.2	0	36.1	1.3	57.4
Ojibwa	28	64.3	3.6	7.1	0	25

gene flow from the Russian gene pool into the Aleuts was primarily paternal.

(4) Sequence analyses indicate that Asian Eskimos and Chukchi cluster closely with the Aleuts, while the populations from Kamchatka are distinct. What this study indicates is that the ancestors of the Aleuts shared common ancestry with the populations of Chukotka and they crossed Beringia and doubled-back towards Siberia. The archaeological evidence shows that the Aleuts did not reach the western islands of the archipelago until about 3,500 years BP.

(5) Regional genetic differences continue to exist in the Aleutian Islands, despite the migrations and relocations of Aleut communities during World War II. AMOVA analyses of Excoffier *et al.* (1992) were conducted in order to test for population structure and patterning of the Aleut communities located in the western, central, and eastern regions of the archipelago. The results of the AMOVA analyses, based on the mother's place of birth, indicated that there were significant differences in variation among the three regional groups (Rubicz, 2001).

Future field research: learning from mistakes

North American Native Tribes, such as the Navajo, Pima, and Papago, because of their proximity to major universities, their tolerance, generosity, and academic interests have at times been besieged by researchers. One Native group jokes that their extended families consist of a mother, father, children and an anthropologist. Some of the early research came at the expense of the Native groups. Whether it was head measurements, blood or tissue specimens, burials, or sacred kin terms, much of this information was published without direct benefit to the community. Why should a particular tribe or individual participate in a research project under those circumstances?

The ethical field research of the future has to be based on a true partnership in science with the Native communities. Scientists and the communities must be *equal partners* in any scientific endeavour. There must be communication, consideration, and sensitivity to each other's culture – i.e. the cultures of science and that of the indigenous peoples. There can be no field investigation without the collaboration of the indigenous community. There are potential health and intellectual payoffs for the participating community. In two of our long-term research projects with the Mennonite communities of Kansas and Nebraska and the study of the origins of Aleuts, I believe that a true *partnership in science (PIS)* was achieved. In both studies, there was communication, discussion and agreement prior to the beginning of the research, followed by continuous dialogue and eventual feedback to the communities. All of these activities are constantly scrutinized by IRBs (institutional review boards) whose responsibility is the protection of research subjects.

In the Aleut project, we discussed with representatives from both the Aleut Corporation and the Aleutian/Priblof Islands Association, followed by presentations and discussions with tribal councils about all aspects of the project. We offered medical or dental care to each community. The Aleuts were not interested in health care, since they received adequate services from the Indian Health Services. They also were sceptical about outside researchers coming and inappropriately characterizing the health conditions in the villages. On the other hand, many of the Aleuts were fascinated by their history and wanted to know where in Siberia did they come from? Who were they most closely related to? These were questions that scientists could answer and the Aleut community wanted to know. After the first three years of the project, the Aleuts invited me to come to the Board of Trustees meeting in Anchorage to summarize the results of the study. More than 200 Aleuts attended this meeting, listened with great attention, and posed numerous well-thought out and probing questions. Such a relationship between a well-educated and a curious community and scientists begins to approach the principles of partnership in science. Obviously, this approach cannot be entirely replicated in a non-literate community, traditional society, where science may have little meaning or interest.

In the Mennonite research programme on the genetics of biological aging, there was direct interest in the health and well-being of the members of the communities. The Mennonites wanted physicians to provide health assessments and screening of blood samples for 25 different clinical chemicals – such as cholesterol, triglycerides, thyroxine, glucose, and BUN. The results of these blood tests were provided to both the participants and their physicians. The Mennonites wanted to improve their health and to better understand the aging process. In addition to the various scholarly publications about the genetics and the aging process in the Mennonites, an edited volume, entitled *Different Seasons,* was completed and copies were presented to the Mennonite community, local libraries, and the Mennonite colleges – Tabor and Bethel. Feedback to the community was provided through a number of presentations made to the Mennonite communities in Goessel and Tabor.

The future of field research portends a close collaboration between the scientists and the communities being studied. No scientists contemplating field research should attempt to collect data, publish it, and disappear from the community without a trace. This sort of behaviour has occurred too many times and has precipitated a crisis in the perception of scientists by native communities. Such unethical behaviour would impact on any possible future contacts between ethical scientists and the communities. We must work together for the greater good of humanity.

References

Almasy, L. and J. Blangero (1998). Multipoint quantitative-trait linkage analysis in general pedigrees. *American Journal of Human Genetics,* **62**(5), 1198–211.

Bamshad, M., A. E. Fraley, M. H. Crawford, R. L. Cann, B. R. Busi, J. Naidu and L. B. Jorde (1996). Mitochondrial DNA variation in caste populations of Andhra Pradesh, India. *Human Biology,* **68**(1), 1–28. **(1)**

Boas, F. (1911). Changes in bodily form in the descendants of immigrants. In *Abstract of the Report on Changes in Bodily Form of Descendants of Immigrants.* Washington DC, Govt. Printing Office.

Brown, M., S. Hosseini, A. Torroni, H. Bandelt, J. Allen, T. Schurr, R. Scozzari, F. Cruciani and D. Wallace (1998). MtDNA haplogroup X: An ancient link between Europe/West Asia and North America? *American Journal of Human Genetics,* **63**, 1852–61.

Cooper, R. S., C. N. Rotimi, S. L. Ataman, *et al.* (1997). Hypertension prevalence in seven populations of African origin. *Amer. J. of Public Health,* **87**(2), 160–8.

Crawford, M. H. (1975). Genetic affinities and origins of the Irish Tinkers. In *Biosocial Interrelations in Population Adaptations,* eds. E. Watts, F. Johnston and G. Lasker. The Hague: Mouten Press, pp. 93–103.

Crawford, M. H., (ed.) (1976). *The Tlaxcaltecans. Prehistory, Demography, Morphology and Genetics.* University of Kansas Publications in Anthropology 7. Lawrence, KS. **(6)**

Crawford, M. H. (1978). Population dynamics in Tlaxcala, Mexico: The effects of gene flow, selection, and geography on the distribution of gene frequencies. In *Evolutionary Models and Studies in Human Diversity,* eds. R. J Meier, C. M Otten, F. Abdel-Hameed. The Hague: Mouton Press, pp. 215–25. **(7)**

Crawford, M. H. (1980a). Genetic epidemiology and anthropology. *Journal of Medical Anthropology,* **4**(3), 415–22. **(8)**

Crawford, M. H. (1980b). The breakdown of reproductive isolation in an Alpine genetic isolate: Acceglio, Italy. In *Population Structure and Genetic Disorders*, eds. A. W. Eriksson, H. Forsius, H. R. Nevanlinna, P. L. Workman and R. K. Norio. New York: Academic Press, pp. 57–71. (9)

Crawford, M. H. (1983). The anthropological genetics of the Black Caribs (Garifuna) of Central America and the Caribbean. *Yearbook of Physical Anthropology*, **25**, 155–86. (10)

Crawford, M. H. (ed.) (1984). *Current Developments in Anthropological Genetics. Volume 3. Black Caribs. A Case Study in Biocultural Adaptation*. New York: Plenum Press.

Crawford, M. H. (1998). *The Origins of Native Americans. Evidence from Anthropological Genetics*. Cambridge: Cambridge University Press.

Crawford, M. H. (2000a). Anthropological genetics in the 21st Century: An introduction. *Human Biology*, **72**(1), 3–14.

Crawford, M. H. (ed.) (2000b). *Different Seasons. Biological Aging among the Mennonites of the Midwestern United States*. University of Kansas Publications in Anthropology, 21. Lawrence, KS. (14)

Crawford, M. H., A. G. Comuzzie, W. R. Leonard and R. I. Sukernik (1994). Molecular genetics, protein variation, and the population structure of the Evenki. In *Isozymes: Organization and Roles in Evolution, Genetics and Physiology*, eds. C. L. Markert, J. G. Scandalios, H. A. Lim and O. L. Serov. River Edge, NJ: World Scientific, pp. 227–41. (15)

Crawford, M. H. and V. B. Encisco (1982). Population structure of circumpolar groups of Siberia, Alaska, Canada, and Greenland. In *Current Developments in Anthropological Genetics*. Vol. **2**, eds. M. H. Crawford and J. H. Mielke. New York: Plenum Press, pp. 51–91. (16)

Crawford, M. H. and G. Gmelch (1974). Human biology of the Irish Tinkers: Demography, ethnohistory, and genetics. *Social Biology*, **21**, 321–31. (17)

Crawford, M. H., T. Koertvelyessy, M. Pap, K. Szilagyi and R. Duggirala (1999). The effects of a new political border on the migration patterns and predicted kinship (Phi) in a subdivided Hungarian agricultural population: Tiszahat. *Homo*, **50**(3), 201–10. (19)

Crawford, M. H. and W. R. Leonard (2002). The biological diversity of herding populations: An introduction. In *Human Biology of Pastoral Populations*, eds. W. R. Leonard and M. H. Crawford. Cambridge: Cambridge University Press, pp. 1–9. (20)

Crawford, M. H., W. C. Leyshon, K. Brown, F. Lees and L. Taylor (1974). Human biology in Mexico, II: A comparison of blood group, serum and red cell enzyme frequencies, and genetic distances of the Indian populations of Mexico. *Amer. J. Phys. Anthrop*, **41**, 251–68. (21)

Crawford, M. H., D. D. Dykes, K. Skradski and H. F. Polesky (1979). Gene flow and genetic differentiation of a transplanted Mexican population: Saltillo. *Amer. J. Phys. Anthrop*, **50**, 401–12.

Crawford, M. H., R. Lisker and R. Perez Briceno (1976). Genetic differentiation of two transplanted Tlaxcaltecan populations. In *The Tlaxcaltecans: Prehistory, Demography, Morphology, and Genetics*. University of Kansas, Publications in Anthropology, 7, pp. 169–75.

Crawford, M. H., J. McComb, M. S. Schanfield and R. J. Mitchell (2002). Genetic structure of pastoral populations of Siberia: The Evenki of central Siberia and the Kizhi of Gorno Altai. In *Human Biology of Pastoral Populations*, eds. W. R. Leonard and M. H. Crawford, Cambridge: Cambridge University Press, pp. 10–49. (22)

Crawford, M. H., J. H. Mielke, E. J. Devor, D. D. Dykes and H. F. Polesky (1981). Population structure of Alaskan and Siberian indigenous communities. *American Journal of Physical Anthropology*, **55**, 167–85. **(23)**

Crawford, M. H., J. T. Williams and R. Duggirala (1997). Genetic structure of the indigenous populations of Siberia. *American Journal of Physical Anthropology*, **104**, 177–92. **(24)**

Curtin, P. D. (1992). The slavery hypothesis for hypertension among African Americans: The historical evidence. *American J. of Public Health*, **82** (12), 1681–6.

Derenko, M. V., T. Grzybowski, B. A. Malyarchuk, J. Czarny, D. Miscicka-Sliwka and I. A. Zakharov (2001). The presence of mitochondrial haplogroup X on Altaians from south Siberia. *American Journal of Human Genetics*, **69**, 237–41.

Duggirala, R., M. Uttley and M. H. Crawford (2002). Genetics of differential biological aging. *J. of Genetic Epidemiology*, **23**(2), 97–109.

Excoffier, L., P. Smouse and J. Quattro (1992). Analysis of molecular variance inferred from metric distances among DNA haplotypes: Application to human mitochondrial DNA restriction data. *Genetics*, **131**, 479–91.

Gonzalez, N. L. (1984). Garifuna (Black Carib) social organization. In *Current Developments in Anthropological Genetics. Vol. 3. Black Caribs. A Case Study in Biocultural Adaptation*. New York: Plenum Press, 51–65.

Halberstein, R. A, M. H. Crawford and H. G. Nutini (1973). Historical-demographic analysis of Indian populations in Tlaxcala, Mexico. *Social Biology*, **20**, 40–50. **(30)**

Hrdlička, A. (1945). *The Aleutian and Commander Islands and Their Inhabitants*. Wistar Institute of Anatomy and Biology, Philadelphia.

Hutchinson, J. and M. H. Crawford (1981). Genetic determinants of blood pressure level among Black Caribs of St. Vincent. *Human Biology*, **53**(3), 453–66. **(32)**

Kamboh, M. I., M. H. Crawford, C. E. Aston and W. R. Leonard (1996). Population distributions of APOE, APOH, and APOA4 polymorphisms and their relationships with quantitative plasma lipid levels among the Evenki herders of Siberia. *Human Biology*, **68** (2), 231–44. **(33)**

Koertvelyessy, T., M. H. Crawford, M. Pap and K. Szilagyi (1993). The influence of religious affiliation on surname repetition (RP) in marriages of Marokpapi, Hungary. *Antropologischen Anzieger*, **51**, 309–16. **(34)**

Krawczak, C. P., E. Devor, M. Zlojutro and M. H. Crawford (2006). mtDNA variation in the Kizhi population of Gorno Altai: A comparative study. *Human Biology* (In press).

Lasker, G. W., B. Chiarelli, M. Masali, F. Fedele and B. A. Kaplan (1972). Degree of human genetic isolation measured by isonymy and marital distances in two communities in an Italian Alpine Valley. *Human Biology*, **44**, 351–60.

Laughlin, W. (1963). *Aleuts: Survivors of the Bering Land Bridge*. New York: Holt, Rinehart Winston.

Marini, A. M., G. Matassi, V. Raynal, B. Andre, J. P. Cartron and B. Cherif-Zahar (2000). The human Rhesus-associated RhAG protein and a kidney homologue promote ammoniu transport in yeast. *Nature Genetics*, **26**, 341–44.

Martin, L. J., M. H. Crawford, T. Koertvelyessy, D. Keeping, M. Collins and R. Huntsman (2000). The population structure of ten Newfoundland outports. *Human Biology*, **72**(6), 997–1016. **(38)**

Martin, L., K.E. North and M.H. Crawford (2000). The origins of the Irish Travellers and the genetic structure of Ireland. *Annals of Human Biology*, 25(5), 453–65. (39)

McComb, J., N. Blagitko, A.G. Comuzzie, M.S. Schanfield, R.I. Sukernik, W.R. Leonard and M.H. Crawford (1995). VNTR DNA variation in Siberian indigenous populations. *Human Biology*, 67, 217–30. (40)

McComb, J., M.H. Crawford, L. Osipova, T. Karafet, O. Posukh and M.S. Schanfield (1996). DNA interpopulational variation in Siberian indigenous populations: The Mountain Altai. *American Journal of Human Biology*, 8, 599–607. (41)

Mitchell, R.J., S. Howlett, N.G. White, L. Federle, S.S. Papiha, I. Briceno, J. McComb, M.S. Schanfield, C. Tyler-Smith, L. Osipova, G. Livshits and M.H. Crawford (1999). Deletion polymorphism in the human COL1A2 gene: Genetic evidence of a non-African populations whose descendants spread to all continents. *Human Biology*, 71, 901–14.

Mitchell, R.J., S. Howlett, L. Earl, N.G. White, J. McComb, M.S. Schanfield, I. Briceno, S.S. Papiha, L. Osipova, L. Livshits, W.R. Leonard and M.H. Crawford (2000). The distribution of the 3' VNTR polymorphism in the human dopamine transporter gene (DAT1) in world populations. *Human Biology*, 72, 295–304.

Nava, L. (1969). *Trascendencia Historia de Tlaxcala*. Editorial Progress, Mexico City.

North, K.E. and M.H. Crawford (1997). Isonymy and repeated pairs analysis: Mating structure of Acceglio, Italy, 1889–1968. *Rivista d'Anthropologia* (Special issue in honour of G.W. Lasker), 74, 93–103. (45)

Roberts, D.F. (1980). Current developments in anthropological genetics: achievement and gaps. In *Current Developments in Anthropological Genetics: Vol. 1. Theories and Methods*, eds. J.H. Mielke and M.H. Crawford. New York: Plenum Press, pp. 419–30.

Rubicz, R. (2001). *Origins of the Aleuts: Molecular perspectives*. Unpublished MA thesis, University of Kansas, Lawrence, KS. (46)

Rubicz, R., T. Schurr, P.L. Bubb and M.H. Crawford (2003). Origins of the Aleuts: Population variation in mt-DNA. *Human Biology*, 75(6), 809–35. (47)

Santos, F.R., A. Pandya, C. Tyler-Smith, S.D.J. Pena, M.S. Schanfield, W.R. Leonard, L. Osipova, M.H. Crawford and R.J. Mitchell (1999). The Central Siberian origin for Native American Y chromosomes. *American Journal of Human Genetics*, 64, 619–28. (48)

Schanfield, M.S., H.H. Fudenberg, M.H. Crawford and K. Turner (1978). The distributions of immunoglobulin allotypes in two Tlaxcaltecan populations. *Annals of Human Biology*, 5, 577–90. (49)

Shields, G.F., A.M. Schmiechen, B.L. Frazier, A. Redd, M.I. Voevoda, J.K. Reed and R.H. Ward (1993). mtDNA sequences suggest a recent evolutionary divergence for Beringian and Northern North American populations. *American Journal of Human Genetics*, 53, 549–62.

Sirajuddin, S.M., R. Duggirala and M.H. Crawford (1994). Population structure of Chenchu and other South Indian tribal groups: Relationship between genetic, anthropometric, dermatoglyphic, geographic and linguistic distances. *Human Biology*, 66(5), 865–84. (51)

Terwilliger, J.D., S. Zollner, M. Laan, *et al.* (1998). Mapping genes through the use of linkage disequilibrium generated by genetic drift: Drift mapping with small populations with no demographic expansion. *Human Heredity*, 48, 138–54.

Thomasson, H. R., M. H. Crawford, D. Zeng, R. Deka, D. Goldman, A. M. Khartonik, K. Mai, Y. M. Ostrovsky, C. Ching, B. Segal and T.-K. Li (2005). Alcohol and aldehyde dehydrogenase genotypic variation among Chinese, Siberian, North Amerindian populations. Unpublished manuscript.

Torroni, A., R. I. Sukernik, T. G. Schurr, Y. B. Starikovskaya, M. F. Cabell, M. H. Crawford, A. G. Comuzzie and D. C. Wallace (1993). mtDNA variation of aboriginal Siberians reveals distinct genetic affinities with Native Americans. *Amer. J. Hum. Genet*, **53**, 591–608. **(55)**

Turner, J. H., M. H. Crawford and W. C. Leyshon (1975). Phenotypic-karyotypic localization of the human RH-locus on chromosome 1. *J. of Heredity*, **66**, 97–9. **(56)**

Weiner, J. S. and J. A. Lourie (1969). *Human Biology. A Guide to Field Methods. IBP Handbook No. 9*, Oxford: Blackwell Scientific Publications.

Zerjal, T., B. Dashnyam, A. Pandya, M. Kayser, L. Roewer, F. R. Santos, W. Schiefenhovel, N. Fretwell, M. A. Jobling, S. Harihara, K. Shimizu, D. Semjidmaa, A. Sajantila, P. Salo, M. H. Crawford, E. K. Ginter, O. V. Evgrafov and C. Tyler-Smith (1997). Genetic relationships of Asians and Northern Europeans, revealed by Y-chromosomal DNA analysis. *American Journal of Human Genetics*, **60**, 1174–83. **(57)**

Chapter 5

The Confluence of Anthropological Genetics and Anthropological Demography

James H. Mielke

Department of Anthropology, University of Kansas, Lawrence, KS

Alan G. Fix

Department of Anthropology, University of California, Riverside, CA

Introduction

The relationship between demography and evolution is close and long-standing. After all, it was by reading Malthus' (1798) essay on population that both Darwin and Wallace achieved their insight into natural selection. The importance of demography for anthropological genetics continues to be strong. Anthropological genetics, concerned with understanding the patterns and causes of genetic variation within and among populations, depends on anthropological demography to provide data on population sizes and fluctuations, mating structure, and migration patterns and histories that are crucial for that understanding.

While demographers study many aspects of human populations (Preston *et al.*, 2001; Siegel and Swanson, 2004), anthropological demography usually focuses on small-scale populations and is often linked with studies of human biology. Anthropological demographic studies have been undertaken expressly to provide information necessary to understand genetic variation.

Demography is the study of human population. More specifically, as the classic definition states: 'Demography is the study of the size, territorial distribution, and composition of population, changes therein, and the components of such change' (Hauser and Duncan, 1959: 31). The size and composition of a population is caused by three fundamental factors: fertility (births), mortality (deaths), and migration (in-migration and out-migration). The discipline of demography has historically emphasized measurement and description of these vital processes, usually at the macro level of the national population. It is generally the population characteristics of countries that are analysed and compared (e.g. see Keyfitz and Fleiger, 1968). Moreover, formal demography has been less focused on the human *behaviours* that affect population size, composition, or changes. Indeed, as a prominent demographer, Samuel Preston (1993: 594) states: 'In theory, the other behavioural and social sciences hold the keys to the relations between demographic events and the behaviour of individuals and social systems'. The methods of demography provide limited insight into the consequences and causes of variation in migration, fertility, and mortality patterns (McNicoll, 1992). Demographers have, however, excelled in description and careful measurement of populations. Anthropological demography combines this quantitative approach to examining vital events and processes with an interest in social and behaviour processes that influence vital events. The strong quantitative empirical basis acts as an ideal starting place for subsequent research and provides anthropological geneticists with analytical tools to measure population dynamics that are relevant to understanding genetic variation.

In the sections that follow, we will describe how data, methods, and theoretical constructs from anthropological demography shape our understanding of the genetics of human populations.

Human populations: size and structure

The concept of population

The term *population* commonly refers to a group of individuals. For demographers, the definition requires that these individuals are of the same species and live within a delimited geographical area. So, according to Murdock and Ellis (1991: 11), population is the most basic of terms in demography and '... consists of the persons living in a specific geographical area at a specific point in time'. In some cases this definition is adjusted to include the idea that the collection of people must meet certain criteria (Preston *et al.*, 2001). Some demographers prefer to make the distinction that these groups are *subpopulations* within the larger population and that population should always refer to the total population. Thus, a demographer may be concerned with the size and age composition of the population of females in a specific region. Demographers are also concerned with the population as it persists over time, collectively changing through both attrition (deaths and out-migration) and accession (births and in-migration).

In much of demography the focus of analysis is on the aggregate characteristics (collectivity) of the individuals living in a region, with little emphasis on individuals or individual variation. For example, the size and composition of a population are aggregate measures rather than individual measures. The goal is to understand the aggregate processes affecting the population. On the other hand, demography also focuses on the individual and the implications that certain processes have for the individual. There are a number of useful indices that demographers employ that take the aggregate-level processes and apply them to randomly chosen individuals (or the 'average person') within the population. These indices include such measures as the total fertility rate and life expectancy at different ages (Preston et al., 2001). Thus, demographers can focus their studies at the macro-level or the micro-level of analysis by examining the consequences those individual behaviours may have on shaping and changing the aggregate processes.

Imbedded within the demographic conception of population is the concept of total population. This concept is not as simple as it may first appear because there are two types of total population for demographers: the de facto and the de jure (Shyrock et al., 1980). The de facto population comprises all the individuals actually present in a given region at the specific time. The de jure population is a bit more fluid and consists of '. . . all the people who "belong" to a given area at a given time by virtue of legal residence, usual residence, or some similar criterion' (Wilmoth, 2004: 65). Within nations or large areas the distinction between these two concepts may not be that great; however, within subdivisions in a population, they may matter greatly. As demographers have learned, these are ideal definitions and in practice one often finds a mix of the two. For an insightful discussion of the concept of population in demography and sociology, see Ryder (1964).

The population concept is also fundamental to anthropological genetics. However, because genetics focuses on the transmission of genes among individuals between generations, some further specifications are added to the basic definition: a population must be a group of sexually interbreeding (or potentially interbreeding) individuals, and as for demography, they must occupy a specific geographical region. Furthermore, since evolution takes place through time, breeding populations (or mendelian populations as Sewall Wright termed them) should show continuity across generations. Ephemeral groupings do not qualify as mendelian or evolving populations.

Mendelian populations can be visualized as levels of a hierarchy from the entire human species that constitutes a single, long-term breeding population, subdivided into continental and regional populations, that in turn are subdivided into myriads of local populations. The term deme is often applied to these small local population units (Gilmour and Gregor, 1939; Wright, 1969). Ideally, all the members of a deme have the same (and equal) chance of mating with any other member in the group of the opposite sex (panmixia). In practice, many behavioural and cultural factors structure mating even within the deme and much interesting work in anthropological genetics has been concerned with identifying and studying the effects of such factors.

Population subdivision can be caused by a variety of factors such as environmental patchiness, the geographical structure of the area (e.g. mountain ranges, rivers), and socio-cultural factors (e.g. preferential mating, linguistic and religious barriers). These demes usually differ genetically (mainly through non-selective mechanisms such as genetic drift) but are not so different as to completely prevent gene flow and the introduction of new genetic variability.

The concept of *population structure* includes both this hierarchical subdivision of populations as well as those factors differentiating individuals within each subdivision (see below and Relethford, Chapter 7). Thus, anthropological genetic studies include populations that range from small relatively isolated groups to those that are large and subdivided into various units based on social, geographical, and cultural features.

As anthropological geneticists, we should be aware of the varying constructs of *population* and population size that are used by both geneticists and demographers. In some cases these are the same or similar and in other cases they are different. Clarity in the use of these concepts is important as both demography and anthropological genetics converge. Further discussion of these issues will continue throughout this chapter as well as further specification of the crucial concept of population structure and introduction of additional notions such as *effective population size*.

Population size and the concept of effective size

At first glance, population size would seem to be a simple quantity. Governments employ census takers to enumerate all the people in a specified area, seemingly a straightforward procedure. Leaving aside the potential issues of de facto and de jure populations mentioned above, and the technical difficulties of actually finding everyone, the demographer's population size seems unproblematic. For the anthropological geneticist, however, the population of interest is the *breeding* population. Not all individuals (or even most) in a given population satisfy this criterion. The total census population must be reduced to the number of breeding individuals by excluding the older and younger members of the group. At the simplest, for females whose reproductive life is tied more closely to age, those women between the ages of 15 and 45 (sometimes less, depending on patterns of fertility in a population) are included. Depending on the age structure of the population, this might reduce the breeding population to only one-third of the total census number with the pre- and post-reproductive portion of the population comprising the other two-thirds.

However, a more complex formulation is required for understanding the consequences of population size for genetic processes such as inbreeding and genetic drift. Sewall Wright (1931) devised the concept of population *effective size* (N_e) to take into account all the deviations from the idealized random mating that take place in real populations. Thus, all the factors that affect the proportion of the census population contributing genes to the next generation are included in this single parameter. These include basic demographic measures such as age structure (pre- and post-reproductive ages as already mentioned), overlapping generations, unequal sex ratios, as well as differences in fertility rates among individuals and consanguineous mating. For evolving populations, changes in population size through time may also have profound effects on N_e (see section on 'Population bottlenecks' below). Ideally, N_e measures the population size that contributes genes to the next generation or in the evolving population, the average population size over a specified number of generations.

The importance of this modification of population size is clear when one remembers that the effect of random genetic drift is critically dependent on it. Thus, the expected variance of gene frequencies for a two allele locus is just

$$\sigma^2 = pq/2N,$$

where p and q are the allele frequencies and N is population size. The appropriate N, however, is not the census size but rather N_e, the effective breeding size. This difference affects any estimate of expected drift and the same logic applies to all the other measures of genetic drift. For instance, as will be discussed below, N_e has even figured in arguments over the 'Out-of-Africa' *versus* 'Multiregional Evolution' debate in human origins research.

Population fluctuations and bottlenecks

Over the long span of human history and prehistory, especially prior to the development of agriculture, we can assume that the growth of the human species population was minimal (see Coale, 1974). Consideration of the basic equation for population growth makes this evident.

$$N_{t+1} = N_t e^{rt}$$

where N is population size, r the annual rate of growth, t time, and e is the base of the natural logarithms. Even for very small values of r, the compounding of N in this exponential equation would have produced an astronomically sized human population. This is not to say, however, that *local populations* experienced no growth (or decline). While the overall species population did not increase markedly until relatively recently, regional and local populations surely increased or decreased, sometimes catastrophically.

A striking example of this phenomenon is the extremely well documented small population of Tristan da Cunha, a tiny island

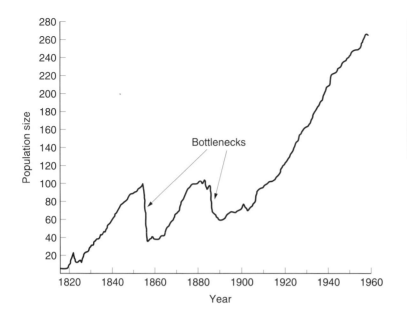

Fig 5.1 Population growth in Tristan da Cunha from 1816 to 1961. Note the two bottlenecks. During the first one, the population dropped from 103 to 33 individuals, and during the second bottleneck the population decreased from 106 to 59 individuals.

in the South Atlantic intensively studied by Derek Roberts and his colleagues (e.g. 1968, 1971). From its founding in 1816, meticulous records of births, deaths, migration, and marriages were kept. These records show the population growing moderately during 'normal' times but periodically plummeting due to accidents and/or large scale emigration. The trajectory of the population over time is shown in Figure 5.1. Each of the troughs in the curve is a *population bottleneck*, a term used for a significant reduction in the size of a population.

Although we should not generalize from one example, much evidence suggests that size fluctuations in small populations have been the normal course of events (e.g. Glass and Eversley, 1965; Kunstadter, 1972; Tilly, 1978). Thus, the often observed moderate growth rates (r < = 0.01 vs. r = 0.2−0.3 in some modern less developed countries, Bentley, 1993) were probably balanced by periodic reductions due to some catastrophic event (e.g. disease, warfare). Indeed, some have suggested that the entire human species was reduced to a mere 10,000 breeding individuals early in our history (Harpending and Rogers, 2000), a point discussed in more detail below.

The genetic effects of population bottlenecks vary depending upon the reduced size of the population and the length of time (number of generations) the population remains small. The usual result of this type of demographic reduction is the loss of genetic variation (and the possible subsequent loss of evolutionary potential). The loss of genetic diversity can seriously threaten the ability of the population to continue and recover. Anthropological geneticists use two primary measures of genetic variation: (1) heterozygosity or the proportion of individuals in the population who are heterozygous at a locus; and (2) allelic diversity or the actual number of different types of alleles present at a locus. These measures can also be used to evaluate the effects of bottlenecks.

In general, population bottlenecks can have a significant effect on allelic diversity. Rare alleles are especially susceptible to loss during bottlenecks. There is a lesser effect on the additive genetic variance of quantitative traits (portion of genetic differences among individuals in a population that can react to natural selection). However, if the bottleneck is prolonged and severe, there can be loss of quantitative genetic variation.

As a simple example, consider a large randomly mating population that experiences a reduction in population size, N, that lasts only one generation. Following this bottleneck, the expected hetero-zygosity (H_e), expressed as a proportion of the original heterozygosity, will be (Allendorf, 1986):

$$H_e = 1 - \frac{1}{2N}$$

As N increases, the term [1/2N] decreases and the proportion of the original heterozygosity remaining also increases. This expectation is valid if we also assume that there is no differential selection among alleles. A short duration bottleneck does not radically affect the heterozygosity. For example, a single generation bottleneck that reduces a population to only two individuals has an expected heterozygosity of 0.75.

It is rather easy to modify the above equation to take into consid-eration a bottleneck lasting more than one generation. The expected proportion of heterozygosity that is retained after a bottleneck of t-generations is (Allendorf 1986):

$$H_e = \left(1 - \frac{1}{2N}\right)^t$$

The calculation of the expected number of alleles is a bit more complicated since it requires knowledge about the number and frequencies of alleles present before the bottleneck; something that is not always possible.

Consider a large randomly mating population with a locus with n alleles at frequencies $p(1)$, $p(2)$, $p(3) \ldots p(n)$. The number of alleles expected after a bottleneck $E(n')$ can be determined by:

$$E(n') = n - \sum_{j=1}^{n} (1 - p_j)^{2N}$$

Using n', one can then calculate the allelic diversity, A, after the bottleneck:

$$A = (n' - 1)/(n - 1)$$

If all the alleles remain, A equals 1.0, and if only one allele remains, $A = 0$.

Heterozygosity usually gives a satisfactory and accurate assess-ment of the loss of genetic variation following a bottleneck.

However, if the population contains many alleles or if the bottleneck is not severe, heterozygosity may underestimate the actual loss of variation. For example, loci with many alleles such as the HLA system illustrate this point. Following a bottleneck, loss of allelic diversity at HLA loci may be great but there may only be a small decrease in the heterozygosity. The reduction in diversity of alleles at HLA loci may thus be detrimental to the survival of the population if disease selection is intense. Knowing the number and frequency distributions of alleles will aid in any interpretation; however, this may not be possible if one is analysing ancient population dynamics.

Computer simulations by Allendorf (1986) show the effect of multiple generations on the heterozygosity and allelic diversity during a bottleneck. As a general result, the allelic diversity is reduced more than the heterozygosity. The effects of bottlenecks are larger if the reduction in population is great and lasts for numerous generations. Heterozygosity is a useful measure of the population's ability to react to natural selection following a bottleneck; while allelic diversity provides a measure of the ability to respond to longer-term natural selection and the survival of the population. Used in conjunction, these two measures provide valuable insights into the genetic effects of population bottlenecks.

Since demographic measures such as the sex ratio, temporal variation in the number of sexually reproductive individuals in the population, and the variability in number of offspring per individual, have an impact on how fast genetic variation is actually lost, the population size, N, used in the above equations must be N_e, the effective population size (see above). These measures appear to work well when calculating heterozygosity, but are less satisfactory when measuring allelic diversity because allelic diversity is largely determined by the number and frequency of alleles.

Many examples of population bottlenecks exist – too many to review exhaustively here. However, to illustrate the genetic effects of bottlenecks, we will consider the following cases.

The population fluctuations of Tristan da Cunha discussed above provide several illustrations of the genetic effects of bottlenecks (Roberts 1968, 1971, 1973). Most notably, each bottleneck changed the composition of the ancestral gene pool. As a large percentage of the previous population was lost (to death or out-migration), their genes were removed from the Tristan gene pool. The genetic ancestry of the population following each bottleneck comprised only those who remained or survived, potentially leading to very different gene frequencies in the descendent populations. Indeed, each bottleneck is the temporal analogue of the spatial phenomenon of *founder effect*. Founder effect occurs when a small group buds off from a population and founds or colonizes a new location. The genetic effect arises from the fact that only a small sample of the parental population's gene pool can be represented in the small set of colonizers. Thus, the newly founded colony may be expected to differ genetically from its parent group since the sample of genes carried by the founders may, by chance, vary greatly. (Founder effect is classified as one component of the broader concept of random genetic drift – see below for a fuller discussion.) In the Tristan case, the genetic composition of the generation after a bottleneck did not represent the entire pre-disaster population's gene pool. Only the survivors contributed genes to the subsequent population.

A further point to make from the Tristan da Cunha example is the effect of kinship on N_e. Human populations are structured by many factors, but in small-scale societies kinship often plays an important role influencing the organization of communities and interaction/cooperation among individuals. Thus, for each bottleneck on Tristan, family ties often determined who would leave the population. Roberts (1968) called this 'booster effect' since he recognized that it augmented the genetic effect of the population reduction. *Kin-structured migration* (Fix, 2004 and below) is an equivalent concept. Since biological kin are likely to share genes by descent from a common ancestor, an emigrating group of families may be even more unrepresentative of the parental gene pool than would a random assemblage of migrants.

The genes shared by relatives (*identical by descent* from a common ancestor) lead to a reduction in heterozygosity through time due to *inbreeding*. Inbreeding is the consequence of kin (or consanguineous) mating. Since each spouse shares a common ancestor with her/his relative, it is more likely that their offspring will be homozygous for these shared alleles. The increase of homozygosity, viewed either as an increased correlation (Wright, 1969) or a probability of identity by descent (Malécot, 1948, 1969), is called the *inbreeding coefficient* or *F*.

The small, bottle-necked population of Tristan da Cunha again nicely illustrates this phenomenon. Records kept from the beginning of the colony show the gradual development of inbreeding as each generation increased the degree of kinship among the whole population. Even while attempting to avoid mating with a relative, a point was reached when it became inevitable (Roberts, 1967). Thus, inbreeding and a consequent reduction in heterozygosity is also to some extent a function of population size and kinship should also be taken into account when calculating N_e.

Turning now from the tiny, relatively isolated Tristan population where our information on population came from excellent demographic records, we examine population bottlenecking at the level of the whole human species and use genetic information to infer the drastic reduction in N_e.

One of the first claims made from the study of worldwide variation in mitochondrial DNA (mtDNA) was that the human species had originated relatively late (*c.* 200,000 years ago) from a small subset of the hominid population living in Africa (Cann *et al.*, 1987). This hypothesis was partially based on the very low levels of diversity in mtDNA across wide geographic regions of our species, variation lower than that found in regional populations of chimpanzees, for instance Morin *et al.* (1994). This view of human origins, sometimes called the 'Out-of-Africa' or 'Garden of Eden' model, contrasts with the multiregional evolution model in positing an initial small ancestral species population rather than evolution from a larger, widely dispersed set of regional archaic populations. (There are many reviews of this continuing debate: see e.g. Relethford, 2001; O'Rourke, Chapter 8; Tishkoff, Chapter 12). Further studies of the non-recombining portion of the Y chromosome (NRY) and some nuclear short-tandem repeat loci seemed to confirm the recent origins of *Homo sapiens* from a small regional population (Harpending *et al.*, 1998), although it should be noted that loci from the nuclear portion of the genome do not all demonstrate this pattern (Harpending and Rogers, 2000).

Notice that the direction of inference in this case is *from* genetics *to* demography. That is, population genetic theory tells us that small N_e should reduce the overall level of genetic diversity in a population. If the current human species is relatively genetically homogeneous, this could be due to a small founding N_e.

A number of investigators have used this logic to estimate the effective population size that would be congruent with human diversity for several genetic systems, including mtDNA and NRY, as approximately 10,000 individuals (Takahata, 1993; Hammer, 1995; Sherry *et al.*, 1997; Harding *et al.*, 1997; Jorde *et al.*, 2001). That is, at some point in the species history for a longer or shorter period of time, the total number of breeding adults would have been relatively small, indeed a bottleneck for the entire species through which only some genes were transmitted to future generations.

The titles of the Harpending *et al.* (1998) paper, 'Genetic traces of ancient demography', and Rogers (1995), 'Genetic evidence for a Pleistocene population explosion,' indicate that there can be a two-way street between demographic and genetic data and inference. Rapid expansions of populations leave different demographic and genetic profiles than those found in populations in which growth and dispersal occurs over a longer time frame at a more constant rate (Tajima, 1989; Slatkin and Hudson, 1991; Rogers and Harpending, 1992; Sherry *et al.*, 1998; Pritchard *et al.*, 1999). After fast demographic expansion, and the often associated geographic isolation, these populations often show substantial genetic diversity.

The nature of demographic expansions (sometimes after a bottleneck) can be examined through mismatch and crossmatch distribution analysis of mtDNA data (Harpending *et al.*, 1993; Harpending, 1994; Sherry *et al.*, 1994; Harpending *et al.*, 1998). For example, using these methods, Shurr and Sherry (2004) examine mtDNA and Y chromosome diversity in relation to the peopling of the New World. They argue that the initial migration occurred between 20,000 and 15,000 years BP along the coastline, reaching South America by at least 14,600 years ago. They also suggest a second, later migration occurred after an ice-free corridor opened. For our purposes here, however, we are more interested in their cautionary notes. As they, and others (e.g. Sherry *et al.*, 1994) suggest, the interpretations of the expansions and their timing must be tempered by the sampling methods and sample sizes collected. Researchers must consider what effect small sample sizes (e.g. 10−25 individuals per group) may have on representing and reflecting the full genetic diversity within the groups being analysed. The demographic size and geographic location of the sampled populations does not always represent the overall sizes of the groups being studied nor does it capture the true extent of genetic diversity present.

Another study illustrates how the amount of genetic diversity reflects the demographic structure of ancestral populations. An examination of *Alu* insertion polymorphisms suggests that the demographic history of African populations may have been different from the demographic history of non-African populations. Stoneking *et al.* (1997) found that all the African populations they sampled showed higher heterozygosity than predicted and higher heterozygosity than non-African populations. Elevated heterozygosity can be the result of higher rates of gene flow. In addition, elevated rates of gene flow should reduce population differentiation. However, this study showed that the African populations had the largest between-population differences as measured by G_{st}. Given these results the researchers concluded that the effective sizes of populations across Africa must have been greater than the non-African populations that showed lower heterozygosity. On the other hand, others (e.g. Cann *et al.*, 1987; Hammer, 1995; Armour *et al.*, 1996) have suggested that the elevated genetic diversity in Africa reflects a greater antiquity of African populations and thus an African origin for modern *Homo sapiens*. Please note that this alternative interpretation does not mean that an African origin of modern humans is incorrect. African populations could have been both larger than non-African populations and the source of modern *Homo sapiens* at the same time.

The argument that our species experienced a severe bottleneck with effective size (N_e) of only 10,000 individuals has been used to undermine the multiregional evolution theory of modern human origins. It would seem very unlikely that a widespread human species population interconnected by gene flow inhabiting the entire old world would comprise only 10,000 breeding individuals (Rogers, 1995). The inferred population bottleneck implies a localized population that 'exploded', colonizing (invading) the territory of archaic hominids and ultimately replacing them.

But this conclusion depends on N_e representing the actual *breeding* size (N_b) of the species during the time span of the bottleneck. N_e may be considerably smaller than N_b due to several factors in addition to the removal of the pre-and post-reproductive members of the population. Depending on the strength of these added factors, N_b might have been sufficient to maintain a transcontinental network of gene flow.

Elise Eller (2002) has pointed out a potentially important cause of sharply reduced N_e in the Pleistocene: extinction and recolonization. Her model proposes that local populations of early humans periodically were wiped out by local catastrophes. Their vacant territories would then be recolonized by small groups. Each such extinction/recolonization event would constitute a founder effect reducing N_e. In fact, she estimates that the total number of breeding individuals (N_b) could have exceeded 300,000 if the local extinction rate was on the order of 10%. This result might also be augmented by kin-structuring of the recolonizing groups (Eller, 2002; see also Fix, 2004 and below).

As these examples show, inferring prehistoric demography from current genetic patterns is not without difficulties. In principle, however, with careful attention to potential pitfalls, it should be able to provide insights not available from other sources.

Population structure

When demographers describe the structure of a population, it is age and sex composition that is presented, often in the form of a *population pyramid*, a visual representation of the numbers (or percentages) of males and females in the population. Specialized studies of human populations by demographers and other social scientists may define other variables of interest in addition to these two basic factors. For example, epidemiologists might characterize a population divided into susceptible or non-susceptible to a particular disease (see 'Disease epidemiology and demography' section below).

Population structure in anthropological genetics is a much broader concept, implying a variety of mechanisms that influence the degree and pattern of genetic variation in a population (Cavalli-Sforza, 1959). Some human geneticists [e.g. Morton and his colleagues (1977)] have defined population structure very specifically as the expected decline of genetic similarity as a function of distance between populations (see Jorde, 1980 and Fix, 1999 for discussion of this work). Most anthropological geneticists, however, include numerous variables in addition to geographic distance as attributes of population structure. In fact, almost any mechanism that influences genetic drift or gene flow can be considered a component of population structure. Thus, as Mielke *et al.* (1976) state, 'the total description of population structure...requires the integration of biological, social, and demographic data set in an ecological framework'. Further, the potentially changing relationships of these variables need to be specified through time in an evolving population.

Demographic variables are an important component of this catalogue. As already noted, the basic variables, age and sex, affect N_e and thereby opportunities for genetic drift. Social and ecological attributes of a population may affect genetic variation indirectly by influencing demographic structure. Population size, density, stability, and migration probabilities and patterns are all determined by socio-ecological conditions. Culture may act more directly in specifying preferential or proscribed marriage practices involving kinship, social class, ethnicity, language, and/or religion.

Migration and gene flow

Migration is one of the basic demographic variables determining changes in population size and composition. In contrast to the other variables, fertility and mortality, migration is not as easily characterized as the unique events (for individuals) of birth and death. Partly for this reason, the study of migration in formal demography is less well developed (Newell, 1988). The numerous studies of world variation in fertility and mortality are not paralleled by world surveys of migration (Livi-Bacci, 1977; Coale *et al.*, 1979; Coale and Watkins, 1986).

Because of the potential ambiguity in characterizing 'migrants' versus 'visitors', the definition of migration in demography relies on *intention*: migration occurs only when people move, remain in the new location for at least a year, and/or declare the intention of remaining in the new location (Newell, 1988). The definition also depends crucially on the units among which migration takes place. Short distance moves may cross political boundaries in some parts of the world while being only the daily commute in others.

Biologists also use migration in several ways, often for the seasonal movements of animals while using *dispersal* for one-time movements, and *range expansion* or *colonization* for occupation of new territories (see Fix, 1999 for discussion of these and other definitions).

For evolutionary genetics, the importance of movement of individuals through migration is its consequences: the transfer of genes between populations or *gene flow*. Anthropological genetics is concerned with the size and structure of migrating groups, the roles of geographic distance and barriers, and social factors which all determine the magnitude, duration, and pattern of gene flow. The study of migration, then, includes the demography of population exchange and the geographic, economic and socio-cultural attributes of populations that structure the flow of genes (see Fix, 1999 for an extended discussion of these topics and many examples of migration in different human societies).

Of these basic variables, geographic distance is foremost in genetic migration models. As noted in the section on population structure, some human geneticists have treated distance as the *sole* factor determining migration (Morton, 1977). These models depend on long-distance migration to stabilize gene frequency variation while isolation by distance produces a characteristic decline of genetic similarity among exchanging populations.

Geographers (e.g. Lewis, 1982) have also emphasized the importance of distance in shaping migration flows. Many empirical studies of human migration have confirmed the central role of geographic distance (e.g. Olsson, 1965; Mielke *et al.*, 1976). However, these studies (and many others) have also identified other significant causal factors. Economic forces have always loomed large in social science thinking about migration. Indeed, a similar causal role of the 'economy' is also central to biological views of migration. Dingle (1996) sees migration as behaviour directed toward improving environmental conditions for organisms. Thus, the numbers and direction of migrants will depend on economic/environmental conditions. For humans, potential movers may be influenced by their knowledge of surrounding places. This insight was formalized in the *neighbourhood knowledge* model of Boyce and his colleagues (1967) in which the distribution of marriage distances is proportional to the amount of visiting (presumably leading to 'knowledge' about the place) among a group of populations.

Geographic barriers may also restrict movement of migrants, although features limiting movement of many animal populations such as rivers (the great apes) and oceans may become channels of communication and movement for boat-possessing people.

This point emphasizes the centrality of socio-cultural factors in structuring human migration. Not only can technology negate geography, economic and technological conditions may determine the mobility of individuals and the frequency and distance of movement. Thus, for many hunting-gathering societies, mobility is the key strategy both in daily foraging but also to resolve conflict and/or escape local shortages. To accomplish this, these populations usually have wide-ranging marriage, kinship and friendship networks. In contrast, high-density agricultural populations are often extremely localized with only very short range marital migration producing local gene flow (see Fix, 1999 for examples).

Culture may also serve as a barrier to intermarriage. Shared language, religion and ethnicity are among numerous factors influencing mate choice. For example, many anthropological genetic studies have been undertaken among American religious communities such as the Amish, Hutterites and Mennonites (e.g. Crawford *et al.*, 1989; Morgan and Holmes, 1982; Olsen, 1987).

Kinship is a major socio-cultural factor affecting marriage and mobility. Incest taboos limit inbreeding in all societies. Moreover, patterns of post-marital residence differentially affect male and female mating distances (matri- versus patrilocal — see below).

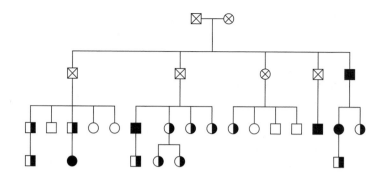

Fig 5.2 Genealogy of a migrant group. Squares are males; females are circles. Hemoglobin E homozygotes (*EE*) are fully darkened, heterozygotes (*AE*) are half-darkened, and homozygotes (*AA*) are not darkened. Crosses indicate deceased individuals.

In some small-scale, low density populations, kinship may also structure the composition of migrating groups, a phenomenon termed *kin-structured migration* (Fix, 1978). For example, among the Semai Senoi, a swidden-farming population of Peninsular Malaysia (Dentan, 1968), family groups often migrate together. The members of these kin groups, being biologically related, share genes. As a consequence, they do not constitute a random sample of the gene pool of their origin population and may deviate sharply in gene frequencies. Figure 5.2 shows a genealogy of such a migrant family group along with their hemoglobin genotypes. As can be seen, this family included many individuals with the hemoglobin *E* allele, one of the several malarial-resistant hemoglobin types (Livingstone, 1985). Since Semai villages are relatively small (ranging between 50 and 250 inhabitants), this large family group made a substantial impact on the hemoglobin *E* frequency in both the recipient and origin populations.

Although migration is generally seen as the major force reducing genetic variation among populations, kin-structured migration may actually *increase* genetic micro-differentiation (Fix, 1978) and may better be classified as one of the components of random genetic drift (Rogers, 1988). (More examples of the causes and consequences of kin-structured migration may be found in Fix, 2004).

Thus, the demographic process, migration, may have genetic consequences ranging from spreading newly arisen mutant alleles through local and regional populations of a species, to mixing gene pools thereby reducing genetic differences among populations, to actually promoting local genetic micro-differentiation when migrant groups are kin-structured.

This example depended on demographic and genealogical measures of migration to infer and examine genetic consequences. As in the case of fluctuating population, we may also use genetic patterns to infer past demographic events such as migration in populations.

An interesting example of such inference is the recent claim that genetic evidence demonstrates a higher migration rate for females during the course of human evolution (Seielstad *et al.*, 1998). The argument is based on the fact that mitochondrial DNA (mtDNA) is transmitted only by females and Y chromosomes only by males. Since migration (gene flow) generally reduces local genetic diversity, a difference in the relative levels of Y chromosome and mtDNA variation could be due to a higher rate of movement of the relevant sex. If males were more mobile, Y chromosome variants would be spread over a wider geographic area; conversely, greater female mobility would diminish mtDNA variation across larger regions. Seielstad and his colleagues (1998) found just such a differential between Y and mtDNA variation in a worldwide sample of human populations. A standard measure of genetic variation, F_{ST}, is 19% for mtDNA and 64% for Y chromosome markers. Similarly, they show that for European populations, there is a more rapid increase in genetic differences with geographic distance for Y chromosome loci than for mtDNA markers. This observation is again consistent with relatively greater movement of females spreading mtDNA variants more widely.

An obvious cause for this difference would be the widespread practice of patrilocality, brides moving to the natal population of their husbands at marriage. But, as Stoneking (1998) points out, some question whether a *local* event could account for the difference in *global* F_{ST} values. Moreover, a more recent study by Wilder *et al.* (2004), based on data from the same set of individuals from ten different world populations, a sampling strategy that removes the potential problem of comparing different molecular data from different populations (apples and oranges, Stoneking 1998), detected no differences between mtDNA and Y chromosome variation, in direct contrast to the Seielstad *et al.* (1998) findings.

Human genes possess enormous potential for inferring past demographic events, including migration. However, the story is not simply read from gene distributions; many factors can shape gene histories and great care must be exercised to evaluate all these forces.

A very active area of molecular anthropological genetics involves reconstructing large-scale movements of human populations in prehistory including the initial spread of *Homo* from Africa as well as various colonizations/invasions of continents or regions. Many of these studies will be presented in subsequent chapters (see Part 4 'The Human Diaspora', Chapters 12–16).

Disease epidemiology and demography

Diseases often play a significant role in influencing the mortality and fertility patterns in populations. Thus, the confluence of demography, genetics and disease is a natural one. The demographic structure of populations often influences disease outbreaks and the disease may, in turn, shape the demographic and genetic structures of the area. Demographic and genetic variations allow some diseases to flourish (McGrath, 1988; Cliff et al., 1981) while preventing others from occurring (e.g. Livingstone, 1984).

The epidemiology of many infectious organisms is linked to the demographic structure of the host populations. The impact and course of many epidemic diseases is directly related to the demographic parameters (e.g. population size, number of births per year) of host populations. For example, the periodic nature of many human epidemic diseases is clearly associated with the number of new susceptible individuals in the population. The early work of Kermack and McKendrick (1927) established the strong relationship between population and epidemic occurrences. If a *threshold* number of susceptible individuals are not present, an epidemic cannot take hold and move through the population. In order for a disease to be sustained in a population there must be a sufficient number of susceptible individuals. The regular reappearance of such epidemic diseases as measles, smallpox, and scarlet fever reflect the dynamic changes in the numbers of susceptible, infectious, and immune individuals in a population. As an epidemic moves through a population, susceptible individuals are eliminated (become infected and then become immune or die). Eventually the number of susceptibles is reduced to the point that the epidemic cannot sustain itself, and it disappears. Over time, births and migration replenish the population of susceptible individuals and once again a *threshold* is attained and a new epidemic can occur (Bailey, 1975; Sattenspiel, 1990, 2003).

It is interesting to note that concept of *threshold* originated as a consequence of the analysis of the Kermack-McKendrick model. Thus, a basic concept, that has practical implications in disease control and understanding of epidemics, is the result of modelling. In general, modelling disease dynamics can provide researchers with insights not gained by other analyses. In some cases, modelling may reveal features that are simply not obvious. Social activities, cultural barriers, membership in particular groups (cultural and biological), geographic separation, genetic makeup and customs are just a few features that can be incorporated into disease models. Models can simultaneously take into consideration epidemic and human behavioural processes and the interaction between the processes. Many diseases are transmitted by contact, direct or indirect, between individuals and since populations are often exposed to various diseases, the genetic and demographic structure may determine the outcome of the contact (i.e. epidemic or not). Thus, models can provide valuable insights into the conditions (genetic and cultural) under which diseases occur and are maintained in populations.

Many basic models of infectious disease spread include four demographic or epidemiological groups: susceptibles, $X(t)$; latents, $H(t)$; infectious, $Y(t)$; and immune, $Z(t)$. For ease of calculation, many of these models, but not all, also assume that the host population size, N, remains constant:

$$N = X(t) + H(t) + Y(t) + Z(t)$$

Quite a number of deterministic and stochastic models of disease transmission and spread have been developed over the years since the initial insight provided by such researchers as Kermack and McKendrick (1927) and Soper (1929) (e.g. see Rushton and Mautner, 1955; Black, 1966; Bailey, 1975; Murray and Cliff, 1975; Rvachev and Longini, 1985; Sattenspeil, 1990; Sattenspiel and Dietz, 1995; Sattenspiel and Herring, 1998).

In some cases the threshold models have given way to the epidemiological concept of *basic reproductive rate* (R_0). This rate is the number of secondary infections that are produced by the initially infected individual (assuming a totally susceptible population). R_0 is related to the epidemic threshold and the initial population of susceptible individuals, and its value is comparable to the *net reproductive rate* used in demography. This rate is very appealing to epidemiologists interested in disease control modelling because if one can break the chain of infection so that an infected individual does not infect more than one other person, the disease outbreak can be controlled. As Sattenspiel (1990) observes, this is similar to decreasing the *effective population size* to a value below the critical threshold in order to sustain an epidemic.

Demographers will be familiar with the *epidemiologic* R_0 and its interpretation because it is analogous to R_0 used in demographic research. The net reproductive rate (R_0 or NRR) was developed by German statistician Richard Böckh in 1884 and written in an approximate formulation as

$$R_o = e^{rT}$$

where e is the base of the natural logarithm, T is average generation length (mean age of mothers at birth of their daughters), and r is the *intrinsic rate of natural increase*. This is a measure of the number of daughters a cohort of newborn females will produce during their lifetime. Two assumptions are made: (1) there is a fixed schedule of age-specific fertility rates; and (2) there is a fixed schedule of age-specific mortality rates. R_0 is thus a measure of the degree to which a cohort of newborn females will replace themselves. The interpretations of the values for the *demographic* R_0 are similar to the interpretation of the *epidemiologic* R_0. Hence, a net reproductive rate of 1.00 means exact replacement, a value below 1.0 indicates that the population is not replacing itself, and a rate above 1.0 means that the population is more than replacing itself. For the *epidemiologic* R_0 the interpretation is similar: if R_0 is greater than 1.0, the population of microorganisms can grow and an epidemic is possible; if R_0 is less than 1.0, the epidemic dies out; and if R_0 is equal to 1.0, the epidemic stays constant in size.

As an interesting addition and to show the confluence of genetics and demography, the *intrinsic rate of natural increase*, r in the equation above, has been used by geneticists to examine the intensity and direction of natural selection in human populations. Demographers measure the rate of increase or decrease in the total size of a population by a quantity known as the intrinsic rate of natural increase (r). This quantity is calculated from the age-specific birth and death rates. In 1930 Fisher suggested combining mortality and fertility rates with factorial inheritance so that the principle of natural selection could be expressed in a mathematical theorem. Increases and decreases in a population are regulated by differences in fertility and mortality rates and these can be used to assess the rate of 'improvement' of a population in relation to its specific environment. This method uses the demographic measure called the intrinsic rate of increase (r). The rate is calculated for both the 'affected' and the 'normal' portions of the population. The rates are then compared to determine the fitness of the two groups:

$$w = \frac{e^{r_1 T}}{e^{r_2 T}}$$

where e is the base of the natural logarithm, T is average generation length for the total sample, r_1 is the intrinsic rate of natural increase for subgroup one and r_2 is the intrinsic rate of natural increase for the subgroup having the largest r_i value. The first study using r to estimate the intensity and direction of natural selection employed an IQ sample from Kalamazoo, Michigan (Bajema, 1963). Not many studies have used this method because of data constraints (see e.g. Bodmer, 1968; Cavalli-Sforza and Bodmer, 1961; Adams and Smouse, 1985). It is often difficult to detect different genotypes and then trace those individuals throughout their entire life.

Demographic features such as the age and sex structure of the host population have an important influence on the behaviour of infectious diseases (e.g. Anderson and May, 1985; Castillo-Chavez *et al.*, 1989). One of the more intriguing questions is how the structure (demographic and genetic) of the host population affects the impact and transmission of diseases (John, 1990). Intimately related to this question is how do the genetic, epidemiological, and demographic characteristics of the disease agent(s) (and the vectors) affect the impact of the disease on the host population.

Spatial, temporal, and individual heterogeneities in susceptibility to infection reflect not only the genetic and demographic structure of human populations but also that of the pathogen and the vector (if the disease is vector borne). As a non-human example shows, with the foot and mouth epidemic that hit the United Kingdom in 2001, factors such as the heterogeneity of farm size and species composition were responsible for the varied dynamics of the epidemic, especially the spatial aggregation of cases (Keeling *et al.*, 2001). As the authors suggest, without good demographic data, the dynamics of these types of outbreaks will not be resolved nor can measures be taken to contain and control disease outbreaks (reduce the reproductive ratio of the disease, R_o). Stochastic models such as those of Keeling *et al.* (2001) which take into account such factors as herd and farm size and composition, spatial aggregation of cases, regional geography, and vaccination differences provide a picture of this disease outbreak that is different from deterministic models of the same event (e.g. see Ferguson *et al.*, 2001). Ideally, one could incorporate the genetic structure of both the host and pathogen into these models.

Changing demographics and shifts in disease patterns

Various diseases, both endemic and epidemic, have contributed to shaping the genetic and demographic structures of human populations over the ages. Conversely, the changing demographic structure of human populations has influenced disease occurrence, transmission, and spread over the same period. Small, mobile populations of hunters/gatherers were subjected to a different set of disease entities than later larger sedentary populations (Armelagos and Dewey, 1970; Cockburn, 1971; Armelagos and McArdle, 1975; Cohen and Armelagos, 1984; Swedlund and Armelagos, 1990; Wilson *et al.*, 1994). These groups were small, dispersed over the landscape, and mobile. These features minimized the effect of many acute communicable diseases, especially those, such as measles, smallpox and mumps, which require a large population base in order to survive and spread. The demographic factors responsible for shaping and maintaining these small populations are debated (Armelagos *et al.*, 1991). Some demographers think that these early forager populations would have had high, maximal, natural fertility coupled with high mortality. On the other hand, these populations may have been sustained by moderate fertility and moderate mortality rates in an environment of low infectious disease pressure (Armelagos *et al.*, 1991).

About 10,000 to 12,000 years ago many of these populations became less mobile and started to produce food. These cultural changes eventually resulted in a dramatic increase in population size and larger population concentrations. These increases probably resulted from maximal natural fertility, moderate mortality, and food surpluses which provided better, and more predictable, nourishment and health. However, empirical evidence suggests that with these demographic changes came an increase in infectious and nutritional disease incidence. This shift has often been referred to as the 'first epidemiological transition' (Omran, 1971, 1983; Barrett *et al.*, 1998). Human interaction with animals intensified, and zoonotic diseases, such as anthrax, tuberculosis, and Q fever probably became important factors. Landscape changes exposed populations to vector borne diseases such as scrub typhus and malaria. In turn, these diseases impacted the human populations and began to mould their genetic structures.

With the rise of civilizations in prehistory, epidemic diseases became important factors in shaping the genetic and demographic structure of human populations. By 5,000 to 6,000 years ago urban centres, some as large as 50,000 individuals, became important contributors to the disease ecology of human groups. Crowd diseases began to play a role in shaping the genetic and demographic structure of populations. During this time, we see the rise of epidemic diseases such as cholera, smallpox, chickenpox and measles. The Black Death eventually courses its way throughout Europe, killing an estimated 25 million people.

Mortality crises in the past (and present) impact both the genetic and the demographic structure of populations. For example, studies of famine induced mortality crises have shown that marked differences in population structure among regions and subdivisions of populations often occur. The effects on population density, household sizes, male/female ratios, and effective population sizes were regionally variable (Bittles *et al.*, 1986; Bittles, 1988; Smith *et al.*, 1990). Given the variable effects, one would expect that genetic changes were likewise variable, enhancing such factors as inbreeding and genetic drift.

The genetic structure of a number of regions of the world appears to reflect the past impact of various diseases both epidemic and endemic. Research by Glass *et al.* (1985) in Bangladesh showed a significant association between infections of *Vibrio cholerae* 01 and O blood. It is suggested that the low frequency of O and B in the Ganges delta region is a result of the endemic presence of cholera. Similarly, Vogel and Motulsky (1997) propose that the high frequency blood group O among native peoples of the Americas may be related to natural selection by syphilis or other treponemal infections (yaws and pinta). The existence of an H-like antigen on the surface of the plague bacterium and the association of smallpox and blood group A suggests that these diseases could have acted as selective agents moulding the ABO blood group frequencies over the centuries (Vogel *et al.*, 1960; Pettenkofer *et al.*, 1962). Indirect evidence for this is the decreased frequency of O and high B found along the major river systems in India and China.

Some historical demographers are now arguing that mortality patterns are much more complex than some thought, reflecting environmental, economic, social, biological, and demographic variables rather than changes in wages and living standards (e.g. Dobson, 1992; Landers, 1992). Mortality varies seasonally and locally, and these historical demographers are starting to focus their work on what they refer to as 'epidemiological landscapes'. The focus has shifted from examining large continental aggregates to exploring the role that spatial variation in population structure (e.g. population size and density, age structure, migration patterns, socio-economic variation) has on influencing the degree and intensity of exposure to infectious agents and hence mortality patterns and changes in demography. Different diseases are influenced by different sets of variables, and anthropological geneticists could contribute by studying the differential genetic susceptibility of various populations to these diseases.

In addition, Dobson (1992: 92) states, 'The role of population movements and their simultaneous passage of vectors, pathogens, carriers and victims, explain many of the dynamic and elusive patterns of infectious disease outbreaks.' And, as Preston (1993: 603) states, 'Demographers' techniques and perspectives are useful elements in the array of approaches used in the social and health sciences. With their careful attention to issues of measurement and data and their orientation to aggregate-level processes, they have provided valuable insights into many issues'. Thus, traditional demographers working with historical demographers and anthropological geneticists could do well to collaborate more often since their goal of understanding the dynamics of population changes is similar.

Summing up

The confluence of anthropological genetics and anthropological demography is clearly seen in many studies that explore human evolution, variation, adaptation, history and prehistory. This confluence has a rich history, with both fields enhancing the other. Theories, methodology, interpretations and data are often shared; and in many cases the different perspectives and goals provide insights that would remain hidden if both fields were separate and not complementary. This chapter has examined a number of these confluences and the shared interests of demographers and geneticists. From an anthropological view, without a demographic perspective and understanding, interpreting many anthropological genetic results would be difficult, if not impossible.

Anthropological genetics is concerned, to a large extent, with understanding the patterns and causes of genetic variation within and among human populations and exploring the patterns of genetic similarity and differences among these populations. By using such data as the mating structure, migration histories and population size fluctuations, anthropological geneticists can often deduce the demographic history of an area. Likewise, anthropological demography is important and sometimes vital to interpreting genetic data. This chapter has explored a number of issues and topics that are important to anthropological geneticists. In addition, we have noted where the two fields overlap and view the same data in different manners. For example, the concept of population is central to both, yet this concept varies and understanding both views enhances and enriches both fields. The concept of effective size of populations is vital to interpretations in anthropological genetics, but it also influences demographic interpretations. The ongoing debate between many paleoanthropologists and molecular geneticists about the origin of modern humans is just one example of how vitally important anthropological demography is to this controversy. The size and structure of early populations and the extent of contemporary genetic variation shows that demography and genetics are clearly linked. The structure of human populations (mating patterns, sex and age composition, migration, population fluctuations) is vital to understanding the genetic structure. The demographic structure of human populations often influences the disease patterns. In turn, these diseases influence and often shape the genetic structure of the population, again linking genetics and demography.

This chapter has not exhausted or explored all the links between genetics and demography. Other confluences of anthropological genetics and anthropological demography can be seen in various chapters in this book.

References

Adams, J., and Smouse, P. E. (1985). Genetic consequences of demographic change in human populations. In *Genealogical Demography*, ed. B. Dyke, and W. T. Morrill. New York: Academic Press, pp. 115–38.

Allendorf, F. W. (1986). Genetic drift and the loss of alleles versus heterozygosity. *Zoo Biology*, **5**, 181–90.

Anderson, R. M., and May, R. M. (1985). Age-related changes in the rate of disease transmission: Implications to the design of vaccination programmes. *Jour. Hygene*, **94**, 365–436.

Armelagos, G. J., and Dewey, J. (1970). Evolutionary response to human infectious disease. *Bioscience*, **20**(5), 271–5.

Armelagos, G. J., and McArdle, A. (1975). Population, disease, and evolution. In ed. A. C. Swedlund, *Population Studies in Archaeology and Biological Anthropology: A Symposium*, Memoir of the Society of American Archaeology, No. 30.

Armelagos, G. J., Goodman, A. H., and Jacobs, K. H. (1991). The origins of agriculture: Population growth during a period of declining health. *Population and Environment*, **13**(1), 9–22.

Armour, J. A. L., Anttinen, T., May, C. A., Vega, E. E., Sajantil, A., Kidd, J. R., Kidd, K. K., Bertranpetit, J., Paabo, S., and Jeffreys, A. J. (1996). Minisatellite diversity supports a recent African origin for modern humans. *Nature Genetics*, **13**, 154–60.

Bailey, N. T. J. (1975). *The Mathematical Theory of Infectious Diseases and Its Applications*. New York: Hafner Press.

Bajema, C. (1963). Estimation of the direction and intensity of natural selection in relation to human intelligence by means of the intrinsic rate of natural increase. *Eugenics Quarterly*, **10**, 175–87.

Barrett, R., Kuzawa, C.W., McDade, T., and Armelagos, G.J. (1998). Emerging and re-emerging infectious diseases: The third epidemiological transition. *Annual Review of Anthropology*, **27**, 247–71.

Bentley, G. (1993). The fertility of agricultural and non-agricultural traditional societies. *Population Studies*, **47**, 269–81.

Bittles, A.H., McHugh, J.J., and Makov, E. (1986). The Irish famine and its sequel: Population structure changes in the Ards Peninsula, Co. Down, 1841–1911. *Annals. Human Biol.*, **13**, 473–87.

Bittles, H.A. (1988). Famine and man: Demographic and genetic effects of the Irish famine, 1846–1851. In *Antropologie et Histoire ou Anthropologie Historique?*, ed. L. Buchet. Actes des Troisiémes Journées Anthropologiques de Valbonne, Notes et monographies techniques No. 24. Paris: Centre National de la Recherche Scientifique, pp. 159–75.

Black, F.L. (1966). Measles endemicity in insular populations: Critical community size and its evolutionary implication. *J. Theoret. Biol.*, **11**, 207–11.

Bodmer, W.F. (1968). Demographic approaches to the measurement of differential selection in human populations. *Proceedings of the National Academy of Sciences*, **59**, 690–9.

Boyce, A.J., Kuchemann, C.F., and Harrison, G.A. (1967). Neighbourhood knowledge and the distribution of marriage distances. *Annals of Human Genetics*, **30**, 335–8.

Cann, R.L., Stoneking, M., and Wilson, A.C. (1987). Mitochondrial, DNA and human evolution. *Nature*, **325**, 31–6.

Castillo-Chavez, C., Hethcote, H.W., Andreasen, V., Levin, S.A., and Liu, W.M. (1989). Epidemiological models with age structure, proportionate mixing, and cross-immunity. *Jour. Mathematical Biology*, **27**, 233–58.

Cavalli-Sforza, L.L. (1959). Some data on the genetic structure of human populations. *Proc. X Internatl. Congress Genetics*, **1**, 389–407.

Cavalli-Sforza, L.L., and Bodmer, W.F. (1961). *The Genetics of Human Populations*. San Francisco, CA: W.H. Freeman.

Cliff, A.D., Haggett, P., Ord, J.K., and Versey, G.R. (1981). *Spatial Diffusion: An Historical Geography of Epidemics in an Island Community*. Cambridge: Cambridge University Press.

Coale, A.J. (1974). The history of the human population. *Scientific American*, **231**(3), 31–51.

Coale, A.J., Anderson, B., and Härm, E. (1979). *Human Fertility in Russia Since the Nineteenth Century*. Princeton, NJ: Princeton University Press.

Coale, A.J., and Watkins, S.C., eds. (1986). *The Decline of Fertility in Europe*. Princeton, NJ: Princeton University Press.

Cockburn, J. (1971). Infectious disease in ancient populations. *Current Anthropology*, **12**(1), 45–62.

Cohen, M.N., and Armelagos, G.J. (1984). *Paleopathology at the Origins of Agriculture*. Orlando: Academic Press.

Crawford, M.H., Dyles, D.D., and Polesky, H.F. (1989). Genetic structure of Mennonite populations of Kansas and Nebraska. *Human Biology*, **61**, 493–514.

Dentan, R.K. (1968). *The Semai: A Nonviolent People of Malaya*, New York: Holt, Rinehart and Winston.

Dingle, H. (1996). *Migration: The Biology of Life on the Move*. Oxford: Oxford University Press.

Dobson, M.J. (1992). Contours of death: Disease, mortality and the environment in early modern England. *Health Transition Review*. Supplementary issue to Volume **2**, 77–96.

Eller, E. (2002). Population extinction and recolonization in human demographic history. *Math. Biosci.*, **177** and **178**, 1–10.

Ferguson, N.M., Donnelly, C.A., and Anderson, R.M. (2001). The foot-and-mouth epidemic in Great Britain: Pattern of spread and impact of interventions. *Science*, **292**, 1155–60.

Fisher, R.A. (1930). *The Genetical Theory of Natural Selection*. London: Clarendon Press.

Fix, A.G. (1978). The role of kin-structured migration in genetic microdifferentiation. *Annals of Human Genetics, London*, **41**, 329–39.

Fix, A.G. (1999). *Migration and Colonization in Human Microevolution*. Cambridge: Cambridge University Press.

Fix, A.G. (2004). Kin-structured migration: causes and consequences. *American Journal of Human Biology*, **16**, 1–8.

Gilmour, J.S.L., and Gregor, J.W. (1939). Demes: A suggested new terminology. *Nature*, **144**, 333.

Glass, D.V., and Eversley, D.E.C., eds. (1965). *Population in History: Essays in Historical Demography*. London: Edward Arnold.

Glass, R.I., Holmgren, J., Haley, C.E., Khan, M.R., Svennerholm, A.-M., Stoll, B.J., Belayet, K.M., Hossain, K., Black, R.E., Yunus, M., and Baru, D. (1985). Predisposition for cholera of individuals with O blood group: Possible evolutionary significance. *Amer. J. Epidemiology*, **121**(6), 791–6.

Hammer, M.F. (1995). A recent common ancestry for human Y chromosomes. *Nature*, **378**, 376–8.

Harding, R.M., Fullerton, S.M., Griffiths, R.C., Bond, J., Cox, M.J., Schneider, J.A., Moulin, D.S., and Clegg, J.B. (1997). Archaic African and Asian lineages in the genetic diversity of modern humans. *Amer. J. Hum. Genet.*, **60**, 772–89.

Harpending, H.C. (1994). Signature of ancient population growth in a low-resolution mitochondrial DNA mismatch distribution. *Hum. Biol.*, **66**, 591–600.

Harpending, H.C., and Rogers, A.R. (2000). Genetic perspectives on human origins and differentiation. *Annu. Rev. Genomics Hum. Genet.*, **1**, 361–85.

Harpending, H.C., Batzer, M.A., Gurven, M.A., Jorde, L.B., Rogers, A.R., and Sherry, S.T. (1998). Genetic traces of ancient demography. *Proc. Natl. Acad. Sci. USA*, **95**, 1961–7.

Harpending, H.C., Sherry, S.T., Rogers, A.R., and Stoneking, M. (1993). The genetic structure of ancient human populations. *Current Anthropology*, **34**, 483–96.

Hauser, P., and Duncan, O.D. (1959). *The Study of Population*. Chicago: Chicago University Press.

John, A.M. (1990). Transmission and control of childhood infectious diseases: Does demography matter? *Population Studies*, **44**(2), 195–215.

Jorde, L.B. (1980). The genetic structure of subdivided human populations: a review. In *Current Developments in Anthropological Genetics*, eds. J.H. Mielke, and M.H. Crawford. New York: Plenum, pp. 135–208.

Jorde, L.B., Watkins, W.S., and Bamshad, M.J. (2001). Population genomics: A bridge from evolutionary history to genetic medicine. *Human Molecular Genetics*, **10**, 2199–2207.

Keeling, M. J., Woolhouse, M. E. J., Shaw, D. J., Matthews, L., Chase-Topping, M., Haydon, D. T., Cornell, S. J., Kappey, J., Wilesmith, J., and Grenfell, B. T. (2001). Dynamics of the 2001 UK foot and mouth epidemic: Stochastic dispersal in a heterogeneous landscape. *Science*, **294**, 813–17.

Kermack, W. O., and McKendrick, A. G. (1927). Contributions to the mathematical theory of epidemics. Part I. *Proceed. Royal Society*, **115**, 700–21.

Keyfitz, N., and Flieger, W. (1968). *World Population: An Analysis of Vital Data*. Chicago: University of Chicago Press.

Kundstadter, P. (1972). Demography, ecology, social structure and settlement patterns. In *The Structure of Human Populations*, eds. G. A. Harrison, and A. J. Boyce. Oxford: Oxford University Press.

Landers, J., ed. (1992). *Historical Epidemiology and the Health Transition. Health Transition Review*. Supplementary issue to Volume **2**.

Lewis, G. H. (1982). *Human Migration: A Geographical Perspective*. London: Croom Helm.

Livi-Bacci, M. (1977). *A History of Italian Fertility During the Last Two Centuries*. Princeton, NJ: Princeton University Press.

Livingstone, F. B. (1984). The Duffy blood groups, vivax malaria, and malaria selection in human populations: A review. *Human Biology*, **56**(3), 413–25.

Livingstone, F. B. (1985). *Frequencies of Hemoglobin Variants: Thalassemia, the Glucose-6-Phosphate Dehydrogenase, G6PD Variants and Ovalocytosis in human Populations*. New York: Oxford University Press.

Malécot, G. (1948). *Les Mathématiques de l'Hérédité*. Paris: Masson.

Malécot, G. (1969). *The Mathematics of Heredity*. San Francisco: WH Freeman.

Malthus, T. R. (1798). An Essay on the Principle of Population, 1st edn. Harmondsworth Penguin Books, 1970.

McGrath, J. W. (1988). Social networks and disease spread in the Lower Illinois Valley: A simulation approach. *Amer. J. Phys. Anthrop.*, **77**, 483–96.

McNicoll, G. (1992). The agenda of population studies: a commentary and complaint. *Population and Development Review*, **18**(3), 399–420.

Mielke, J. H., Workman, P. W., Fellman, J., and Eriksson, A. W. (1976). Population structure of the Aland Islands, Finland. *Advances in Human Genetics*, **6**, 241–321.

Morgan, K., and Holmes, T. M. (1982). Population structure of a religious isolates: the Dariusleut Hutterites of Alberta. In *Current Developments in Anthropological Genetics. Vol. 2. Ecology and Social Structure*, eds. M. H. Crawford, and J. H. Mielke. New York: Plenum Press, pp. 429–48.

Morin, P. A., Moore, J. J., Chakraborty, R., Jin, L., Goodall, J., and D. S. Woodruff (1994). Kin selection, social structure, gene flow and the evolution of chimpanzees. *Science*, **265**, 1193–1201.

Morton, N. E. (1977). Isolation by distance in human populations. *Annals of Human Genetics, London*, **40**, 361–5.

Murdock, S. H., and Ellis, D. R. (1991). *Applied Demography: An Introduction to Basic Concepts, Methods, and Data*. Boulder, CO: Westview Press.

Murray, G. D., and Cliff, A. D. (1975). A stochastic model for measles epidemics in a multi-region setting. *Trans. Inst. Brit. Geograph.*, ns **2**, 158–74.

Newell, C. (1988). *Methods and Models in Demography*. New York: Guilford Press.

Olsen, C. L. (1987). The demography of colony fission from 1878–1970 among the Hutterites of North America. *American Anthropologist*, **89**, 823–37.

Olsson, G. (1965). Distance and human interaction: a migration study. *Geografiska Annaler*, **47**, 3.43.

Omran, A.R. (1971). The epidemiological transition: A theory of the epidemiology of population change. *Milbank Memorial Fund Quarterly*, **49**(4), 509–37.

Omran, A.R. (1983). The epidemiological transition theory: A preliminary update. *J. Tropical Pediatrics*, **29**(6), 305–16.

Pettenkofer, H.J., Stöss, B., Helmbod, W., and Vogel, F. (1962). Severe smallpox scars in A+AB. *Nature*, **193**, 445–46.

Preston, S.H. (1993). The contours of demography: estimates and projections. *Demography*, **30**(4), 593–606.

Preston, S.H., Heuveline, P., and Guillot, M. (2001). *Demography: Measuring and Modeling Population Processes*. Malden, MA: Blackwell Publishers.

Pritchard, J.K., Seielstad, M.T., Perez-Lezaun, A., and Feldman, M.W. (1999). Population growth of human Y chromosomes: A study of Y chromosome microsatellites. *Mol. Biol. Evol.*, **16**, 1791–8.

Relethford, J.H. (2001). *Genetics and the Search for Modern Human Origins*. New York: Wiley-Liss.

Roberts, D.F. (1967). The development of inbreeding in an island population. *Ciencia e Cultura*, **19**, 78–84.

Roberts, D.F. (1968). Genetic effects of population size reduction. *Nature*, **220**, 1084–8.

Roberts, D.F. (1971). The demography of Tristan da Cunha. *Population Studies*, **25**, 465–79.

Roberts, D.F. (1973). Anthropological genetics: Problems and pitfalls. In *Methods and Theories of Anthropological Genetics*, eds. M.H. Crawford, and P.L. Workman, A School of American Research Book. Albuquerque: University of New Mexico Press, pp. 1–17.

Rogers, A.R. (1988). Three components of genetic drift in subdivided populations. *American Journal of Physical Anthropology*, **77**, 435–50.

Rogers, A.R. (1995). Genetic evidence for a Pleistocene population explosion. *Evolution*, **49**, 608–15.

Rogers, A.R., and Harpending, H. (1992). Population growth makes waves in the distribution of pairwise genetic differences. *Mol. Biol. Evol.*, **9**, 552–69.

Rushton, S., and Mautner, A.J. (1955). The deterministic model of a simple epidemic for more than one community. *Biometrika*, **42**, 126–32.

Rvachev, L.A., and Longini, I.M. Jr. (1985). A mathematical model for the global spread of influenza. *Mathematical Biosciences*, **75**, 3–22.

Ryder, N.B. (1964). Notes on the concept of a population. *American Journal of Sociology*, **69** (5), 447–63.

Sattenspiel, L. (1990). Modeling the spread of infectious disease in human populations. *Yearbook Phys. Anthrop.*, **33**, 245–76.

Sattenspiel, L. (2003). Infectious diseases in the historical archives: A modeling approach. In *Human Biologists in the Archives*, eds. D. Ann Herring, and A.C. Swedlund. Cambridge: Cambridge University Press, pp. 234–65.

Sattenspiel, L., and Dietz, K. (1995). A structured epidemic model incorporating geographic mobility among regions. *Mathematical Biosciences*, **128**, 71–91.

Sattenspiel, L., and Herring, D.A. (1998). Structured epidemic models and the spread of influenza in the Central Canadian subarctic. *Human Biology*, **70**, 91–115.

Seielstad, M.T., Minch, E., and Cavalli-Sforza, L.L. (1998). Genetic evidence for a higher female migration rate in humans. *Nature Genetics*, **20**, 278–80.

Sherry, S. T., Batzer, M. A., and Harpending, H. C. (1998). Modeling the genetic architecture of modern populations. *Annual Review of Anthropology*, **27**, 153–69.

Sherry, S. T., Harpending, H. C., Batzer, M. A., and Stoneking, M. (1997). *Alu* evolution in human populations: Using the coalescent to estimate effective population size. *Genetics*, **147**, 1977–82.

Sherry, S. T., Rogers, A. R., Harpending, H., Soodyall, H., Jenkins, T., and Stoneking, M. (1994). Mismatch distributions of mtDNA reveal recent human population expansions. *Hum. Biol.*, **66**, 761–75.

Shurr, T. G., and Sherry, S. T. (2004). Mitochondrial DNA and Y chromosome diversity and the peopling of the Americas: Evolutionary and demographic evidence. *Amer. J. Human Biology*, **16**, 420–39.

Shyrock, H. S., Siegel, J. S. and Associates (1980). *The Methods and Materials of Demography*. Volume **1**, US Department of Commerce, Bureau of the Census. Washington, DC: US Government Printing Office.

Siegel, J. S., and Swanson, D. A., eds. (2004). *The Methods and Materials of Demography*, 2nd edn. London: Elsevier Academic Press.

Slatkin, M., and Hudson, R. R. (1991). Pairwise comparisons of mitochondrial DNA sequences in stable and exponentially growing populations. *Genetics*, **129**, 555–62.

Smith, M. T., Williams, W. R., McHugh, J. J., and Bittles, A. H. (1990). Isonymic analysis of post-famine relationships in the Ards Peninsula, N.E. Ireland: Effects of geographical and politico-religious boundaries. *Amer. J. Human Biol.*, **2**, 245–54.

Soper, H. E. (1929). Interpretation of the periodicity in disease prevalence. *Jour. Royal Statistical Society*, **92**, 34–73.

Stoneking, M. (1998). Women on the move. *Nature Genetics*, **20**, 219–20.

Stoneking, M., Fontius, J. J., Clifford, S. L., Soodyall, H., Arcot, S. S., Saha, N., Jenkins, T., Tahir, M. A., Deininger, P. L., and Batzer, M. A. (1997). *Alu* insertion polymorphisms and human evolution: Evidence for a larger population size in Africa. *Genome Research*, **7**, 1061–71.

Swedlund, A. C., and Armelagos, G. J. (1990). *Disease in Populations in Transition: Anthropological and Epidemiological Perspectives*. New York: Bergin and Garvey.

Tajima, F. (1989). The effects of change in population size on DNA polymorphism. *Genetics*, **123**, 597–601.

Takahata, N. (1993). Allelic genealogy and human evolution. *Mol. Biol. Evol.*, **10**, 2–22.

Tilly, J. (ed.) (1978). *Historical Studies of Changing Fertility*. NJ: Princeton University Press.

Vogel, F., and Motulsky, A. G. (1997). *Human Genetics: Problems and Approaches*. Berlin: Springer-Verlag.

Vogel, F., Pettenkofer, H. J., and Helmbold, W. (1960). Über die Populationsgenetik der ABO-Blutgruppen. *Acta Genet.*, **10**(2), 267–294.

Wilder, J. A., Kingan, S. B., Mobasher, Z., Pilkington, M. M., and Hammer, M. F. (2004). Global patterns of human mitochondrial DNA and Y-chromosome stsructure are not influenced by higher migration rates of females versus males. *Nature Genetics*, **36**, 1122–5.

Wilmoth, J. (2004). Population size. In eds. Siegel, J. S., and D. A. Swanson (2004). *The Methods and Materials of Demography*, 2nd edn. London: Elsevier Academic Press, pp. 65–80.

Wilson, J. E., Levins, R., and Spielman, A., eds. (1994). *Disease in Evolution: Global Changes and Emergence of Infectious Diseases*. Annals of the New York Academy of Sciences, Vol. **740**.

Wright, S. (1931). Evolution in Mendelian populations. *Genetics*, **16**, 97–159.

Wright, S. (1969). *Evolution and the Genetics of Populations*, Volume **2**, *The Theory of Gene Frequencies*. Chicago: The University of Chicago Press.

Chapter 6

Molecular Markers in Anthropological Genetic Studies

Rohina C. Rubicz, Phillip E. Melton and
Michael H. Crawford

Laboratory of Biological Anthropology, University of Kansas, Lawrence, KS

Introduction

In 1973, a chapter entitled 'The use of genetic markers of the blood in the study of the evolution of human populations', was published in the first volume that attempted to synthesize the field of anthropological genetics (Crawford, 1973). This chapter defined genetic markers as 'discrete segregating, genetic traits which can be used to characterize populations by virtue of their presence, absence, or high frequency in some populations and low frequency in others' (Crawford, 1973: 38). This definition similarly applies to molecular markers, which are segregating regions of DNA, present in some populations but absent or infrequent in others. The 1973 chapter summarized the available genetic markers of the blood that could be used for the measurement of evolutionary processes and the characterization of human population structure. The list of available polymorphic loci included 16 blood groups, 11 red blood cell proteins, 10 serum proteins and 3 white cell and platelet systems. These 'riches' of available variation of the blood followed 70 years of research on the blood group systems (since Karl Landsteiner's original work in 1900), and Oliver Smithies (1955) development of zone electrophoresis for the separation of specific proteins from mixtures such as the serum of the blood (Landsteiner and Levine, 1927). At the time the first volume in anthropological genetics was compiled, the physiological functions of blood groups were unknown, other than their involvement in blood transfusion and some suspect statistical associations with disease. The role of Duffy blood group as a resistance factor to *Plasmodium vivax* in Africans was not known

This research was sponsored by NSF grants OPP-990590 and OPP-0327676.

until 1976 (Miller *et al.*, 1976). In 1993, Horuk *et al.* demonstrated that Duffy blood group antigen was a chemokine receptor that the malarial parasite utilized for its entry into the erythrocyte. The function of the Rhesus blood group system and the homology between the *mep* gene in yeast and the ammonia transport system across cell membranes was unknown until the molecular genetic research by Marini and Urrestarazu in 1997. In the 1970s, from an evolutionary perspective, we knew little about genetic markers and their functions. Our genome was thought to consist of 100,000 genes with 30,000 estimated to be polymorphic with fewer than 100 loci had been adequately documented.

Methodological developments

The past three decades have seen a number of major technological developments that graduate students in the 1960s could have only dreamed about. Variation in primary gene products (proteins) or secondary gene products (such as blood groups) were used as proxies for the DNA that synthesized them. The technological changes that eventually resulted in the direct manipulation of DNA included:

(1) Rapid methods of DNA extraction. The first isolation of DNA occurred in the late 1860s by Friedrich Miescher, who was attempting to investigate the characteristics of proteins in pus cells. He observed that alkaline cellular extracts, when neutralized, yielded a precipitate that he termed 'nuclein'. Classical techniques for extracting DNA required from 3 to 24 hours to complete and the use of expensive equipment and caustic and toxic chemicals. Early extraction techniques also required large quantities of cellular materials, with placentae and blood being the preferred tissues. Now, modern protocols bundled into extraction kits can purify DNA in as little as 30 minutes and do not require toxic chemicals. Most of these methods are based on cell lysis, protein degradation with the destruction of the nucleases, nucleic acid precipitation and fractionation.

(2) The discovery that restriction enzymes produced by bacteria cleaved DNA at or after specific sequences and could be applied to solving problems in molecular genetics won the 1978 Nobel Prize in Physiology or Medicine for Hamilton O. Smith and Dan Nathan. This discovery enabled early molecular specialists to identify mutations and variants in specific regions of the genome. This was a less labour-intensive method than sequencing for the characterization of restriction fragment length polymorphisms (RFLPs) in mitochondrial and nuclear DNA.

(3) DNA hybridization techniques (annealing two homologous DNA strands) that allowed comparisons of DNA between different species. This method compares single strands of

two different DNA molecules that are re-associated and their thermal stability is ascertained. The original experiments by Sibley and Ahlquist (1984) demonstrated that chimpanzees and humans were more similar than either species was to gorillas and placed the chimpanzees as our closest phylogenetic relative.

(4) Polymerase Chain Reaction (PCR) revolutionized molecular genetics by making copies of DNA sequences through flanking a particular region with primers, denaturing the DNA, and annealing to the target sequence. The method of alternate heating and cooling allows the synthesis of specific DNA regions in geometric progression. Development of the PCR method and the thermocycler in the late 1980s won for Kary Mullis the 1993 Nobel Prize in Chemistry. This methodological breakthrough has had profound effects on anthropological genetics and forensic sciences. Researchers in these fields deal with minute quantities of DNA that must be amplified for either genotypic characterization or sequencing of regions of the genome.

(5) Automated DNA sequencing, followed by the development of high throughput sequencing in the late 1990s resulted in the rapid characterization of the human genome. The earliest methods of sequencing were extremely time intensive and were based on either chemical identification of the purines and pyrimidines or the use of dideoxyribose chain terminators (Maxam and Gilbert, 1977; Sanger et al., 1977). Both methods required the identification of specific DNA fragments on polyacrylamide gels followed by radio-labelling (see Devor, Chapter 10 of this volume for a discussion of the development of sequencing methodologies). The high throughput methods using fluorescence dye-terminators and capillary electrophoresis sped up the sequencing exponentially and made possible the completion of a draft of the human genome sequences in 2001.

The remainder of this chapter provides: (1) a description of the available DNA markers that can be utilized in anthropological genetic investigations; (2) a discussion of the most common and informative analytical tools that can be applied to molecular marker distributions in populations to answer evolutionary and historical questions; (3) the applications of specific DNA markers and analytical techniques to the study of evolutionary processes.

Molecular markers

In humans, DNA is packaged into 22 pairs of autosomes and a pair of sex chromosomes (XX or XY), but is also present outside the nucleus in the form of mitochondrial DNA (mtDNA). Based on function, nuclear DNA can be subdivided into coding and non-coding regions. Coding DNA, i.e. genes or exons, are defined as sequences that carry instructions for synthesizing proteins. Non-coding regions of the DNA make up the largest portion of our genome with an estimated 98.5% of the total nuclear DNA. They are not under the functional constraints of genes, therefore allowing them to exhibit greater variation, and are generally considered to be selectively neutral. Previously described as 'junk' DNA, it is now evident that some non-coding regions may function in providing structural support for DNA molecules and may be involved in gene regulation.

Several types of polymorphisms are present in the human genome. The most basic of these is the substitution of one base for another, called a single nucleotide polymorphism (SNP). Insertions and deletions, 'indels', of single bases are sometimes included in the first category, although the mechanism by which they arise differs. Larger indels also occur throughout the genome. Another category, that of repetitive DNA, comprises approximately 45% of the genome (Lander *et al.*, 2001) and includes sequences repeated in tandem arrays, such as micro- and minisatellites, and retroelements (DNA sequences that have been inserted into the genome through reverse transcriptase).

Autosomal markers

Anthropological genetics studies have utilized molecular markers characterizing polymorphisms occurring at various locations throughout the genome to investigate questions concerning the history of a population, population structure, and events such as migration and gene flow, including the time frame in which they occurred. Studies using autosomal markers have an advantage over those using Y chromosome and mtDNA markers in that they sample a larger portion of the gene pool and may be more representative of a population as a whole. Autosomal polymorphisms are biparentally transmitted, therefore they are not limited to either the paternal (in the case of the Y chromosome) or the maternal (in the case of mtDNA) side, but rather can provide information about both sexes. Alleles at different loci may undergo recombination and assort independently of one another as they are passed down each genera-tion. This reshuffling generates novel combinations of genetic material, although depending on the particular research question, it is not always a desirable characteristic. The effective population size or N_e, an estimate of the breeding size of a population, is largest for the autosomal loci since they are present in two copies for both males and females. In contrast, there are three-quarters the number of X chromosomal loci (2 X's for each female, and 1 X for each male), quarter the number of Y chromosomal loci in the same population assuming equal numbers of males and females, and quarter the number of mtDNA loci. Markers present on the X chromosome and pseudo-autosomal regions of the Y chromosome (small areas at the tips that recombine with the X chromosome) are otherwise similar to autosomal markers in their analysis and potential for resolving anthropological questions such as population history and relationships between populations.

Of the estimated 32,000 genes present in the human genome, the majority are located on the autosomes, while the X chromosome has approximately 1,500 genes, and the Y contains approximately 78 (Venter *et al.*, 2001; Skaletsky *et al.*, 2003). These coding regions are under functional constraint, since mutations that disrupt protein synthesis (such as indels that shift the DNA reading frame) can result in genetic diseases. Sometimes these mutations are associated with lowered fitness or even death of the individual. In these cases the variant is likely to be quickly eliminated from the population. Some mutations may have no effect on the phenotype, for example a base substitution in the third position of a codon often codes for the same amino acid. Still other mutations may produce variants of a gene that are advantageous to a human population living in a particular environment.

Collagen genes (COLIA2)

One example of an autosomal coding marker found to be informative for human phylogenetic research is the human α2 (1) collagen gene (COL1A2) used by Mitchell *et al.* (1999). COL1A2 is one of two genes that code for peptides in type 1 collagen (a component of skin, bone, blood vessels, ligaments, and dentin, among other tissues). In Mitchell *et al.*'s (1999) study, the worldwide distribution of a 38 base pair deletion in this gene among human populations (which appears to be neutral) was characterized in order to test the recent out-of-Africa model. This model, supported by previous molecular studies (Vigilant *et al.*, 1991; Stoneking and Soodyall, 1996; Hammer *et al.*, 1997, 1998), states that modern *Homo sapiens* originated in sub-Saharan Africa and rapidly spread throughout the rest of the world, replacing earlier hominid forms. The collagen deletion was determined to be present in high frequencies in non-African populations, but was completely absent in sub-Saharan groups. These results indicate the marker likely arose just before or shortly after modern humans left Africa, but before their spread throughout the rest of the world. A similar distribution as found in the collagen deletion was noted for the immunoglobulin GM locus. The GM*A,X,G haplotype is absent in sub-Saharan Africa but occurs worldwide in all other human populations. Thus, the population out-of-Africa experienced at least two distinct mutations before it spread throughout the world. Mitchell *et al.*'s (1999) study lends further support to the recent out-of-Africa model, and demonstrates that COL1A2 is a useful marker in reconstructing phylogenetic relationships between human populations.

Single nucleotide polymorphisms (SNPs)

A variety of markers located in non-coding autosomal regions are useful for anthropological genetics studies. Of these, single nucleotide polymorphisms (SNPs) tend to evolve at a slow rate of approximately 2.3×10^{-8} (Nachman and Crowell, 2000). These simple base substitutions may result from the action of mutagenic agents, such as radiation, or be due to nucleotide misincorporation during DNA replication. Transitions, where a purine is substituted for another purine (i.e. A→G) or pyrimidine for another pyrimidine (i.e. T→C), occur at a higher rate than transversions, where there is a change in nucleotide class (i.e. A→T). SNPs (including deletions or insertions of single bases) can be identified through RFLP analysis, whereby the target DNA is amplified by PCR, digested by restriction enzymes, electrophoresed, and scored. A single nucleotide change in the DNA sequence recognized by the enzyme will prevent its cleavage. While this method is widely used for the detection of SNPs, direct sequencing of the nucleotides is also common. The low mutation rate of these markers makes it unlikely that they have reoccurred during the evolution of modern humans, and therefore individuals sharing the same marker can be assumed to share common ancestry, or be identical by descent. The ancestral state for a particular locus can be inferred by comparing the sequence to that of our closest relatives, the great apes. Humans and chimpanzees are estimated to have diverged around 5 million years ago. SNPs, also described as binary markers or unique mutational events (UMEs), may be of use for investigating phylogenetic questions of great time depth.

Tandem repeats

Short tandem repeats (STRs)

Tandemly repeated DNA sequences evolve at a faster rate than SNPs and are common throughout the human genome. Small repeat units called microsatellites or short tandem repeats (STRs) range from 1–6 base pairs (e.g. CA-, CAT-, CCG-, CAG-) with the total repeat unit usually less than 350 base pairs (Guarino et al., 1999). These occur approximately every 6 to 10 kilobases. They are believed to typically mutate according to the gain or loss of single repeat units at a time, although they may undergo larger 'jumps' whereby several repeats are simultaneously inserted or deleted. The mutation rate varies among STR loci, partly due to size and composition of the repetitive unit, and the number of repeats (alleles with larger numbers of repeats tend to mutate more rapidly than those with fewer repeats).

DNA Marker	Mutation rate per locus per generation
Some expanded polymorphic microsatellites	$< 10^0$
Minisatellites	10^{-2} to 10^{-1}
Microsatellites	10^{-4} to 10^{-3}
Some structural polymorphisms	10^{-5} to 10^{-4}
Base substitutions (SNPs)	10^{-8} to 10^{-7}
Retroelement insertions	10^{-11} to 10^{-10}

Table 6.1. Average mutation rate for DNA markers (after Jobling *et al.*, 2004).

STR mutation rate is estimated between 10^{-4} and 10^{-3} mutations per locus per generation (see Table 6.1), making them ideal for investigations into recent human history, such as the recent separation of human groups. While a number of studies have indicated that microsatellites tend to be selectively neutral, there is evidence implicating several trinucleotide repeats in human diseases. For example, 'CCG' repeat in fragile-X syndrome, 'CAG' in Huntington's Disease, and 'CTG' in myotonic dystrophy (Fu *et al.*, 1991; The Huntington's Disease Collaborative Research Group, 1993; Brook *et al.*, 1992). The alleles become increasingly unstable with larger numbers of repeats, until they reach a threshold at which the disease manifests.

Variable number of tandem repeat (VNTRs)

Another class of repetitive DNA sequences consists of minisatellites or variable number of tandem repeat (VNTR) polymorphisms. Some investigators use an alternative nomenclature system in which VNTRs refer to all tandem repeats; microsatellites, minisatellites, and satellites. According to the more specific definition used here, VNTRs consist of core repeat units of approximately 10 to 100 base pairs, strung together up to 1,000 base pairs in length. They are usually GC-rich, and exhibit sequence variation in addition to variability in the number of repeats. VNTR mutation rate is higher than that of STRs, at approximately 10^{-2} to 10^{-1} per locus per generation (see Table 6.1), and their mutation process is more complex, with a bias toward repeat gains over losses. VNTRs exhibit a great amount of variation between individuals which makes them particularly useful for forensic studies. Their application to individual identification, using a method called 'DNA fingerprinting', such as for paternity testing, is described in greater detail by Schanfield, Chapter 9 of this volume.

VNTRs have been successfully used in anthropological genetic research to explore the Siberian origins of Native American populations. In studies by the University of Kansas research group (1989–1995), five VNTR loci (D7S104, D11S129, D18S17, D20S15, and D21S112) were used to characterize the population structures and genetic affinities of Siberian groups (McComb *et al.*, 1995, 1996). Siberia is of evolutionary importance to the prehistory of human populations as a crossroads between Europe, Asia, and the Americas. In the first season of fieldwork, a comparison was made between two Evenki villages, Surinda and Poligus, and the Ket village of Sulamai (McComb *et al.*, 1995). The Evenki are Tungusic-speaking reindeer herders who are widely dispersed throughout the taiga of central Siberia and are relatively isolated. The Kets subsist on hunting and fishing, speak a language unrelated to the three major linguistic phyla of Siberia (Altaic, Uralic, and Paleoasiatic), and are of unknown origin. VNTR frequencies were significantly different between Sulamai and Surinda at the D11S129 locus. Sulamai was shown to have experienced more gene flow from Russians, a finding in agreement with the available ethnographic information, and Poligus and Surinda were relatively isolated and experienced little gene flow from outside groups. Overall, the VNTR data demonstrated the distinctiveness of the Kets and Evenki. In a second year of field investigations, a Kizhi population from Gorno-Altai was compared to the Kets and Evenki (McComb *et al.*, 1996). The Kizhi of the Gorno-Altai are a group of Turkic-speaking pastoralists, previously shown to be genetically more diverse than other Siberians and they apparently exhibited the so-called Asian-specific 9 bp deletion in mtDNA. The presence of the 9-bp deletion, which defined the B mtDNA haplogroup, made the Gorno-Altai unique in Siberia with the presence of A, B, C, and D haplogroups (Shields *et al.*, 1993). More recently the Altai became of great interest to anthropological geneticists because of the reported presence of haplogroup X, shared with Native American populations (Derenko *et al.*, 2003). The VNTR data confirmed that the Altai were indeed genetically distinct from the other Siberian groups, although not to as great an extent as previously indicated. In both studies, the Siberians were shown to cluster together when compared to outside populations, indicating their relative genetic homogeneity, and they were closer to Native Americans than to other American populations (African Americans and European Americans) reflecting their recent common evolutionary history. Thus, VNTR markers (DNA fingerprints) can be useful for discriminating between populations, measuring population affinities, and examining recent events such as human migrations.

Telomeric arrays and satellites

Telomeric arrays and satellites are repeat polymorphisms that have not been widely used in anthropological genetics. Telomeric arrays are DNA-protein structures with tandem repeats, which are located at the ends of the chromosomes, and are sometimes classified as minisatellites. They play a functional role in preventing chromosomes from fusing together and preventing chromosomal degradation. Satellites are very large tandem repeats of hundreds of thousands to millions of bases long. Some have been shown to provide structural support, such as the alpha satellite which forms part of the centromeres. Satellites are difficult to work with because of their tremendous size.

Retroelements

Retroelements are DNA elements that transpose into genomic locations after their transcription to RNA from an active genomic copy, and then reverse transcription into DNA. One example is the *Alu* family, consisting of short interspersed repetitive units (SINEs) which are only found among primates. An *Alu* consists of two approximately 150 base pair units comprising a highly repetitive sequence of around 300 nucleotides, which may be randomly inserted hundreds to thousands of times. In humans there are an estimated 500,000 copies per haploid genome (Deininger and Batzer, 1993; Novick *et al.*, 1996). Of these, a few thousand are thought to be polymorphic in humans, and therefore useful in anthropological genetics studies. Insertion events of these elements are rare, and once inserted, they are extremely stable. The ancestral state of any *Alu* repeat is its absence, while its presence is considered to be the derived state. There is no known mechanism for the complete removal of an *Alu* from its insertion site. Other retroelements include LINEs (long interspersed nuclear elements), such as the L1 element, and HERVs (human endogenous retroviruses). These retroelements also show promise for understanding the etiology of certain diseases, as well as for phylogenetic studies of human populations.

Randomly amplified polymorphic DNA (RAPD)

Another type of marker, randomly amplified polymorphic DNA (RAPD), will only be mentioned briefly here because it has had limited use in studies of human diversity. RAPDs are generated by using short primers (8–12 bases) to randomly amplify segments of DNA. Variation is seen in the presence or absence of bands, or in different lengths of bands. Although early research on humans using RAPDs was promising in that phylogenetic reconstruction of populations of known histories gave accurate results (McComb, 1999; Melvin, 2001), this method has fallen out of favour for human studies because of criticism that the majority of fragments are non-specific and cannot be reproduced.

Y chromosome DNA markers

The Y chromosome (see Figure 6.1), composed of large amounts of chromatin and few genes (related to male sex determination), is passed exclusively from father to son. Only a small portion of the Y, the pseudoautosomal region located at the tips of its chromosomal arms, recombines with the other sex chromosome, the X.

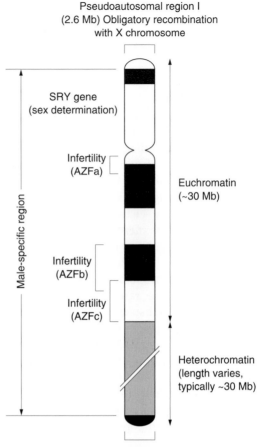

Fig 6.1 Human G-banded Y chromosome (after Jobling et al., 2004).

Pseudoautosomal region I
(2.6 Mb) Obligatory recombination
with X chromosome

SRY gene
(sex determination)

Infertility
(AZFa)

Euchromatin
(~30 Mb)

Infertility
(AZFb)

Infertility
(AZFc)

Heterochromatin
(length varies,
typically ~30 Mb)

Male-specific region

Pseudoautosomal region II
(0.32 Mb) Non-obligatory recombination
with X chromosome

Approximately 95% of the Y chromosome does not recombine, and it is this portion, referred to as the MSY (male-specific region of the Y), that is targeted for population studies. The MSY is said to be haploid, present in only one copy, because it does not share a corresponding region on the X chromosome. Being uniparentally inherited, MSY markers have at most a quarter of the effective population size (N_e) of autosomal markers. This may be even further reduced because the Y chromosome is subject to genetic drift since not all males will contribute equally to the gene pool of future generations, and some Y chromosomes may be lost through stochastic processes.

Y chromosome DNA polymorphisms are useful for the phylogenetic reconstruction of populations because in the absence of recombination, any mutations that occur will be passed on to future generations. In this way, long-lasting patrilineages can be identified and traced back to a common male ancestor. At a local level, higher female than male migration rates due to cultural practices (including patrilocality, the practice of women moving to their husbands' place of residence after marriage) may lower Y diversity in comparison to that of mtDNA (Salem *et al.*, 1996; Oota *et al.*, 2001). Although some studies indicated this pattern extended to a global scale (Seielstad *et al.*, 1998; Romualdi *et al.*, 2002) this no longer appears to be the case, as it is likely that longer-distance migrations were more often made by males (Wilder *et al.*, 2004).

The most widely used MSY polymorphisms for phylogenetic studies are binary markers and STRs. The binary markers, consisting mainly of SNPs and indels, evolve slowly and are considered to be unique mutational events (UMEs). They are used to define the major Y chromosome lineages, or haplogroups. Recently, the haplogroup nomenclature was revised in order to construct a single system that reflects the phylogenetic relationship between haplogroups and has the flexibility for the incorporation of new mutations (Y chromosome consortium 2002). Previous to this, seven different systems of nomenclature were in use. Y chromosome STRs are highly polymorphic and are used to characterize variation within the haplogroups. A Y specific minisatellite (MSY1) is even more variable and has been similarly used by researchers (Jobling *et al.*, 1998; Brion *et al.*, 2003). The characterization of new Y chromosomal markers in populations worldwide has contributed to their increasing usefulness for anthropological genetic studies, which until recently lagged behind mtDNA studies (see the chapters by Arredi *et al.*, Tishkoff and Gonder, and Salzano in their application of Y chromosome markers to the peopling of Africa, Europe and the Americas).

Mitochondrial DNA markers

The other non-recombining portion of the human genome is the mitochondrial DNA (mtDNA). MtDNA is a double-stranded, circular molecule believed to be of bacterial origin (Margulis 1981), and located outside the nucleus in the energy-producing mitochondria of the cell (see Figure 6.2). MtDNA is maternally inherited, meaning that it is passed from a mother to all of her children, but only her daughters will pass it on to subsequent generations. Similar to the Y chromosome, the mtDNA has an effective population size equal to a quarter that of the autosomes, and it is also subject to genetic drift. MtDNA is present in multiple copies per cell, hundreds to thousands, depending on the tissue. This characteristic is desirable for ancient DNA studies, where the material recovered from deceased individuals is often degraded (see Chapter 8 by O'Rourke on ancient DNA). There are approximately 16,569 base pairs in the mtDNA molecule, which consists of a coding region with 37 genes and two non-coding hypervariable regions (HVS-1 and HVS-2). The molecule mutates approximately ten times faster than nuclear DNA because it lacks the nuclear repair mechanisms, and the hypervariable region or D-loop

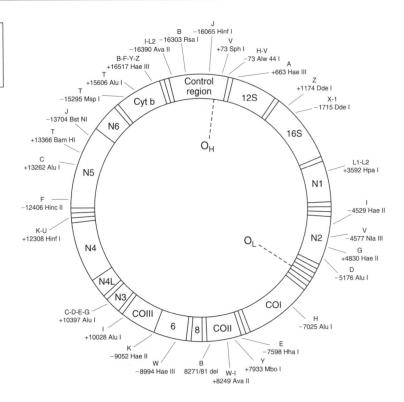

Fig 6.2 Human mitochondrial DNA molecule with haplogroup specific restriction sites (after Francalacci *et al.*, 1999).

has an even faster evolutionary rate, making it useful for studies of human population history dating back 100,000 years or less.

MtDNA markers used in population studies mainly consist of RFLPs (restriction fragment length polymorphisms) and sequencing of HVS-1 and HVS-2. RFLPs (and increasingly sequencing) are used to characterize SNPs and a 9-base pair deletion located in the coding region of the molecule. These polymorphisms define the major mtDNA haplogroups, which are considered to be relatively stable. HVS-1 and HVS-2 sequences have been used to characterize diversity within the haplogroups. Because recurrent mutations occur in the hypervariable region, shared polymorphisms in this segment cannot be assumed to be identical by descent. Some researchers have even sequenced the entire mtDNA genome, providing a detailed picture of individual maternal lineages.

The Aleuts provide an informative example of the application of mtDNA markers to the study of human populations (Rubicz, 2001; Rubicz *et al.*, 2003). The Aleuts are an indigenous Alaskan population located near the entry point to the Americas (see Figure 6.3), who represent one of the final migrations of humans into previously unoccupied territory. Like other Native Americans, they are proposed to have originated in Asia, and crossed over to Alaska by route of the Bering Land Bridge, which was exposed during the last Ice Age when sea levels were lower. Aleut mtDNA RFLP and HVS-1 sequencing variation was characterized in order to investigate their origins and role in peopling of the New World. Of the four haplogroups common

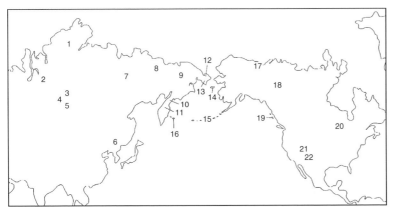

Fig 6.3 Map of population locations: 1) Nganasan; 2) Sel'kup; 3) Evenki; 4) Ket; 5) Altai; 6) Udehe; 7) Even; 8) Yukagir; 9) Reindeer Chukchi; 10) Koryak; 11) Itel'men; 12) Coastal Chukchi; 13) Asian Eskimo; 14) Savoonga Eskimo; 15) Aleut; 16) Bering Aleut; 17) Inuit; 18) Dogrib; 19) Haida; 20) Ojibwa; 21) Navajo; 22) Apache.

Table 6.2. MtDNA frequencies among indigenous Siberian and American populations (Rubicz, 2001).

Population	n	MtDNA*A	mtDNA*B	mtDNA*C	mtDNA*D	mtDNA*OT	Ref
Aleut	179	28.5%	0%	0%	71.5%	0%	1
Altai	28	3.6%	3.6%	35.7%	14.3%	42.8%	5
Asian Eskimo	50	80.0%	0%	0%	20.0%	0%	7
Coastal Chukchi	46	23.9%	0%	21.7%	8.7%	45.7%	5
Dogrib	154	90.9%	0%	2.0%	0%	7.1%	3
Savoonga Eskimo	49	93.9%	0%	0%	2.0%	4.1%	3
Evenki	51	3.9%	0%	84.3%	9.8%	2.0%	7
Even	43	0%	0%	58.1%	7.0%	34.9%	7
Haida	25	96.6%	0%	0%	3.4%	0%	6
Inuit	30	96.7%	0%	0%	3.3%	0%	2
Itel'men	47	6.4%	0%	14.9%	0%	78.7%	4
Koryak	155	5.2%	0%	36.1%	1.3%	57.4%	4
Nganasan	49	2.1%	0%	38.8%	36.7%	22.4%	5
Ojibwa	28	64.3%	3.6%	7.1%	0%	25.0%	6
Reindeer Chukchi	24	37.5%	0%	16.7%	16.7%	29.1%	5
Sel'kup	20	0%	0%	35.0%	0%	65.0%	7
Udehe	45	0%	0%	17.8%	0%	82.8%	5
Yukagir	27	0%	0%	59.3%	33.3%	7.4%	7

1 = Rubicz et al., 2003; 2 = Lorenz and Smith, 1996; 3 = Merriwether et al., 1995; 4 = Schurr et al., 1990; 5 = Sukernik et al., 1996; 6 = Torroni et al., 1993a; 7 = Torroni et al., 1993b.

among Native Americans (A, B, C and D), the Aleuts had only two, A and D. Their high frequency of haplogroup D (71.5%) set them apart from Eskimos, Athapaskans, and other Northern Amerindians (see Table 6.2). The Aleuts shared several HVS-1 sequences with other circumarctic populations, but did not have the 16265G mutation which is specifically found among Eskimo groups. They also significantly differ from the Koryak and Even populations of Kamchatka. The Aleuts were closest genetically to the Chukchi and

Siberian Eskimo populations of Chukotka. These results indicate that the Aleuts likely migrated across the Bering Land Bridge and settled the Aleutian Islands from the east, rather than island-hopping from the Kamchatka Peninsula to settle the western Aleutians. This lends further support to the hypothesis that multiple migrations were responsible for the peopling of the New World.

Analytical tools

Understanding the underlying biological and social processes that occur within and between human societies allow anthropological geneticists to make inferences concerning two areas of inquiry that have shaped modern and ancient human populations. The first of these processes focuses on population structure, which may be defined as any number of factors that impact the amount of gene flow or genetic drift within a group or population (Crawford, 1998). The second area of inquiry is population history, which focuses on the extent of biological similarity among groups or populations, reflecting either common ancestry or gene flow (Harpending and Jenkins, 1973). The advent of molecular genetic markers has led to a number of analytical techniques that have either been adopted from classical genetic studies or generated to deal specifically with molecular data. The following is not a comprehensive review of these techniques, but rather offers examples of some of these methods that address questions regarding population structure or history.

Population structure

Several different definitions of population structure exist. A few researchers define population structure as all factors that cause deviation from Hardy-Weinberg equilibrium (Cavalli-Sforza and Bodmer, 1971). Others restrict their definition exclusively to population subdivision, based on cultural or geographic factors (Schull and MacCluer, 1968). The relationship between genetic elements (genes, genotypes, and phenotypes) has also been used to define the structure of human populations (Workman and Jorde, 1980). Population structure can be further subdivided into intrapopulation (variation within a population or subdivisions of a population) and interpopulation (variation between populations) differences. This division may be limited because the boundaries between populations and their subdivisions are often arbitrary and all human populations may be considered a subdivision of a single ancestral population (Crawford, 1998).

Intrapopulation variation

The majority of human populations exhibit some form of internal subdivision due to geographic, linguistic or cultural factors. These subdivisions include language families, clans, tribes, castes, religions, socio-political units or other classifications. Frequently, these aggregates are hierarchical and may serve as effective barriers to gene flow between groups, with a varying degree of success. A number of analytical methods are available for evaluating the effects of evolutionary forces on population subdivision. A few techniques that may be useful in quantifying intrapopulation variation and applicable to molecular data are nucleotide diversity statistics, tests of selective neutrality, mismatch distribution, and analysis of molecular variance (AMOVA).

Measures of nucleotide diversity

According to the neutral theory of molecular evolution, the majority of nucleotide substitutions have no effect on fitness of an individual (i.e. are neutral) and most polymorphisms are transient, awaiting fixation due to drift (Kimura, 1968a,b). Therefore, assuming mutation-drift equilibrium it is possible to determine the expected level of diversity (θ) in a population or its subdivision using the mutation rate (μ) and effective population size (N_e) for diploid loci, with the equation:

$$\theta = 4N_e\mu (2N_e\mu \text{ for haploid data}). \tag{1}$$

The parameter θ is an important factor in several different molecular statistical techniques and is often compared to the nucleotide diversity measure (π) and the number of nucleotide variant sites, generally shown as θ_S (S/a_n, where a_n is $\sum_{i=1}^{n-1} 1/i$) (Waterson, 1975). The π statistic is a measure independent of sample size and analogous to Nei's gene diversity measure (Nei, 1987). This diversity measure describes the probability that two copies of the same nucleotide drawn at random from the same set of sequences will differ and is represented using the equation:

$$\pi = n(x_i x_j \pi_{ij})/(n-1) \tag{2}$$

where n equals the number of sampled sequences, x_i and x_j are the frequencies of ith and jth sequences and π_{ij} is the proportion of nucleotide differences between them.

Measures of selective neutrality

In order to determine whether or not populations are being influenced by evolutionary forces other than selection it is important to ascertain whether the amount of genetic diversity exhibited by these populations deviates from neutrality. Several different neutrality tests exist, including Tajima's D, HKA, McDonald-Kreitman, Fu's Fs, Fu and Li's D as well as others. These statistics have been recently reviewed in detail by Kreitman (2000) and we will only briefly discuss two common neutrality measures (Tajima's D, Fu's Fs) used in the anthropological literature. The underlying genetic structure of a population may play a key role in the detection of positive selection (Przeworski, 2002) but factors other than selection such as expansion, bottlenecks, or background selection may also have an impact (Aris-Brosou and Excoffier, 1996; Schneider *et al.*, 2000; Tajima, 1993; Foy and Wu, 1999).

In anthropological genetics, two common measures of selective neutrality, Tajima's D (Tajima, 1989) and Fu's Fs (Fu, 1997), are often applied with the assumptions of the neutral theory of evolution. These statistics are appropriate in distinguishing population expansion from constant population size. Population growth generates an excess of mutations in the external branches of the genealogy and therefore an excess of substitutions are present in only one sampled sequence (Ramos-Onsins and Rozas, 2002). This leads to a star-like phylogeny that includes a large central node with several radiating spokes each represented by a single individual.

Tajima's D (Tajima, 1989) uses information from the sample mutation frequency and is based on the infinite-sites model without recombination. This statistic is appropriate for short DNA sequences or RFLP haplotypes. Tajima's D compares two estimators of the mutation parameter θ. The test statistic D is estimated as:

$$D = \frac{\theta\pi - \theta s}{\sqrt{Var(\theta\pi - \theta s)}} \tag{3}$$

where $\theta\pi$ is equivalent to the mean number of pairwise differences between sequences (π) and θ_S is based on the number of nucleotide variant sites. Negative scores are indicative of larger values for θ_S relative to $\theta\pi$ signifying the potential effects of population expansion. Positive or statistically non-significant negative scores may indicate the effects of genetic bottlenecks on a population, which tend to create highly fragmented phylogenies and represent inflated values for $\theta\pi$ relative to θ_S.

Fu's Fs (1997) is also based on the infinite-site model without recombination but utilizes data from the haplotype distribution. This test statistic is based on the equation:

$$Fs = \ln\left(\frac{S'}{1 - S'}\right) \tag{4}$$

where S' is the probability of observing a random neutral sample and defined as $S'=PR(K \geq k_{obs}|\theta=\theta\pi)$, where (k) is equal to the number of alleles similar or smaller than the observed value given $\theta\pi$ and Fs is the logit of S'. Statistically significant negative scores indicate an excess of alleles, a signature of population expansion. This test is considered less conservative than Tajima's D and is more sensitive to large population expansions expressed as large negative numbers whereas positive numbers indicate populations impacted by genetic drift (Schneider et al., 2000; Fu, 1997).

Mismatch distribution

Another common method for representing molecular data is through the distribution of pairwise differences, also known as mismatch distribution. This method is applicable to molecular data where differences between alleles can be counted and includes nucleotide substitutions, RFLPs, VNTRs, or STRs (Jobling et al., 2004). The mismatch distribution is constructed by counting the number of

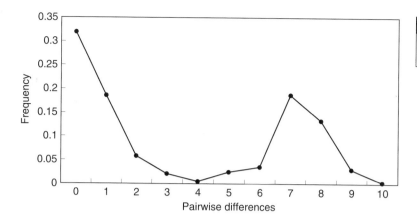

Fig 6.4 Mismatch distribution for all Aleut mtDNA HVS-I sequences (after Zlojutro et al., 2006).

differences between each pair of subjects and then using histograms or scatter plots to display the frequencies of sites that differ (Rogers *et al.*, 1996). This measure of diversity summarizes the discernible amount of genetic variation within a population. The shape of the mismatch distribution is also highly informative. A unimodal distribution is indicative of population expansion whereas a multimodal distribution indicates constant population size over a long time period (Rogers and Harpending, 1992).

NO – Forster (1997)

Zlojutro *et al.* (2006) used mtDNA diversity and a number of these techniques in order to assess intrapopulation variation in Aleutian islanders. Figure 6.4 shows a mismatch distribution for the total Aleut population. This distribution is bimodal with peaks at 0 and at 7, generally interpreted as a signature of long-term population stability. However, others have argued that population substructure and mutation rate heterogeneity may account for multimodal mismatch distributions (Aris-Brosou and Excoffier, 1996; Marjoram and Donnelly, 1994). In order to determine whether or not population structure was influencing the overall mismatch distribution, Zlojutro *et al.* (2006) separated out three mtDNA haplotypes A3 and D2 (shared by other Native American and Siberian populations), and A7 (characterized by a 16212A transversion) specific to the Aleut population (Figure 6.5). Mismatch distributions for all three haplotypes are unimodal and support evidence for differential population expansion within their respective haplogroups. The A3 haplotype possibly indicates an older demographic event than both the A7 and D2 haplotypes, due to its higher peak at two mutational differences. This latter mismatch distribution is indicative of two population expansions (one for A3 and a later expansion for A7 and D2), which may not have been detected if the underlying mtDNA haplotype structure had not been investigated independently of the total population.

Further support for a dual population expansion in Aleuts is provided through nucleotide diversity scores and neutrality test statistics (Zlojutro *et al.*, 2006). Table 6.3 shows values for θ_π, θ_S, Tajima's D, and Fu's Fs values within the total Aleut population and

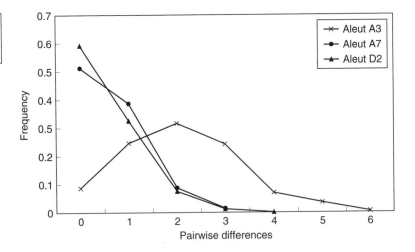

Fig 6.5 Mismatch distributions for select mtDNA haplotypes (after Zlojutro *et al.*, 2006).

Table 6.3. Aleut Diversity and neutrality test statistics for population structure as inferred through mtDNA haplogroups (after Zlojutro *et al.*, 2006).

	θ_π (SD)	θ_S (SD)	Tajima's D	Fu's Fs
Aleut total	3.338 (1.904)	3.582 (1.080)	−0.234	−6.678
Subhaplogroup A3	2.099 (1.358)	2.743 (1.216)	−0.822	−6.866
Subhaplogroup A7	0.607 (0.549)	1.223 (0.636)	−1.337	−3.093
Haplogroup D	0.363 (0.390)	1.085 (0.496)	−1.421*	−4.799*

two mtDNA subhaplogroups (A3, A7) and mtDNA haplogroup D. The Aleuts are characterized by low mtDNA diversity levels with θ_π equal to 3.338 and θ_s equal to 3.582. The values are lower for all three of the other haplogroups with the lowest θ_π (0.363) and θ_s (1.085) diversity being located in haplogroup D. Fu's Fs statistics are significant for all three haplotypes, while only one, haplogroup D, was found to be statistically significant for Tajima's D. This is indicative of the greater statistical sensitivity of Fu's Fs to Tajima's D, which some researchers consider overly conservative (Ramos-Onsins and Rozas, 2002).

AMOVA

AMOVA is an alternate to allele frequency methods and takes into account the molecular relationship of alleles. AMOVA is analogous to a nested analysis of variance (ANOVA) derived from a matrix of squared distances among all pairs of haplotypes. This in turn produces variance estimates and F-statistic analogues designated as Φ-statistics that reflect the correlation of haplotypic diversity at different hierarchical levels of population subdivision. The following equation is used to calculate the total sum of squared deviations (SSD):

$$SSD_{(Total)} = \frac{1}{2N} \sum_{j=1}^{N} \sum_{k=1}^{N} \delta_{jk}^2 \qquad (5)$$

Table 6.4. Aleut AMOVA for mother's place of birth (Rubicz, 2001).

Source of variation	Df	SSD	MSD	Variance component	% Total	p	Φ-st
Among groups	2	1.37	0.70	0.04	27.22%	<0.001	$\Phi_{ct}=0.272$
Among communities within groups	4	0.50	0.13	0.00	2.00%	0.366	$\Phi_{sc}=0.027$
Within communities	43	4.73	0.11	0.11	70.79%	<0.001	$\Phi_{st}=0.292$

where N equals the number of haplotypes, δ_{jk}^2 is the Euclidean distance between haplotypes j and k. This allows for the hierarchical partition of the haplotypes into SSD within populations, SSD within regional groups, and SSD among populations within regional groups. The mean squared deviation (MSD) is obtained by dividing the corresponding SSD by the appropriate degrees of freedom (Excoffier et al., 1992). This method is appropriate for any data where genetic distances between alleles can be calculated.

Rubicz (2001) used AMOVA on mtDNA control region data to determine whether population structure was present between Aleut communities located in western, central, and eastern regions of the Aleutian island chain. Table 6.4 shows the results of an analysis that investigated mother's place of birth for 47 individuals. This AMOVA indicates significant differences exist between the aforementioned groups and accounted for 27.22% of the variation present in these data. No significant difference was found within groups and the greatest amount of variation was located within communities (70.79%). According to this analysis, Aleuts appear to demonstrate population structure along an east-west axis and these AMOVA results are concordant with both linguistic and archaeological evidence.

Interpopulation variation

Comparisons between populations have long been of interest to anthropological geneticists due to the information they provide regarding stochastic processes impacting human population structure and phylogeny. These differences may be based on either geographic location or linguistic affiliation of cultural groups and are informative in regards to evolutionary forces affecting genetic variation in these populations. Analytical methods used to study these differences have largely been adopted from classical genetic studies, with a few exceptions such as microsatellite genetic distances and phylogenetic networks. Population relationships are often established through the comparison of genetic distances and displayed either through genetic maps, phylogenetic trees, or in select cases, phylogenetic networks.

Genetic differences and affinities within human populations or subgroups are often measured through allele or haplotype frequencies. It is difficult to measure the resulting similarities or diversity between groups by solely viewing the matrices of several alleles or populations. Therefore, a number of genetic distance measures have been developed for comparing variation between populations through the use of summary statistics (Crawford, 1998). A number of genetic distances measures are applicable to molecular data and include those that compute standardized Euclidian squared distances, angular transformations, gene substitutions, and coancestry coefficients (Jorde, 1985). Two commonly used classical genetic distance measures designed for measuring genetic differentiation in subdivided populations are Wright's F_{st}, appropriate for sequence or microsatellite data (Reynolds *et al.*, 1993) and Nei's standard genetic distance, D (Nei, 1987).

In addition to these techniques, several distance measures have been developed that deal exclusively with microsatellite loci, which follow a stepwise mutational model (Goldstein *et al.*, 1995). Measures applicable to STR or VNTR data include $\delta\mu^2$ (Goldstein *et al.* 1995), R_{ST} (Slatkin, 1995), and D_{SW} (Shriver *et al.*, 1995). However, microsatellite distance measures that rely primarily on mutation ($\delta\mu^2$ and R_{ST}) are less effective then methods that grant greater weight to genetic drift (F_{st} and D_{SW}) for recognizing population associations (Destro-Bisol *et al.*, 2000; Perez-Lezaun *et al.*, 1997).

Three common techniques for graphically displaying genetic distance data are: genetic maps, phylogenetic trees, or phylogenetic networks. Genetic maps reduce distance matrices into a more manageable two or three-dimensional graphical representation based on various matrix algebra methods such as principal coordinate analysis (Jorde, 1985). This method is mathematically sound and distortions common to phylogenetic tree building are often erased when using gene maps. Phylogenetic trees not only contain information about the relationships between populations but also provide data regarding the fissioning of groups and the time of divergence (Crawford, 1998). Several different methods of tree construction utilizing different assumptions exist and include: maximum likelihood, maximum parsimony, neighbour-joining, and unweighted paired group method (UPGMA). Statistical error in tree building can be high and it is recommended that over 30 loci be used along with bootstrapping methods, which may allow for an estimation of confidence limits (Jorde, 1985).

An alternative to phylogenetic trees for certain types of molecular data (mtDNA RFLP, mtDNA control region, and male specific Y-chromosome STRs) are phylogenetic networks. These networks offer an advantage over traditional tree building methods that utilize maximum parsimony or maximum likelihood, because networks can distinguish between irresolvable and resolvable character conflict errors that may occur due to homoplasy. Networks represent 'all most parsimonious trees' by highlighting conflicts in the form of reticulations (equally possible mutation routes between nodes in the network) and interpreted as homoplasy (parallel mutations or

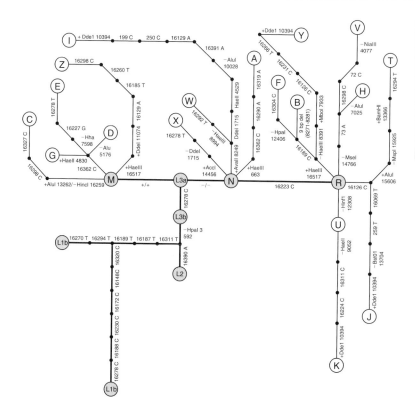

Fig 6.6 Generalized phylogenetic network of worldwide human mtDNA haplogroups: $+/+$, $-/- =$ *DdeI 10394/AluI 10397*, Gray Circles = macrohaplogroups, White Circles = haplogroups, characteristic RFLP sites shown in italics, mtDNA Control Region SNPs shown in plain text.

reversals), recombination, or sequence errors (Bandelt *et al.*, 1995). The network is sequentially constructed through the addition of consensus points (median vectors) to three mutually close sequences at a time. These median vectors are then inferred as either extinct sequences or extant unsampled sequences within the population. Four different types of networks exist and include: minimum spanning networks (MSN), reduced median networks (RM), median joining networks (MJ), and quasi-median spanning networks (QSN). The underlying assumptions of these networks are that ambiguous states are infrequent and that recombination is absent. These assumptions are met for aforementioned molecular data types (Bandelt *et al.*, 1999).

Figure 6.6 illustrates a generalized phylogenetic network of worldwide human mtDNA haplogroups and macrohaplogroups along with their characteristic RFLP, HVS-I and HVS-II sites. Macrohaplogroups consist of a number of haplogroups shared by several populations. A number of these haplogroups are considered to be continent-specific and integration between geographic regions is rare. Sub-Saharan African populations are all characterized by the L-macrohaplogroups (L1, L2, and L3). The macrohaplogroup L3 is thought to have diverged into the macrohaplogroups M and N, which arose in northeastern Africa and subsequently spread throughout the rest of the world. The final macrohaplogroup R then diverged from N. European populations belong almost exclusively to N and R as the haplogroups

H, I, J, N, T, U, V, W and X make up 98% of the total mtDNA variation (Mishmar *et al.*, 2003). Asian populations consist of all three non-African mtDNA macrohaplogroups (M, N and R) and haplogroups (A, B, C, D, E, F, G, X, Y and Z), and, as previously mentioned, Native Americans are almost exclusively composed of haplogroups (A, B, C, D and X).

Genes and language

A research area that has long been of interest to anthropological geneticists is the relationship between linguistic affiliation and genetics (Cavalli-Sforza *et al.*, 1994). Languages are often considered to be potential barriers to reproductive success of a population and may influence the population structure found within a region. The nature of the language–gene relationship is often controversial and the results are frequently dependent on both the linguistic classification and analytical method utilized. Some areas of the world demonstrate a clear relationship between genetic and linguistic distances, while for other regions the association is more ambiguous or absent (Crawford, 1998). One example of a research study that examined this relationship between American Na-Dene and Central Siberian Yeniseian speakers was conducted by Rubicz *et al.* (2002) in order to test Ruhlen's (1998) hypothesis that these two linguistic families were closely related. This relationship was considered controversial because of the large geographic distance separating the speakers of these two language families.

Rubicz *et al.* (2002) investigated this genetic-linguistic relationship with several Native American and Siberian populations using blood group polymorphisms and four mtDNA haplogroups (A, B, C and D). A common genetic map technique known as R-matrix analysis was utilized in order to visualize the association between the two regions. R-matrix is a relational statistical technique that allows for the representation of population structure in two or three dimensional space. The first step is to calculate a genetic distance matrix using the equation:

$$R_{ij} = (p_i - \bar{p})(p_j - \bar{p})/\bar{p}(1 - \bar{p}) \tag{6}$$

where R_{ij} is the kinship coefficient for every allele, p_i and p_j are the allele frequencies in populations i and j and $\bar{p}$ is the weighted gene mean frequency of allele p in the matrix. The genetic distances are then averaged into a variance-covariance R-matrix. The R-matrix is reduced into eigenvectors that correspond to the percentage of variation observed in the matrix and plotted using principal component analysis (Harpending and Jenkins, 1973).

Figure 6.7 shows the plot of the first two principal components of an R-matrix using 17 populations and the four mtDNA haplogroups (A, B, C and D) from Rubicz *et al.* (2002). The first two eigenvectors account for 78.5% of the variation found within the sample populations. The first axis separates Native American groups from Siberian populations, with the exception of Asian Eskimo who group with

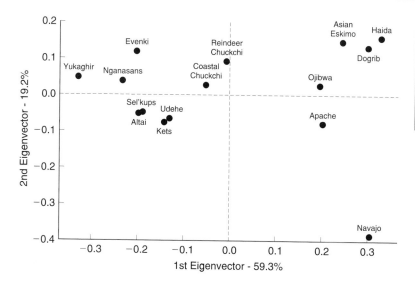

Fig 6.7 R-matrix of ten Siberian and seven North American indigenous population using four mtDNA haplogroups. First two eigenvectors account for 78.5% of total variance. (Adapted from Rubicz et al., 2002).

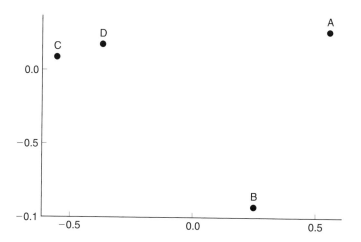

Fig 6.8 S-matrix of four mtDNA haplogroups used in the R-matrix. (Adapted from Rubicz et al., 2002).

northern Native Americans. The second axis divides northern and southern Na-Dene speakers. Figure 6.8 shows the plot of the mtDNA haplogroup and illustrates that the single extant Yeniseian speaker population (Kets) are separated from Na-Dene speakers (Haida, Dogrib, Apache, and Navajo) by the presence of mtDNA haplogroup C and the absence of haplogroup A. Along with evidence provided by additional analytical techniques, Rubicz et al. (2002) concluded that the available genetic data demonstrated that Na-Dene speaking populations clustered with other Native American groups, while the Kets resemble surrounding Siberian populations and that spatial patterning accounts for the majority of variation present in these populations.

Genes and geography

The relationship between genetics and geographic subdivision is a second area that allows anthropological geneticists to generate inferences regarding interpopulation structure. This relationship reflects the 'isolation-by-distance model' and is based on the assumption that populations proximal geographically will demonstrate a higher genetic affinity than those groups found to be at greater spatial distances. Crawford *et al.* (2002) investigated the relationship of two Siberian populations, the Evenki from central Siberia and the Kizhi of Gorno Altai from southern Siberia, to surrounding populations using a number of classic, coding and non-coding molecular markers including ADH, ALDH, COL1A2, mtDNA, Y-chromosome, VNTRs and STRs. Figure 6.9 shows a phylogenetic tree of four autosomal STR loci (TPOX, CSF1PO, THO1, and vWA) among three Siberian populations (Ket, Altai, Evenki) along with four other populations (Asian, Caucasian, Javan, and Amerinds). The genetic distances for this tree were calculated using the Shriver's D_{SW} distance (Shriver *et al.*, 1995) and a dendogram was constructed utilizing the Fitch-Margoliash method (Fitch and Margoliash, 1967). The first bifurcation of the tree separates the Evenki from the other sample populations. The Altai separate out next and are distinct from the other Siberians, Asians, and Amerinds. The Amerinds and Caucasians cluster together as do the Asians and Javans. Based on this evidence, along with results from other markers, it was concluded that the Evenki have maintained their genetic uniqueness as opposed to the Altai who demonstrate a closer relationship to European populations.

A problem for genetic distance measures and their association with other distance matrices (such as geography) is their inability to provide measures of statistical significance to the relationships demonstrated between populations. A commonly used method that does allow for comparisons between matrices and testing of statistical significance is the Mantel test (Mantel, 1967). This method tests

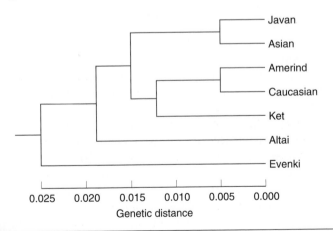

Fig 6.9 Phylogenetic tree based on four STRs (TPOX, CSFIPO, THOI, and vWA) for three Siberian population (Ket, Altai, Evenki), Asian, Native American and European population. Genetic distances calculated using Shriver's method (Shriver *et al.*, 1995) and tree was constructed using the Fitch-Margoliash method (Fitch-Margoliash, 1967). (Adapted from Crawford *et al.*, 2002).

the association between the elements of two matrices using the equation:

$$Z_{AB} = A_{ij}B_{ij} \tag{7}$$

where A_{ij} and B_{ij} are elements of row i and column j of matrices A and B, resulting in an unnormalized correlation coefficient. The statistical significance of the correlation is tested by comparing the observed correlations against a sample distribution of Z based on randomized B matrix (Crawford and Duggirala, 1992).

Other factors besides geographic distance may have an impact on the relationship between genetics and geography. These factors include past geological and ecological events that may have isolated human populations and contributed to present day genetic makeup of these groups. Crawford *et al.* (1997) used Mantel's test to investigate potential ecological events that occurred during the Pleistocene and early Holocene in Siberia and may have contributed to the genetic isolation of these groups. Based on Mantel's test results they found that 30.6% of genetic variation in the region could be explained through the joint effects of geography and language. However, partial correlations between all three matrices (while holding geography constant) demonstrated that the relationship between genetics and linguistics was statistically insignificant. Crawford *et al.* (1997) concluded that most of the genetic differentiation in Siberia is due to geographic patterning. This may be due to the location of Pleistocene glaciers of northern Siberia as well as the presence of Lake Mansi, which played active roles in the geographic isolation and linguistic differentiation in the region.

Population history

Population history focuses on the genetic resemblance of populations, reflected through common descent or gene flow (Harpending and Jenkins, 1973). These two aspects of genetic similarity are normally impossible to differentiate and are closely related to interpopulation variation. However, population history is concerned with past evolutionary events and their timing, which may have impacted the current genetic makeup of populations, whereas interpopulation variation may reflect the existing biological relationships between living populations. Molecular markers and the advent of coalescent theory (Hudson, 1990; Kingman, 1982) have enhanced the ability to determine both the effects of stochastic evolutionary events that shaped a population and the relative time that they occurred. A number of coalescent dating techniques have been developed and aid in providing a context for relating genetic data with other forms of prehistoric data. Three dating methods that may be applicable to anthropological genetic research are mismatch distributions (Rogers and Harpending, 1992), ρ (Forster *et al.*, 1996; Saillard *et al.*, 2000), and the averaged square distance (ASD) between a root microsatellite haplotype and all other haplotypes (Goldstein *et al.*, 1995).

Techniques for estimating population history from sequence data are frequently based on a group of mathematical models collectively referred to as coalescent theory (Kingman, 1982). This theory is useful for the characterization of statistical properties located within intrapopulation phylogenies and is used to estimate effective population size, recombination, and migration rates from sampled sequence data. The mutual history of the sampled sequences creates a genealogy, the lineages of the sample sequences converge or 'coalesce' backwards in time, until the most recent common ancestor of the sample is reached (Pybus and Rambaut, 2002). The coalescent process is an approximation of classic population genetic models and is valid when effective population size is large (Strimmer and Pybus, 2001). Coalescent theory has several useful applications for anthropological geneticists including mathematical modelling, simulation tool for hypothesis testing, inferential tool, and exploratory data analysis (Rosenberg and Nordborg, 2002).

Chronometric techniques

In order to make sense of population history, anthropological geneticists are challenged to place it within a broader context, which is most readily achieved through chronometric techniques. The dating of molecular polymorphisms is dependent on the theory of the molecular clock. According to this theory, genetic variation (consisting of mutation, recombination, and genetic drift) accumulates at a predictable rate (Jobling et al., 2004). The measurement of this rate can be thought of as the speed at which the clock ticks. This can be done directly, through the observation of mutations occurring in rapidly mutating markers such as microsatellites, or indirectly by comparison of closely related species in conjunction with their estimated time of divergence (based on the fossil record). Typically this method assumes that natural selection has not been operating on the loci under investigation. Non-recombining portions of the genome, the MSY and mtDNA, have been extensively used for dating because in these cases mutation alone is assumed to drive the molecular clock. The number of mutations between lineages can be directly counted and related to the time of the most recent common ancestor, without the confounding effects of recombination. After estimating the age of a particular node, it may be related to a particular event (such as the split between two human populations) in association with a larger anthropological question such as the timing of the peopling of the New World by migrations originating in Asia (see Salzano, Chapter 15). In this way, molecular data can be used to establish a chronological record of events in human prehistory. Three molecular dating techniques common in the anthropological literature are mismatch analysis, ρ, and ASD.

Mismatch analysis is also a useful statistical approach for estimating time of expansion for recently established populations. Rogers and Harpending (1992) demonstrated that pairwise differences between nucleotide sequences increased by a rate of 2μ (where μ equals the mutation rate) for each generation during population growth. From this estimated substitution rate, it is possible to estimate N (population size) of a sample prior to population expansion. It is also possible to use coalescent data (Hudson, 1990) to estimate the initial timing of population growth in mutational units using the equation:

$$\tau = 2\mu t \tag{8}$$

where t is time in generations and μ is the mutation rate. Taking the parameters θ_0 and θ_1 as the population estimates before and after expansion, respectively, and fitting these and τ through the least squares method to the observed mismatch distribution permits an estimate of expansion time in mutational units over divisions of time (Rogers and Harpending, 1992). Using the mutation rate of the molecular marker as an estimate, an absolute date can be calculated and then be used to test hypotheses about the history of a population by comparing data from other sources such as archaeology or ecology. Some researchers have cautioned that underlying intrapopulation structure may have a profound impact on the chronometric dates provided by mismatch analysis, and that phylogenetic dating should take into account the root phylogenetic structure of the population (Zlojutro et al., 2006).

Two commonly used chronometric techniques that require a specified root haplotype are ρ (Forster et al., 1996; Saillard et al., 2000) and ASD (Goldstein et al., 1995). Both statistics are expressed as being equal to μ (mutation rate) multiplied by t (time in generations), however the methods are appropriate for different categories of molecular markers. ASD is restricted to microsatellite haplotypes, but ρ is applicable to any marker that produces haplotype data. The ρ statistic is an intra-allelic diversity measure that requires the construction of a phylogeny, most often a network, and is representative of the average number of mutations between the root haplotype and every individual in the sample. These mutations are quantified from the network itself in order to account for mutation rate heterogeneity or homoplasy occurring with the sample data. A phylogenetic reconstruction is not necessary for ASD as the equation that it is derived from corrects for homoplasy and mutation rate heterogeneity (Goldstein et al., 1995). Different methods of chronometric dating often result in temporal dates that are not always concordant with each other and some caution should be exercised in extrapolating inferences solely from molecular data.

Forces of evolution

The modern synthesis of evolutionary theory indicates that there are four forces of evolution, all resulting from deviations of the underlying assumptions of Hardy-Weinberg-Castle genetic equilibrium.

These forces of evolution change the frequencies of alleles in a gene pool, over time. The concept of genetic equilibrium, as first verbalized by Castle and shortly thereafter by Hardy and Weinberg, assumes: (1) populations of infinite size; (2) panmixis (random mating); (3) equal genetic contribution of each genotype to the next generation. Given these conditions, the frequencies of genes remain constant from generation to generation. Changes in the frequencies of genes over time constitute evolution.

A discussion of the forces of evolution played a prominent role in the 1973 volume on Methods and Theories of Anthropological Genetics with at least four chapters covering various facets of selection, gene flow, and genetic drift. Although mutations were not specifically covered by a chapter, the generation of variation as observed in genetic markers was discussed in one chapter. However, in this volume on Anthropological Genetics, the remainder of this chapter focuses on all four of the forces of evolution and how the molecular revolution has affected their study and measurement.

Natural selection

Darwin's major contribution to science was the development of the theory of evolution, guided through the actions of natural selection. He stressed survival of the fittest and the action of mortality as the driving force of evolution. However, he did not understand either the sources of normal genetic variation or the mechanisms of heredity. Ronald Fisher in 1930 helped create a new synthesis in evolutionary theory with the publication of his tome 'Genetical Theory of Natural Selection'. Like Darwin, Fisher emphasized natural selection as the engine of evolution.

Selection operates entirely through differential mortality and fertility. It is not merely the survival of the individual but the successful reproduction and the transmission of the genes. In a given environment the individuals who are the fittest, survive, reproduce and make up the subsequent generations of the population.

Until recently there were few documented cases of natural selection detected in human populations. Prior to the availability of molecular data, in order to demonstrate the action of natural selection in human populations significant differences in fertility had to be ascertained when comparing specific genotypes or genes. Since fitness differences between genotypes in a given generation were often minute, except for rare deleterious or fatal mutations, studies of selection required exceptionally large sample sizes. The best two examples of selection operating on humans involve genes associated with resistance to malaria and with birth weight.

Selection and extremes in birth weight

Karn and Penrose (1951), based upon 6,693 English female babies, showed that selection operates against both the very small (particularly premature) and very large births. They demonstrated that only 41% of infants born weighing less than 4.5 lbs survived 28 days after birth. The additional mortality due to selection acting on birth weight was computed by subtracting the fitness of the optimal set of phenotypes ($So = 0.985$) from the overall fitness ($S = 0.959$) which equals 0.026 (Spuhler, 1973). However, during the last five decades with the technological developments associated with treatment and maintenance of premature births, the survivorship of small infants has increased exponentially.

Adaptation to malaria

The best example of selection operating on human populations is the association between malaria and a series of mutations that alter the environment of the erythrocyte and interfere with the life cycle of the malarial parasite. According to World Health Organization, approximately 1 to 2 million people (mainly children) die from malaria each year. A disease with such an ancient evolutionary history, high mortality and lowered fertility is likely to be acted upon by natural selection. A number of hemoglobinopathies (hemoglobin S, C, D, E, α- and β-thalassemia), the absence of a chemokine receptor (Duffy null), and low levels of the RBC enzyme glucose-6-phosphate dehydrogenase (G-6-PD deficiency) affect the life cycle of the *Plasmodium* organism and increase the fitness of the carriers in malarial environments. However, the specific mutations that provide resistance to malaria vary by geographical region where the chance mutations (particularly in the β-globin region of the genome) had occurred. For example, a suite of mutations in Africa include: S (sickle cell anemia), C hemoglobin, α-thalassemia, G-6-PD deficiency, and Duffy null blood group. While in the Mediterranean region, β-thalassemia, G6PD deficiency and hemoglobin D provide resistance to malaria. The dramatic spread of hemoglobin S mutation in Africa accompanied the introduction of slash-and-burn agriculture from Malaysia and demonstrated how culture and genetics interact in human populations.

Applications of molecular markers to selection

Molecular genetics offers a new dimension to the study of selection: with the sequencing of the coding regions of the genome it is possible to detect the signature and possibly the type of selection. By comparing sequences in a coding region of the genome to a region that is neutral selectively it is possible to determine whether balancing, directional selection or no selection may have been involved. In balancing selection, the region shows a higher level of sequence diversity, an excess of intermediate-frequency variants and a positive

Table 6.5. A summary of molecular regions of coding DNA that have been affected by selection (after Olson, 2002).

Region	Type of selection	Selection agent
β-globin (Hb S)	Balancing selection	Malaria, Sickle Cell
CCR5	Balancing selection	Smallpox, Plague
Lactase	Directional selection	Ability to drink milk
G6PD	Directional selection	Malaria

value for Tajima's D (Bamshad and Wooding, 2003). With directional selection a lower level of sequence diversity, an excess of low-frequency variants are coupled with a negative value of Tajima's D. A beneficial gene spreads rapidly and widely reducing genetic variation. Table 6.5 summarizes from the recent literature regions of the genome that selection may have acted upon and the type of possible selection.

Table 6.5 summarizes the signatures of natural selection in four regions of the genome that have been associated with factors that may affect fertility and mortality in human populations. As expected, the β-globin region and G-6-PD deficiency involve the signature of selection, but through different forms of adaptation. The β-globin region contains mutations for hemoglobin S, which is maintained as a balanced polymorphism with the heterozygotes having a selective advantage over the homozygotes. On the other hand, directional selection appears to be operating in favour of lower levels of G-6-PD, with fewer deleterious effects in homozygous females or hemizygous males. However, individuals who are designated as genetically deficient for G-6-PD, still synthesize some enzyme, although at a lower level. A recent review article by Kwiatkowski (2005) examines the relationship between common erythrocytic variants that affect resistance to malaria. CCR5 gene encodes a cell surface receptor for mediator molecules (chemokines) for HIV-1 virus and bears the signature of selection (Bamshad et al., 2002). The sequence variation in the CCR5 region is substantially higher than expected and deviates significantly from tests of neutrality. One mutation in the CCR5 region, Δ32, has been linked with bubonic plague and prevents both HIV and possibly *Yersina pestis* from entering human macrophages. However, a recent study has questioned whether bubonic plague could extract the intensity of selection necessary to account for the observed signature and suggests that smallpox is a more likely agent of selection at the CCR5 region (Galvani and Slatkin, 2003).

Mishmar et al. (2003) have recently presented an argument that the colonization of Central Asia and Siberia by humans from Africa resulted in a five-fold expansion of mtDNA diversity. They also found that the Siberian and Native American mtDNA haplogroups, A,C,D,G,Z,Y and X showed significant deviation from neutrality for Fu and Li (D*) test but not Tajima's D. Under conditions of extreme cold, Mishmar *et al.* have argued that selection had operated on humans by favouring the increase of heat production. Variants of mtDNA reduce ATP production but increase heat production and require a higher caloric intake provided by a high-fat diet. This dietary-ATP link would provide an advantage under cold stress. Although higher basal metabolic rates are observed in arctic populations it is unclear whether the elevated metabolism is due to variants of mtDNA and oxidative phosphorylation, higher caloric intake, or increased levels of thyroxine, or the interaction of all of these variables. With the molecular genetic tools that are now available, such hypotheses can be tested and the relationship between human molecular characteristics and physiological function can be fruitfully explored.

Mutation

Mutation is the random introduction of new alleles into a population which occurs when the DNA structure of a chromosome is changed through the insertion, replacement, deletion, or inversion of genetic material. These genetic changes may be caused by factors such as radiation, extreme temperatures, or chemical mutagens. On a species level, germ cell mutation is the source of all new genetic material and in order for changes to be passed from one generation to the next, transformations must occur in the gametes' DNA. Therefore, mutation is the creative basis of evolution and is the foundation on which all other forces operate.

The Dutch botanist, Hugo De Vries (1901–1903), was the first to suggest the importance of mutation in evolutionary theory. Through the experimental investigation of the evening primrose plant (*Oenothera lamarkiana*), De Vries suggested that speciation occurred through single mutations. This hypothesis was subsequently challenged by Renner (1914) and Cleland (1923) who demonstrated that evening primrose is a permanent heterozygote and that De Vries' plants were segregates of this rare genetic form. However, this hypothesis led Morgan (1925, 1932) to propose a more gradual role of mutation in evolution. Morgan *et al.* (1915) conducted extensive breeding experiments on *Drosophila melangaster* and showed that mutations occur in small but quantifiable frequencies. Morgan concluded that evolution occurs through the substitution of existing genes by more beneficial mutations in a population and that the role of natural selection was to preserve useful mutations and remove unfit genotypes.

Morgan's mutation-selection theory began to lose popularity with the formation of the synthetic theory of evolution as proposed by Fisher (1930), Wright (1931), and Haldane (1932). These three individuals performed extensive mathematical formulations on mutation, genetic drift, and natural selection and concluded that selection was much more effective in changing gene frequencies than mutation. In the 1960s the synthetic theory was adopted by the majority of biological scientists who characterized mutation as: (1) being random and recurring at a reasonably high frequency; (2) the principal source of variation; (3) natural populations contain a large amount of genetic variability due to previous mutations; (4) due to this high degree of genetic variability, populations have a certain degree of evolutionary flexibility and do not have to evolve in response to changes in the environment; (5) the majority of advantageous mutations should remain fixed due to the high mutation rate; and (6) evolution of new species occurs gradually through the accumulation of advantageous mutations that are acted on by natural selection (Nei, 1987).

In the late 1960s, Kimura (1968a,b) and King and Jukes (1969) began to question some of the basic assumptions of the synthetic theory of evolution through the experimental investigation of newly discovered techniques for the identification of protein polymorphisms. Kimura (1968a, 1983) investigated amino acid substitution rates and allozyme polymorphisms in a number of different species and argued that their genetic load would be too great if explained only by selection. He concluded that these protein polymorphisms were more easily explained through the interactions of genetic drift and neutral mutation rather than solely by natural selection. These findings led researchers to propose the neutral theory of evolution based on the principle that nearly all mutations are selectively neutral and that the majority of molecular variation is driven through genetic drift rather than natural selection. The basic tenet of this theory is that through genetic drift, new neutral alleles become more frequent within a population. Over time these neutral mutations decline or disappear, or in rare cases become fixed within a population or species. When fixation of an allele occurs they accumulate within a previously fixed sequence of the genome. The advent of DNA sequencing technology in the 1980s provided further support for the neutral theory and it now forms the foundation of molecular DNA analysis. Neutral theory has led to a number of advances in molecular biology, including the ability to test null hypotheses using neutrality test statistics and the formation of the molecular clock hypothesis, both of which have been previously discussed in this chapter.

Applications of molecular markers to mutation

Mutations form the foundation of molecular anthropology as both sequence differences and protein polymorphisms are the starting points for drawing inferences regarding the structure or history of a population. The ability of molecular technology to resolve single mutations within the entire human genome has greatly enhanced the ability to formulate hypotheses about the other evolutionary forces impacting human populations. Molecular technology also allows for the study of the effects of different causes of mutation such as those genetic differences caused by long-term exposure to radiation.

Dubrova *et al.* (2002) investigated germ-line mutations at eight autosomal human minisatellite loci (D2S90, D1S172, D10S180, D10S473, D1S7, D7S21, D1S8 and B6.7 located on chromosome 20q13) in rural areas of the Ukraine and Belarus that were heavily contaminated by radiation from the Chernobyl nuclear accident. Using control and exposed groups composed of families with children conceived before and after the Chernobyl accident they found a significant 1.6-fold increase in the germ-line mutation rate of exposed fathers. They did not find an elevated mutation rate in exposed mothers. They suggested that this resulting sexual dimorphism may be to due to differences in the timing of spermatogenesis and oogenesis. Spermatogenesis is a continuous process of mitotic and meiotic cell divisions occurring in males at the onset of puberty, whereas female oocytes are formed in the late stages of embryogenesis and remain inactive until puberty. All females used in this study were at least eight years old at the time of the Chernobyl accident, indicating that ionizing radiation may be unable to affect the stability of minisatellite loci in the maternal germline. Based on these results they concluded that for the males of this population the elevated mutation rates of minisatellites were caused by post-Chernobyl radioactive exposure. They also indicated the need for future research on the mechanisms of radiation-induced mutation at human minisatellites.

Gene flow

This force of evolution has been defined by Workman (1973: 117) as 'the intermixture among individuals from different populations'. It can also be viewed as the addition of new genes into a gene pool. Migration rates and hybridization determine the degree of gene flow into and out of specific populations. Gene flow increases genetic variability within populations (i.e. increases heterozygosity) and decreases genetic differences among populations. Gene flow makes it possible for mutations arising in one population to be exposed to a new environment with its unique selective factors.

Estimates of admixture in human populations can be traced to the original formulation of Bernstein (1931) which was based on the proportionate contribution of one population to the genetic makeup of the hybrid, expressed as:

$$m = q_x - Q/q - Q, \tag{9}$$

where m is the proportion of admixture, Q and q are the allelic frequencies in the two parental populations, and q_x is the frequency of the same allele in the hybrid population. This method was first applied to Brazilian populations (Ottensooser, 1944; DaSilva, 1949) and then to Afro-American US populations by Glass and Li (1953). Since the use of single loci gave vastly different estimates of m,

Roberts and Hiorns (1962), Pollitzer (1964) and Elston (1971) used multiple alleles and various multivariate methods to estimate m in tri-racial populations (see Crawford *et al.*, 1976 for a discussion of multivariate approaches, such as maximum likelihood, true least squares, multiple regression to the measure of admixture).

Application of molecular data to the measurement of admixture

Molecular markers offer greater precision than classic markers (such as blood groups and protein) for the measurement of admixture because they are more informative and maternal migration (using mtDNA) can be distinguished from paternal (male) migration (using MSY). However, autosomal microsatellites have been effectively utilized to measure admixture in various populations. For example, Destro-Bisol *et al.* (1999), using ten unlinked microsatellites plus a linked parital *Alu* deletion, estimated admixture by means of a maximum likelihood procedure in an African American population. The percentage of European admixture was 26% ± 2%, a narrow confidence interval indicating that the allele frequency data from the microsatellite frequencies for the hybrid and parental populations provided a reliable estimate.

An elegant study by DiBenedetto *et al.* (2001) applying autosomal markers, MSY and mtDNA, permitted the testing of specific historically derived models concerning the nature of the gene flow from Central Asia into the Anatolian peninsula approximately 1,000 years ago. They tested three models: (1) elite dominance — i.e. a small portion of an invading army imposes the Turkic language on the inhabitants of Anatolia but do not contribute genetically to the Anatolian gene pool. This model was contradicted by the substantial Central Asian contribution, with a median rate of admixture of 31% based on Y-chromosome markers; (2) massive instantaneous population movement in the form of a military invasion. This model predicts greater incidence of Central Asian haplotypes at Y-chromosome markers than mtDNA. This prediction was not confirmed; (3) small numbers of Oghuz Turks invaded Anatolia, followed by continuous gene flow from Central Asia, at about 1% for 40 generations. The introduction of the Turkic language into the population of Anatolia facilitated long-term, continuous gene flow. This model best fits all of the available data.

Genetic drift
Genetic drift is the random fluctuation of gene frequencies from one generation to the next, resulting in unpredictable evolutionary outcomes. It is the accidental loss (or fixation) of alleles due to sampling error, the effects of which are most pronounced in small populations. Within a population, genetic drift acts to reduce heterozygosity and

causes the loss of alleles. The long-term effects of genetic drift are the reduction of variation within a population and the increase of variation between populations, leading to their divergence.

Gulick was the first to note the importance of random evolutionary processes in small isolated populations (Gulick 1872); however, it was Sewall Wright who developed the theory of genetic drift. In the 1930s and 1940s he worked out a series of mathematical models to deal with the effects of random processes in populations of finite size (detailed in Wright, 1969). The characteristics of random processes can be summarized as follows: (1) it is not possible to predict the direction of change; (2) the magnitude of change depends on the size of the population, with smaller populations experiencing larger changes; and (3) two samples taken from a population will differ not only from the parental population, but also from each other. According to Wright, the repeated random sampling of a population may cause genetic drift to occur, and this can have an evolutionary impact on a species.

Genetic drift can cause the loss of alleles from a population and reduce its heterozygosity (Wright, 1931). Alleles that are initially present in small frequencies in a population are more likely to be lost, while those present at higher frequencies have a greater chance of going to fixation. Most new mutations that arise in a population are transitory. However, if the population is small and isolated, a new mutation has an increased likelihood of becoming fixed through genetic drift. Fewer generations are needed for the fixation or elimination of alleles than in large, panmixtic populations. In addition, individuals are more likely to be identical by descent in small populations, meaning they share certain alleles through common ancestry. This contributes to the reduction of heterozygosity, and can result in a decrease in fitness. Recessive alleles that are normally rare in other populations will have a greater chance of being homozygous and expressing genetic diseases.

Unique historical events can have an impact on human populations, making them more susceptible to the actions of genetic drift. Two examples of situations that cause a reduction in population size include population bottleneck, in which there is decrease in size of an existing population, and founder effect, where a subset of the original population establishes a new colony. These events are themselves categorized as types of genetic drift.

In 1931, Wright introduced effective population size as a means by which the magnitude of genetic drift can be measured (Wright, 1931). The relationship between effective population size (N_e) and the census size of the population (N) is expressed as:

$$N_e = N + 1/2. \tag{10}$$

In the absence of natural selection and mutation, N_e allows for the calculation of the probability that a new allele will become fixed, and the rate at which this may occur. The smaller the effective population size, the greater the effect drift will have on it. N_e is nearly always smaller than N, and is difficult to measure because it is affected by a number of parameters, including an uneven number of males and females, and variation in family size. Fluctuations in the overall size of a population will also affect its evolution, and reductions will be disproportionately important because N_e is shown to approximate the harmonic mean of the population (Wright, 1938; Crow and Kimura, 1970). Most large populations are subdivided into smaller groups (demes) that are partially isolated from each other, and whose members are more closely related to each other than to the population as a whole. Thus, population structure also acts to reduce N_e.

In the 1950s and 60s there was a surge in studies of reproductively isolated, inbred human populations in order to better understand the effects of genetic drift and transmission of rare genetic diseases. Examples include studies of the populations of Tristan da Cunha, Parma Valley, St Bartholemew, the Habbanite Jews, and various Anabaptist groups such as the Amish, Dunkers, and Hutterites (Goldschmidt, 1963). Among the Amish, McKusick *et al.* (1964, 1971) described four rare recessive genetic disorders, each of which was present at a relatively high frequency in a different community: Ellis-van Creveld syndrome in Lancaster County (Pa.) Amish; Pyruvate Kinase deficient hemolytic anemia in Mifflin County (Pa.) Amish; Hemophilia B in Holmes County (Ohio) Amish; and limb-girdle muscular dystrophy in Amish of Adams and Allen Counties, Indiana.

Application of molecular data to the measurement of genetic drift

Today, molecular markers facilitate investigations into the effect of random genetic drift on human populations because allele frequencies can be assessed directly rather than relying on observations of phenotypic changes, such as rare recessive disorders, or the quantification of gene products.

One example of the application of molecular data to assess the effects of genetic drift in a population is the use of mtDNA and MSY markers in investigating the population history of the Tristan da Cunha islanders by Soodyall *et al.* (1997, 2003). Tristan da Cunha is a small volcanic island in the southern Atlantic Ocean, discovered by the Portuguese in 1506. In 1816 the British placed a garrison on the island to prevent the French from attempting to rescue Napoleon Bonaparte from exile on St Helena, 1,300 miles to the North. The British garrison was removed a year later, at which point Corporal Glass, along with his wife, their two children, and two other men, stayed behind to establish the first permanent residence on Tristan. Genealogical records, going back to those first settlers, have been carefully reconstructed for all the island's inhabitants. The current population of 278 is described as having descended from 15 known individuals, all of western European ancestry. There is also historical documentation of two population bottlenecks which occurred on Tristan, one in the 1850s when the original founder died and 25 of his descendants left, followed by the emigration of a pastor and 45 additional members of the community (Roberts, 1968). The second occurred in 1885 due to a boat capsize that killed 15 adult males, leaving only four on the island (one of which was mentally retarded and two were elderly). In 1997, Soodyall *et al.* compared the genealogical information with mtDNA data, and found both lines of evidence indicated the current Tristan population was derived from only five females. Documentation of two sister pairs among the original female founders was not supported by the mtDNA data, however, which only found evidence for one pair. Y-chromosome analysis of Tristan males demonstrated that while seven surnames were in use in the living population, more than seven males had contributed to the genetic makeup of the living population (Soodyall *et al.*, 2003). Two additional Y-chromosome types were present that did not match with the recorded surnames, one of which could be accounted for by a single-step microsatellite mutation from a legitimate Y haplotype, while the other was not indicated in the records and was of obvious non-island origin. In these studies, when molecular markers were compared with the written records of Tristan da Cunha, they pointed out errors in the genealogical information, and provided additional documentation of the action of founder effect and bottlenecks in this small island population.

In another study, molecular markers were used to compare the genetic makeup of the Aleuts of Bering Island, Russia with that of their founding population in the Aleutian Islands and to determine the genetic consequences of unique historical events on this population (Rubicz *et al.*, 2005). Between 1824 and 1828, Aleuts were forcibly relocated from their homeland in the Aleutian Islands to the Commanders (Bering and Medni Islands) to work for the Russian-American Company. Families from Atka, in the central Aleutians, were relocated to Bering, and Medni was settled by Aleuts from Attu, in the western Aleutians. In 1825 only 45 Aleuts resided on Bering, in addition to 15–30 original Russian settlers. By 1826 the number of Aleuts increased to 110. The Commander Aleut population sizes fluctuated over time, with an all-time high of 626 (330 on Bering) in 1892. Depopulation due to epidemics and relocations to the Kamchatka Peninsula reduced this number to 364 (204 on Bering) by 1923. In 1969 the Medni Aleuts were relocated to Bering. Today the Bering population consists of approximately 300 individuals. Rubicz *et al.* (2005) found that Bering Island Aleut mtDNAs belonged to a single haplogroup, D, whereas the founding Aleut population had both haplogroup D (72%) and A (28%) mtDNAs. While Derbeneva *et al.* (2002) suggest the fixation of haplogroup D in the Bering population is due to intergenerational drift, Rubicz *et al.* favour the explanation that it was the result of a founder effect. There is historical documentation that Aleut families, not a random sample of individuals from the parental population, were originally brought to the Commander

Islands. It is more likely that haplogroup A mtDNAs were absent from the Aleut founders of Bering Island, rather their being lost over the few generations since the establishment of the community, considering they were originally present at a frequency of nearly 30%. This study provides further details of the effect of random processes such as historic events on small isolated communities.

Conclusion

This chapter demonstrates the essential nature of molecular markers to the study of human evolution and the out-of-Africa diaspora, topics central to the field of anthropological genetics. These markers, mtDNA, STRs, VNTRs, SNPs, MSY and the sequences of the coding regions of the genome, are much more informative than our early markers, such as blood groups and proteins. The molecular markers characterize the gene pool (total genetic makeup) of a population and bear witness to the indelible signatures of the forces of evolution. The molecular markers have provided researchers with a unique set of tools that can be used to reconstruct sex specific population history and a molecular chronology of events. We have at our disposal a more powerful array of tools for mapping genes and QTLs by placing SNPs or STRs equidistantly throughout the genome and locating their positions through linkage analyses and association studies. We are now experiencing the repercussions of a molecular revolution that has forever altered the field of anthropological genetics.

References

Aris-Brosou, S. and Excoffier, L. (1996). The impact of population expansion and mutation rate heterogeneity on DNA sequence polymorphism. *Molecular Biology and Evolution*, **13**, 494–504.

Bamshad, M., Mummidi, S., Gonzalez, E., Ahuja, S.S., Dunn, D.M., Watkins, W.S., Wooding, S., Stone, A.C., Jorde, L.B., Weiss, R.B. and Ahuja, S.K. (2002). A strong signature of balancing selection in the 5′ cis-regulatory region of CCR5. *Proceedings of the National Academy of Sciences, USA*, **99**, 10539–44.

Bamshad, M. and Wooding, S.P. (2003). Signatures of natural selection in the human genome. *Nature Reviews*, **4**, 99–111.

Bandelt, H.J., Forster, P. and Rohl, A. (1999). Median-joining networks for inferring intraspecific phylogenies. *Molecular Biology and Evolution*, **16**, 37–48.

Bandelt, H.J., Forster, P., Sykes, B.C. and Richards, M.B. (1995). Mitochondrial portraits of human populations using median networks. *Genetics*, **141**, 743–53.

Bernstein, F. (1931). Die geographische verteilung der blutgruppen und ihre anthropologische bedeutung. *Comitato Italiano per Studio dei Problemi della Populazione*. Istituto Poligraficao dello Stato, Rome, pp. 227–43.

Brion, M., Salas, A., Gonzalez-Neira, A., Lareu, M.V. and Carracedo, A. (2003). Insights into Iberian population origins through the construction of highly informative Y-chromosome haplotypes using biallelic markers, STRs, and the MSY1 minisatellite. *American Journal of Physical Anthropology*, **122**, 147–161.

Brooke, J., McCurrah, M., Harley, H., Buckler, A., Church, D., Aburatani, H., Hunter, K., *et al.* (1992). Molecular basis for myotonic dystrophy: expansion of trinucleotide (CTG) repeat at the 3′ end of a transcript encoding a protein kinase family member. *Cell*, **68**, 799–808.

Cavalli-Sforza, L.L. and Bodmer, W.F. (1971). *The Genetics of Human Populations*. San Francisco: W.H. Freeman.

Cavalli-Sforza, L.L., Menozzi, P. and Piazza, A. (1994). *The History and Geography of Human Genes*. Princeton, NJ: Princeton University Press.

Cleland, R.E. (1923). Chromosome arrangements during meiosis in certain oenotheras. *American Naturalist*, **57**, 562–66.

Crawford, M.H. (1973). The use of genetic markers of the blood in the study of the evolution of human populations. In *Methods and Theories of Anthropological Genetics*, eds. M.H. Crawford & P.L. Workman. Albuquerque: University of New Mexico Press, pp. 19–38.

Crawford, M.H. (1998). *The Origins of Native Americans*. Cambridge, UK: Cambridge University Press.

Crawford, M.H. and Duggirala, R. (1992). Digital dermatoglyphic patterns of Eskimo and Amerindian populations: relationships between geographic, dermatoglyphic, genetic, and linguistic distances. *Human Biology*, **64**, 683–704.

Crawford, M.H., McComb, J., Schanfield, M. and Mitchell, R. (2002). Genetic structure of pastoral populations of Siberia: the Evenki of central Siberia and Kizhi of Gorno Altai. In *Human Biology of Pastoral Populations*, eds. W. Leonard & M.H. Crawford. Cambridge, UK: Cambridge University Press, pp. 10–49.

Crawford, M.H., Williams, J.T. and Duggirala, R. (1997). Genetic structure of the indigenous populations of Siberia. *American Journal of Physical Anthropology*, **104**, 177–92.

Crawford, M.H., Workman, P.L., McLean, C. and Lees, F.C. (1976). Admixture estimates and selection in Tlaxcala. In *The Tlaxcaltecans. Prehistory, Demography, Morphology and Genetics*, ed. M.H. Crawford. Lawrence: University of Kansas Publications in Anthropology, pp. 161–8.

Crow, J. and Kimura, M. (1970). *An Introduction to Population Genetics Theory*. New York: Harper and Row.

DaSilva, E.M. (1949). Blood groups of Indians, whites and white-Indian mixtures in southern Mato Grosso. Brazil. *American Journal of Physical Anthropology*, **7**, 575–86.

Deininger, P. and Batzer, M. (1993). Evolution of retroposons. In *Evolutionary Biology*, eds. M. Hect, R. MacIntyre & M. Clegg. New York: Plenum Press, pp. 157–96.

Derbeneva, O., Sukernik, R., Volovko, N., Hosseini, S., Lott, M. and Wallace, D. (2002). Analysis of mitochondrial DNA diversity of the Aleuts of the Commander Islands and its implications for the genetic history of Beringia. *American Journal of Human Genetics*, **71**, 415–21.

Derenko, M.V., Grzybowski, T., Malyarchuk, B.A., Dambueva, I.K., Denisova, G.A., Czarny, J., Dorzhu, C.M., Kakpakov, C.M., Miscicka-Sliwaka, D., Wozniak, M. and Zakharaov, I.A. (2003). Diversity of mitochondrial DNA lineages in South Siberia. *Annals of Human Genetics*, **67**, 391−411.

Destro-Bisol, G., Spedini, G. and Pascali, V.L. (2000). Application of different genetic distance methods to microsatellite data. *Human Genetics*, **106**, 130−2.

Destro-Bisol, G., Maviglia, R., Caglia, A., Boschi, I., Spedini, G., Pascali, V., Clark, A. and Tishkoff, S. (1999). Estimating European admixture in African Americans by using microsatellites and a microsatellite haplotype (CD4/Alu). *Human Genetics*, **104**, 149−57.

De Vries, H. (1901−1903). *Die Mutationstherorie*. Leipzig: Von Veit. (109−10 English translation). The Mutation Theory, trans. J.B. Farmer and A.D. Darbishire. Chicago: Open court.

DiBenedetto, G., Erguven, A., Stenico, M., Castri, L., Bertorelle, G., Togan, I. and Barbujani, G. (2001). DNA diversity and population admixture in Anatolia. *American Journal of Physical Anthropology*, **115**, 144−56.

Dubrova, Y.E., Grant, G., Chumak, A.A., Stezhka, V.A. and Karakasian, A.N. (2002). Elevated minisatellite mutation rate in the post-Chernobyl families from Ukraine. *American Journal of Human Genetics*, **71**(4), 801−9.

Elston, R.C. (1971). The estimation of admixture in racial hybrids. *Annals of Human Genetics*, **35**, 9−17.

Excoffier, L., Smouse, P.E. and Quattro, J.M. (1992). Analysis of molecular variance inferred from metric distances among DNA haplotypes: application to human mitochondrial DNA restriction data. *Genetics*, **131**, 479−91.

Fisher, R.A. (1930). *Genetical Theory of Natural Selection*. Oxford: Clarendon Press.

Fitch, W.M. and Margoliash, E. (1967). Construction of phylogenetic trees. *Science*, **155**, 279−84.

Forster, P., Harding, R., Torroni, A. and Bandelt, H.J. (1996). Origin and evolution of Native American mtDNA variation: a reappraisal. *American Journal of Human Genetics*, **59**, 935−45.

Foy, J.C. and Wu, C.I. (1999). A human population bottleneck can account for the discordance between patterns of mitochondrial versus nuclear DNA variation. *Molecular Biology and Evolution*, **16**, 1003−5.

Francalacci, P., Montiel, R. and Malgosa, A. (1999). A mitochondrial DNA database. In *Genomic Diversity: Applications in Human Population Genetics*, eds. S. Papiha, R. Deka & R. Chakraborty. New York: Kluwer Academic/Plenum Publishers, pp. 103−19.

Fu, Y.X. (1997). Statistical tests of neutrality of mutations against population growth, hitchhiking and background selection. *Genetics*, **147**, 915−25.

Fu, Y., Kuhl, D., Pizzuti, A., Pieretti, M., Sutcliffe, J., Richards, S., Verkerk, A., *et al.* (1991). Variation in the CGG repeat at the fragile X site results in genetic instability: resolution of the Sherman paradox. *Cell*, **67**, 1047−58.

Galvani, A.P. and Slatkin, M. (2003). Evaluating plague and smallpox as historical selective pressures for the CCR5- 32 HIV-resistance allele. *Proceedings of the National Academy of Sciences, USA*, **100**, 15276−9.

Glass, B. and Li, C.C. (1953). The dynamics of racial intermixture−an analysis based on the American Negro. *American Journal of Human Genetics*, **5**, 1−20.

Goldschmidt, E. (1963). *The Genetics of Migrant and Isloate Populations*. New York: Williams and Wilkins.

Goldstein, D. B., Ruiz Linares, A., Cavalli-Sforza, L. L. and Feldman, M. W. (1995). An evaluation of genetic distances for use with microsatellite loci. *Genetics*, **139**, 463–71.

Guarino, F., Federle, L., van Oorschot, R., Briceno, I., Bernal, J., Papiha, S., Schanfield, M. and Mitchell, R. (1999). Genetic diversity among five Native American tribes of Columbia: Evidence from nine autosomal microsatellites. In *Genomic Diversity: Applications in Human Population Genetics*, eds. S. Papiha, R. Deka & R. Chakraborty. New York: Kluwer Academic/Plenum Publishers, pp. 33–51.

Gulick, J. (1872). On the diversity of evolution under one set of external conditions. *Linnean Society Journal of Zoology*, **11**, 496–505.

Haldane, J. B. S. (1932). *The Causes of Evolution*. London: Longmans and Green.

Hammer, M. F., Karafet, T., Rasanayagam, A., Wood, E. T., Altheide, T. K., Jenkins, T., Griffiths, R. C., Templeton, A. R. and Zegura, S. L. (1998). Out of Africa and back again: Nested cladistic analysis of human Y-chromosome variation. *Molecular Biology and Evolution*, **15**, 427–41.

Hammer, M. F., Spurdle, A. B., Karafet, T., Bonner, M. R., Wood, E. T., Novelletto, A., Malaspina, P., Mitchell, R. J., Horai, S., Jenkins, T. and Zegura, S. L. (1997). The geographic distribution of human Y-chromosome variation. *Genetics*, **145**, 787–805.

Harpending, H. and Jenkins, T. (1973). Genetic distance among South African populations. In *Methods and Theories in Anthropological Genetics*, eds. M. H. Crawford & P. L. Workman. Albuquerque: University of New Mexico Press, pp. 177–99.

Horuk, R., Chitnis, C. E., Darbonne, W. C., Colby, T. J., Rybicki, A., Hadley, T. J. and Miller, L. H. (1993). A receptor for the malarial parasite Plasmodium vivax: The erythrocyte chemokine receptor. *Science*, **261**(5125), 1182–4.

Hudson, R. (1990). Gene genealogies and the coalescent process. *Oxford Surveys in Evolutionary Biology*, **9**, 1–44.

Jobling, M. A., Bouzekri, N. and Taylor, P. G. (1998). Hypervariable digital DNA codes for human paternal lineages: MVR-PCR at the Y-specific minisatellite, MSY1 (DYF155S1). *Human Molecular Genetics*, **7**, 643–53.

Jobling, M., Hurles, M. and Tyler-Smith, C. (2004). *Human Evolutionary Genetics: Origins, Peoples & Disease*. New York: Garland Science.

Jorde, L. B. (1985). Human genetic distance studies: present status and future prospects. *Annual Review of Anthropology*, **14**, 343–73.

Karn, M. N. and Penrose, L. S. (1951). Birth weight and gestation time in relation to maternal age, parity, and infant survival. *Annals of Eugenics*, **15**, 206–33.

Kimura, M. (1968a). Genetic variability maintained in a finite population due to mutational production of neutral and nearly neutral isoalleles. *Genetic Research*, **11**, 247–69.

Kimura, M. (1968b). Evolutionary rate at the molecular level. *Nature*, **217**, 624–26.

Kimura, M. (1983). *The Neutral Theory of Evolution*. Cambridge: Cambridge University Press.

King, J. L. and Jukes, T. H. (1969). Non-Darwinian evolution. *Science*, **164**, 788–98.

Kingman, J. (1982). The coalescent. *Stochastic Processes and their Applications*, **13**, 335–48.

Kreitman, M. (2000). Methods to detect selection in populations with application to the human. *Annual Review of Genomics and Human Genetics*, 1, 539–59.

Kwiatkowski, D.C. (2005). How malaria has affected the human genome and what human genetics can teach us about malaria. *American Journal of Human Genetics*, 77, 171–92.

Lander, E., Linton, L., Birren, B., Nusbaum, C., Zody, M., Baldwin, J., Devon, K., et al. (2001). Initial sequencing and analysis of the human genome. *Nature*, 409, 860–921.

Landsteiner, K. and Levine, P. (1927). Further observations on individual difference of human blood. *Proceedings of the Society of Experimental Biology*, 24, 941–2.

Lorenz, J. and Smith, D. (1996). Distribution of the four founding haplogroups among Native North Americans. *American Journal of Physical Anthropology*, 101, 307–23.

Mantel, N. (1967). The detection of disease clustering and a generalized regression approach. *Cancer Research*, 27, 209–20.

Margulis, L. (1981). *Symbiosis in Cell Evolution: Life and its Environment on the Early Earth*, New York: W. H. Freedman & Co.

Marjoram, P. and Donnelly, P. (1994). Pairwise comparisons of mitochondrial DNA sequences in subdivided populations and implications for early human evolution. *Genetics*, 136, 673–83.

Marini, A.-M. and Urrestarazu, A. (1997). The Rh (Rhesus) blood group polypeptides are related to NH4+ transporters. *Trends in Biochemical Sciences*, 22, 460–1.

Maxam, A.M. and Gilbert, W. (1977). A method for sequencing DNA. *Proceedings of the National Academy of Sciences, USA*, 74, 560–4.

McComb, J. (1999). The development of dual-primer randomly amplified polymorphic DNA (RAPD) and the application to the study of anthropological genetics. Unpublished Dissertation. Lawrence, Kansas: University of Kansas.

McComb, J., Crawford, M., Osipova, L., Karaphet, T., Posukh, O. and Schanfield, M. (1996). DNA interpopulational variation in Siberian indigenous populations: The mountain Altai. *American Journal of Human Biology*, 8, 559–607.

McComb, J., Blagitko, N., Comuzzie, A., Schanfield, M., Sukernik, R., Leonard, W. and Crawford, M.H. (1995). VNTR DNA variation in Siberian indigenous populations. *Human Biology*, 67, 217–29.

McKusick, V., Egeland, J., Eldridge, R. and Krusen, D. (1964). Dwarfism in the Amish. I. The Ellis-van Creveld syndrome. *Bulletin Johns Hopkins Hospital*, 115, 306–36.

McKusick, V., Hostetler, J., Egeland, J. and Eldridge, R. (1971). The distribution of certain genes in the Old Order Amish. In *Human Populations, Genetic Variation, and Evolution*, ed. L.N. Morris. San Francisco: Chandler Publishing Company, pp. 358–80.

Melvin, K. (2001). Genetic diversity among the chuvash using randomly amplified polymorphic DNA (RAPD) markers. Unpublished thesis. Lawrence, Kansas: University of Kansas.

Merriwether, D., Rothhammer, F. and Ferrell, R. (1995). Distribution of the four founding lineage haplotypes in Native Americans suggests a single wave of migration for the New World. *American Journal of Physical Anthropology*, 98, 411–30.

Miller, L.H., Mason, S.J., Clyde, D.F. and McGinniss, M.H. (1976). The resistance factor to Plasmodium vivax in blacks. The Duffy-blood-group genotype, FyFy. *New England Journal of Medicine*, **295**, 302–4.

Mishmar, D., Ruiz-Pesini, E., Golik, P., Macaulay, V., Clark, A.G., Hosseini, S., Brandon, M., Easley, K., Chen, E., Brown, M.D., Sukernik, R.I., Olckers, A. and Wallace, D.C. (2003). Natural selection shaped regional mtDNA variation in humans. *Proceedings of the National Academy of Sciences, USA*, **100**, 171–6.

Mitchell, R., Howlett, S., White, N., Federle, L., Papiha, S., Briceno, I., McComb, J., Schanfield, M., Tyler-Smith, C., Osipova, L., Livshits, G. and Crawford, M.H. (1999). Deletion polymorphism in the human COL1A2 gene: Genetic evidence of a non-African population whose descendants spread to all continents. *Human Biology*, **71**, 901–14.

Morgan, T.H. (1925). *Evolution and Genetics*. Princeton, NJ: Princeton University Press.

Morgan, T.H. (1932). *The Scientific Basis of Evolution*. New York: Norton.

Morgan, T.H., Muller, H.J., Sturtevant and Bridges, C.B. (1915). *The Mechanisms of Mendelian Heredity*. New York: Holt.

Nachman, M.W. and Crowell, S.L. (2000). Estimate of the mutation rate per nucleotide in humans. *Genetics*, **156**, 297–304.

Nei, M. (1987). *Molecular Evolutionary Genetics*. New York: Columbia University Press.

Novick, G., Batzer, M., Deininger, P. and Herrera, R. (1996). The mobile genetic element *Alu* in the human genome. *Bioscience*, **46**, 32–44.

Olson, S. (2002). Seeking the signs of selection. *Science*, **298**, 1324–5.

Oota, H., Settheetham-Ishida, W., Tiwawech, D., Ishida, T. and Stoneking, M. (2001). Human mtDNA and Y chromosome variation is correlated with matrilocal versus patrilocal residence. *Nature Genetics*, **29**, 20–21.

Ottensooser, F. (1944). Calculo do gran de mistura racial atranes dos grupos sanguineos. *Review of Brasilian Biology*, **4**, 531–7.

Perez-Lezaun, A., Calafell, F., Mateu, E., Comas, D., Ruiz-Pacheco, R. and Bertranpetit, J. (1997). Microsatellite variation and the differentiation of modern humans. *Human Genetics*, **99**, 1–7.

Pollitzer, W.S. (1964). Analysis of a tri-racial isolate. *Human Biology*, **36**, 362–73.

Przeworski, M. (2002). The signature of positive selection at randomly chosen loci. *Genetics*, **160**, 1179–89.

Pybus, O.G. and Rambaut, A. (2002). GENIE: estimating demographic history from molecular phylogenies. *Bioinformatics*, **18**, 1404–5.

Ramos-Onsins, S.E. and Rozas, J. (2002). Statistical properties of new neutrality tests against population growth. *Molecular Biology and Evolution*, **19**, 2092–100.

Renner, O. (1914). Befruchtung und Embroylobilidng bei *Oenothera lamrckiana* und einigen Verwandten Arten. *Flora*, **107**, 115–50.

Reynolds, J., Weir, B.S. and Cockerham, C.C. (1993). Estimation of the coancestry coefficient for a short term genetic distance. *Genetics*, **105**, 767–79.

Roberts, D. (1968). Genetic effects on population size reduction. *Nature*, **220**, 1084–8.

Roberts, D.F. and Hiorns, R.W. (1962). The dynamics of racial intermixture. *American Journal of Human Genetics*, **14**, 261–77.

Rogers, A. R., Fraley, A. E., Bamshad, M. J., Watkins, W. S. and Jorde, L. B. (1996). Mitochondrial mismatch analysis is insensitive to the mutational process. *Molecular Biology and Evolution*, **13**, 895–902.

Rogers, A. R. and Harpending, H. (1992). Population growth makes waves in the distribution of pairwise genetic differences. *Molecular Biology and Evolution*, **9**, 552–69.

Romualdi, C., Balding, D., Nasidze, I. S., Risch, G., Robichaux, M., Sherry, S., Stoneking, M., Batzer, M. and Barbujani, G. (2002). Patterns of human diversity, within and among continents, inferred from biallelic DNA polymorphisms. *Genome Research*, **12**, 602–12.

Rosenberg, N. A. and Nordborg, M. (2002). Genealogical trees, coalescent theory and the analysis of genetic polymorphisms. *Nature Reviews Genetics*, **3**, 380–90.

Rubicz, R. (2001). Origins of the Aleuts: Molecular perspectives. Unpublished thesis. Lawrence, Kansas: University of Kansas.

Rubicz, R., Melvin, K. L. and Crawford, M. H. (2002). Genetic evidence for the phylogenetic relationship between Na-Dene and Yeniseian speakers. *Human Biology*, **74**, 743–60.

Rubicz, R., Schurr, T., Babb, P. and Crawford, M. H. (2003). Mitochondrial DNA variation and the origins of the Aleuts. *Human Biology*, **75**, 809–35.

Rubicz, R., Sun, G., Devor, E., Spitsyn, V., Deka, R. and Crawford, M. H. (2005). *Genetic architecture of a small, isolated population: Bering Island, Russia.* Poster presented at the 30th annual meeting of the Human Biology Association, Wisconsin.

Ruhlen, M. (1998). *A Guide to the World's Languages.* Stanford: Stanford University Press.

Salem, A., Badr, F., Gaballah, M. and Paabo, S. (1996). The genetics of traditional living: Y-chromosomal and mitochondrial lineages in the Sinai Peninsula. *American Journal of Human Genetics*, **59**, 741–743.

Saillard, J., Forster, P., Lynnerup, N., Bandelt, H. J. and Norby, S. (2000). mtDNA variation among Greenland Eskimos: the edge of the Beringian expansion. *American Journal of Human Genetics*, **67**, 718–26.

Sanger, F., Micklen, S. and Coulson, A. R. (1977). DNA sequencing and chain-terminating inhibitors. *Proceedings of the National Academy of Sciences, USA*, **74**, 5463–7.

Schneider, S., Roessli, D. and Excoffier, L. (2000). Arelquin ver 2.000: A software for population genetic data analysis. Geneva, Switzerland: Genetics and Biometry Laboratory, University of Geneva.

Schull, W. and MacCluer, J. (1968). Human genetics: structure of populations. *Annual Review of Genetics*, **2**, 279–304.

Schurr, T., Ballinger, S., Gan, Y., Hodge, J., Merriwether, A., Lawrence, D., Knowler, W., Weiss, K. and Wallace, D. (1990). American mitochondrial DNAs have rare Asian mutations at high frequencies, suggesting they derived from four primary maternal lineages. *American Journal of Human Genetics*, **46**, 613–23.

Seielstad, M., Minch, E. and Cavalli-Sforza, L. (1998). Genetic evidence for a higher female migration rate in humans. *Nature Genetics*, **20**, 278–280.

Shields, G. F., Schmiechen, A. M., Frazier, B. L., Redd, A., Voevoda, M. I., Reed, J. K. and Ward, R. H. (1993). mtDNA sequences suggest a recent evolutionary divergence for Beringian and Northern North American populations. *American Journal of Human Genetics*, **53**, 549–62.

Shriver, M. D., Jin, L., Boerwinkle, E., Deka, R., Ferrell, R. E. and Chakraborty, R. (1995). A novel measure of genetic distance for highly polymorphic tandem repeat loci. *Molecular Biology and Evolution*, **12**, 914–20.

Sibley, C. G. and Ahlquist, J. E. (1984). The phylogeny of the hominid primate, as indicated by DNA-DNA hybridisation. *Journal of Molecular Evolution*, **20**, 2–15.

Skaletsky, H., Kuroda-Kawaguchi, T., Minx, P., Cordum, H. S., Hillier, L., Brown, L. G., Repping, S., *et al.* (2003). The male-specific region of the human Y chromosome is a mosaic of discrete sequence classes. *Nature*, **423**, 825–37.

Slatkin, M. (1995). A measure of population subdivision based on micro-satellite allele frequencies. *Genetics*, **139**, 457–62.

Smithies, O. (1955). Zone electrophoresis in starch gels: Group variations in serum proteins of normal human adults. *Biochemistry Journal*, **61**, 629.

Soodyall, H., Jenkins, T., Mukherjee, A., du Toit, E., Roberts, D. and Stoneking, M. (1997). The founding mitochondrial DNA lineages of Tristan da Cunha islanders. *American Journal of Physical Anthropology*, **104**, 157–66.

Soodyall, H., Nebel, A., Morar, B. and Jenkins, T. (2003). Genealogy and genes: tracking the founding fathers of Tristan da Cunha. *European Journal of Human Genetics*, **11**, 705–9.

Spuhler, J. N. (1973). Anthropological genetics: An overview. In *Methods and Theories of Anthropological Genetics*, eds. M. H. Crawford & P. L. Workman. Albuquerque: University of New Mexico Press, pp. 423–51.

Stoneking, M. and Soodyall, H. (1996). Human evolution and the mitochon-drial genome. *Current Opinion in Genetics & Development*, **6**, 731–6.

Strimmer, K. and Pybus, O. G. (2001). Exploring the demographic history of DNA sequences using the generalized skyline plot. *Molecular Biology and Evolution*, **18**, 2298–305.

Sukernik, R., Schurr, T., Starikovskaya, E. and Wilson, A. (1996). Mitochondrial DNA variation in Native Siberians, with special reference to the evolu-tionary history of American Indians. *Genetika*, **32**, 432–9.

Tajima, F. (1989). Statistical method for testing the neutral mutation hypo-thesis by DNA polymorphism. *Genetics*, **123**, 585–95.

Tajima, F. (1993). Measurement of DNA polymorphisms. In *Mechanisms of Molecular Evolution: Introduction to Molecular Paleopopulation Biology*, eds. N. Takahata & A. Clark. Tokyo: Japan Scientific Society Press, pp. 37–59.

The Huntington's Disease Collaborative Research Group (1993). A novel gene containing a trinucleotide repeat that is expanded and unstable on Huntington's disease chromosomes. *Cell*, **72**, 971–983.

Torroni, A., Schurr, T., Cabell, M., Brown, M., Neel, J., Larsen, M., Smith, D., Vullo, C. and Wallace, D. (1993a). Asian affinities and continental radiation of the four founding Native American mtDNAs. *American Journal of Human Genetics*, **53**, 563–90.

Torroni, A., Sukernik, R., Schurr, T., Starikovskaya, Y., Cabell, M., Crawford, M. H., Comuzzie, A. and Wallace, D. (1993b). Mitochondrial DNA variation of aboriginal Siberians reveal distinct genetic affinities with Native Americans. *American Journal of Human Genetics*, **53**, 591–608.

Venter, J. C., Adams, M. D., Myers, E. W., Li, P. W., Mural, R. J., Sutton, G. G., Smith, H. O., *et al.* (2001). The sequence of the human genome. *Science*, **291**, 1304–51.

Vigilant, L., Stoneking, M., Harpending, H., Hawkes, K. and Wilson, A. C. (1991). African populations and the evolution of human mitochondrial DNA. *Science*, **253**, 1503–07.

Waterson, G. (1975). On the number of segregating sites in genetical models without recombination. *Theoretical Population Biology*, **7**, 256–76.

Wilder, J.A., Kingan, S.B., Mobasher, Z., Pilkington, M.M. and Hammer, M.F. (2004). Global patterns of human mitochondrial DNA and Y-chromosome structure are not influenced by higher migration rates of females versus males. *Nature Genetics* S36(10), 1122–1125.

Workman, P.L. (1973). Genetic analyses of hybrid populations. In *Methods and Theories of Anthropological Genetics* eds. M.H. Crawford & P.L. Workman. Albuquerque, New Mexico: University of New Mexico, pp. 117–50.

Workman, P.L. and Jorde, J.B. (1980). The genetic structure of the Aland islands. In *Population Structure and Genetic Disease,* eds. A. Erikson, H. Forsius, H. Nevanlinna & P. Workman. New York: Academic Press, pp. 287–508.

Wright, S. (1931). Evolution in Mendelian populations. *Genetics*, **16**, 97–159.

Wright, S. (1938). Size of population and breeding structure in relation to evolution. *Science*, **87**, 430–1.

Wright, S. (1969). *Evolution and the Genetics of Populations II. The Theory of Gene Frequencies.* Chicago: University of Chicago Press.

Y Chromosome Consortium (2002). A nomenclature system for the tree of human Y-chromosomal binary haplogroups. *Genome Research*, **12**, 339–48.

Zlojutro, M., Rubicz, R., Devor, E., Spitsyn, V.A., Wilson, K. and Crawford, M.H. (2006). Genetic structure of the Aleuts and circumpolar populations based on mitochondrial DNA sequences: A Synthesis. *American Journal of Physical Anthropology*, **129**, 446–64.

Chapter 7

The Use of Quantitative Traits in Anthropological Genetic Studies of Population Structure and History

John H. Relethford

Department of Anthropology, State University of New York College at Oneonta, Oneonta, NY

Introduction

This chapter may seem to some to be a fond remembrance of things past. Given the increasing wealth of new data on genetic variation in humans, it might seem rather old-fashioned to deal with quantitative traits. Long before the discovery of blood groups, protein and enzyme polymorphisms, and DNA sequences, the only way anthropologists could describe human variation was in terms of quantitative traits, such as anthropometrics. In the broadest sense, a quantitative trait is one whose genotypic and/or phenotypic distribution is continuous, such as stature or head length, rather than discrete, such as a blood type or a DNA haplotype. Quantitative traits are often referred to as 'complex traits' because the phenotype is a reflection of both genetic and non-genetic influences, the latter of which can include age, diet and climate, as well as many other factors. Some quantitative traits, such as the level of phenylalanine in the blood, are due to a single gene whose phenotypic expression is affected by environmental (non-genetic) influences resulting in a continuous distribution. Other traits, such as cleft palate, are phenotypically discrete, but reflect an underlying continuous genotypic distribution under a threshold model. Most quantitative traits examined in the course of anthropological research are assumed to be polygenic, where a continuous distribution (often reflecting a normal distribution) is a function of multiple genes and environmental influences (Cavalli-Sforza and Bodmer, 1971). The primary focus of this chapter is on polygenic quantitative traits and their application to questions of population structure and history.

In terms of anthropological research, there are several classes of quantitative traits that have frequently been studied. Perhaps the most commonly measured quantitative traits in living populations are anthropometrics, which are measures of the body, skull and face. Similar measures are available for skeletal data, including those of the cranium (craniometrics). Other commonly used quantitative traits are dental measures (odontometrics), which can be measured on both living subjects and from skeletal data, as well as finger and palm prints (dermatoglyphics) and measures of skin colour. Quantitative traits in the broadest sense have been used to investigate a variety of questions in human biology, including studies of sexual dimorphism, growth, aging, physiologic adaptation and others. Such traits have also been applied to questions concerning population structure and history, key concerns of anthropological genetics.

Why study quantitative traits?

Before getting into some of the approaches used with quantitative traits, it is worthwhile examining a more fundamental question: why bother? When William Boyd (1950) wrote his classic book, *Genetics and the Races of Man*, he argued that the accumulating data on genetic markers (primarily blood groups at that time) were superior to physical traits, such as anthropometrics, because the former did not change during an individual's life and provided a closer window on underlying genetic differences. Over the next several decades, additional data became available due to developments of electrophoretic techniques that allowed direct observations of the genotypes of many red blood cell proteins and enzymes. Anthropological investigations of these, and other, genetic markers provided a wealth of information on human population structure and history (Crawford, 1973). The utility of such markers for examination of anthropological questions of populational relationships in terms of geography and history is clear from Cavalli-Sforza *et al.*'s (1994) *magnum opus*.

Yet even as these genetic markers supplanted earlier phenotypic measures, they too were later supplanted to some extent by other, more direct, measures of underlying genetic variation. Today, the revolution in molecular biology has provided an even better window on genetic diversity, right down to the level of direct comparisons of DNA sequences. Every month brings forth the results of new DNA markers, and an increasing sophistication in assessing population structure and history. As such, the red and white blood cell polymorphisms that were considered state of the art a short while ago are now considered 'classical' genetic markers. Given this shift, a focus on quantitative traits must seem positively archaic. As someone who has devoted much of his career to using quantitative traits in research on anthropological genetics, I wonder at times whether my work is a matter of being in the right place at the wrong time. However, instead of characterizing quantitative traits as an interesting approach worth a short note in the intellectual history of anthropological genetics and then moving on, I suggest that there are still valuable uses for such traits in our investigations of population structure and history. Although classical marker and newer DNA-based measures are indeed better suited to many questions in anthropological genetics, this does *not* mean that quantitative traits are worthless. Far from it.

Obviously, one reason to study quantitative traits is to find out how population structure and history have affected such traits. The primary emphasis here is on the relative effect that genetic and environmental forces had on quantitative trait variation. Such an emphasis ties in closely with long-standing debates over 'nature' and 'nurture'. For example, when one examines population differences in a quantitative trait, such as cranial shape, the question arises as to whether these populational differences reflect underlying genetic differences or phenotypic plasticity. Additionally, if populational comparisons reflect genetic differences, then we are interested in whether such differences reflect neutral variation (a balance between gene flow and genetic drift), or the action of natural selection.

Another, more pragmatic, reason for studying quantitative traits is that in many cases they are the best (and often the only) source of information on human populations. Even though more and more genetic markers are continually being discovered, in many cases we have limited geographic and cultural coverage from contemporary human populations. Data on quantitative traits have been collected for a large number of populations. Even though classical genetic and DNA markers are better measures of underlying genetic variation, we still ought to consider what can be learned from these vast amounts of data on quantitative traits, rather than simply throw them out. As Chakraborty (1990: 149) notes, 'It is therefore only appropriate to ask how best to utilize quantitative traits and to determine their limitations and not to ignore them altogether'.

Another reason to study quantitative traits is that they often provide us with data for historic and prehistoric populations that we might otherwise lack. Part of the problem with classical genetic markers and DNA markers is that they have only been developed in recent times, and therefore were not measured directly in generations past, a time when data on quantitative traits (such as anthropometrics) were routinely collected. Even the oldest known classical genetic marker, the ABO blood group, has been known for only a bit over a century. Most of the other genetic markers have been known for considerably less time. The rapid demographic changes taking place in today's world, including rapid population growth and migration, can lead to rapid changes in the pattern of genetic variation among human populations. As I have noted elsewhere, 'The ability to peer back in time even a few decades offers us an opportunity to look at a world that no longer exists' (Relethford, 2003a: 33).

The problem of available data is even more critical as we attempt to assess variation in prehistoric populations, where skeletal and dental evidence are our primary source of information. To some extent, direct measures of genetic variation can now be obtained from some prehistoric data using ancient DNA analysis. However, ancient DNA analysis is still limited in many cases and tends to focus primarily on mitochondrial DNA (O'Rourke, 2000). This is not to downplay the obvious importance of ancient DNA analysis, but simply to note that such data may never be as abundant as skeletal or dental data. It seems more reasonable to use *both* data on ancient DNA and skeletal/dental data to examine questions of prehistoric variation.

Background on quantitative genetics

Before examining how quantitative traits can be used in anthropological studies of population structure and history, it is useful to examine some of the more basic principles of quantitative genetics, starting with the simplest underlying model known as the equal and additive effects model. This model is central to much of quantitative genetics and it answers a basic question concerning the relationship of Mendelian genetics to quantitative traits — how can discrete units of inheritance (alleles) result in a continuous trait?

The equal and additive effects model

The starting point here is a simple model of genetic variation where there is a single locus with two codominant alleles, A and a, which have different effects on a quantitative trait, such as height, where the A allele adds one unit to the trait, and the a allele adds nothing. According to this simple model, individuals with genotype AA will have two units of height added by this gene (because they have two A alleles, each adding $+1$). Individuals with genotype Aa will have one unit of height ($A = 1$, $a = 0$, such that $Aa = 1 + 0 = 1$). Finally, individuals with genotype aa have zero units of height added by this gene. Assuming no environmental effects, the phenotypes will be $AA = 2$, $Aa = 1$, and $aa = 0$ (Konigsberg, 2000).

Now, consider another unlinked locus that also has two alleles, which we call B and b, and assume that the effect of this locus is the same as for the A locus; that is, the B allele adds one unit of height and the b allele adds zero units of height. If we then consider all the possible genotypes for both loci, we get the list of genotypes and their respective phenotypes shown in the following table.

Genotype	Phenotype
AABB	4
AABb	3
AAbb	2
AaBB	3
AaBb	2
Aabb	1
aaBB	2
aaBb	1
aabb	0

For the two-locus model above there are nine genotypes and five different phenotypes. This simple model can be extended to more than two loci. For example, imagine three loci, each with two alleles, that affect a quantitative trait (A and a, B and b, and C and c). There will be 27 different genotypes (e.g. *AABBCC*, *AABBCc*, and so on) and seven different phenotypes, ranging from a value of 0 for genotype *aabbcc* to a value of 6 for genotype *AABBCC*. For a four-locus model, there are 81 genotypes and nine phenotypes, and for a five-locus model there are 243 genotypes and 11 phenotypes. In general, if there are L loci each with two codominant alleles, there will be 3^L genotypes and $2L+1$ phenotypes.

The basic principle of the equal and additive effects model is that the phenotype reflects the net effect of polygenic inheritance, with each locus having an *equal* effect on the genotype, and where these effects are *additive*. As the number of loci increases, the phenotypic distribution has more and more categories and more closely resembles a continuous distribution. In addition, environmental (non-genetic) influences can further alter the phenotype, resulting in a true continuous distribution (Falconer, 1989). The result is that a quantitative trait can take on a range of possible values and is not limited to a finite number of discrete classes. Consider, for example, the length of the human skull. Some people may have a head length of 190 mm, others may have a head length of 191 mm, and still others may have a head length of 190.5 mm. The underlying variation is continuous, and our ability to measure this variation is limited only by the precision of our measuring instruments.

A continuing concern is the extent to which the equal and additive effects model is appropriate for the type of traits of interest to anthropologists. Genetic epidemiologic studies have found some traits that are best described by a major gene model, where there are different and non-additive effects of loci. Konigsberg (2000) notes that further study is needed on traditional anthropological traits to determine if their patterns of inheritance deviate significantly from the equal and additive effects model and whether the models are robust to such deviations.

Statistical aspects of quantitative variation

For the above model, if we wanted to know the proportion of individuals with a given genotype or phenotype, we would need to know the frequencies of the different alleles. An interesting property of the phenotypic distribution of most quantitative traits is that they form the familiar 'bell-shaped' curve, or normal distribution. The normal distribution will always occur if the number of loci contributing to a quantitative trait is large (Konigsberg, 2000). As such, the basic principles of statistical inference for normally distributed variates apply to most quantitative traits.

Distributions of quantitative traits are described primarily in terms of their mean and variance. The mean is simply the average of all observations in the sample. Although the mean of a quantitative trait is a useful and necessary statistic, it is also clear that the mean by itself does not describe the variation present in a sample; not everyone in the sample will have the same value for a given quantitative trait. For example, the distribution of height in any population will show individuals somewhat shorter than average, and those that are somewhat taller than average. There will also be some individuals that are quite a bit shorter or taller than on average. The variance is the statistical measure of this variability; it is simply an average of the squared differences from the mean, and represents the area under the curve for the normal distribution. A related measure, the standard deviation, is simply the square root of the variance and measures variability along the horizontal axis of the curve. A trait that fits a normal distribution has certain well-established properties: roughly 68% of all cases lie within 1 standard deviation (plus or minus) of the mean, and roughly 95% of all cases lie within 2 standard deviations of the mean (Sokal and Rohlf, 1995).

Heritability

The observed variation of a quantitative trait is the net effect of genetic and environmental influences on variation. So far, we have only considered additive genetic effects, but a more comprehensive model also considers genetic variation due to dominance. Specifically, we can partition the total phenotypic variance (V_p) into three components: (1) The additive genetic variance (V_a) resulting from the effects of the equal and additive effects model; (2) The non-additive genetic variance resulting from dominance effects (V_d); and (3) Environmental (non-genetic) variance (V_e). This partitioning is expressed mathematically as

$$V_p = V_a + V_d + V_e$$

Heritability (h^2) is the measure of the relative proportion of total phenotypic variation that is due to genetic variation. There are two ways of doing this, depending on whether we are interested only in the additive genetic effects or the total genetic effects (additive plus dominance variation). Narrow-sense heritability is a measure of the proportion of the total phenotypic variance that is due to additive genetic variance, or

$$h^2 = \frac{V_a}{V_p}$$

while broad-sense heritability includes additive and dominance variation, as

$$h^2 = \frac{V_a + V_d}{V_p}$$

Although broad-sense heritability gives us an idea of the overall impact of genetic variation on phenotypic variation, dominance effects are not inherited, such that narrow-sense heritability actually gives us a better idea of what is actually transmitted from parent to offspring. Heritability can be estimated by comparing phenotypic traits across relatives, such as comparing parents and offspring, twins and other siblings, as well as more complex pedigrees.

As the concept of heritability is frequently misunderstood, it is worth noting briefly that it is a relative measure and not an absolute measure. The level of heritability is not fixed. If, for example, the amount of environmental variation in a population declines for one reason or the other, the relative amount of genetic variation increases by definition. In addition, heritability is a measure that applies to variation within a population, and does not apply to individuals. If, for example, we estimate that heritability of height is 0.8, this does *not* mean that 80% of any individual person's height is due to genetics and 20% to environment.

The above discussion of statistical and genetic aspects of quantitative genetics has been deliberately brief. Further details and elaboration of the concepts lies beyond the scope of this chapter, but can be found in a number of sources (e.g. Falconer, 1989; Konigsberg 2000).

How can we analyse quantitative traits in studies of population structure and history?

Given that quantitative traits are influenced in part by genetics, and given that genetic differences between populations form the basis of analysis of population structure and history, it seems reasonable to assume that quantitative traits can, in principle, be used to shed light on anthropological genetic questions. However, the nature of quantitative variation also makes it clear that specific methods are needed to analyse patterns of variation.

Indirect and direct application of population genetic models

As with any biological data, models of analysis can rely on indirect or direct application of population genetic models. Take for example the isolation by distance model, which predicts that the genetic distance between pairs of populations will increase as a function of the geographic distance between populations. This model is based on the limiting effect of geography on gene flow. The further apart geographically two organisms are, the less likely they will mate. Now, assume we have a set of populations for which we have estimates of both genetic and geographic distance between all pairs of populations. An *indirect* test of the isolation by distance model might simply involve a graph of genetic versus geographic distance, or more formally the computation of a measure of correlation between the two distance measures. In either case, the general prediction of the isolation by distance model is that genetic distance and geographic distance are related. On the other hand, a *direct* application of the isolation by distance model would involve a specific mathematical function relating genetic and geographic distance and one of the goals of such an analysis would be estimation of the parameters of the model. In an earlier review, my colleague Frank Lees and I referred to these two different approaches as 'model-free' and 'model-bound' (Relethford and Lees, 1982), but I now think this is a bit misleading, as indirect models still rely on some underlying model.

Indirect methods have a long history in anthropology, and include virtually any type of comparison of quantitative traits between populations. Such comparisons include traditional univariate and multivariate statistics, such as analysis of variance and discriminant analysis, as well as correlation of distance measures. These types of analyses, and some examples of each, have been reviewed elsewhere (Relethford and Lees, 1982) and will not be presented here. Instead, this section deals more specifically with the *direct* methods of analysis that have been developed for quantitative traits over the past 15 years. These methods are particularly important because they allow comparison with other genetic data in terms of underlying measures and parameters, and not simply indirect correspondence.

The R matrix

There are many different measures of genetic similarity that have been developed for allele and haplotype frequency data. One method in particular which has been frequently in the anthropological literature, and is easily extended to quantitative traits, is the R matrix method developed by Harpending and colleagues (Harpending and Jenkins, 1973, 1974; Workman *et al.*, 1973). When g is the number of populations being studied, R is a matrix with g rows and g columns, where each element in the matrix represents the genetic similarity between a pair of populations. For a given allele, the genetic similarity between population i (row) and population j (column) is defined as

$$r_{ij} = \frac{(p_i - \bar{p})(p_j - \bar{p})}{\bar{p}(1 - \bar{p})}$$

where p_i and p_j are the frequencies of the allele in populations i and j respectively, and $\bar{p}$ is the mean allele frequency over all populations in the analysis, ideally a weighted mean where weighting is by population size. The values of the R matrix are then averaged over all alleles. The R matrix has certain interesting properties. The average of all matrix elements (weighted by population size) is by definition equal to zero, a positive r_{ij} value indicates a pair of populations more similar to each other than on average, and a negative r_{ij} value indicates a pair of populations less similar to each other than on average. Most often, the diagonal elements of the matrix ($i = j$) are positive because genetic similarity between pairs of individuals is typically greater than between individuals in different populations.

The R matrix was originally developed for allele/haplotype frequency data, but it can be extended to quantitative traits. Under the equal and additive effects model the mean of a quantitative trait is proportional to the underlying mean allele frequency and the variance is proportional to heterozygosity (Falconer, 1989). Given data on means and estimates of heritability (or an estimate of average heritability), an R matrix can be estimated from quantitative traits. The basic method was developed by Williams-Blangero and Blangero (1989) and Relethford and Blangero (1990), and later refined by Relethford *et al.* (1997) (see also Rogers and Harpending, 1983).

F_{ST} and minimum F_{ST}

Once an R matrix has been obtained, there are several further computations that provide useful comparative measures of population structure and history. One common measure is F_{ST}, a measure of differentiation among populations that for neutral traits reflects a balance between gene flow, genetic drift and mutation. F_{ST} can be compared across studies. For example, suppose you analyse allele frequency data for two sets of populations and estimate an F_{ST} value of 0.05 for one set and a value of 0.01 for the second set. The first set of populations shows more differentiation between populations than the second, which is likely due to differences in population size (and, hence, genetic drift) and/or gene flow. Further demographic, historical, and cultural information could provide greater insight into the specific factors responsible for differences in the level of differentiation. F_{ST} can also be interpreted as a measure of variation among populations relative to the amount of variation expected under panmixia (where all individuals from all populations are equally likely to mate with each other). For example, an F_{ST} value of 0.10 shows that 10% of the total variation exists *among* populations, with the remainder ($100 - 10 = 90\%$) of the variation existing *within* populations.

F_{ST} can be estimated from the R matrix as the average diagonal of the matrix ($i = j$), which is the average amount of genetic similarity within populations. In order to estimate F_{ST} from quantitative traits we need estimates of heritability for the traits used in the analysis or at least an estimate of average heritability. A heritability estimate is needed because our analysis is based on phenotypic variation, which in turn reflects both genetic variation and environmental variation; in order to get an estimate of genetic differentiation, we need to adjust phenotypic variation by the heritability ratio. Ideally, we would have heritability estimates derived from the same set of populations that we are studying. In practice, this is rare and we more often have to use a general estimate of heritability based on other studies (e.g. values of h^2 are often around 0.5 for a number of measures of the head and face in a number of studies).

In some cases we can simply look at the phenotypic variation without a heritability estimate. If F_{ST} is computed in this way, where we are assuming that $h^2 = 1$, we are computing what is known as the 'minimum F_{ST}' (Williams-Blangero and Blangero, 1989). This estimate is a *minimum* possible value, because the smaller the heritability, the greater the actual value of F_{ST}. For an average heritability, F_{ST} and minimum F_{ST} ($MinF_{ST}$) are related as

$$F_{ST} = \frac{MinF_{ST}}{MinF_{ST} + h^2(1 - MinF_{ST})}$$

If, for example, you compute a minimum F_{ST} of 0.05, this means that the true value of F_{ST} cannot be less than 0.05, but it actually could be larger. If h^2 is actually 0.5, then the true F_{ST} would be 0.095. Minimum F_{ST} therefore has some limited comparative use by showing the lowest possible limit of the actual level of genetic differentiation. Care must be exercised in comparing minimum F_{ST} values across different studies and should be restricted to situations where a similar level of average heritability seems reasonable. Otherwise, comparison of minimum F_{ST} values could be misleading.

Genetic distance and minimum genetic distance

Although F_{ST} provides information on the *degree* of differentiation, it does not tell us anything about the *pattern* of variation. Most often, our interests in population structure and history require information on which populations are most similar to each other and whether there is any particular patterning, such as geographical or historical, to population relationships. Genetic distances (as discussed in other chapters) can be used to construct graphic 'maps' or in correlation analyses to illustrate patterns of relationship. The starting point in all cases is computation of a measure of genetic distance between all pairs of populations, which are then easily interpreted – the larger the genetic distance between two populations, the less similar they are genetically. For example, imagine we have genetic data for three populations (A, B, C) and compute the genetic distance between each pair of populations and find the following distances: A and B = 0.03,

A and C $= 0.06$, B and C $= 0.06$. These values show immediately that populations A and B are more similar to each other (with a lower genetic distance) than either is to population C.

There are many different types of genetic distance measures for allele and haplotype frequency data, many of which are strongly correlated. The approach taken for quantitative traits is to derive the genetic distances from the R matrix. Harpending and Jenkins (1973) showed that the genetic distance between populations i and j can be computed from the R matrix as

$$d_{ij}^2 = r_{ii} + r_{jj} - 2r_{ij}$$

As with F_{ST}, the same approach can be used with R matrices from either allele/haplotype frequencies or quantitative traits. When applied to quantitative traits, the exact magnitude of genetic distance requires an estimate of average heritability. If phenotypic variation is used without an adjustment for heritability, then the distances are 'minimum genetic distances' subject to the same caveats as minimum F_{ST} (Williams-Blangero and Blangero, 1989). However, even here the minimum genetic distances are proportional to the true genetic distances and can be used for producing distance maps, correlation analysis, and other methods of distance analysis.

The Relethford-Blangero method

Studies of genetic distance are not always easy to interpret. Consider, for example, a case where a given population is rather divergent from others in the analysis. Is this divergence due to genetic drift, gene flow with populations outside of the area of analysis, a difference in population history, or some other factor(s) making that population genetically more dissimilar? Correlations of genetic distance with geography, history and demography often provide clues for answering such questions. Such approaches rely on making inferences about population relationships from external data, such as geographic distance, population size, the history of the populations, or any other type of data that might shed light on the patterns of genetic distances that we observe.

Another approach, developed for allele/haplotype frequency data by Harpending and Ward (1982), provides additional information about patterns of genetic variation from the observed data. The method essentially compares two different measures of variation within populations. One way of assessing variation in a population is to compute the average level of heterozygosity within each population from the allele/haplotype frequencies. The average per-locus heterozygosity in population i is

$$H_i = 1 - \frac{\sum p_k^2}{l}$$

where p_k is the frequency of allele k in population i, l is the number of loci, and summation is over all loci and alleles. In general, the level of heterozygosity increases with mutation and gene flow, and decreases with genetic drift. The quantity defined here is the *observed* heterozygosity. Harpending and Ward (1982) showed that the *expected* level of heterozygosity in a population could be derived from the level of heterozygosity for the total population (allele frequencies from all populations pooled together), H_T, and the genetic distance, r_{ii}, of population i to the set of mean allele frequencies. This genetic distance is the diagonal element of the R matrix. Given these values, Harpending and Ward showed that the expected level of hetero-zygosity in population i is

$$E[H_i] = H_T(1 - r_{ii})$$

The *observed* and *expected* values of heterozygosity can be compared, which in turn can tell us something about the level of *external* gene flow into populations (that is, gene flow from outside the set of populations that we are analysing). Specifically, observed and expected heterozygosity will be the same if the level of external gene flow is the same across all populations in our analysis. If observed heterozygosity is greater than expected heterozygosity in a given population, then greater than average external gene flow is likely the cause of the excess heterozygosity. On the other hand, if observed heterozygosity is less than expected heterozygosity, then that population appears to have been more isolated and has received less external gene flow. The Harpending-Ward method thus provides us with yet another way of drawing inferences about population relationships, and has proven to be useful in a number of analyses based on allele frequencies.

This method has been discussed here because it has also been applied directly to the analysis of quantitative traits. Relethford and Blangero (1990) showed that because of the proportional rela-tionship between heterozygosity and phenotypic variation, that the Harpending-Ward method could be extended to quantitative traits as

$$E[V_i] = \frac{V_w(1 - r_{ii})}{1 - F_{ST}}$$

where V_i is the average phenotypic variance over all traits in population i (after conversion to standardized scores), V_w is this average phenotypic variance averaged over all groups, and r_{ii} and F_{ST} are as estimated from quantitative traits. Relethford and Blangero (1990) showed that their method successfully identified patterns of differential external gene flow in two different cases – dermato-glyphic traits from Nepal and anthropometric traits from Ireland. Further, they found that the method was very robust to different values of average heritability. Thus, even when only crude estimates of average heritability are available, the method still works well.

What have we learned from quantitative traits?

Although the analysis of quantitative traits has a long history in anthropology for addressing questions about population structure and history in an indirect manner, the methods developed over the past 15 years and described above have provided the ability to make direct estimation of model parameters which can be compared with genetic marker data. The methods outlined above, focusing on estimation of F_{ST}, estimation of genetic distances, and the Relethford-Blangero model for detecting differential gene flow, have now been applied to quantitative traits (primarily anthropometric and craniometric) from contemporary, historic, and prehistoric populations. In most cases, these methods have proven useful in addressing specific questions about population structure and history. Table 7.1 lists a sampling of these studies from recent years. Judging from this list, it appears that direct methods of incorporating quantitative traits have been particularly useful in studies of prehistoric populations. These studies can be consulted for additional detail on the extent to which quantitative traits have been useful in addressing specific questions about population structure and history, and will not be reviewed here. Instead, I shift the focus in the remainder of this chapter to more general questions regarding the nature of quantitative variation in human populations.

Does plasticity erase or obscure population history?

As originally noted by Boyd (1950) and others, a major advantage of genetic polymorphisms is that they provide a more direct assessment of underlying genetic variation, a point that is even more true for the many DNA markers used today. Quantitative traits such as anthropometrics and dental measures are by their very nature further removed from the underlying genetic code. Although we can make inferences about genotypic variation, what we actually observe is phenotypic variation that includes the joint effects of both genetic and non-genetic influences. Given many potential non-genetic influences, it has often been felt that quantitative traits therefore have very limited use in studies of genetic relationships within and between populations.

The view that non-genetic factors played a key role in structuring phenotypic variation came to greater popularity in the early twentieth century, replacing previous notions of 'racial' fixity. A key figure in this change was the anthropologist Franz Boas, whose work on environmental changes on cranial shape supported the notion of phenotypic plasticity. If the cranial measures used to define and support racial classifications were subject to environmental changes during an individual's lifetime, then they were clearly poor measures of population affinity. In his landmark study, Boas (1912) examined cranial measures of thousands of European immigrants and their children, comparing children born in Europe and those born in the

Table 7.1. Some examples of recent studies that have used direct application of population-genetic models to studies of quantitative variation in human populations.[1]

Population	Data	Reference
New World (Iroquois)	Anthropometrics	Langdon (1995)
New World (Algonquian speakers)	Anthropometrics	Jantz and Meadows (1995)
New World (comparative)	Anthropometrics	Ousley (1995)
New World (Sioux and Assiniboine)	Anthropometrics	Wescott and Jantz (1999)
New World (prehistoric Spanish Florida)	Odontometrics	Stojanowski (2004)
New World (prehistoric Mississippians)	Craniometrics	Steadman (2001)
New World (prehistoric Ohio)	Craniometrics	Tatarek and Sciulli (2000)
New World (comparative prehistoric samples)	Craniometrics	Powell and Neves (1999)
New World (comparative prehistoric samples)	Craniometrics	González-José et al. (2001)
New World (prehistoric Chile)	Craniometrics	Varela and Cocilovo (2002)
Easter Island (protohistoric/prehistoric)	Craniometrics	Stefan (1999)
Nepal	Anthropometrics	Williams-Blangero and Blangero (1989)
Nepal	Dermatoglyphics	Relethford and Blangero (1990)
India (east coast)	Anthropometrics Dermatoglyphics	Reddy and Chopra (1999)
India (southern Andhra Pradesh)	Anthropometrics Dermatoglyphics	Reddy et al. (2001)
Ireland (19th century)	Anthropometrics	Relethford and Blangero (1990); Relethford (1991)
Ireland (20th century)	Anthropometrics	Relethford (2003a); Relethford and Crawford (1995); Relethford et al. (1997)
Global analyses	Craniometrics	Relethford (1994, 2001, 2002, 2004a, 2004b); Relethford and Harpending (1994, 1995)

[1]In addition to studies on human populations, the quantitative-genetic methods described in this chapter have also been applied to craniometric data on gorilla subspecies (Leigh et al., 2003).

United States. This was one of the first uses of a migrant study to test the hypothesis of developmental plasticity, the idea that the phenotype will change during growth depending on environmental conditions. After comparing cranial measures in foreign-born and US-born children from seven different ethnic groups, Boas concluded that the environment did indeed have an impact. These results influenced the further development of American anthropology in the early twentieth century, coming at a time when genetic/racial fixity of quantitative traits was the dominant ideology (Harris, 1968).

Boas's study has recently attracted new attention following the work of two separate groups that reanalysed the original data with modern statistical methods, but came to rather different initial conclusions. Sparks and Jantz (2002, 2003) suggested that the Boas data provided very little evidence of developmental plasticity, whereas Gravlee *et al.* (2003a, 2003b) argued that Boas *did* get it right, and that there *was* evidence of plasticity. Closer examination of the results of these studies shows that there are some important methodological differences as well as some subtle differences in interpretation. My own reading of these studies suggests that craniometric traits *do* show some evidence of developmental plasticity, but that the magnitude of these changes is not sufficient to erase patterns of populational relationships. In the analysis by Gravlee *et al.* (2003b), for example, *both* the foreign-born and US-born analyses show a consistent pattern where two ethnic groups (Scottish, Sicilian) are distinct from other ethnic groups. As I have noted elsewhere,

> To some, the primary focus is on developmental plasticity, in which case the question is a test of the null hypothesis of no plasticity. This is what Gravlee *et al.* have demonstrated. But, the question of whether plasticity exists is different from the question of whether it is the major influence on cranial variation. In a test of developmental plasticity, genetic and/or environmental differences between ethnic groups may represent something to control for, in this case by comparing foreign-born and U.S.-born children separately by ethnic group. However, one person's noise may be another person's signal. If the primary interest is on underlying group differences, then plasticity is noise, and the question is not so much whether it represents a significant influence, but rather the extent to which it affects other analyses
>
> (Relethford, 2004a: 381).

The bottom line here is that although plasticity does exist for craniometric traits, it does not necessarily obscure underlying genetic differences between populations, suggesting continued potential for such traits in the study of human population structure and history.

Does natural selection erase or obscure population history?

When using genetic distances to make inferences regarding population structure and history, we must always be aware that genetic differences between populations could reflect natural selection rather than common ancestry. This caveat applies to any type of genetic data; use of a given gene or trait in a study of population history makes the assumption that the gene or trait is neutral, or if not, then that the key environmental factors do not vary appreciably between the populations being studied. This problem might seem particularly acute when dealing with quantitative traits, many of which show evidence of natural selection and adaptation to the environment, such as skin colour (Jablonski and Chaplin, 2000), cranial size and shape (Beals, 1972; Beals *et al.*, 1983, 1984), and body size and shape (Roberts, 1978; Ruff, 1994). A simple example of this type of problem is making an inference based on skin colour that the dark pigmentation shared by native populations in sub-Saharan Africa and New Guinea reflects common ancestry, when genetic evidence shows a closer relationship of New Guineans to Southeast Asians, who are phenotypically lighter. In this case, skin colour would be a poor choice of trait for analysis of population history.

Even quantitative traits that show evidence of having been affected by natural selection might still be useful in studies of population structure and history, provided that there are small environmental differences between the populations being studied. On a broader geographic scale, the assumption of environmental homogeneity will clearly be violated, such as any global analysis of human variation that includes populations from different latitudes and climates. In general, we might expect that the larger the geographic scale of an analysis (i.e. global or regional rather than local), the more profound this effect, and the less likely an analysis of quantitative traits will provide us with information on population structure and history.

The idea that natural selection will distort patterns of population structure and history on a global level is an assumption that can be tested. Recently, two studies have examined global patterns of craniometric variation and have found that they reflect *both* population affinities and natural selection, but that natural selection does not obscure the underlying patterns of population relationships (Relethford, 2004a; Roseman, 2004). In both cases, genetic distances were derived from a global database of craniometric traits originally collected by W. W. Howells (1989, 1996) and compared with distance measures reflecting natural selection due to climatic adaptation, based on the squared difference in temperature for the locations of each pair of populations. Under a model of natural selection due to heat loss, larger cranial size is expected to be selected for in colder climates because larger crania lose heat less rapidly than smaller crania, whereas smaller cranial size is expected to be selected for in warmer climates because smaller crania lose heat more rapidly. Thus, a model of climatic selection predicts greater craniometric distances between populations that live in populations with dissimilar average temperature.

One of the studies (Relethford, 2004a) compared craniometric distance with temperature distance and geographic distance, the latter used as a proxy for gene flow. Under a neutral model of gene flow, the craniometric distance between pairs of populations is expected to increase as the geographic distance between them increases because geographic distance mediates the effects of gene flow. Both geographic and temperature distances showed significant correlations with craniometric distance, suggesting that *both* gene flow and natural selection due to climatic adaptation has shaped global patterns of craniometric variation. However, the pattern of relationship between craniometric and geographic distance was scarcely changed after the craniometric traits were statistically adjusted for climatic differences, showing that although natural selection *is* a significant component of global craniometric variation, it does not erase or greatly obscure underlying patterns of geographic relationship, which themselves presumably tell us something about global population structure and history.

Roseman's (2004) study took a different and more direct approach to the question of neutral and selective models of craniometric variation. Instead of comparing craniometric distance with geographic distance, he used actual genetic distances, based on microsatellite DNA markers, to approximate the genetic distance expected between populations under a neutral model. Both population structure and history (as reflected by neutral DNA markers) and natural selection (as reflected by temperature distances) were significantly correlated with craniometric distance. Although global craniometric variation fit a neutral model to a large extent, the influence of natural selection appeared to be particularly strong for analyses that included a Siberian population which lives in a particularly cold climate. Roseman also found that some classes of craniometric traits (particularly measures of cranial breadth) showed stronger associations with temperature distance than others.

Do global patterns of genetics and craniometrics agree?

Roseman's (2004) study also contributes to a growing number of studies that have looked at the degree to which global analyses of genetic variation correspond with craniometric variation (see also González-José *et al.*, 2004). The availability of global data on craniometric traits (Howells, 1996), classical genetic polymorphisms (e.g. Roychoudhury and Nei, 1988), and DNA markers (e.g. Rosenberg *et al.*, 2002) has allowed broad comparisons with a large number of populations. Perhaps more significantly, the extension of population-genetic models to quantitative traits, as reviewed earlier in this chapter, has allowed *direct* comparison of a number of model parameters, which in turn has provided us with a better understanding of the relative influence of neutral influences versus natural selection on quantitative traits.

One approach of comparative studies has been the examination of how global genetic diversity is partitioned. Starting with the pioneering work of Lewontin (1972), a number of studies have examined our species' genetic diversity among and within major geographic regions (e.g. sub-Saharan Africa, Europe, East Asia, and so forth). Total species diversity is made up of two major components: variation *among* geographic regions and variation *within* geographic regions, with the latter further broken down into variation *among* local populations and variation *within* local populations. Studies of classical genetic markers and nuclear DNA markers are generally in close agreement, showing roughly 10% variation among geographic regions, 5% among local populations within geographic regions, and 85% within local populations. Thus, most neutral genetic loci show the overwhelming amount of variation occurs within local populations (Barbujani *et al.*, 1997).

Methods of estimating F_{ST} and related statistics from quantitative traits have allowed direct comparison of the results of neutral loci with quantitative traits. Global craniometric data shows very similar values; using an average heritability of 0.55 for craniometrics, the apportionment of diversity is roughly 13% among geographic regions, 6% among local populations within geographic regions, and 81% within local populations (Relethford, 2002). Although such estimates vary depending on the number of geographic regions used and the specific populations used within each, as well as the choice of average heritability, the general agreement with results from classical genetic marker and DNA marker studies suggests that the *overall* partitioning of craniometric variation is close to those for neutral loci. These results in turn support the idea that craniometric traits can provide useful inferences regarding global population structure history, at least when used in a multivariate analysis incorporating many traits simultaneously.

Not all quantitative traits show this pattern of agreement with genetic markers. Skin colour shows an opposite pattern, with considerably more variation *among* geographic regions (about 85%) than within local populations (9%) (Relethford, 2002). The atypical nature of skin colour variation makes sense given that skin colour appears very strongly related to natural selection. Given the wide range of latitude encompassed in global analysis, the major differences between populations are between those that are found in different geographic regions. It remains to be seen how other quantitative traits (e.g. odontometrics) compares with studies of neutral genetic marker loci. It is worth noting that a preliminary study of global dermatoglyphic variation showed regional F_{ST} values similar to those found from genetic studies (DiGangi and Jantz, 2004).

Another approach comparing genetic marker loci and quantitative traits is to examine the correspondence of each type of data with geographic distance in global samples. Studies of classical genetic markers and DNA markers have found a strong relationship between genetic distance and geographic distance across the world (Cavalli-Sforza *et al.*, 1994; Eller, 1999; Relethford, 2003b). It is not clear whether the global correlation between genetic distance and geographic distance reflects the limiting effect of geographic distance on gene flow between populations (the classic isolation by distance model), the genetic signature of global dispersion following an African origin of modern human genetic diversity, or both (Relethford, 2004b). Regardless of the underlying factors, correspondence between genetic distance and geographic distance surely represents neutral variation, and it is this pattern that allows comparison with quantitative traits. Relethford (2004b) examined measures of genetic similarity for global datasets for classical genetic markers, microsatellite DNA markers, and craniometrics, finding the same rate of distance decay in all three types of data. As with the studies of apportionment of genetic diversity, this close correspondence again suggests that multivariate patterns of craniometric variation mirror those of neutral genetic variation to a large extent.

Closing thoughts

The issues raised in this chapter all relate to a single question – are quantitative traits useful in studying population structure and history? The short answer is a definitive 'Yes'. Of course, the specific answer depends in large part on the nature of the samples, the specific type of quantitative trait being studied, and the number of variables incorporated in the analysis. Craniometric traits have been the most widely studied in direct application of population genetic models. These traits do show some influence of natural selection, at least at a global level, but the results of several avenues of research suggest that this selection does not necessarily erase or obscure underlying patterns of population history. Other traits, such as skin colour, show a much greater influence of natural selection, and are less useful in regional or global studies of population history.

The history of increasing sophistication of genetic markers has led many researchers away from the study of quantitative traits as indices of population history (excepting, of course, those whose primary interest is in quantitative variation). Both classical genetic markers and DNA markers *are* preferred in any study of population history, in part because they are not affected by the environment during a individual's life, and in part because many of the kinds of historical questions of ancestry and descent are best addressed using traits such as mitochondrial DNA and Y chromosome DNA, which do not recombine. Thus, in most cases a researcher starting a new study would attempt, where possible, to collect DNA samples and not bother with taking anthropometric or other quantitative measures.

This does not mean, however, that quantitative traits are without value. I have attempted to argue here that quantitative traits can be useful in studies of population history, and should not be ignored. There are cases where an analysis of quantitative variation might provide an additional window on patterns of population history, particularly where large datasets of such measures already exist (e.g. my studies in Ireland, and Jantz and colleagues' studies of Native Americans – see Table 7.1). Further, the success in using craniometric data in studies of population structure and history means that a wide amount of skeletal data from prehistoric populations can be used in population genetic analysis. Indeed, it is my feeling that the methods that John Blangero and I developed have had their greatest impact in biodistance studies of skeletal biologists (see Table 7.1). Although ancient DNA analysis is one way in which to assay prehistoric variation, it cannot be used for all samples, and even in a best case scenario provides essentially a single-locus perspective on the past. Analysis of craniometric traits provides another way to examine our past.

References

Barbujani, G., Magagni, A., Minch, E. and Cavalli-Sforza, L. L. (1997). An apportionment of human DNA diversity. *Proceedings of the National Academy of Sciences USA*, **94**, 4516–519.

Beals, K. L. (1972). Head form and climatic stress. *American Journal of Physical Anthropology*, **37**, 85–92.

Beals, K. L., Smith, C. L. and Dodd, S. M. (1983). Climate and the evolution of brachycephalization. *American Journal of Physical Anthropology*, **62**, 425–37.

Beals, K. L., Smith, C. L. and Dodd, S. M. (1984). Brain size, cranial morphology, climate, and time machines. *Current Anthropology*, **25**, 301–30.

Boas, F. (1912). *Changes in the Bodily Form of Descendants of Immigrants*. New York, NY: Columbia University Press.

Boyd, W. C. (1950). *Genetics and the Races of Man: An Introduction to Modern Physical Anthropology*, Boston, MA: Little, Brown and Company.

Cavalli-Sforza, L. L. and Bodmer, W. F. (1971). *The Genetics of Human Populations*. San Francisco, CA: W. H. Freeman.

Cavalli-Sforza, L. L., Menozzi, P. and Piazza, A. (1994). *The History and Geography of Human Genes*. Princeton, NJ: Princeton University Press.

Chakraborty, R. (1990). Quantitative traits in relation to population structure: Why and how are they used and what do they imply? *Human Biology*, **62**, 147–62.

Crawford, M. H. (1973). The use of genetic markers of the blood in the study of the evolution of human populations. In *Methods and Theories of Anthropological Genetics*, ed. M. H. Crawford and P. L. Workman. Albuquerque, NM: University of New Mexico Press, pp. 19–38.

DiGangi, E. A. and Jantz, R. L. (2004). Dermatoglyphic ridge counts compared to short tandem repeats as measures of population distance. *American Journal of Physical Anthropology* Supplement, **38**, 87–88 (abstract).

Eller, E. (1999). Population substructure and isolation by distance in three continental regions. *American Journal of Physical Anthropology*, **108**, 147–59.

Falconer, D. S. (1989). *Introduction to Quantitative Genetics*, 3rd edn. New York, NY: John Wiley & Sons.

González-José, R., Dahinten, S. L., Luis, M. A., Hernández, M. and Pucciarelli, H. M. (2001). Craniometric variation and the settlement of the Americas: Testing hypotheses by means of R-matrix and matrix correlation analyses. *American Journal of Physical Anthropology*, **116**, 154–65.

González-José, R., Van der Molen, S., González-Pérez, E. and Hernández M. (2004). Patterns of phenotypic covariation and correlation in modern humans as viewed from morphological integration. *American Journal of Physical Anthropology*, **123**, 69–77.

Gravlee, C. C., Bernard, H. R. and Leonard, W. R. (2003a). Boas's *Changes in bodily form*: the immigrant study, cranial plasticity, and Boas's physical anthropology. *American Anthropologist*, **105**, 326–32.

Gravlee, C.C., Bernard, H.R. and Leonard, W.R. (2003b). Heredity, environment, and cranial form: a reanalysis of Boas's immigrant data. *American Anthropologist*, **105**, 125–38.

Harpending, H. and Jenkins, T. (1973). Genetic distance among Southern African populations. In *Methods and Theories of Anthropological Genetics*, ed. M.H. Crawford and P.L. Workman. Albuquerque, NM: University of New Mexico Press, 177–99.

Harpending, H. and Jenkins, T. (1974). !Kung population structure. In *Genetic Distance*, ed. J.F. Crow and C. Denniston. New York, NY: Plenum Press, pp. 137–65.

Harpending, H. and Ward, R. (1982). Chemical systematics and human evolution. In *Biochemical Aspects of Evolutionary Biology*, ed. M. Nitecki. Chicago, IL: University of Chicago Press, pp. 213–56.

Harris, M. (1968). *The Rise of Anthropological Theory*. New York, NY: Harper and Row.

Howells, W.W. (1989). Skull Shapes and the Map: *Craniometric Analyses in the Dispersion of Modern Homo. Papers of the Peabody Museum No. 79*. Cambridge, MA: Peabody Museum, Harvard University.

Howells, W.W. (1996). Howells' craniometric data on the Internet. *American Journal of Physical Anthropology*, **101**, 441–2.

Jablonski, N.G. and Chaplin, G. (2000). The evolution of human skin coloration. *Journal of Human Evolution*, **39**, 57–106.

Jantz, R.L. and Meadows, L. (1995). Population structure of Algonquian speakers. *Human Biology*, **67**, 375–86.

Konigsberg, L.W. (2000). Quantitative variation and genetics. In *Human Biology: An Evolutionary and Biocultural Perspective*, ed. S. Stinson, B. Bogin, R. Huss-Ashmore and D. O'Rourke. New York, NY: Wiley-Liss, pp. 135–62.

Langdon, S.P. (1995). Biological relationships among the Iroquois. *Human Biology*, **67**, 355–74.

Leigh, S.R., Relethford, J.H., Park, P.B. and Konigsberg, L.W. (2003). Morphological differentiation of *Gorilla* subspecies. In *Gorilla Biology: A Multidisciplinary Perspective*, ed. A.B. Taylor and M.L. Goldsmith. Cambridge: Cambridge University Press, pp. 104–31.

Lewontin, R.C. (1972). The apportionment of human diversity. *Evolutionary Biology*, **6**, 381–98.

O'Rourke, D.H. (2000). Genetics, geography, and human variation. In *Human Biology: An Evolutionary and Biocultural Perspective*, ed. S. Stinson, B. Bogin, R. Huss-Ashmore and D. O'Rourke. New York, NY: Wiley-Liss, pp. 87–133.

Ousley, S.D. (1995). Relationships between Eskimos, Amerindians, and Aleuts: Old data, new perspectives. *Human Biology*, **67**, 427–58.

Powell, J.F. and Neves, W.A. (1999). Craniofacial morphology of the first Americans: Pattern and process in the peopling of the New World. *Yearbook of Physical Anthropology*, **42**, 153–88.

Reddy, B.M. and Chopra, V.P. (1999). Biological affinities between migrant and parental populations of fisherman on the east coast of India. *Human Biology*, **71**, 803–22.

Reddy, B.M., Pfeffer, A., Crawford, M.H. and Langstieh, B.T. (2001). Population substructure and patterns of quantitative variation among the Gollas of Southern Andhra Pradesh, India. *Human Biology*, **73**, 291–306.

Relethford, J.H. (1991). Genetic drift and anthropometric variation in Ireland. *Human Biology*, **63**, 155–65.

Relethford, J.H. (1994). Craniometric variation among modern human populations. *American Journal of Physical Anthropology*, **95**, 53–62.

Relethford, J.H. (2001). Global analysis of regional differences in craniometric diversity and population substructure. *Human Biology*, **73**, 629–36.

Relethford, J.H. (2002). Apportionment of global human genetic diversity based on craniometrics and skin color. *American Journal of Physical Anthropology*, **118**, 393–98.

Relethford, J.H. (2003a). Anthropometric data and population history. In *Human Biologists in the Archives: Demography, Health, Nutrition and Genetics in Historical Populations*, ed. D.A. Herring and A.C. Swedlund. Cambridge: Cambridge University Press, pp. 32–52.

Relethford, J.H. (2003b). *Reflections of Our Past: How Human History is Revealed in Our Genes*. Boulder, CO: Westview Press.

Relethford, J.H. (2004a). Boas and beyond: migration and craniometric variation. *American Journal of Human Biology*, **16**, 379–86.

Relethford, J.H. (2004b). Global patterns of isolation by distance based on genetic and morphological data. *Human Biology*, **76**, 499–513.

Relethford, J.H. and Blangero, J. (1990). Detection of differential gene flow from patterns of quantitative variation. *Human Biology*, **62**, 5–25.

Relethford, J.H. and Crawford, M.H. (1995). Anthropometric variation and the population history of Ireland. *American Journal of Physical Anthropology*, **96**, 25–38.

Relethford, J.H., Crawford, M.H. and Blangero, J. (1997). Genetic drift and gene flow in post-famine Ireland. *Human Biology*, **69**, 443–65.

Relethford, J.H. and Harpending, H.C. (1994). Craniometric variation, genetic theory, and modern human origins. *American Journal of Physical Anthropology*, **95**, 249–70.

Relethford, J.H. and Harpending, H.C. (1995). Ancient differences in population size can mimic a recent African origin of modern humans. *Current Anthropology*, **36**, 667–74.

Relethford, J.H. and Lees, F.C. (1982). The use of quantitative traits in the study of human population structure. *Yearbook of Physical Anthropology*, **25**, 113–32.

Roberts, D.F. (1978). *Climate and Human Variability*, 2nd edn. Menlo Park, CA: Cummings Publishing.

Rogers, A.R. and Harpending, H.C. (1983). Population structure and quantitative characters. *Genetics*, **105**, 985–1002.

Roseman, C.C. (2004). Detecting interregionally diversifying natural selection on modern human cranial form using matched molecular and morphometric data. *Proceedings of the National Academy of Sciences USA*, **101**, 12824–9.

Rosenberg, N.A., Pritchard, J.K., Weber, J.L., Cann, H.M., Kidd, K.K., Zhivotosky, L.A. and Feldman, M.W. (2002). Genetic structure of human populations. *Science*, **298**, 2381–5.

Roychoudhury, A.K. and Nei, M. (1988). *Human Polymorphic Genes: World Distribution*. New York, NY: Oxford University Press.

Ruff, C.B. (1994). Morphological adaptations to climate in modern and fossil hominids. *Yearbook of Physical Anthropology*, **37**, 65–107.

Sokal, R.R. and Rohlf, F.J. (1995). *Biometry: The Principles and Practice of Statistics in Biological Research*, 3rd edn. New York, NY: W.H. Freeman.

Sparks, C.S. and Jantz, R.L. (2002). A reassessment of human cranial plasticity: Boas revisited. *Proceedings of the National Academy of Sciences USA*, **99**, 14636–9.

Sparks, C. S. and Jantz, R. L. (2003). Changing times, changes faces: Franz Boas's immigrant study in modern perspective. *American Anthropologist*, **105**, 333–7.

Steadman, D. W. (2001). Mississippians in motion? A population genetic analysis of interregional gene flow in west-central Illinois. *American Journal of Physical Anthropology*, **114**, 61–73.

Stefan, V. H. (1999). Craniometric variation and homogeneity in prehistoric/protohistoric Rapa Nui (Easter Island) regional populations. *American Journal of Physical Anthropology*, **110**, 407–19.

Stojanowksi, C. M. (2004). Population history of native groups in pre- and postcontact Spanish Florida: Aggregation, gene flow, and genetic drift on the Southeastern U.S. Atlantic coast. *American Journal of Physical Anthropology*, **123**, 316–22.

Tatarek, N. E. and Sciulli, P. W. (2000). Comparison of population structure in Ohio's Late Archaic and Late Prehistoric periods. *American Journal of Physical Anthropology*, **112**, 363–76.

Varela, H. H. and Cocilovo, J. A. (2002). Genetic drift and gene flow in a prehistoric population of the Azapa Valley and coast, Chile. *American Journal of Physical Anthropology*, **118**, 259–67.

Wescott, D. J. and Jantz, R. L. (1999). Anthropometric variation among the Sioux and the Assiniboine. *Human Biology*, **71**, 847–58.

Williams-Blangero, S. and Blangero, J. (1989). Anthropometric variation and the genetic structure of the Jirels of Nepal. *Human Biology*, **61**, 1–12.

Workman, P. L., Harpending, H., Lalouel, J. M., Lynch, C., Niswander, J. D. and Singleton, R. (1973). Population studies on southwestern Indian tribes. VI. Papago population structure: a comparison of genetic and migration analyses. In *Genetic Structure of Populations*, ed. N. E. Morton. Honolulu, HI: University Press of Hawaii, pp. 166–94.

Chapter 8

Ancient DNA and its Application to the Reconstruction of Human Evolution and History

Dennis H. O'Rourke

Department of Anthropology, University of Utah, Salt Lake City, UT

Introduction

The use of ancient nucleic acids to infer population history and phylogeny is now entering its third decade, with the initial demonstration of the possibility and utility of the approach pioneered by Higuchi *et al.* (1984) on museum specimens of the extinct quagga, and by Pääbo (1985) on preserved soft tissue from Egyptian mummies. Now uniformly termed ancient DNA (aDNA) studies, the approach has exploded in the past decade to encompass studies of modern human origins, regional history and dynamics of prehistoric human populations, as well as phylogenetic studies of nonhuman organisms. A full review of this vast and rapidly growing literature is beyond the scope of this chapter, and interested readers are directed to several excellent and recent reviews of the field from a variety of disciplinary perspectives (e.g. Wayne *et al.*, 1999; O'Rourke *et al.*, 2000a; Hofreiter *et al.*, 2001a; Kaestle and Horsburgh, 2002, Pääbo *et al.*, 2004, Cipollaro *et al.*, 2005).

The study of contemporary patterns of human genetic variation has proven a powerful approach to inferring human population history and evolution, although such approaches are bound by assumptions of evolutionary rates in the markers under study, effective population sizes over time, rates of population movement, levels of admixture, etc. The use of aDNA analyses in conjunction with such modern genetic studies affords a temporal perspective on human genetic variation that is, to some degree, independent of model assumptions. As such, it provides an approach to more directly study the dynamics of migrating and colonizing populations in prehistory, and occasionally, as a test of assumptions made using other analytical approaches. To date, aDNA studies have focused on variation in the mitochondrial genome, in large measure due to the high copy number present in most cells, thereby increasing the likelihood of recovery of small target sequences from ancient degraded samples.

Ancient DNA analyses are not without their own limitations, however. Foremost among these is documenting the authenticity of typing ancient nucleic acids; providing convincing evidence that results are not the product of modern contaminants. Ancient DNA analyses, then, are highly dependent on the quality of analytical methods and procedures used, and these play a central role in documenting aDNA results.

The centrality of method

DNA may be recovered from a variety of organic materials, including bone, organ tissue, skin, teeth and hair (reviewed in O'Rourke *et al.*, 2000a; Kaestle and Horsburgh, 2002; Pääbo *et al.*, 2004). Different aDNA analysts prefer alternative sources for DNA extraction when they are available. Some prefer hair, especially if hair bulbs are present, on the grounds that it is easier to remove external con-taminants from hair, and thus minimize spurious results due to adhering modern molecules (Gilbert *et al.*, 2004a). Others prefer teeth since the encapsulated nature of the nucleic acid bearing tissue in teeth is also thought to minimize contamination. This may be true in unworn and intact teeth, but wear, cracks in the enamel or root may compromise contamination control in teeth (Gilbert *et al.*, 2005). However, both hair and teeth provide low DNA yields on extraction and, at least for teeth, replication is difficult or impossible without access to additional samples. The latter is sometimes not possible from rare archaeological materials. Alternatively, some analysts prefer bone as a DNA source. In well preserved skeletal series, small bone samples may be plentiful, and since only ~0.25 gm of dry bone [or less] is needed to extract adequate volumes of DNA for analysis, multiple extractions and replications are possible. However, bone is porous, and contamination with modern DNA is common, and difficult if not impossible to remove (Gilbert *et al.*, 2005; Hofreiter *et al.*, 2001a; Serre *et al.*, 2004). The most likely source of contamination is by individuals studying the skeletal material, archaeologists, skeletal biologists, and technicians working on the samples in the molecular lab. The most efficient method of minimizing contamination is to 'clean collect' the samples directly from their archaeological context as soon as they are exposed. Immediate collection and sequestration of a sample for later molecular analysis minimizes handling and exposure to contami-nants. This makes it more likely that any modern contaminants will be introduced in the lab, and thus, the range of contaminants is reduced, and more easily identified, since everyone with access to the lab should have a genetic profile on record. Contamination issues with museum specimens and other archaeologically recovered samples are discussed by Yang and Watt (2005) and Gilbert *et al.* (2005).

Irrespective of source material, aDNA is usually present in low copy number, uniformly fragmented to small molecular size (typically <500 bp), characterized by modified bases, and often associated with undetermined molecules (both organic and inorganic) that inhibit amplification of target sequences by the polymerase chain reaction (PCR). As a consequence of the condition of ancient DNA samples, any contamination with modern DNA molecules, even in very low concentration, will likely result in preferential amplification of the modern contaminant over the ancient target template. Thus, a variety of methods have been developed to minimize modern contaminants and to detect their presence when they occur. Many of these contamination control measures span extraction techniques as well as PCR protocols and have been reviewed by others (O'Rourke *et al.*, 2000a; Hofreiter *et al.*, 2001a; Kaestle and Horsburgh, 2002; Gilbert *et al.*, 2005).

Extraction

Standard quality control measures need not be exhaustively reviewed here as they have been discussed by many authors previously, and many such protocols are now becoming standard procedure in most experienced labs. A variety of extraction methods are available and selection of an extraction method often is determined by trial and error, determining which method yields best results given a sample set. Early methods relied on standard phenol/chloroform methods, although these methods are known to lose DNA during extraction, and the toxicity of phenol makes it less desirable as an extraction method if other methods prove efficacious. Guanidinium based extraction methods are often preferred due to their high affinity for nucleic acids. This results in higher DNA yields per unit extract and, unlike the phenol based methods, minimizes coextraction of molecules that may impede PCR. However, the high affinity of guanidinium based methods for DNA means that any modern DNA molecule present in the sample may well be preferentially extracted in higher titers as well. Moreover, the guanidinium as well as the glass milk used in the extraction process must be carefully cleaned from the extract since both are powerful inhibitors to the enzyme used in PCR. Simple spin-column filtration extraction methods may be adequate for relatively clean, well preserved samples. In general, the fewer steps involved in the extraction process the better, since each additional step is an opportunity for the introduction of contamination. Finally, use of multiple negative controls is essential during extraction so as to detect contamination when it occurs. While this is much easier said than done, routine use of negative, or blank, controls at all stages of aDNA analysis is essential. One or two such controls is typically inadequate since low concentration contaminants may only be apparent in one or two of several such controls. Standards vary, but we rarely perform extractions on more than six samples at a time (usually less), and never run less than four negative controls per extraction.

PCR preparation

While extreme care must be taken when extracting nucleic acids from source tissue, the most likely time for contamination to occur is during preparation of sample extracts for PCR amplification. Thus, extensive contamination preventive protocols are associated with PCR preparation of samples. PCR set-up should be conducted in a room separate from all other laboratory work, preferably one which has a positive pressured and HEPA filtered air supply (Yang and Watt, 2005). Importantly, the air filtration of this work space should be separate from other lab spaces to prevent contamination via airflow patterns between work areas. In addition, PCR preparation is often conducted within a workstation within this 'clean' room, where the workstation has its own positive pressured HEPA filtered airflow as well as a UV lighting system for cross-linking the interior of the workstation. All instruments and reagents possible should be autoclaved, cross-linked or otherwise sterilized to minimize the possibility of contamination. While none of these procedures can guarantee the absence of contamination, they do help increase the probability of a contaminant free amplification. Prior to introduction of any reagents or samples, the immediate work environment (e.g. PCR workstation) should be thoroughly cleaned with bleach. This seems to be more effective at removing known contaminants from surface areas than cross-linking (Yang and Watt, 2005), and suggests that the standard practice of soaking tissue samples (e.g. bone fragments) in bleach prior to processing and extraction may be an efficient method of removing surface contaminants from samples.

As more investigators have begun to work with ancient DNA samples, a variety of protocols have emerged to minimize or eliminate modern contaminants, or to remove coextracted molecules that inhibit the enzymatic reaction driving PCR amplification. Eshleman *et al.* (2001) suggest using DNase to remove modern DNA contaminants in reagents prepared for use in PCR. DNase degrades any DNA molecules present, but is inactivated by high temperature prior to the introduction of enzyme or aDNA sample during PCR preparation. Kiesslich *et al.* (2002) advocate the use of dialysis through nitrocellulose membranes for DNA purification, followed by ethanol or isopropanol precipitation (Hänni *et al.*, 1995). Pruvost and Geigl (2004) advocate the use of real-time quantitative PCR to assess aDNA authenticity and quantify initial target volumes and identify presence of contaminants. Others suggest the use of additional molecular compounds, e.g. PTB (Poinar *et al.*, 1998), in order to overcome the presence of enzymatic inhibitors in sample extracts, and to realize amplification from samples that have previously been recalcitrant to PCR. Most of these newer methods are not yet in wide usage, but many will undoubtedly prove useful in broad contexts, and the standard laboratory procedures employed in aDNA labs will continue to evolve and improve.

Authentication

The risk of contamination, and hence report of false positives, is sufficiently great that a number of methods have been proposed to authenticate aDNA results. Over a decade ago, Handt *et al.* (1994a) proposed six criteria for the evaluation of aDNA authenticity which have become the basic standard in the field. The six criteria recommended by Handt *et al.* were: (1) pre- and post-PCR activities should be spatially separate, perhaps even in different laboratories; (2) strict laboratory protocols should be adopted and rigorously followed to both prevent contamination with modern DNAs, and to monitor and detect contaminants when they occur; (3) the routine use of controls; (4) replicate experiments are required to confirm initial results; (5) the results should make phylogenetic sense; and (6) there should be an inverse relationship between fragment size and PCR efficiency. In general, these guidelines are closely followed by most experienced workers. However, with increasing experience with aDNA methods, and as methods have changed, some of these authenticity criteria have also been altered.

Today, not only do most workers spatially segregate pre- and post-PCR activities, but there is an increasing trend to do extractions, PCR prep, amplifications, and post-PCR work (e.g. electrophoresis, sequencing, etc.) in separate spaces. Further, some of these spaces (extraction, PCR prep) are now expected to be HEPA filtered and positively pressured, and to contain PCR workstations with additional levels of contamination control. This makes the space and laboratory requirements much more substantial (and expensive) than was true a decade ago. Routine use of negative controls is still considered important, indeed many researchers use more negative controls per experiment than was common only a few years ago. Few workers any longer use positive controls, as the latter are simply a source for cross-over contamination. Replication is still a valuable criterion, although its nature has changed over time (see below). Finally, it is true that there should be a negative correlation between fragment size and PCR efficiency, but even a cursory review of the literature reveals that it is a vastly underutilized criterion in many studies. Few authors who are able to amplify their aDNA target of interest report experiments with longer fragments to determine if this relationship holds in their samples.

A decade ago, much aDNA research, especially human aDNA research, focused on screening for discrete markers in mitochondrial DNA that defined haplotypes or haplogroups. The latter are simply groups of haplotypes that share a common (set of) marker(s).

In this environment, replication of results from independent extractions from a single sample were usually adequate to confirm an initial result, especially if the investigators' haplotype or haplogroup differed from that of the sample. In recent years, increasing attention has focused on generating sequence data from ancient samples, and the criterion for replication and authenticity has changed. One of the major changes was first introduced by Krings *et al.* (1997) in their initial genetic characterization of the Feldhofer Neandertal. It is known that cytosine deamination may result in sequence artefacts when PCR amplicons are directly sequenced (Höss *et al.*, 1996; Hansen *et al.*, 2001; Hofreiter *et al.*, 2001b). To accommodate this source of error, Krings *et al.* (1997) cloned PCR fragments from multiple independent extracts and sequenced multiple clones to identify the authentic ancient sequence, sequence artefacts, and cases of modern contamination. Increasingly, this strategy has become the standard practice for aDNA studies relying on sequence data. The importance of this approach can not be overstated. Gilbert *et al.* (2003) demonstrated that postmortem alterations to mtDNA sequences was not randomly distributed throughout the molecule. Rather, postmortem damage clustered in 'hotspots' that correlated with areas of high *in vivo* mutation rates, at least in HVR1. Moreover, higher levels of postmortem damage were observed in regions of HVR1 that have no structural function. This result suggests that future aDNA studies might profitably target not only the mtDNA hypervariable region, but also coding regions of the molecule (e.g. Adachi *et al.*, 2004).

As was the case with extraction and PCR protocols, protocols for sequencing ancient templates have also been proposed to minimize contamination, or identify it when it occurs. Following earlier suggestions of Longo *et al.* (1990) and Hofreiter *et al.* (2001b), Pruvost *et al.* (2005) suggest using dUTP and uracil-N-glycosylase, quantitative real-time PCR, and an *E. coli* strain deficient in UNG and dUTPase for cloning the uracil containing PCR product to reduce sequence artefacts due to deamination. Treating ancient DNA extracts with *E. coli* UNG acts to eliminate uracil containing sequences that derive from the deamination of cytocine (Pääbo, 1989; Gilbert *et al.*, 2003). Alternative enzymatic treatments that act to reduce the number of postmortem damaged templates in ancient DNA extracts are available, e.g. EndoIV (Pääbo, 1989); Pol 1 in conjunction with T4 ligase (Pusch *et al.*, 1998; Di Bernardo *et al.*, 2002), and alkyl-adenine glycosylase (Lau *et al.*, 2000). While reducing the number of templates exhibiting postmortem damaged bases, such treatment also reduces the number of starting templates for PCR in an extract, possibly eliminating some samples from further analysis (Gilbert *et al.*, 2003). The protocol presented by Pruvost *et al.* (2005) is also designed to dramatically reduce carry-over contamination from earlier PCR products or from plasmid DNA. Although not yet in wide use (and the use of quantitative, real-time PCR will be a hindrance in many anthropological genetics labs), innovations such as these are likely to become a standard component of the array of protocols that increasingly target contamination control as central to the methodology.

Another authenticity criterion that has been proposed is that replication of initial results, especially sequence results, take place in a separate, independent laboratory by a second set of laboratory personnel (Cooper and Poinar, 2000). In many cases this may be appropriate, but in others perhaps unnecessary. Independent replication is maximally useful if modern contaminants are introduced by laboratory workers, and the genetic profile of experimenters in the different labs are distinctively different. Moreover, independent replication is most likely to be useful if the replicating laboratory is permitted to do their own independent DNA extraction. Sharing extracts to simply replicate sequence data is not as powerful a technique for contamination control and detection. Replication of discrete marker data may be accomplished without use of a separate lab, if the target markers are clearly distinct from laboratory workers (i.e. lab technicians are characterized by haplotypes or haplogroups that are not observed in the study population). In such instances, contamination detection is straightforward. This is often not the case with DNA sequence data due to template jumping during amplification, sequence artefacts due to template modification, etc. In such cases, not only are multiple sequences from cloned PCR products essential, but independent replication may also be required, especially when the investigators may be characterized by sequences common to the ancient sample.

Fortunately, with the tremendous growth of aDNA studies, the number and diversity of laboratory protocols aimed at increasing success rates in generating ancient nucleic acid samples for study while simultaneously minimizing undetected contaminants and the reporting of false positives has also burgeoned. As these are fully tested in the scientific market-place, a new and continuously improving set of methods will define the field. Despite the state of constant flux in methodology, numerous examples of aDNA studies that have contributed to our understanding of human evolution and population history exist.

Ancient DNA studies and human evolution — whither the Neandertals?

Few controversies in the study of human origins have remained so current for so long as the debate regarding the place of Neandertals in human evolution. In 1997 Krings *et al.* addressed the debate using aDNA data by demonstrating that the mitochondrial control region sequence of the Neandertal type specimen from Feldhofer, Germany was not present in a large series of modern human sequences. Comparing the Feldhofer Neandertal mtDNA sequence to 994 modern mitochondrial lineages, Kringsa *et al.* (1997) found an average of 27.2 substitutions between the modern and Neandertal lineages (range = 22–36). In comparison, the modern human sequences exhibited an average difference of only eight nucleotides. The maximal difference between any two modern human sequences was 24 substitutions, only two more than the minimal difference between the Feldhofer Neandertal and the large modern mtDNA series. For comparison, the authors also examined 16 chimp mtDNA lineages and found that the mean difference between the chimp sequences and modern humans was 55 nucleotide positions. Thus, while the sequence divergence between the Neandertal sequence and a large series of modern human sequences was triple that found among modern populations, the difference was only one-half that observed between modern human and chimpanzee sequences.

Perhaps of equal interest was the fact that the mean number of nucleotide differences was not correlated with geography. If European Neandertals were ancestral to modern European populations, it would be predicted that they would share more genetic material in common than Neandertals would with modern populations of other regions, for whom they would be unlikely direct ancestors. In fact, the mean nucleotide difference between the Feldhofer sequence and modern Europeans was 28.2 ± 1.9, while the mean differences between the Neanderal sequences and modern mtDNA lineages in other regions was comparable (27.1 ± 2.2 substitutions with African sequences, 27.7 ± 2.1 with Asian lineages, 27.4 ± 1.8 with American Indian sequences, and 28.3 ± 3.7 with mtDNA sequences observed in Australia and Oceania). The authors concluded that such a dramatic hypervariable mtDNA sequence difference between Neandertal and modern human mitochondrial lineages strongly indicated that Neandertals were not ancestral to modern humans.

Nordborg (1998) was the first to demonstrate that despite the substantial difference in mtDNA sequences between the Feldhofer Neandertal and modern humans, some contribution from Neandertals to the modern human gene pool could not be statistically rejected. Using a coalescent approach, he demonstrated that the mtDNA data on the Feldhofer Neandertal and modern humans was inconsistent with random mating of Neandertals and early anatomically modern humans, but did not permit rejection of other interbreeding scenarios. Nordborg (1998) notes the difficulty of obtaining adequate samples, especially ancient samples, to obtain sufficient power to reject intermediate models of interbreeding under different demographic models. Wall (2000) suggested that sufficient statistical power would be obtained if something on the order of 200 independent nuclear loci were typed in modern populations, but, of course, this approach is not possible for ancient samples given current technology. A number of authors (e.g. Hawkes *et al.*, 2001; Relethford, 2001; Gutiérrez *et al.*, 2002) also noted the lack of power to unequivocally reject a Neandertal contribution to the modern gene pool based on the single Feldhofer mtDNA data.

Additional relevant data soon appeared, however. Krings *et al.* (1999) subsequently obtained sequence data from the second mtDNA hypervariable region (HVR2) of the Feldhofer specimen. Combining these new data with the sequence data generated earlier on HVR1 resulted in a mean substitution difference between the Neandertal and modern humans of over 35, while the mean difference among modern humans was just under 11 (Krings *et al.*, 1999), concordant with the earlier results. Moreover, these authors estimate the time to the most recent common ancestor (TMRCA) between Neandertals and modern humans based on mtDNA sequence data to be over 400,000 years ago (range = 317,000 − 741,000) long prior to the appearance of anatomically modern humans in Europe, thus lending little support to the argument of Neandertal ancestry for modern humans.

Ovchinnikov *et al.* (2000) obtained mtDNA sequence data from a young Neandertal specimen recovered from the Mezmaiskaya Cave site in the northern Caucasus. The HVR1 sequence from this Neandertal individual differed from the Feldhofer HVR1 sequence by 12 substitutions and from the Cambridge reference sequence (CRS, Anderson *et al.*, 1981) by 22 differences, and a mean difference from 300 randomly selected modern sequences of >23 pairwise differences, depending on geographic origin of the modern samples (Ovchinnikov *et al.*, 2000). These authors also estimated the divergence of modern human and Neandertal mtDNA lineages to between 365,000 and 853,000 years ago, the time to most recent common ancestor of the two Neandertal samples to be ~150,000−350,000 BP, and the TMRCA of the modern human sequences to between 106,000−246,000 years ago. Like the estimates of Krings *et al.* (1999), these coalescent dates would seem to preclude Neandertal ancestry in the modern human mtDNA gene pool.

Although these early reports were consistent in demonstrating substantial sequence divergence between Neandertal samples and modern humans (but not between the temporally and spatially divergent Neandertal specimens), it was not clear when the sequence divergence originated. For example, early anatomically modern humans might also be genetically divergent from modern mtDNA sequences, and thus force a reconsideration of the relationship between archaic and modern human mtDNA sequences. Caramelli *et al.* (2003) sequenced the mtDNA HVR1 of two early anatomically modern forms from Italy dating to 23,000−25,000 BP. One sequence was identical to the CRS while the other differed from the CRS by a single substitution. This result was the first to demonstrate that the modern HVR1 sequence was present early in the expansion of modern humans, and diminished the time frame possible if the Neandertal and modern human HVR1 sequence diversity arose after their contact.

Serre *et al.* (2004) provide mtDNA data on four additional Neandertal specimens (Vindija 77 & 80 from Croatia, Engis 2 from Belgium, and La Chapelle-aux-Saints from France), as well as five additional early anatomically modern humans (Mladeč 2 & 25c from the Czech Republic, Cro-Magnon, Abri Pataud, and La Madeleine from France). Concerned that modern contamination might bias results, and that Neandertal sequences similar to modern human ones might be discarded erroneously as modern contaminants, the authors relied on a novel analytical approach. Taking advantage of the earlier work of Krings *et al.* (2000) and Schmitz *et al.* (2002), these workers employed PCR primers that were specific for Neandertal mtDNA sequences, and which failed to amplify modern human DNA. Similarly, more general primers were constructed that amplified both Neandertal and modern human DNA, as well as mtDNA from the African apes (Serre *et al.*, 2004). Thus, rather than generating sequence data from the expanded fossil series, the experimental design was to test whether Neandertal specific primers would amplify DNA extracts from early anatomically modern samples as well as from Neandertals. The result was that all of the Neandertal samples tested amplified with the Neandertal specific primers, but none of the early anatomically modern samples did. Similarly, both Neandertal and early anatomically modern samples amplified with the generalized 'hominoid' primers. This result also was taken as evidence that Neandertals had contributed little, if any, to the modern human gene pool.

Most recently, Lalueza-Fox *et al.* (2005) reported a short (47 BP) fragment of HVR1 from a 43,000 year old Neandertal from the El Sidrón Cave in northern Spain that exhibited Neandertal specific substitutions. Comparing substitutions in this sample with the other available Neandertal sequences led to a TMRCA estimate of 245,5000 BP, although with a very wide standard error. Interestingly, Lalueza-Fox *et al.* (2005) estimated the female effective size of the Neandertal population at between 5,000 and 9,000, similar to the estimates for modern humans based on mtDNA sequence data. The authors infer that Neandertals and modern humans are characterized by similar demographic parameters, but that the Neandertal data are inconsistent with a recent bottleneck prior to their extinction (e.g. during the last glacial maximum in Europe ~130 Kya when populations were hypothesized to have declined and coalesced in refugia in southern Europe (Lalueza-Fox *et al.*, 2005)). They also suggest that the Neandertal mtDNA sequence data suggest a speciation event for the origin of Neandertals in Europe coincident with the appearance of the distinctive Neandertal morphology approximately 250,000 years ago (Lalueza-Fox *et al.*, 2005; Rightmire, 2001).

Given the importance attached to quality control and the problem of contamination identification in ancient DNA studies cited above, some comment on the methods used in the Neandertal aDNA work is warranted. When modern contaminants might be contributed from researchers who may share a genetic heritage with the samples under study, it is critically important to be extra vigilant in attempts to detect modern contamination. This was initially the case with Neandertal aDNA, which derived from European fossil specimens and was analysed by investigators of European ancestry. Several methodological innovations were employed to ensure the authenticity of the results. In virtually every study mentioned above, initial sequence results were replicated in an independent laboratory by other researchers. Moreover, PCR products to be sequenced were generally cloned and multiple sequences from each clone were obtained in order to identify consensus ancient target sequence from artefacts or contamination. These procedures have proven most useful in many aDNA analyses, especially those focusing on mtDNA sequence variation, and for many applications have become standard procedures.

Finally, the large sample sizes needed to resolve alternative Neandertal phylogenies are not yet possible. However, the accumulating evidence on Neandertal mtDNA sequence variation is increasingly consistent with substantial genetic divergence between Neandertals and early modern human populations, providing little support for a significant Neandertal contribution to the modern human gene pool. It should be emphasized, however, that all the aDNA analyses brought to bear on this question rely on mitochondrial DNA. As a single molecule that is uniparentally inherited, mtDNA can only provide one window to our understanding of early human evolution. Data from the Y-chromosome and genetic variation in the autosomes would be most helpful in discriminating between alternative demographic and evolutionary scenarios. Such data must come from modern population studies rather than ancient DNA analyses given current methods. However, one observation regarding Neandertal mtDNA does seem relevant in the context of recent population genetic studies. A number of researchers (e.g. Batista dos Santos *et al.*, 1999; Bosch *et al.*, 2003) have noted that in instances where we know the historical context of expanding, colonizing populations, the uniform pattern of genetic variation observed is a preponderance of mtDNA lineages deriving from the 'recipient' population while a majority of Y lineages derive from the 'migrant' or 'colonizing' population. Neandertals existed in Europe for tens of millennia before the appearance of anatomically modern humans. If the latter migrated into the region, and admixture between them and Neandertals occurred, the Neandertals would be considered the 'recipient' population, while the more recent, anatomically modern human population the 'migrant' or 'colonizing' population. Yet it is the Neandertal mtDNA lineages that are absent from modern populations – exactly the opposite pattern expected based on more recent cases of migration and admixture. Thus, if there was substantial admixture between Neandertals and anatomically modern humans in Europe, it must have been of a demographic form very different than recent evidence and experience would indicate.

Archaeological and temporal context

Although a significant contribution to human evolutionary studies, the analysis of Neandertal and early anatomically modern aDNA constitutes only a small portion of the field. A larger effort, conducted by a series of scholars, has been to use aDNA to study regional population histories. In this approach, multiple samples that may be viewed as prehistoric 'populations' are analysed for either discrete marker data or sequence variation in order to help clarify prehistoric population dynamics or ancestral/descendant relationships. A number of such anthropological genetic aDNA studies from around the world have been summarized in recent reviews (e.g. O'Rourke *et al.*, 2000b; Kaestle and Horsburgh, 2002) and need not be recounted here. It will suffice to say that aDNA analyses of archaeologically recovered skeletal samples have been used to track prehistoric population movements, colonization events, and clarify ancestral/descendant relationships between temporally distinct populations in Europe, Africa, Asia, Oceania and the Americas.

It is useful to recount some issues in aDNA analyses relating to prehistoric populations that bear on historical inference. The first issue of concern is the definition of population. By the very nature of sampling in the prehistoric record, it is unlikely that prehistoric samples used in aDNA analyses actually represent a 'population' in the traditional sense of that term. That is, most ancient samples cover a time span of several hundred, perhaps thousands of years, and therefore the individuals comprising the sample are clearly not members of a breeding population. As such, use of standard population genetic analytical approaches that assume samples derive from a random mating population should be used with caution. We have suggested elsewhere (O'Rourke, *et al.*, 2005) that archaeologically recovered human samples might be considered members of a continuous local or regional population if they: (1) derive from a uniform and continuous cultural context, (2) are recovered from a small, proscribed geographic area, and (3) the timeframe spanned by the samples is relatively short. While such criteria are hopefully useful guidelines, they are short on specificity. Uniform cultural context may be clear and easily documented archaeologically, but how small a geographic area or time span is sufficient to treat individual prehistoric samples as members of a temporally continuous population? The answer to this question, of course, is highly dependent on the nature of the question being addressed, and the nature of the available samples. However, an example regarding treatment of prehistoric samples as populations may illustrate several issues in this regard.

In an effort at testing Hrdlicka's (1945) proposal of a population replacement in the Aleutian Islands approximately 1,000 years ago, we examined mtDNA lineage frequencies in Aleutian skeletal material archaeologically recovered from three sites (Kagamil,

Chaluka, and Shiprock) in the eastern Aleutian Islands (Hayes, 2002). From a larger series of samples, 36 were initially selected for aDNA analysis. Amplification success rates from these samples varied by primer set between 81% and 94%. Variable success rates for individual samples across primer sets is not unexpected in ancient samples. Thirty of the samples could unambiguously be classified into one of the Native American founding haplogroups. In fact, all the samples proved to be either haplogroup A or D, with the frequency of the latter predominating. Since the material culture across the Aleutian chain has generally been considered to be continuous in time over the past 4,000 years (e.g. McCartney, 1984) the first criterion above is met for these samples. The samples were excavated from three adjacent islands in the eastern Aleutians that span a geographic area of less than 125 kilometres, fulfilling criterion (2) above. Archaeological dating for the sites suggested the time span of occupation was a few hundred years, perhaps a bit more than 1,000 years. Although a longer time frame than hoped, those were the data, and we assumed it sufficient for criterion (3). Hayes (2002) showed that the haplogroup frequencies across the sites, while somewhat divergent, were not statistically significantly different, nor were there significant differences between the 'early' and 'late' groups proposed by Hrdlicka (1945). Taken as a whole, the haplogroup frequencies characterizing the ancient Aleuts were virtually identical to those observed in contemporary Aleut communities (Rubicz et al., 2003), leading to the inference that the Aleutian genetic data spanning the last one to two millennia were consistent with a continuous population model rather than a replacement event.

However, the initial aDNA analyses were performed by taking a sample from each of the three archaeological sites without the benefit of knowing directly their temporal distribution. Subsequent to the initial sampling and aDNA haplogroup typing, all of the samples available were directly dated using AMS ^{14}C methodology (Coltrain et al., 2005) and the remaining samples screened for mtDNA haplogroup markers. Several unexpected results emerged from these continued analyses of the Aleutian samples. First, the temporal distribution proved to be nearly 3,500 years, over twice the temporal span initially assumed (the oldest samples were dated to 3,800 BP; Coltrain et al., 2006). Second, the initial samples characterized genetically proved not to be a random sample from throughout the temporal range, but clustered in the more recent half of the range (Hayes, 2002; Smith, 2006). Third, the temporal distribution of haplotypes was altered when all 80 samples were tested and placed in temporal order. That is, in the earlier analysis there was minimal and nonsignificant evidence for temporal change in haplogroup frequency; approximately 75% of the samples were haplogroup D and 25% haplogroup A throughout the time range. Expanding the sample size from 36 to 61 (i.e. 61/80 yielded haplogroup destinations), and having a direct date for each specimen demonstrated that while only haplogroups A and D were present in the full sample, as had been the case in the initial analyses, haplogroup A predominated in the earlier time periods, with the increase in haplogroup D not being observed until later in time (Smith, 2006). Curiously, if all the ancient samples are taken together, and considered a 'temporal' population, the preponderance of haplogroup D is still observed, and is concordant with the frequency observed in the modern Aleut population (Rubicz et al., 2003; Smith et al., 2006), which is consistent with a view of continuity of population in the Aleutians. However, the temporal change in haplogroup frequency in this small geographic area is equally consistent with an incursion of new people into the region, and a corresponding change in haplogroup frequencies (Smith, 2006). Indeed, the latter view, dependent on the molecular analysis and dating of the full skeletal series from this region is now more consistent with the fuller analysis of the modern genetic data, as well (Zlojutro et al., 2005).

There are several lessons to be learned here. First, dating each specimen analysed is essential in aDNA research if multiple samples are to be used to infer population history. A reliable temporal context for aDNA data should be considered equally important in such studies as the use of rigorous laboratory protocols to prevent and detect contamination (see discussion above). Incorrect assumptions about temporal context of samples are as likely to lead to spurious inferences as undetected contaminants. Second, sample size matters. Sample sizes of 30–40 are generally considered large in most aDNA studies, with only a few studies reporting larger sample sizes (e.g. Stone and Stoneking, 1998). Increasing the sample size from 36 to 61 in the Aleutian data changed the perception of temporal change and possible population history dramatically. This should come as no surprise. When sampling even a continuous population through time, sample sizes need to be larger to adequately capture not only the inherent variability in the population, but to document any temporal trends in frequencies in genetic characters. This is as important for sequence data as it is for simple haplogroup frequencies determined from discrete markers. This is not to denigrate studies of single, individual samples (e.g. the early work on Neandertals discussed above, or the Tyrolean Ice Man (Handt et al., 1994b)). Interesting and important questions in human evolution and population history can only be addressed using aDNA methods by studying the available relevant samples, and sometimes this means studying a single sample. However, if the aim is to assess population variability over time, and distinguish between alternative population histories, larger samples, if available, are much more powerful than smaller ones.

These observations are not new, and have been emphasized before (e.g. Nordborg, 1998; O'Rourke et al., 2000a; Kaestle and Horsburgh, 2002), but they do have some practical ramifications beyond simple sampling strategies. One of the most important is fiscal. In planning aDNA projects it is important to anticipate the need to obtain ^{14}C dates on all specimens, which can dramatically increase the cost of individual projects. Additionally, large sample sizes require more laboratory analyses, reagents, lab tech time, etc., adding to the cost of such projects. These realities are not meant as an agument against conducting aDNA research. Rather, they are meant as a guide to realistic planning and experimental design when such analyses can genuinely contribute to the testing of specific hypotheses.

Properly designed and executed aDNA studies can make significant contributions, indeed have made significant contributions, to long-standing problems in human evolution and population history. Such approaches provide a level of resolution to many questions and hypotheses that have proven difficult with other types of data with lower levels of resolution. Combining ancient DNA analyses, stable isotope determinations, and archaeological inference provides a powerful methodology for studying past human populations (Coltrain *et al.*, 2004; Hayes *et al.*, 2005).

Additional directions in aDNA research

Ancient DNA analyses have proven most helpful in several other areas of inquiry. Phylogenetic analyses of a number of species, including brown bear (Leonard *et al.*, 2000; Barnes *et al.*, 2002); bison (Nielsen-Marsh *et al.*, 2002); and penguins (Lambert *et al.*, 2002; Ritchie *et al.*, 2004), among several others (see review in Pääbo *et al.*, 2004). The brown bear data are particular interesting in that the geographic pattern of ancient mitochondrial variation did not reflect the geographic distribution in modern populations of bears. This result is in contrast to the results of O'Rourke *et al.* (2000b) who examined mtDNA haplogroup frequency variation in modern and ancient human populations of North America, and found that the well documented geographic structure in modern Native American mtDNAs was reflected in the ancient mtDNA data as well. This result was somewhat surprising since disruption of indigenous patterns of genetic variation would have been expected as a result of dramatic population declines at the time of European contact. This was not observed in the mtDNA haplogroup data, suggesting a rather longer term stability to geographic patterns in humans than expected. The disparity between the brown bear and human result may be attributed to not only the different evolutionary and demographic histories of bears and people, but also the different timescales over which genetic variation was examined in the two studies. Curiously, relatively little aDNA research has informed primate phylogenetic studies. Exceptions are the recent reports by Yoder and colleagues (Yoder *et al.*, 1999; Karanth *et al.*, 2005) on subfossil lemurs from

Madagascar, where aDNA analyses have helped clarify the phylogeny and origin of Malagasy primates.

In addition to phylogenetic studies, much may be learned about human behaviour from analysis of human prey species (e.g. Yang *et al.*, 2004, 2005b) or commensals (e.g. Matisoo-Smith *et al.*, 1997; Savolainen *et al.*, 2004) in archaeological faunal collections. Paxinos *et al.* (2002) used aDNA methods to document dramatic population declines in the Hawaiian goose (Nene) both at the time of colonization of the islands by early Polynesians, as well as earlier. Given the availability of large faunal collections from many archaeological sites, and the fact that contamination issues are less severe in working with nonhuman target sequences (given proper primer design such that human sequences will not be amplified), the study of archaeological faunal collections is an underutilized resource, and we may expect an increase in such studies in the future.

Finally, there is a growing literature in aDNA applications to the study of disease evolution. Like mtDNA, many infectious pathogens (i.e. viruses and bacteria) may be present in high copy number in individuals who were infected with the pathogen at time of death. Identification of the pathogen through recovery of fragments of its genome is therefore feasible. Examples of this approach abound in the literature, including study of the historic or prehistoric presence of tuberculosis (e.g. Spigelman and Lemma, 1993; Salo *et al.*, 1994; Arriaza *et al.*, 1995; Donoghue *et al.*, 1998; Zink *et al.*, 2003); plague (Drancourt *et al.*, 1998; Raoult *et al.*, 2000); Chagas Disease (Guhl *et al.*, 1999); and syphilis (Kolman *et al.*, 1999). Reid *et al.* (1999) recovered portions of the RNA genome of the virus that caused the Spanish flu pandemic of 1918 from a victim of that pandemic, expanding aDNA applications in this area to ancient RNA as well (see also Basler *et al.*, 2001). If it proves possible to recover RNA fragments from truly ancient samples, it will dramatically expand the possibilities for studying a wider array of prehistoric pathogens, and possibly permit more detailed study of the co-evolution of human hosts and our pathogens over time. However, there are reasons to remain cautious about such results, since replication of results in different laboratories has not been as readily forthcoming as has been the case with ancient human or animal DNA (Gilbert *et al.*, 2004b; Bouwman and Brown, 2005).

The ancient DNA literature is growing rapidly as more and more investigators realize the inherent power of molecular analysis of prehistoric material to address questions that have proven recalcitrant to traditional methods of inquiry. There is little evidence that the increase in use of molecular methods will diminish anytime soon. That makes it all the more important to rationally assess not only the power, but also the pitfalls, of such approaches. It also means that training and preparation in anthropological genetics needs to emphasize advanced training in molecular laboratory methods, in addition to genetics, in order to prepare practitioners for the rigours of conducting ancient DNA research. With proper background, training, and adherence to the emerging standards of laboratory protocols, the future looks bright for continued aDNA contributions in anthropological genetics.

References

Adachi, N., K. Umetsu, W. Takigawa and K. Sakaue (2004). Phylogenetic analysis of the human ancient mitochondrial DNA. *J. Archaeol. Sci.*, **31**, 1339–48.

Anderson, S., A. Bankier, B. Arrell, M. de Bruijn, A. Coulson, J. Drouin, I. Eperon, D. Nierlich, B. Roe, F. Sanger, P. Schreier, A. Smith, R. Staden and I. Young (1981). Sequence and organization of the human mitochondrial genome. *Nature*, **290**, 457–65.

Arriaza, B. T., W. Salo, A. C. Auferheide and T. A. Holcomb (1995). Pre-Columbian tuberculosis in Northern Chile – molecular and skeletal evidence. *Amer. J. Phys. Anthropol.*, **98**, 37–45.

Barnes, I., P. Matheus, B. Shapiro, D. Jensen and A. Cooper (2002). Dynamics of Pleistocene population extinctions in Beringian brown bears. *Science*, **295**, 2267–70.

Basler, C. F., A. H. Reid, J. K. Dybing, T. A. Janczewski, T. G. Fanning, H. Zheng, M. Salvatore, M. L. Perdue, D. E. Swayne, A. García-Sastre, P. Palese and J. K. Taubenberger (2001). Sequence of the 1918 pandemic influenza virus nonstructural gene (SN) segment and characterization of recombinant viruses bearing the 1918 NS genes. *PNAS*, **98**, 2746–51.

Batista dos Santos, S. E., J. D. Rodrigues, A. K. C. Ribeiro-Dos-Santos and M. A. Zago (1999). Differential contribution of indigenous men and women to the formation of an urban population in the Amazon Region as revealed by mtDNA and Y-DNA. *Amer. J. Phys. Anthrop.*, **109**, 175–80.

Bosch, E., F. Calafell, Z. H. Rosser, S. Norby, N. Lynnerup, M. E. Hurles and M. A. Jobling (2003). High level of male-biased Scandinavian admixture in Greenlandic Inuit shown by Y-chromosomal analysis. *Hum. Genet.*, **112**, 353–63.

Bouwman, A. S. and T. A. Brown (2005). The limits of biomolecular paleopathology: ancient DNA cannot be used to study venereal syphilis. *J. Archaeol. Sci.*, **32**, 703–13.

Caramelli, D., C. Lalueza-Fox, C. Vernesi, M. Lari, A. Casoli, F. Mallegni, B. Chiarelli, I. Dupanloup, J. Bertranpetit, G. Barbujani and G. Bertorelle (2003). Evidence for a genetic discontinuity between Neandertals and 24,000-year-old anatomically modern Europeans. *PNAS*, **100**, 6593–97.

Cipollaro, M., U. Galderisi and G. DiBernardo (2005). Ancient DNA as a multidisciplinary experience. *J. Cell Physiol.*, **202**, 315–22.

Coltrain, J. B., M. G. Hayes and D. H. O'Rourke (2004). Sealing, whaling and caribou: the skeletal isotope chemistry of Eastern Arctic foragers. *J. Arch. Sci.*, **31**, 39–57.

Coltrain, J. B., M. G. Hayes and D. H. O'Rourke (2006). A radiometric evaluation of Hrdlicka's Aleutian replacement hypothesis: population continuity and morphological change. *Current Anthropology*, in press.

Cooper, A. and H. Poinar (2000). Ancient DNA: do it right or not at all. *Science*, **289**, 1139.

Di Bernardo, G., S. Del Gaudio, M. Cammarota, U. Galderisi, A. Cascino and M. Cipollaro (2002). Enzymatic repair of selected cross-linked homo-duplex molecules enhances nuclear gene rescue from Pompeii and Herculaneum remains. *Nucleic Acids Res.*, **30**, e16.

Donoghue, H. D., M. Spigelman, J. Zias, A. M. Gernaey-Child and D. E. Minnikin (1998). *Mycobacterium tuberculosis* complex DNA in calcified pleura from remains 1400 years old. *Let. Appl. Microbiol.*, **27**, 265–9.

Drancourt, M., G. Aboudharam, M. Signoli, O. Dutour and D. Raoult (1998). Detection of 400-year old *Yersinia pestis* DNA in human dental pulp: an approach to the diagnosis of ancient septicemia. *Proc. Natl. Acad. Sci., USA*, **95**, 12637–40.

Eshleman, J. A. and D. G. Smith (2001). Use of DNase to eliminate contamination in ancient DNA analysis. *Electrophoresis*, **22**, 4316–19.

Gilbert, M. T. P., E. Willerslev, A. J. Hansen, I. Barnes, L. Rudbeck, N. Lynnerup and A. Cooper (2003). Distribution patterns of postmortem damage in human mitochondrial DNA. *Amer. J. Hum. Genet.*, **72**, 32–47.

Gilbert, M. T. P., A. S. Wilson, M. Bunce, A. J. Hansen, E. Willerslev, B. Shapiro, T. F. G. Higham, M. P. Richards, T. C. O'Connell, D. J. Tobin, R. C. Janaway and A. Cooper (2004a). Ancient mitochondrial DNA from hair. *Curr. Biol.*, **14**, R463–R464.

Gilbert, M. T. P., J. Cuccui, W. White, N. Lynnerup, R. W. Titball *et al.* (2004b). Absence of *Yersinia pestis*-specific DNA in human teeth from five European excavations of putative plague victims, *Microbiol. Sgm.*, **150**, 341–54.

Gilbert, M. T. P., L. Rudbeck, E. Willerslev, A. J. Hansen, C. Smith, K. E. H. Penkman, K. Prangenberg, C. M. Nielsen-Marsh, M. E. Jans, P. Arthur, N. Lynnerup, G. Turner-Walker, M. Biddle, B. Kjølbye-Biddle and M. J. Collins (2005). Biochemical and physical correlates of DNA contamination in archaeological human bones and teeth excavated at Matera, Italy. *J. Archaeol. Sci.*, **32**, 785–93.

Guhl, F., C. Jaramillo, G. A. Valleho, R. Yockteng, F. Cárdenas-Arroyo *et al.* (1999). Isolation of Trypanosoma cruzi DNA in 4,000-year-old mummified human tissue from Northern Chile. *Amer. J. Phys. Anthrop.*, **108**, 401–07.

Gutiérrez, G., D. Sánchez and A. Marín (2002). A reanalysis of the ancient mitochondrial DNA sequences recovered from Neandertal bones. *Mol. Biol. Evol.*, **19**, 1359–66.

Handt, O., M. Höss, M. Krings and S. Pääbo (1994a). Ancient DNA: methodological challenges. *Experientia*, **50**, 524–9.

Handt, O., M. Richards, M. Trommsdorff, C. Kilger, J. Simanainen *et al.* (1994b). Molecular genetic analyses of the Tyrolean ice man. *Science*, **264**, 1175–8.

Hänni, C., T. Brousseau, V. Laudet and D. Stehelin (1995). Isoproponaol precipitation removes PCR inhibitors from ancient bone extracts. *Nucleic Acid Res.*, **23**, 881–2.

Hansen, A., E. Willerslev, C. Wiuf, T. Mourier and P. Arctander (2001). Statistical evidence for miscoding lesions in ancient DNA templates. *Mol. Biol. Evol.*, **18**, 262–5.

Hawkes, J., K. Hunley, S. H. Lee and M. Wolpoff (2000). Population bottlenecks and Pleistocene human evolution. *Mol. Biol. Evol.*, **17**, 2–22.

Hayes, M. G. (2002). Paleogenetic Assessments of Human Migration and Population Replacement in North American Arctic Prehistory. Ph.D. Dissertation, University of Utah.

Hayes, M.G, J.B. Coltrain and D.H. O'Rourke (2006). Genetic signature of human population replacement coincident with paleoclimatic change in the North American Arctic. Submitted.

Higuchi, R., B. Bowman, M. Freiberger, O.A. Ryder and A.C. Wilson (1984). DNA sequences from the quagga, an extinct member of the horse family. *Nature*, **312**, 282–4.

Hofreiter, M., D. Serre, H.N. Poinar, M. Kuch and S. Pääbo (2001a). Ancient DNA. *Nature Rev. Genet.*, **2**, 353–9.

Hofreiter, M., V. Jaenicke, D. Serre, A. von Haeseler and S. Pääbo (2001b). DNA sequences from multiple amplifications reveal artifacts induced by cytosine deamination in ancient DNA. *Nucleic Acids Res.*, **29**, 4793–9.

Höss, M., P. Jaruga, T. Zastawny, M. Dizdaroglu and S. Pääbo (1996). DNA damage and DNA sequence retrieval from ancient tissue. *Nucleic Acids Res.*, **24**, 1304–7.

Hrdlicka, A. (1945). *The Aleutian and Commander Islands and Their Inhabitants*. The Wistar Institute of Anatomy and Biology, Philadelphia.

Kaestle, F.A. and K.A. Horsburgh (2002). Ancient DNA in anthropology: methods, applications and ethics. *Yrbk Phys. Anthrop*, **45**, 92–130.

Karanth, K.P., T. Delefosse, B. Rakotosamimanana, T.J. Parsons and A.D. Yoder (2005). Ancient DNA from giant extinct lemurs confirms single origin of Malagasy primates. *PNAS*, **102**, 5090–5.

Kiesslich, J., M. Radacher, F. Neuhuber, H.J. Meyer and K.W. Zeller (2002). On the use of nitrocellulose membranes for dialysis-mediated purification of ancient DNA from human bone and teeth extracts. *Ancient Biomol.*, **4**, 79–87.

Kolman, C.J., A. Centurion-Lara, S.A. Lukehart, D.A. Owsley and N. Tuross (1999). Identification of Treponema pallidum subspecies pallidum in a 200-year-old skeletal specimen. *J. Infect. Dis.*, **180**, 2060–3.

Krings, M., A. Stone, R.W. Schmitz, H Krainitzki, M. Stoneking and S. Pääbo (1997). Neandertal DNA sequences and the origin of modern humans. *Cell*, **90**, 19–30.

Krings, M., H. Geisert, R.W. Schmitz, H. Krainitzki and S. Pääbo (1999). DNA sequence of the mitochondrial hypervariable region II from the Neandertal type specimen. *PNAS*, **96**, 5581–5.

Krings, M., C. Capelli, F. Tschentscher, H. Geisert, S. Meyer *et al.* (2000). A view of Neandertal genetic diversity. *Nat. Genet.*, **26**, 144–6.

Lalueza-Fox, C., M.L. Sampietro, D. Caramelli, Y. Puder, M. Lari, F. Calafell, C. Martínez-Maza, M. Bastir, J. Fortea, M. de la Rasilla, J. Bertranpetit and A. Rosas (2005). Neandertal evolutionary genetics: mitochondrial DNA data from the Iberian peninsula. *Mol. Biol. Evol.*, **22**, 1077–81.

Lambert, D.M., P.A. Ritchie, C.D. Miller, B. Holland, A.J. Drummond and C. Baroni (2002) Rates of evolution in ancient DNA from Adelie penguins. *Science*, **295**, 2270–3.

Lau, A., M. Wyuatt, B. Glassner, L. Samson and T. Ellenberger (2000). Molecular basis for discrimination between normal and damaged bases by the human alkyladenine glycosylase, AAG. *PNAS*, **97**, 13575–8.

Leonard, J.A., R.K. Wayne and A. Cooper (2000). Population genetics of Ice Age brown bears. *Proc. Natl. Acad. Sci., USA*, **97**, 1651–4.

Longo, M.C., M.S. Berninger and J.L. Hartley (1990). Use of uracil DNA glycosylase to control carry-over contamination in polymerase chain reactions. *Gene*, **93**, 125–8.

Matisoo-Smith, E., R. M. Roberts, J. S. Allen, G. J. Irwin, D. Penny and D. M. Lambert (1997). Patterns of human colonization in Polynesia revealed by mitochondrial DNA from the Polynesian rat. *PNAS*, **95**, 15145–50.

McCartney, A. P. (1984). Prehistory of the Aleutian Region. In *Handbook of North American Indians*, Vol. 5 The Arctic, ed. D. Damas. Washington, DC: Smithsonian Institution, p. 119–35.

Nielsen-Marsh, C. M., P. H. Ostrom, H. Gandhi, B. Shapiro, A. Cooper, P. V. Hauschka and M. J. Collins (2002). Sequence preservation of osteo-calcin protein and mitochondrial DNA in Bison bones older than 55 ka. *Geology*, **30**, 1099–1102.

Nordborg, M. (1998). On the probability of Neanderthal ancestry. *Amer. J. Hum. Genet.*, **63**, 1237–40.

O'Rourke, D. H., M. G. Hayes and S. W. Carlyle (2000a). Ancient DNA studies in physical anthropology. *Ann. Rev. Anthropology*, **29**, 217–42.

O'Rourke, D. H., M. G. Hayes and S. W. Carlyle (2000b). Spatial and temporal stability of mtDNA haplogroup frequencies in Native North America. *Hum. Biol.*, **72**, 15–34.

O'Rourke, D. H., M. G. Hayes and S. W. Carlyle (2005). The consent process and aDNA research: contrasting approaches in North America. In *Biological Anthropology and Ethics*, ed. T. R. Turner. SUNY Press, Albany, pp. 231–40.

Ovchinnikov, I. V., G. Götherström, G. P. Romanova, V. M. Kharitonov, K. Lidén and W. Goodwin (2000). Molecular analysis of Neanderthal DNA from the northern Caucasus. *Nature*, **404**, 490–3.

Pääbo, S. (1985). Preservation of DNA in ancient Egyptian mummies. *J. Archaeol. Sci.*, **12**, 411–17.

Pääbo, S. (1989). Ancient DNA: extraction, characterization, molecular cloning, and enzymatic amplification. *PNAS*, **86**, 1939–43.

Pääbo, S., H. Poinar, D. Serre, V. Jaenicke-Després, J. Hebler, N. Rohland, M. Kuch, J. Krause, L. Vigilant and M. Hofreiter (2004). Genetic analyses from ancient DNA. *Ann. Rev. Genet.*, **38**, 645–79.

Paxinos, E. E., H. F. James, S. L. Olson, J. D. Ballou, J. A. Leonard and R. C. Fleischer (2002). Prehistoric decline of genetic diversity in the Nene. *Science*, **296**, 1827.

Poinar, H., M. Hofreiter, G. Spaulding, P. Martin, A. Stankiewicz, H. Bland, R. Evershed, G. Possnert and S. Pääbo (1998). Molecular coproscopy: dung and diet of the extinct ground sloth Nothrotheriops shastensis. *Science*, **281**, 402–6.

Pusch, C., I. Giddings and M. Scholz (1998). Repair of degraded duplex DNA from prehistoric samples using Escherichia coli, DNA polymerase I and T4 DNA ligase. *Nucleic Acids Res.*, **26**, 857–9.

Pruvost, M. and E.-M. Geigl (2004). Real-time quantitative PCR to assess the authenticity of ancient DNA amplification. *J. Archaeol. Sci.*, **31**, 1191–7.

Pruvost, M., T. Grange and E.-M. Geigl (2005). Minimizing DNA contamination by using UNG-coupled quantitative real-time PCR on degraded DNA samples: application to Ancient DNA studies. *BioTechniques*, **38**, 569–75.

Raoult, D., G. Aboudharam, E. Crubezy, G. Larrouy, B. Ludes and M. Drancourt (2000). Molecular identification by 'suicide PCR' of *Yersinia pestis* as the agent of Medieval Black Death. *PNAS*, **97**, 12800–3.

Reid, A. H., T. G. Fanning, J. V. Hultin and J. K. Taubenberger (1999). Origin and evolution of the 1918 'Spanish' influenza virus hemagglutinin gene. *PNAS*, **96**, 1651–6.

Relethford, J. H. (2001). Absence of regional affinities of Neandertal DNA with living humans does not reject multiregional evolution. *Amer. J. Phys. Anthrop.*, **115**, 95–8.

Rightmire, G. P. (2001). Patterns of hominid evolution and dispersal in the Middle Pleistocene. *Quaternary Int.*, **75**, 77–84.

Ritchie, P. A., C. D. Millar, G. C. Gibb, C. Baroni and D. M. Lambert (2004). Ancient DNA enables timing of the Pleistocene origin and Holocene expansion of two Adelie penquin lineages in Antarctica. *Mol. Biol. Evol.*, **21**, 240–8.

Rubicz, R., T. G. Schurr, P. L. Babb and M. H. Crawford (2003). Mitochondrial DNA variation and the origins of the Aleuts. *Hum. Biol.*, **75**, 809–35.

Salo, W. L., A. C. Aufderheide, J. Buikstra and T. A. Holcomb (1994). Identification of *Mycobacterium tuberculosis* DNA in a Pre-Columbian Peruvian mummy. *PNAS*, **91**, 2091–4.

Savolainen, P., T. Leitner, A. N. Wilton, E. Matisoo-Smith and J. Lundeberg (2004). A detailed picture of the origin of the Australian dingo, obtained from the study of mitochondrial DNA. *Proc. Natl. Acad. Sci., USA*, **101**, 12387–90.

Schmitz, R. W., D. Serre, G. Bonani, S. Feine, F. Hillgruber *et al.* (2002). The Neandertal sype site revisited: Interdisciplinary investigations of skeletal remains from the Neander Valley, Germany. *PNAS*, **99**, 13342–7.

Serre, D., A. Langaney, M. Chech, M. Teschler-Nicola, M. Paunovic, P. Mennecier, M. Hofreiter, G. Possnert and S. Pääbo (2004). No evidence of Neandertal mtDNA contribution to early modern humans. *PLoS Biology*, **2**, 313–17.

Smith, S. E., M. G. Hayes, J. B. Coltrain and D. H. O'Rourke (2006). Inferring population continuity versus replacement with aDNA: a cautionary tale in the Aleutians. *Amer. J. Phys. Anthrop.*, **129**(S42), 167.

Spigelman, M. and E. Lemma (1993). The use of the polymerase chain reaction to detect *Mycobacterium tuberculosis* in ancient skeletons. *Int. J. Osteoarch.*, **3**, 137–43.

Stone, A. and M. Stoneking (1998). mtDNA analysis of a prehistoric Oneota population: implications for the peopling of the New World. *Amer. J. Hum. Genet.*, **62**, 1153–70.

Wall, J. (2000). Detecting ancient admixture in humans using sequence polymorphism data. *Genetics*, **154**, 1271–9.

Wayne, R. K., J. A. Leonard and A. Cooper (1999). Full of sound and fury: the recent story of ancient DNA. *Ann. Rev. Ecol. Syst.*, **30**, 457–77.

Yang, D. Y., A. Cannon and S. R. Saunders (2004). DNA species identification of archaeological salmon bone from the Pacific Northwest Coast of North America. *J. Archaeol. Sci.*, **31**, 619–31.

Yang, D. Y. and K. Watt (2005). Contamination controls when preparing archaeological remains for ancient DNA analysis. *J. Archaeol. Sci.*, **32**, 331–6.

Yang, D. Y., J. R. Woiderski and J. C. Driver (2005). DNA analysis of archaeological rabbit remains from the American Southwest. *J. Archaeol. Sci.*, **32**, 567–78.

Yoder, A. D., B. Rakotosamimanana and T. J. Parsons (1999). Ancient DNA in subfossil lemurs: methodological chanllenges and their solutions. In *New Directions in Lemur Studies*, eds. B. Rakotosamimanana, *et al*. New York: Kluwer Academic/Plenum Press.

Zink, A. R., W. Grabner, U. Reischl, H. Wolf and A. G. Nerlich (2003). Molecular study on human tuberculosis in three geographically distinct and time delineated populations from ancient Egypt. *Epidemiol. Infect.*, **130**, 239–49.

Zlojutro, M., R. Rubicz, E. J. Devor, V. A. Spitsyn, S. V. Makarov, K. Wilson and M. H. Crawford (2006). Genetic structure of the Aleuts and circumpolar populations based on mitochondrial DNA sequences: a synthesis. *Amer. J. Phys. Anthrop.*, **129**, 446–64.

PART 3

General Applications

Michael H. Crawford

Chapter 9

Applications of Molecular Genetics to Forensic Sciences

Moses Schanfield

Department of Forensic Sciences, George Washington University

Introduction

There are many parallels between anthropological genetics and forensic science. Both have evolved as new useful human polymorphisms were found. Both have evolved as technology has advanced. See the excellent coverage of this in Chapter 7. The approaches of anthropological genetics to the study of human populations is described in Chapter 7. In contrast to anthropological genetics which usually looks at the population, forensic science uses the genetically useful markers found in populations to characterize or in forensic terminology 'individualize' evidence and individuals. In the ideal sense both would like to have genetic markers that will either individualize populations or evidence to a given population of geographical origin. In forensic science the area that tests biological testing of evidence is called forensic genetics. In forensic genetics there are two general ways in which evidence is individualized. The first is done by direct comparison of two genetic profiles; the second is done using Mendelian inheritance to establish the genetic relationship.

Forensic science uses a process of comparison in which an unknown sample is always compared to a known or reference sample. In the case of identification of drugs a white powder suspected of being a controlled substance is seized by a police officer and submitted to the crime laboratory for identification. The laboratory will do a presumptive test to see if the powder could be a controlled substance. If it is positive for the screening test it will be tested by a method that will definitively identify the compound, such as a gas chromatograph with a mass spectra detector (GC-MS), which provides a spectrum of ions that characterizes each compound such as cocaine, methamphetamine, marihuana, heroin, etc. In the case of drug testing, the spectra of the unknown drug are compared to known samples which are stored in the computer of the GC-MS. This is referred to as a drug identification. In the case of biological evidence there is no standardized profile of everyone on the earth in the computer. Thus biological materials collected at a crime scene such as blood stains, semen stains, vaginal swabs, or body fragments are considered as 'unknown' or 'questioned' samples, sometimes referred to as 'Q's, because the genetic profile of the item is unknown or the object of the question. Once a genetic profile of the Q is generated it has to be compared to a known sample, or 'K', also referred to as a reference sample. This can be done in two ways. The first is a direct comparison to a known reference sample from either a victim or suspect to the evidence. For instance, in a homicide or sexual assault the best type of evidence is referred to as a two-way transfer. Blood or body fluid from the suspect is left on the victim, which often happens in a sexual assault case in which the perpetrator leaves semen in the victim's vaginal vault. Strengthening this evidence would be finding of the victim's blood or body fluid on the suspect or his clothing. The latter type of evidence may occur in the absence of the former. This is known as a 'direct' comparison identification. In this case there has to be an exact match between the genetic profile in the evidence when compared to the known sample. If the known sample matches at many, or shares one allele at many loci it is possible that the genetic profile originated with a direct relative or the suspect. If the evidence does not match the genetic profile at one or more locations the known sample could not have been the source of the biological material. The usefulness of forensic testing is determined by the ability of a genetic profile to exclude people. The larger the number of individuals that can be excluded the more powerful the test. In the ideal case a genetic profile that only occurs once in all of humanity or in genetically identical individuals (monozygous twins) would maximally exclude all evidence from individuals that were not the donor. Ultimately, all forensic testing is exclusionary. Thus, if you test an item of evidence, and it is the same as the known, depending on the uniqueness of the genetic profile, that can indicate that no one except the known source can be the donor. When testing the female component of a vaginal swab in a sexual assault case it simply confirms the origin of the sample to the victim. On the other hand if it matches the suspect, unless there is some other reason why it could be there, such as consensual sex, it is a compelling argument that the suspect committed a sexual assault.

A second type of identification is that of parentage testing; this is most commonly in the form of a civil paternity test, in which the question is whether the alleged father can be excluded as the biological father of the child in question. This is most often performed because a local department of social services is trying to determine who is responsible for the support of a minor child. At the present time some 300,000 of these are done annually (AABB, Parentage Testing Report, 2002). Parentage testing does not have the same level of individualization as comparison identification because in parentage testing the frequency of non-exclusion will be frequency of all possible donors of a specific allele which all of those individuals homozygous for the allele (p^2) and all of those individuals heterozygous for the allele ($2p[1-p]$) versus the joint likelihood of a specific one or two match in a comparison identification (either $p_1{}^2$ or $2p_1p_2$). Parentage testing can also be involved in forensic cases when it comes to the case of sexual assaults leading to pregnancy, child abandonment cases or other criminal actions in which a child of a suspect or victim is involved. However, unfortunately, in recent years a new application of parentage or relatedness testing has come to the fore because of the necessity of identifying remains from mass casualty events such as the war in the Balkans or the World Trade Center, or the identification of remains found subsequent to a homicide. Examples of both types of cases will be given below.

Testing in forensic genetics paralleled the development of new markers, such that when enzyme variation was found it was quickly put to use in testing forensic evidence and as other protein markers were discovered they were often incorporated into the test battery. Many of the markers used in forensic genetics were used anthropologically as described in Chapter 7. However, genetic markers to be used in the testing of forensic evidence had to be detectable in forensic evidence which consists of blood stains, semen stains, saliva, urine, faeces and other tissues. Often genetic markers found in whole blood, the original sample of choice of anthropologists, could not be detected with the technology available in forensic evidence. Prior to the introduction of DNA testing, the topic of this review, sexual assault evidence was the most common biological evidence submitted for testing and was often problematic; because the forensic scientist was testing genetic markers in the vaginal secretions mixed with semen there was the possibility of masking the semen contribution by the victim's secretions. For example, if the victim is blood group O Secretor and the semen donor is blood group 'A' Secretor the vaginal/semen mixture would be informative that the semen donor was an A Secretor. However, if the victim is an A Secretor and the semen donor is anything but a B Secretor or A,B Secretor the swab would not be informative for ABO type. Also, unlike blood which could generate highly individualizing profiles, semen and vaginal secretions contain a very limited number of genetic markers. The introduction of RFLP technology and the finding of highly polymorphic RFLP loci and the creation of modern PCR opened an entirely new possibility of characterizing vaginal and sperm components.

History of DNA and forensic DNA

The nature of DNA variation is described excellently in Chapter 7, making it unnecessary to repeat that information. However, to understand the present state of forensic DNA testing it is necessary to understand what has happened in the field. I will try to emphasize areas that are relevant to forensic science, though some of the applications had uses in anthropological genetics.

Restriction Fragment Length Polymorphism (RFLP)

Polymorphism in length of DNA segments detected by cutting intact DNA with a sequence specific restriction enzyme, followed by separation using electrophoresis and detecting the separated DNA fragments which differ by size after transferring them onto a membrane and interrogating the membrane with a labelled piece of single stranded DNA specific for the region of interest, referred to as a probe, is described in Chapter 7 for anthropological genetics. However, until highly polymorphic regions were identified, starting in 1980, RFLP was of little use to forensic scientists or individuals involved in paternity testing. The search began for areas of high polymorphism, with the idea of mapping the human genome. In 1980, the reporting of the first hypervariable DNA polymorphism using the probe PAW101 (D14S1) by Wyman and White (1980) opened a new era. Figure 9.1 is an example of the types of bands found with D14S1 from cases of disputed paternity. Laboratories in the United States and elsewhere began searching for hyper variable RFLP loci. Ray White's laboratory at the University of Utah was at the forefront of this research and quickly detected a series of loci that would become useful in forensic testing (White *et al.*, 1985). Other groups such as Collaborative Research in Boston also identified many highly polymorphic loci detected by RFLP; though these were not used forensically, many were used in paternity testing and anthropological studies (Dykes *et al.*, 1986; Dykes *et al.*, 1990a; McComb *et al.*, 1995; McComb *et al.*, 1996) (see Chapter 7 for specific examples).

Role of standardization and quality assurance

A major difference between research testing, such as that performed by anthropological geneticists and forensic scientists, is the requirement that the results of the testing are admitted in a court of law. Historically there have been two legal standards for the admission of novel scientific evidence: the lowest standard is known as the relevancy standard; this is used in the United States in states that did not adopt the Federal Rules of Evidence. The relevancy standard states that evidence is admissible if the testifying expert is duly qualified, the expert's opinion is relevant and will assist the fact

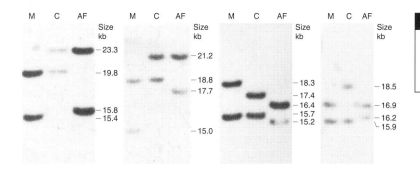

Fig 9.1 Cases of disputed paternity tested using the probe PAWI0I detecting the locus DI4SI. (Picture courtesy of Dale Dykes, Memorial Blood Center of Minneapolis.)

finder (jury), and the testimony is not so prejudicial as to outweigh the probative value (Flatal, 2005). Those jurisdictions that accepted the Federal Rules of Evidence as their guiding law used the Frye standard until recently. The Frye standard held sway in its various derivations until the Daubert decision. The Frye decision in 1923 was on the admissibility of a primitive form of deception detector measuring changes in blood pressure and galvanic skin response, or what is now known as a polygraph or lie detector. Though the novel evidence was not admitted it set the standard for the admissibility of novel scientific evidence for over 70 years. The Frye standard states, 'Just when a scientific principle or discovery crosses the line between the experimental and demonstrable stages is difficult to define. Somewhere in this twilight zone the evidential force of the principle must be recognized, and while courts will go a long way in admitting expert testimony deduced from a well-recognized scientific principle or discovery, the thing from which the deduction is made *must be sufficiently established to have gained general acceptance in the particular field in which it belongs* (emphasis added) (Frye v. United States, 1923). In 1993 the Frye standard was superseded by the more stringent Daubert *et al.* v. Merrell Dow Pharmaceuticals, Inc (1993), which placed a larger burden on the interpretation in the hands of the judges. The Daubert standard as it affected the Federal Rules of evidence affected all states under the Federal Rules of Evidence, except for states that had specifically accepted the Frye Standard as the state standard for the admission of novel evidence. These standards of admissibility played a major role in the admission of DNA evidence in the courts, especially prior to 1993. Another area of difference between research based testing and forensic and clinical laboratories is the requirement of regulation. Clinical testing came under regulation when third party carriers (insurance companies and government agencies) pay for testing and wanted guarantees of the accuracy and specificity of testing. This is normally done through several venues including the accreditation of laboratories or certification of analysts or both; along with standardization of technology and quality assurance procedures. These have existed for a long time in the area of clinical testing. Accreditation of parentage testing laboratories was begun in early 1990s by the American Association of Blood Banks, Parentage Testing Committee.

For a very long time accreditation of forensic laboratories was voluntary under the American Society of Crime Laboratories-Laboratory Accreditation Board (ASCLD-LAB). However, with changes in the Federal Law in 1994, DNA laboratories had to be accredited to participate in CODIS (see below).

Though private companies in the United States had begun testing forensic evidence by RFLP technology in 1985, the FBI did not begin testing cases until late 1988/early 1989. In the meantime, in England, Alex Jeffries, used a different approach, using low stringency RFLP, a series of probes that identified multibanded patterns. Alex Jeffries coined the term 'DNA fingerprinting' to describe these barcode like patterns (Jeffreys *et al.*, 1985). Although these regions proved to be highly informative for parentage testing and some initial success forensically they did not have the sensitivity needed for routine forensic testing and were not used routinely for parentage or forensic testing. The private laboratories operating in the United States used at least two different restriction enzymes and a large array of probes to detect different regions. As the power of forensic DNA testing developed it was obvious that a DNA testing system needed to be developed for publicly funded forensic laboratories in the United States. Thus in 1987 the FBI began developing a DNA testing system for use by the FBI and ultimately to be transferred to the municipal, county and state crime laboratories. All of the available probes from the Ray White, Alex Jeffreys and other sources were evaluated as were a number of different restriction enzymes. The idea was to have highly polymorphic systems, with the smallest fragments that could be detected such that relatively short gels could be used with good resolution, to minimize the amount of band shifting. The FBI settled on the restriction enzyme Hae III, a four base cutter with the restriction site of GG↓CC. The FBI originally proposed a very short 10 cm format; however after initial testing, it was determined that a 14 cm gel provided better resolution and it became the standard. Figure 9.2 is an example of a 14 cm gel FBI gel. To help the FBI set up DNA testing a technical working group on DNA analysis and methodology (TWGDAM) was set up to develop uniform standards for the forensic community. The FBI started training forensic scientists from municipal, county and state crime laboratories in the TWGDAM/FBI forensic DNA protocol. For the USA and Canada the TWGDAM/FBI protocol was the standard for publicly funded crime laboratories, leading to all private forensic laboratories to switch to the FBI based system. However, since paternity testing laboratories were separately regulated a large array of probes and enzymes were used (AABB, Parentage Testing Report, 2002).

Presumptive testing

Another difference between anthropological testing and forensic testing is the requirement to identify what you are testing.

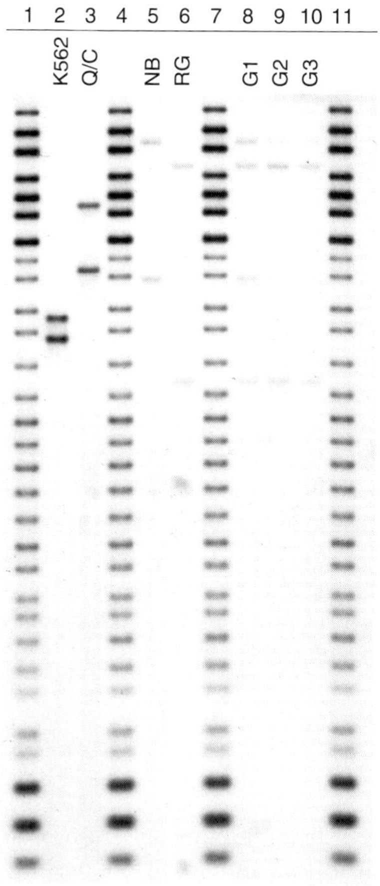
1 2 3 4 5 6 7 8 9 10 11
K562
Q/C
NB
RG
G1
G2
G3

Fig 9.2 A long format FBI gel from the OJ Simpson case. Lanes 1, 4, 7 and 11 are sizing ladders. Lane 2 is the K562 genomic DNA standard. Lane 3 is a quality control sample, that is not known to the analyst, but must be within measurement criteria for the gel interpretation to be valid. Lanes 5 and 6 are known samples from Nicole Brown and Ron Goldman. Lanes 8, 9 and 10 are stains from a glove found at the scene. The stains from the glove (G1, G2 and G4) represent different proportions of DNA from Nicole Brown and Ron Goldman. G1 represents an approximately equal mixture of Nicole Brown and Ron Goldman, while stains G2 and G4 are predominantly from Ron Goldman. At the time of the OJ Simpson case the technology had changed and this gel is not an autoradiograph but rather a lumigraph generated by chemiluminescence (light emission) rather than radioactivity. See the section below on the chemiluminescent detection of DNA fragments. Note: The original image was enhanced using Photoshop to improve contrast. (Source: G. Mathews, Los Angeles Sheriff Department Crime Laboratory, Los Angeles, CA.)

Anthropologists in general know the origin of the sample they are testing, whether it is from a living volunteer donor or from a mummified individual found in a cave. Forensic evidence is an unknown material; the fact something is a dried red stain on an item of clothing does not mean that it is blood, and if it is blood it does not mean that it is human blood. Historically, forensic scientists have spent a great deal of time identifying stains to determine if the colourless stains are semen, saliva, urine or some other identifiable fluid, and whether the red stains are blood. This is done at two levels. The first is a presumptive or screening test. Presumptive tests are for compounds that have a high concentration in a specific fluid, but may not be unique to that fluid. For example, there is a high concentration of the enzyme acid phosphatase in semen. Thus, if a stain is positive for acid phosphatase there is a high likelihood that it is semen. On the other hand if a stain is negative for acid phosphatase there is a low likelihood that it is semen. Body fluids that have good presumptive tests are semen (acid phosphatase), blood (catalase activity) and saliva (amylase), other fluids such as urine are not as well detected but can be with tests for urea or urease. Once a stain has been found to be positive for a presumptive test the results are confirmed using a 'confirmatory' test. For semen this can be done by looking for sperm, and if no sperm is found testing for P30, a prostate specific protein found in all semen including that of vasectomized males. (Note: P30 is the same as the PSA protein used to detect prostate cancer.)

DNA extraction

Unlike the testing of blood groups or serum proteins and enzymes which required little preparation before testing could be begun, the testing for DNA based polymorphisms requires the purification of the DNA. Forensic evidence typically consists of sexual assault evidence, of semen stains on cloth, but more often swabs collected during a sexual assault examination of the vaginal, rectal and other areas of a living or deceased sexual assault victim, often a mixture of semen and vaginal or oral secretion. Other evidence can be blood stains which consist of blood dried on clothing or surfaces, or other body fluids including skin cells, oral cells (saliva), bladder and kidney cells (urine). Removing DNA from dried swabs or clothing always includes a rehydration step in which a portion of the dried vaginal swab or stained cloth is soaked in deionized water or buffer for a period of time. After rehydration the cellular debris is pelleted, the hydrating liquid discarded. The need to work with mixed stains is rarely a problem in anthropological genetics unless one is working with a co-mingled grave. One enormous advantage of DNA technology over enzymes, proteins and blood groups in sexual assault evidence is that the male (sperm) component can be separated from the female (vaginal cell) component due to the differential lysis properties of sperm. Vaginal cells, like white cells in peripheral blood or cheek cells on a buccal swab, are relatively easy to lyse to remove the DNA. In contrast, the head of the sperm (DNA bearing portion) is heavily cross-linked, probably as an adaptation to penetrating the exterior cell wall of the egg. This allows for a process of differential separation of female and male DNA components for the testing of forensic evidence that can be applied to all DNA based testing.

DNA quantitation

Whether RFLP or PCR based testing is performed it is necessary to know how much DNA is present. For RFLP testing it was important to know the total amount of DNA present for two reasons. The first is so that the correct amount of restriction enzyme could be added to get complete digestion of the DNA present. The second is that DNA migration in the RFLP technology was affected by the amount of DNA present and how degraded it was. This was normally detected using a yield gel which could measure the relative amount of DNA in an unknown sample by comparing it to known samples. The overall quality of the DNA could also be measured. An example of a forensic yield gel is given in Figure 9.3. Ultimately, the National Institute of Standards and Technology developed a standardized yield gel for forensic applications (SRM 2390). Paternity laboratories and other applications could quantitate DNA with a spectrophotometer or other technology due to the large amount of high quantity DNA available.

PCR based testing also required the quantitation of the DNA but for a different reason. Since PCR is relatively species specific it was critical to know how much human DNA was present. This was usually detected with a slot blot in which a human specific probe was used and compared to known quantity human DNA standards. In recent years that technology is being replaced by real-time quantitative PCR.

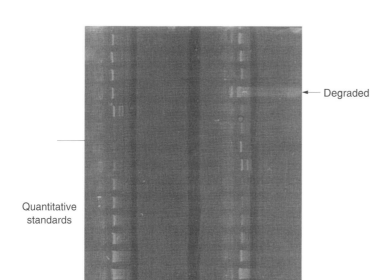

Quantitative standards

Degraded

Fig 9.3 Yield gel of extracted DNA stained with ethidium bromide and visualized with a 360 nm UV light box. The originating well is to the left of the DNA. Quantitative standards are on the left. Kit was part of NIST SRM 2390. (Image courtesy of AGTC, Denver, CO.)

Destruction of DNA

Unlike the testing of fresh samples or those that have been well preserved, forensic evidences are going to be affected by their environment. DNA has been found to be relatively robust when it is in the form of dry stains. Initial environmental studies indicated some of the limitations of DNA based on the material it is deposited upon and the environmental conditions. Environmental insult to DNA does not change the results of testing; you will either obtain results, or if the DNA has been too badly affected by the environment (i.e. the DNA is degraded) you may not get either RFLP or PCR results. The rate limiting difference is the amount of degradation. Because RFLP testing required larger fragments of DNA than the PCR based testing it was more susceptible to non-detection due to degradation than PCR based typing. Thus, the amount of time necessary for stains to dry, exposure to moisture, ultra violet light and the substrate could all affect the ability to detect DNA. This can be seen by a simple experiment performed at AGTC using known concentration DNA solutions. 100 µl samples of a standard DNA solution were exposed to fluorescent light in the laboratory, early sunset light in January, mid day sunlight in January, and a UV germicidal light (254 nanometer), in 15 minute increments, up to one hour. The results are presented in Figure 9.4. There is a linear decrease in high molecular weight DNA with the UV germicidal light exposure and increasing exposure time ($r = 0.996$, $p = 0.003$) such that after an hour about 96% of the high molecular weight DNA has been lost. Even in the weak mid day light in January, over 60% of the high molecular weight DNA was lost. If the aberrant zero time result is removed there is a significant correlation of DNA loss and time ($r = 0.962$, $p = 0.018$) for exposure to noonday sunlight. In contrast, the fluorescent lighting in the laboratory and the after sunset light had no effect on the amount of high molecular weight DNA. Though this was not a rigorous experiment the effects are dramatic enough to demonstrate the effect of ultra violet light exposure to DNA before a stain dries.

The environmental insulting of DNA in forensic evidence does not exist in the research or routine paternity testing environment. The closest area that has this problem is the study of ancient DNA by anthropological geneticists. Thus, tools used in forensic science for the testing of DNA are often also used in the study of ancient DNA.

Forensic RFLP DNA cases 1985–1990

The current acceptance and use of PCR based DNA technology in the forensic arena is based on the challenges created by RFLP testing. Prior to the introduction of standardized RFLP testing by the FBI

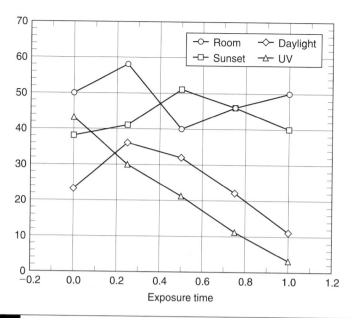

Fig 9.4 Effect of UV light exposure on high molecular weight of DNA as detected by fluorescence of stained yield gels. 100 ul samples of a single DNA sample were exposed to 'room' light (fluorescent light) at counter level, 'sunset' in Denver, CO in January, 'daylight' at 1 pm in January, in Denver, CO, 258 nm UV sterilization light. Exposure times were 0, 0.25, 0.50, 0.75 and 1.0 hours. Regression for UV light is highly significant ($p = 0.0003$), regression for daylight is significant after aberrant time 0 is removed ($p = 0.018$), there is no significant relationship between exposure time and DNA degradation at room light or sunset.

and other publicly funded laboratories, testing in the United States was done by two private laboratories (Lifecodes and Cellmark) both using different restriction enzymes and technology that differed in detail. Lifecodes technology for many reasons was flawed and led to markedly different gel migrations when the samples were not of a similar DNA concentration or degraded. The phenomenon was known as band shifting and led to many challenges of this technology. The FBI developed technology was designed to minimize the problems of band shifting. In addition to the problems with band shifting, there were many issues to be addressed. The sizing of DNA fragments on the same or different gels was not precise. The same fragments sized multiple times did not yield identical results, nor did the running of the same DNA sample on multiple gels. Thus, band sizing was to be considered to be a point estimate with a standard deviation. In paternity testing the community working through the American Association of Blood Banks Parentage Testing Committee decided that, to make sure that the two bands were different between a father and child, required that there would be three lanes run, minimally the child, the alleged father's lane, and a mixture of the alleged father and child. If the bands shared between the father and child are the same there will be one, two or three bands in the mixed lane depending on the phenotype of the alleged father and child (one band if both the father and child have a single band, two bands if the father and/or child has two bands and the other only one band, and three bands if they both have two band patterns). If the bands do not match there will be two to four bands depending on the phenotypes in the father and child. In forensic testing there was virtually never enough sample to use a mixed lane. Thus, it was necessary to determine a scientifically rigorous method of declaring a match or mismatch based on the normal variation in migration of DNA in this technology. This was usually estimated based on the standard deviation of the band size, as determined by testing the same DNA sample multiple times on different gels. For example, using non-isotopic detection at my laboratory in Denver the standard deviation was approximately 0.6% of the fragment size. We chose a conservative 2% match criteria. That is to say, in the absence of a mixed lane if two bands were more than 2% apart they were considered to be different bands; if less than 2% they were considered to be indistinguishable. The forensic community under TWGM chose a very conservative 5% window.

As DNA cases became more common in the courts, and the population frequencies of the multi-locus genotypes became extremely damning, the defence community started mounting challenges to early forensic DNA technology. Note: Using only the four original Ray White loci (D2S44, D10S28, D14S13 and S17S79) the average match likelihood was 1 in 12,800,000 for European Americans. The challenges came in three areas. The first was a challenge under the Frye standard, that the new technology was unproven and not accepted by the scientific community as reliable. The second had to do with match criteria and band shifting as seen primarily in Lifecodes' cases, and the last one had to do with the databases and whether the loci were in Hardy-Weinberg equilibrium and unlinked so that the product rule could be applied. Most of these challenges were not terribly successful; however most of the initial DNA cases were not reviewed by individuals with enough expertise to be adequately critical.

In 1989, shortly after the FBI started reporting out cases under the TWGDAM standards, things changed radically. During the trial NY v. Castro, in which the DNA testing had been performed by the Lifecodes Corporation, the defence took the autorads from the case to Eric Lander of the Whitehead Institute at MIT. In turn, Eric Lander took the autorads to a meeting at Stony Brook where he and other molecular biologists looked at the DNA results from the case as shown in the autorads. This case, as in many other Lifecodes' cases, had significant band shifting, placing bands clearly outside of the match criteria that Lifecodes used but were reported to be within criteria. Figure 9.5a and 9.5b is an example of a Lifecodes autorad from the period of the Castro case. The scientists

D14S13

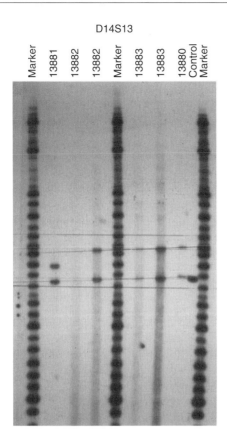

Fig 9.5a Long format Lifecodes gel detecting the locus D14S13, showing band shifting as indicated by the lines diagonally connecting RFLP bands as opposed to the rectangular lines created by the sizing ladders. (Autorad from TX v. Trimbolli, courtesy AGTC, Denver, CO).

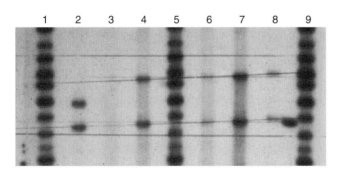

Fig 9.5b Close up of Lifecodes' D14S13 autorad from TX v. Trimbolli. Evidence items are in lanes 4, 6, and 7. The reference sample from Mr Trimbolli is in lane 8. Size of standards are in lanes 1, 5 and 9. Parallel lines at right angles connect all three size ladders indicating that the DNA size standards are migrating in a parallel fashion. However, the lines connecting the items of evidence form a parallelogram, indicating that they are not migrating at the same rate. The evidence in lane 4 is at approximately 10 sd from the reference standard, lane 7 is greater three to four sd. By any reasonable standard of matching these results would have been reported as inconclusive; however, Lifecodes declared that they matched the reference sample. These results are similar to those observed in the Castro case.

wrote a position paper stating that the evidence was not acceptable and the Judge ruled that the inclusionary evidence was not to be admitted, but the evidence that excluded an individual as the source could be used.

Because of the Castro case a period of serious review of forensic DNA testing in the United States began, culminating in the first and second reports of the National Academy of Sciences (NRC, 1992 and NRC, 1996). The first report was directed primarily at RFLP testing as PCR based testing was in its infancy. The aim of the report issued in 1992 summarized many of the issues:

> DNA typing for personal identification is a powerful tool for criminal investigation and justice. At the same time, technical aspects of DNA typing are vulnerable to error, and the interpretation of results requires appreciation of the principles of population genetics. These considerations and concerns arising out of the felon DNA databanks and the privacy of DNA information made it imperative to develop guidelines and safeguards for the most effective and socially beneficial use of this powerful tool.

Though TWGDAM had begun to make quality assurance guidelines in 1989 and 1991, the 1992 report provided the impetus for the development of formal standards for DNA testing in publicly funded crime laboratories with regard to quality assurance programmes and the use of external review.

There were many issues not handled well in the 1992 NRC report, leading to William Session, the Director of the FBI, to request a follow-up study in April 1993. The second NRC report revisited several issues of the first NRC, and clarified several others.

Areas included corrections for possible substructuring of populations based on current population genetic theory, and tests for substructuring that were difficult to implement in RFLP testing, but were readily amenable to PCR based testing. The report spent more time with the new PCR based testing as STR tests were beginning to appear. Note that a concise review of the material on the development of DNA quality standards can be found in Presley (1999).

With the passage and funding of the Violent Crime Control and Law Enforcement Act of 1994, which included the DNA Identification Act of 1994 which mandated the FBI to set up a convicted offender database of individuals involved in sexual assaults and homicides, authorized grants to state and local publicly funded crime laboratories for establishing and improving DNA laboratories, and establish standards for forensic DNA testing through a national DNA Advisory Board (DAB). The DAB was funded by the DNA Identification Act of 1994, and went to work in 1995. Using the large amount of work done by TWGDAM the 'Quality Assurance Standards for Forensic Testing Laboratories' were created and approved by the Director of the FBI. These standards took effect 1 October 1998 and included material from the second NRC report. After 1998 the DAB developed standards for the forensic DNA databases were completed in the form of the 'Quality Assurance Standards for Convicted Offender DNA Databasing Laboratories'.

It should be noted that by the time standards for forensic DNA testing became law in 1998 there were standards in place for RFLP testing in paternity cases by the American Association of Blood Banks Committee on Parentage Testing, which had been providing standards since the early 1980s in the United States and the International Society of Forensic Haemogenetics (ISFHG) now the International Society of Forensic Genetics. In Europe EDNAP (European DNA Proficiency) was created to do proficiency testing of DNA programmes, while the ISFHG had established standards. Further, working groups for various countries or language groups had been established to develop local guidelines. Countries outside of Europe had also set up guidelines for forensic DNA testing.

The development of standards took ten years between 1988 when TWGDAM was formed to the implementation of the DAB guidelines. Over this period many things happened. The issue of band shifting, quality control and standardized testing for the scientific community was only one area. There were issues of databases to be litigated, as well as issues of accuracy of DNA sizing which led to developing binning criteria for allele frequencies and how one declared a match and a mismatch (see above). Ironically many of these issues were long gone by the time the DAB guidelines were implemented in 1998. However, by 1998 many of the issues relating to RFLP were dead issues since RFLP was being replaced by PCR based STR testing. At AGTC in Denver by mid 1997 all RFLP based testing in paternity cases had had been replaced by PCR based STR testing.

The death of RFLP in forensic science

With the introduction of the Polymerase Chain Reaction (PCR) the fate of isotopic RFLP was sealed, only waiting on the discovery of enough useful markers. The end of RFLP was delayed by the development of chemiluminescent technology. It was found the 1–2 dioxetane, when triggered by enzymes such as alkaline phosphatase decomposed and generated photons, would do the same thing as isotopes and x-ray film. However, they had higher energy so that detection time could be cut down from days and weeks to hours. Figure 9.2 is an example of forensic RFLP detected using chemiluminescent technology. The sensitivity was further improved by the creation of oligonucleotide probes labelled with dioxatane. These probes aimed at the repeats had a much higher efficiency of binding creating a very rapid hybridization with high sensitivity. These probes could detect genomic DNA in the range of 25–50 nanograms of DNA. Note: This level of sensitivity was usually obtained by diluting genomic DNA which had not be environmentally abused. Evidence was rarely detected at that level of sensitivity.

Though there was good individualization with five to seven probes, ultimately the speed, simplicity and sensitivity of PCR would win. With sensitivity of one nanogram or less PCR was at least 25–50 times more sensitive than the best RFLP technology and could be done in several days rather than weeks. PCR only awaited the development of high informative markers. This meant that many types of evidence that were previously not amenable to DNA testing could be used as evidence.

However, as fluorescently labelled multiplexes of PCR based short tandem repeats were developed along with the technology to detect them, the death knell was sounded for forensic RFLP technology. Thus by 1995 PCR was replacing RFLP as the main forensic testing tool, even though PCR had been around for ten years.

PCR

At the same time as the revolution in RFLP was beginning in the mid 1970s, an early version of copying DNA using repair enzymes, called polymerases, was being explored. In early uses, the polymerase had to be replenished after the denaturation step because heating the polymerase to near boiling temperatures destroyed it. After a few cycles the system no longer worked because of the build up of denatured proteins. It would not be until the early 1980s when the modern Polymerase Chain Reaction (PCR) tests was developed (Mullis *et al.*, 1986). The discovery of high temperature polymerase and the existence of microprocessors allowed the Cetus Company to develop and patent automated PCR as we know it now. See the review in Chapter 7.

PCR based technology was much less expensive to implement than RFLP based testing since it did not require a laboratory capable of handling radioactive isotopes. It has higher throughput since each worker can do more cases in the same amount of time. PCR by its nature works with smaller amounts of DNA and with DNA that has been environmentally abused. Finally, since the DNA product can be identified as different alternative forms the statistical manipulation of data was similar to that for other genetic polymorphisms and more amenable to genetic analysis.

All of the DNA extraction systems described previously worked with PCR based testing. Additionally, relatively rapid, non-organic extraction methods specific for PCR such as Chelex were developed which only worked with PCR based testing (Walsh *et al.*, 1991).

As stated previously, forensic science is an applied science. In general technology is not specifically developed for forensic science. The initial genetic tests created for PCR were developed by the Cetus Corporation, patent holders on the PCR process.

The interest of Dr Henry Ehrlich, one of the principles, was transplantation genetics. He developed a test to detect genetic variation at the DQA1 locus then called DQα. Thus the first forensic application by Dr Edward Blake used a test for DQA1 occurred in 1985 (Von Beroldingen *et al.*, 1989). However, at that time there were no standardized test kits or anything useful for forensic scientists. In fact from 1985 until 1991, when AGTC in Denver began forensic PCR testing, there were no other laboratories doing PCR based testing. Further, the initial kits that were developed were of limited information content, such that though they were more sensitive with degraded DNA samples than RFLP they did not have individualizing power of RFLP. The earliest PCR based genetic test detected SNP polymorphisms at the DQA1 locus. DQA1 had many advantages: it was relatively quick and it would work on samples that RFLP could not get results. In my experience at AGTC in the early days of PCR, using DQA1, results were obtained in about 70% of the cases that RFLP did not have results. These results were also reported at a meeting by Forensic Science Services in England (David Werrett, personal communication). However, due to the low discrimination power of DQA1 (see Table 9.1), if there were multiple victims, suspects or combinations it was not uncommon for two or more to have the same DQA1 type, leading to the need for additional markers.

In the late 1980s in an effort to improve the resolution of some RFLP loci they were converted to PCR based systems. The locus Apolipoprotein B was the first published locus to be converted from RFLP to PCR based testing. Two laboratories independently in the same year. [Boerwinkle *et al.*, (1989) and Ludwig *et al.*, (1989)], indicating the importance of that locus as a possible risk factor in coronary artery diseases, published articles on the conversion of the APOB 3′ hypervariable region to a PCR based test. The term amplified fragment length polymorphism, or AFLP, was coined by Boerwinkle *et al.* (1989) to indicate that it was a PCR generated length polymorphism. We started using the acronym AFLP in 1992 and published using the term in publications in 1993 (Latorra *et al.*, 1994). Note: The acronym AFLP was subsequently patented inappropriately as a technique for creating a PCR barcode used to identify plants and bacteria, so the acronym may or may not be correct in a given citation. To further complicate the issue Bruce Budowle of the FBI coined the term AmpFLP for the same amplified fragment length polymorphisms. Rapidly, other RFLP systems were published, including D17S5(30) (Horn *et al.*, 1989), D1S80 (Kasai *et al.*, 1990) and PAH (Goltsov *et al.*, 1992). These AFLP systems all had large numbers of alleles, and were characterized by the presence of relatively large tandem repeats (LTRs) between 15 (D1S80 and APOB) and 70 (D17S5) nucleotides, the same size as the RFLP repeats. However, the products were much shorter with in general the longest being around 1,000 bp, were relatively easy to amplify and, more importantly, did not have the electrophoretic quirks of RFLP testing. Adding these AFLP

Table 9.1. | Increase in discrimination power by the addition of PCR based systems to the original DQA1 system, based on 20 European Americans. Numbers in the table represent the likelihood of a coincidental match for the most common profile found, the median profile and the least common profile in these 20 individuals. PCR VNTR loci include APOB, D1S80 and PAH systems tested at AGTC. Frequency data for APOB, D1S80 and PAH from Latorra, Stern and Schanfield (1994), DQA1 data from Helmuth et al. (1990), PM frequencies taken from *dna-view.com/Cellmark.htm*. Five RFLP loci were also tested on the same 20 individuals. They are D7S107, D11S129, D17S34, D18S17.2 and D21S112 using the enzyme *Pst*I. They were used extensively in paternity testing but are less polymorphic than the *Hae*III systems used forensically in the United States.

Profile	DQAI	PM	DQ+PM	DQ+PM +PCR VNTR	RFLP Loci
Most Common	9	87	882	8,000,000	740,000
Median	16	302	4,400	520,000,000	16,000,000
Least Common	59	68,000	1,300,000	116,000,000,000	1,200,000,000

loci to DQA1 significantly improved the individualization of PCR based typing in the early 1990s. This is seen in Table 9.1.

Case Example I: South Dakota v. DM

As stated previously, the problem of DQA1 was that, with regularity, a victim and suspect or multiple suspects would have the same DQA1 type, making the PCR based testing uninformative. I became involved in such a case in 1992. At 6 o'clock in the evening in May of 1991 in Sioux Falls, SD, a seven-year-old female went to a local Seven–Eleven for milk and did not return. Police were notified and search begun. That evening there was a heavy rainfall. The next day a naked body of a young female was found in a field in Lincoln County, just south of Sioux Falls. The child had been brutally murdered after being sexually assaulted. There was significant tearing of the rectal and vaginal vaults. The body had been exposed to the rain all night long. Rectal and anal swabs had weak acid phosphatase (presumptive test for semen) activity and P30 (a confirmatory test for semen) was not detected. Sperm were detected in the swabs. There was no blood group substance detected. The PGM1 (a polymorphic enzyme found in semen and blood) matched the victim.

A suspect DM was developed and after apprehension in the state of Washington was returned to SD for trial. Unfortunately, there was insufficient DNA for RFLP testing. Samples were referred to Dr Edward Blake for DQA1 typing. The results gave a limited level of individualization (6–7%); further, another possible suspect, the victim's stepfather, could not be excluded.

AGTC was requested to do APOB typing on the evidence to see if the stepfather could be eliminated, while not excluding the suspect. AGTC had recently completed forensic validation of the APOB AFLP, the first laboratory to validate an AFLP locus. The results of testing the victim (RO), the sperm fraction from the rectal/anal swab, the suspect (DM) and the stepfather (KC) are presented in Table 9.2.

The request for APOB testing came late, such that the report was not issued until trial was under way. The prosecution attempted to introduce the results of APOB testing at trial; however, the results were not admitted, but DM was convicted. The conviction was overturned because of prosecutorial misconduct. At the second trial the APOB evidence was introduced at trial.

At the second trial the introduction of the APOB VNTR was extremely useful for several reasons. The results of the APOB typing clearly indicated that KC was excluded as a possible semen donor while DM could not be excluded as the semen donor (Table 9.2). The frequencies of DQA and APOB in European-Americans, African-Americans and SW Hispanics are presented in Table 9.3. APOB 30, 38 is not common in any reference populations indicating that the likelihood of a coincidental match for both DQA1 and APOB is

Table 9.2. Results of DQA and APOB testing on the sperm fraction of the rectal swab collected from the victim RO and compared to RO, suspect DM and stepfather KC. Note: DM and KC cannot be differentiated on the basis of DQA1, however, they both have different APOB genotypes. Only suspect DM cannot be excluded. Results courtesy of AGTC, Denver, CO, SD v DM.

	RO	Sperm	DM	KC
DQAI	4	1.2,3	**1.2,3**	**1.2,3**
APOB	36,46	30,38	30,38	34,36
Conclusion			Not-Excluded	Excluded

Table 9.3. This table presents the coincidental match frequencies for possible donors of the sperm fraction obtained after testing the anal/rectal swab taken from RO in three US populations (European American, African American and SW Hispanics). Based on the observed results less than 1 in 1,950 individuals would match by chance.

		European American	African American	SW Hispanic
DQAI	1.2,3	0.068	0.064	0.06
	1 in	15	16	17
APOB	30,38	0.003	0.008	0.001
	1 in	329	120	1,304
Combined		0.02%	0.05%	0.01%
	1 in	4,900	1,950	16,700

between one in 2,000 and one in 17,000, or in other words with these two loci alone between 99.95% and 99.994% of the population would be eliminated. (See Example 2 for specific discussion of mathematical treatments.)

In this case we have a simple example of a forensic comparison. The evidence in this case, the results of PCR based DNA testing on the sperm fraction from the vaginal/rectal swabs, is being compared to the victim, the primary suspect DM and the possible suspect. From the results shown in Table 9.2 the sperm fraction profile DQA1 1.2,3, APOB 30,38 does not match the victim RO, nor the secondary suspect KC, excluding them as a source of the sperm fraction. However, the primary suspect DM cannot be excluded. Though the primary suspect cannot be excluded the question is, how significant is this match? This is referred to as the weight of the evidence. If the evidence had been the ABO blood types and the victim was type O and the semen blood group A, then approximately 40% of the population would match by chance. This would not have been terribly useful. In this case as we can see in Table 9.3, based on the DQA1 typing, between 6% and 7% of the population would match by chance, or between 93% and 94% of unrelated individuals would be excluded. This clearly has more weight than an ABO match. The APOB population frequency is even less with only 0.1% to 0.8% of the population matching by chance, or over 99% of the population excluded.

During the second trial DM was convicted a second time and given the death penalty. Note: After extensive appeals DM was tried a third time (I do not remember the reason). In the meantime STR technology had been developed and could be applied to the case (see discussion of STR loci below). At the third trial STR profiles were added to the DQA1 and APOB. The ultimate individualization reached the hundreds of millions and DM was convicted of the murder of OR for the third time and given the death penalty.

Note: This case was one of the first eight PCR cases upheld on appellate review and the first case in which AFLP evidence was upheld at the appellate level. At the time of writing I am still reviewing Habeus application from DM.

About the time of the DM case the FBI began evaluating PCR based AFLP loci available for inclusion in the forensic battery of PCR based tests available. For various technical reasons the FBI only chose D1S80 and this was ultimately turned into a commercially available forensic test kit. Unfortunately, even with the addition of D1S80 the PCR based tests for forensic applications did not yield the level of individualization seen with RFLP (Table 9.1); further, they were labour intensive, created biohazard waste and did not lend themselves to automation. The SNP based tests had additional problems. Though the amplified products were quite small, and the products amplified readily, they did not discriminate mixed samples very well, such that the power of discrimination of a mixture, when the heterozygous type was an A, B was 0 as no one could be excluded from a mixture in a two allele system. Further, the SNPs were not useful in parentage testing under most circumstances, though we used them with limited success in the identification of remains in Croatia in the early 1990s (Primorac *et al.*, 1996), as these were the only tests we had at the time.

Further, a drawback of the AFLP PCR based markers was their size; just as RFLP failed because of the degraded nature of forensic evidence, though smaller, AFLP fragments were still in the area of 1,000 bp. Thus, though more sensitive than RFLP the large repeat AFLP loci would lose alleles due to sample degradation.

Though the first generation PCR based SNP technology and the second generation AFLP was useful, they were limited by the technology. The replacement technology for DQA1, PM and the AFLP began at the same time that AFLP were being developed. Driven by the human genome project a search had been started for new markers (Weber and May, 1989). The new polymorphisms being found had much smaller repeats consisting of two, three, four or five bases to a repeat unit. These new markers were called 'short tandem repeats' or STR for short or microsatellites. Unfortunately, the Taq polymerase cannot accurately copy two base pair repeats leading to artefacts called 'stutter bands' because though highly polymorphic is not useful forensically. For technical reliability it turned out that the four base pair or tetra nucleotide repeats became the genetic markers of choice to map the human genome. As more and more tetra nucleotide repeat loci were discovered it became possible to pick sets of these markers to make highly discriminatory multiplexes. The first commercial kits to detect tetra nucleotide repeats were developed by the Promega Corporation, initially for single loci, and latter multiplexes containing three loci amplified in a single reaction. The initial multiplex kits made by the Promega Corporation were detected by silver staining. The need for silver staining for detection of bands limited the use of the technology because of the hazards and labour intensive nature of the technology. However, some forensic laboratories, including AGTC, used these kits for forensic, paternity and anthropological genetic studies.

A major technological breakthrough came with the development of primers with fluorescent dyes that could be detected on gel scanners. Initially, the fluorescent labels were excited by laser and the different coloured emitted light detected using filters. Applied Biosystems (ABI) patented a rotating filters system that allowed for the detection of four colours on the ABI DNA analysers. This allowed for rapid four colour DNA sequencing as well as the fluorescent detection of DNA fragments for genotyping purposes. It became evident that these automated testing procedures would be much more useful for forensic and parentage testing purposes. With the availability of fluorescent labels for PCR primers an array of detection devices came on the market using slightly different technologies to bypass the ABI patents. Some of these devices used infrared spectra detection (Li-Core, Beckman), others used simple filter scanning (Beckman, Hitachi, ELF). At the present time, though some laboratories are still using Hitachi slab gel scanners, virtually all forensic laboratories are using ABI slab gel or capillary electrophoresis systems to detect sequencing products or genotyping results for paternity or forensic application. Capillary electrophoresis of CE systems have the advantage that they can be implemented with automated sample loading, eliminating the labour intensive gel loading step and largely replacing slab gel systems.

The first commercial kits for STR testing were made by the Promega Corporation and consisted of single colour kits labelled with the fluorescent label fluorescein. As ABI and other scanners became available, such as those made by Hitachi and Beckman, Promega began making kits in initially two different colours and ultimately made large multiplex kits with multiple colours (Powerplex 16). An example of an electropherogram of Powerplex 16 from Promega is seen in Figure 9.6. Note that a critical ingredient of this test system is the use of the internal size standard original described by Dykes *et al.* (1990) as a solution for band shifting during RFLP testing. ABI was later into the market with kits, initially in single colours but ultimately in four (Profiler, Profiler Plus, Co-Filer and SGM for Europe) and ultimately five colour multiplexes (Identifiler).

One of the extremely useful characteristics of fluorescent imaging devices is that the amount of light read by the detection device is quantified. The electropherograms produced by the real-time scanners is quantitative. However, the CCD image can also be used as a scanning densitometer to determine the amount of light in each peak. Since there is one fluorescent molecule per band the amount of fluorescence is linear with the number of molecules in a band. This allows for many different types of analysis to be performed on the data generated. One of the more important of these is the ability to detect mixtures (see below).

STR loci for forensic applications

With the passage of the DNA Identification Act of 1994 there was a need for a standardized panel in the United States, and a need for there to be at least some sharing of loci with forensic counterparts in Canada, England and Europe. The Technical Working Group on DNA Analysis Methods (TWGDAM) implemented a multi-laboratory evaluation of those STR loci available in kits in the United States. The loci chosen would be the PCR based core of a national sex offender file required under the 1994 DNA Identification Act. The national programme is called *C*ombined *D*NA *I*ndexing *S*ystem or CODIS for short.

The TWGDAM/CODIS loci were announced at the Promega DNA Identification Symposium in the fall of 1997 and at the American Academy of Forensic Science meeting in February 1998. The following loci were chosen to be part of what was originally called the CODIS 13 loci: CSF1PO, D3S1358, D5S818, D7S820, D8S1179, D13S317, D16S539, D18S51, D21S11, FGA, THO1, TPOX and VWA03. These loci overlapped with the Forensic Science Services multiplexes and the Interpol multiplexes. The 13 loci can be obtained in two amplifications using Profiler Plus and Cofiler or in a single amplification using a kit in development called Identifiler from PE Biosystems (ABI), or in two amplifications using Powerplex 1 and Powerplex 2 from Promega, or in a single reaction with Powerplex 16 by Promega. The 13 CODIS loci are present in Figure 9.6. One of the effects of developing a national DNA database is the need to compare DNA profiles from no suspect cases to convicted offenders. It is expected that the National DNA Indexing System (NDIS) could have as many as five million profiles. One of the reasons for the 13 CODIS loci is that the level of individualization should be high enough that, after correction for making five million comparisons, there is still sufficient power of individualization. Table 9.4 compares the individualization of five RFLP loci, DQA1/PM/AFLP and the 13 CODIS loci. The first two are taken from Table 9.2 or 9.3, but reformatted into the exponential format needed for the CODIS loci. It is clear that even the most common profile is less frequent than the least

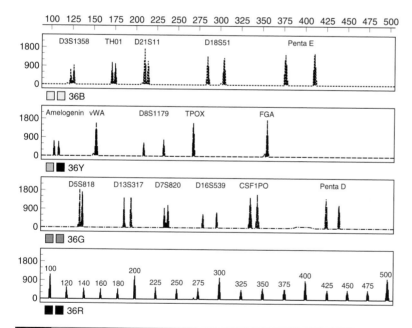

Fig 9.6 Electropherogram of Promega Powerplex 16® using four colour detection. Blue lane from smallest to largest is D3S1358, THOI, D2ISII, DI8S5I, and Penta E (a proprietary penta nucleotide repeat), Yellow lane (shown in black) Amelogenin, vWA (incorrect should be VWA03), D8SII79, TPOX, and FGA, Green lane D5S8I8, DI3S3I7, D7S820, DI6S539, CSFIPO and Penta D (another proprietary penta nucleotide). The red lane is the internal size standard which allows the system to size each fragment with an accuracy of less than one base pair.

Table 9.4. Levels of individualization generated testing 20 European American individuals with three different test batteries in. Details of RFLP and DQA1/PM/AFLP loci are found in Table 9.1. The 13 CODIS loci are described in the text and in Figure 9.6.

Profile	RFLP Loci	DQ+PM+PCR VNTR	13 CODIS Loci
Most Common	7.40E+05	8.00E+06	1.918E+11
Median	1.60E+07	5.02E+08	5.028E+15
Least Common	1.20E+09	1.16E+11	3.788E+25

common profile in the other two test batteries. One of the purposes of the test battery is to make sure that false matches do not occur when screening multi-locus genotype profiles against large data bases. Based on the most common European American profile, it is unlikely that there would be a coincidental match even with the most common profile, when compared to say five million individuals. The estimated likelihood of a coincidental match after taking into account the five million comparisons would be five million times the likelihood of a coincidental match. Using the data in Table 9.4 the corrected likelihood for the most common phenotype is approximately 1 in 40,000 while the median results is approximately 1 in a billion. Thus, using the 13 CODIS loci there is a low likelihood of getting a false match on a database search.

Individualization or sole source attribution

In the United States the FBI has started releasing reports indicating that biological material originated from a specific source (sole source attribution) to a scientific certainty. Much as fingerprint and firearms examiners have done for many years, the FBI has decided that if the population frequency exceeds 1 in 270 billion the sample is individualized. Other laboratories have chosen higher threshold levels such as 1 in 500 billion. Whatever the level chosen it is estimated that the average power of exclusion for these 13 CODIS loci exceeds 1 in a million billion, and though it is possible to obtain a frequency more common than those required for individualization it will occur infrequently.

Interpreting mixtures

Routine testing of evidence in forensic cases normally consists of sexual assault evidence, blood and other body fluids. However, from the very beginning of the use of PCR based technology individuals have been pushing the envelope so to speak by trying to test more and more extreme evidence. As stated above, one valuable characteristic of the fluorescent technology is the ability to quantitate the amount of DNA such that it is often possible to determine major and minor contributors.

When there are many alleles and no clear major contributor or more than two donors the interpretation gets complicated. In Table 9.5 taken from an FBI case it is clear that there are at least four donors including males and females. Though the suspect cannot be excluded there is not a great deal of evidence including him since there is a high likelihood that many people would match. Further, because the forensic community operating under the DAB/FBI guidelines has to have rigid rules for interpreting evidence the suspect could only be included in locus D8S1179, as that is the only one that both alleles of the suspect exceed 200 RFU used by the FBI. An indication of the exclusionary power of the technology is the fact that if the suspect had been identifiable in the mixture for the locus D8S1179 alone, only one in 75 individuals would match by chance compared to virtually all of the population for the mixture, using the average of $2p_ip_j$ for three populations. However,

Table 9.5. The table contains the Relative Fluorescent Units (RFU) from a Profiler Plus green channel in a profile from an FBI case with a mixed sample. There is no dominant component that matched the suspect. The bold faced bands are above 200 RFU, which is the lower limit of stating that a band is present using the FBI criteria for the interpretation of mixtures. The highlighted alleles are those found in one of the suspects. Those less than 200 RFU could not be used to include a suspect but they could be used to exclude some one. With eight alleles detected there are at least four DNA donors in this mixture, including both males and females based on the excess of the X band of Amelogenin. Due to the large numbers of bands there are large number of individuals that would not be excluded. Using these three loci approximately 70% of the population could not be excluded from this mixture. The suspect could only be included in D8S1179, since the other profiles one allele is less than 200 RFU, making the result inconclusive. If the suspect had been a major component for that single locus approximately 44% of the population could not be excluded or about 1 in 2 individuals would match by chance. This indicates how much information is lost in this type of complex mixture.

Amelogenin			D8S1179			D2S11			D18S51		
allele	RFU	%	allele	RFU	%	allele	RFU	%	allele	RFU	%
X	6363	64.15%	10	220	3.98%	28	476	19.44%	13	87	7.30%
Y	3556	35.85%	11	1294	23.40%	29	881	35.99%	14	59	4.95%
			12	518	9.37%	30	187	7.64%	15	342	28.72%
			13	688	12.44%	31	90	3.68%	16	64	5.37%
			14	550	9.95%	31.2	287	11.72%	17	107	8.98%
			15	1525	27.58%	32.2	115	4.70%	18	262	22.00%
			16	734	13.28	33.2	412	16.83%	19	174	14.61%
									20	96	8.06%
RFU	9919			5529			2448			1191	
Alleles			7			7			8		

the information content in the mixture is calculated using two of the proposed formulas. The first is from the NRC (1996, 129) report:

$$\text{NRC 2 mixture formula match probability} = 2\left(\sum p_i^* p_j\right) \qquad (9.1)$$

where $p_i p_j$ represents all of the possible two allele combinations. Such that if the mixture consisted of four alleles D8S1179 10, 11, 12 and 13, then the term $\Sigma p_i p_j$ would consist of the sum of the individual allele frequency products times two

$$2(10,11 + 10,12 + 10,13 + 11,12 + 11,13 + 12,13).$$

For the example in Table 9.5 on average 44% of the population could not be excluded. The FBI has chosen an even more conservative approach:

$$\text{FBI mixture formula} = \left(\sum p_i\right)^2 \qquad (9.2)$$

where the sum of the alleles identified in the mixture is squared. In the example above approximately 93% of the population could not be excluded using the FBI equation (FBI Laboratory Protocols). Indicating that with complex mixtures, even with the power of multiplex PCR based testing, it may not be possible to identify individuals.

Modern forensic PCR testing

As of this point in time the 13 CODIS Loci are the law of the land in the United States and similar numbers of loci are being used on a worldwide basis to test sexual assault evidence, homicide, assault and personal crime evidence. Forensic Science Services of the UK was the first law enforcement agency to put in large-scale DNA screening of individuals from all major crimes. Thus they were profiling individuals convicted of committing burglaries and other similar crimes. The net effect of this was the inclusion of many individuals not originally included under the DNA Identification Act of 1994 which required the data basing of homicides and sexual assaults. However, this also meant that they had to start testing evidence from crimes involving non-violent crimes. Though it is certainly possible that burglars cut themselves and leave blood at a crime scene, most of the evidence left are going to be fingerprints or other skin contact transfers.

It should be noted that parentage laboratories in the US and worldwide rapidly adopted multiplex STR typing as the method of choice, as the testing could be highly informative and amenable to high throughput automation. It should be noted that the information content for parentage testing is much less than that seen in forensic identity testing. This is demonstrated in the example below.

Low copy number DNA testing

The body sheds thousands of skin cells a day. Most of these are nucleated. However, the number in any given transfer are few. This type of evidence does not yield large amounts of DNA. A single cell contains approximately 6 picograms of DNA, the normal lower limit for forensic DNA is considered to be about 250 picograms or the equivalent of about 40 cells. To deal with what is referred to a 'low copy number DNA' Forensic Science Services changed the parameters. Typically the number of cycles was increased from 27 to 34 to increase the amount of PCR product. Low copy number DNA may or may not be successful and has many problems. Because of the low copy number initial amplification may not amplify all of the alleles leading to what is called allelic dropout due to stochastic events. Further, interpretation guidelines usually state that for a heterozygote to be called the ratio of the highest to the lowest peak must be within a certain range usually greater than 75%; however, often in low copy number DNA this ratio is not met. Thus, revalidation of the entire procedure under low copy number protocols is necessary for the evidence to be accepted in court. Finally, low copy number evidence has often been handled by multiple individuals leading to mixtures. A general rule in forensic science is the last donor has the most DNA present; however, a recent study by students at the Department of Forensic Sciences at George Washington University indicates that under laboratory conditions only about 72% of the time did the last person handling the simulated evidence have the most DNA present (Lewis and Lyons, unpublished results, GWU student research project). In part this is due to variation in how many skin cells individuals shed; this is variously referred to as 'shedders' or 'sloughers', and it turns out that there is a great deal of variation in how many cells are sloughed. Many laboratories are in the process of setting up low copy number DNA protocols. Forensic Science Services has been reporting on low copy number testing since 1999. The outcome of their work has been the identification of significant numbers of career burglars that were responsible for many burglaries. One burglar was attributed with 80 burglaries (personal communication, David Warrett, Forensic Science Services). The testing of low copy number DNA testing was recently reviewed by Gill (2001).

Example 2: Identification of remains from Croatia

With the fall of the Soviet Union in 1991 Slovenia and Croatia withdrew from Yugoslavia to form independent democratic countries. This did not sit well with the Serbians within Slovenia and Croatia and more importantly in Belgrade. By the end of 1991 a state of armed conflict existed between Croats and Serbs with the Serbs initially militarily capturing significant Croatian territory, both within Croatia and the Croatian area of Herzegovina. In the process Serbs would kill both military and civilians and bury their remains in mass graves of varying sizes. In 1995 with the tide of war turning such that Croatian territory was being liberated, the reports of atrocities had come in and mass graves were being identified in both Croatia and in Croatian Herzegovina.

In October 1995 I participated in my first mass grave exhumation in the town of Kupres in Herzegovina. At that time 27 new remains had been located of the 120 civilians that had disappeared during the Serbian occupation between 1991 and 1995. The case below is not one of those. These remains (NNIV) were located in 1997 but were not profiled until 1999. The case is treated in two ways, first as a missing person case with no reference samples in what is called a reverse paternity, and the second as a missing persons case with reference samples and thus a forensic identification. Both of these scenarios can occur when an individual goes missing. The issue is whether reference samples are available from the individuals that have disappeared, or only genetic references from relatives. This has been a major problem in the 9/11 World Trade Center identification. The individuals that have not been identified have neither reference samples nor relatives to test.

Note on reference samples: To do DNA identification it is necessary to have samples of the victim or suspect to compare to evidence. The evidence may be the individual's remains if they are missing, or it can be blood stains, semen stains or other biological evidence. In any case it is necessary to have a reference or 'known' to compare to the questioned material. We are used to using blood or tissue from the victim or blood specimens, buccal swabs or hair from the suspect. However, when the victim is missing these items are not available. Even blood stains found at the crime scene would have to be checked against other sources to make sure that they are from the victim. Sometimes idealized reference sources do not turn out to be unmixed, such as tooth brushes, hair brushes and articles of clothing. Often multiple 'references' have to be collected and checked against each other for consistency. The military collects reference samples on all active military personnel, such that should they go missing a reference sample is available to compare to recovered remains. These have been referred to as 'biological dog tags'. For example these were used when the Air Force plane went down in Croatia with Ron Brown on board.

Unlike forensic identification where we are looking at the direct comparison of the genotype/phenotype of the reference sample and the evidence, in forensic paternity testing we look at the evidence, which could be a child, or a parent or suspected remains and see if the possible relative has the allele that is required by Mendelian inheritance. In the data presented below (Table 9.6) we have a mother, a child and a bone suspected to be from the missing

Table 9.6. Results of testing Wife of suspected deceased father, Child and bone NNIV with Profiler Plus STR kit (Applied Biosystems). The alleles are indicated by the number of repeats at each locus. The maternal allele is underlined. The 'obligate allele' is the allele that had to originate in the biological father is in italics, 'p' is the frequency of the obligate allele, 'RMNE' is the frequency of the population that cannot be excluded, 'product' is the running likelihood of getting a Croatian male to match by chance. Allele frequencies are based on Schanfield *et al.* (2002).

Locus	Wife	Child	Obligate Allele	NNIV	p	RMNE	Product
D3S1358	14,16	15,16	*15*	15,17	0.267	0.463	0.463
VWA031	15,16	16,19	*19*	19	0.098	0.186	0.086
FGA	21,22	21,22	*21,22*	21,22	0.354	0.583	0.0502
THOI	7,8	7,8	*7,8*	8,9	0.302	0.513	0.0258
TPOX	11	10,11	*10*	9,10	0.05	0.097	0.0025
CSFIPO	10,12	10	*10*	10,11	0.223	0.396	0.001
D5S818	11,12	10,12	*10*	10,12	0.044	0.085	<0.0001
D13S317	8,12	8,11	*11*	11,12	0.379	0.614	<0.0001
D7S820	8,11	8	*8*	8,10	0.178	0.325	1.68E-05

father of the child and husband of the child's mother. Looking at the first locus D13S1358 we see the following:

	Wife	Child	Obligate	NNIV
D3S1358	14,16	*15*,16	*15*	*15*,17

The Wife is D3S1358 14,16 and the Child is D3S1358 15,16. The Mother and Child share D3S135816*, therefore the biological father of the child has to have contributed the *D3S1358*15* allele. This is called the 'obligate allele'. The bone NNIV has *D3S1358*15* and therefore cannot be excluded. How many other people won't be excluded? This is the parentage equivalent of the population frequency described briefly in Example 1 but in greater detail below. Unfortunately, in parentage testing the segment of the population that cannot be excluded is almost always a larger population segment than that in forensic identification, because we are looking for all of the people that can contribute a specific allele, which means all of those individuals with a single D3S1358*15 (heterozygotes) and all of those individuals with two alleles (homozygotes). This population frequency is called the 'Random Man Not Excluded or RMNE' and it represent the proportion of the population that cannot be excluded. It is expressed in the following formula:

$$RMNE = 2^*p_i(1 - p_i) + p_i^2 \tag{9.3}$$

$$= 2^*p_i - p_i^2 \tag{9.4}$$

In this case the frequency of *D3S1358*15* is 0.267, thus the RMNE is as follows:

$$\text{RMNE} = 2^*0.267 - (0.267)^2$$
$$= 0.534 + 0.071$$
$$= 0.463$$

Thus, for this allele at this locus we are not able to exclude 46.3% of the population. In contrast, to match NNIV directly the frequency of D3S1358 15,17 is 0.09 (see below). Thus, forensic identification is about five times as efficient as forensic parentage.

Combining RMNE

To combine all of the genetic information from the nine loci tested, presented in Table 9.6, we are asking the 'and' question: What is the chance that a single person will have *D3S1358*15* and *VWA031*19?* etc. In probability theory this is calculated by multiplying all of the individual values. Thus, the general formula for combining individual RMNE is as follows:

$$\text{Combined RMNE} = \prod \text{RMNEi} \tag{9.5}$$

where $\prod$ is the product operator indicating that the individuals RMNE terms are multiplied times each other in a string.

In Table 9.6, the individual RMNE values are presented, followed by the running product. Even though the individual values are not very informative after nine loci are tested only 0.00168% of the Croatian population would match by chance, or 99.998% of Croatian males could be excluded, indicating that it is highly likely that NNIV belongs to the biological father of the child in question. Bayes theorem is a method of calculating a posterior probability for an event that cannot be directly measured. It does require the insertion of a prior probability of the events occurrence. The general equation for Bayes estimate of a posterior probability is:

$$\text{Post Pr X} = \frac{X^*\text{pr}}{X^*\text{pr} + Y^*(1 - \text{pr})} \tag{9.6}$$

where 'X' is the likelihood of the hypothesized event, 'Y' is the likelihood of the alternative event and 'pr' is the prior probability of the event occurring. In the specific case of calculating the probability that a non-excluded man is the father of a child, assumes a prior probability of one half (0.5) as it is assumed to be a neutral probability. This reduces to a specific equation for the likelihood of paternity of a non-excluded father:

$$\text{Pr Paternity} = \frac{1}{2 - P_E} \tag{9.7}$$

where 'P_E' is the Power of Exclusion which is equal to 1 – the combined RMNE.

This is taken from Weiner, 1976. Note: This equation applies to all non-excluded men, and not just NNIV, that have this array of obligate alleles. In the NNIV example the Bayesian probability of paternity based on the non-exclusion of the bone, that the bone came from the father of the child is 99.9983%, proving to a reasonable degree of scientific certainty the bone was that of the father of the child. Note: This is the probability of paternity based on non-exclusion; it is not the same as the probability of paternity based on the paternity indices, which are not allowed in the United States in forensic cases since they assume a likelihood of guilt on the part of the alleged fathers. In this case additional data also came from Y-STR data, further supporting the likelihood (data not presented, see below for data on Y STR typing).

Example of forensic identification using NNIV

Should an adequate reference sample had been stored on NNIV then a direct forensic identification can be made. The results of this hypothetical comparison are shown in Table 9.7. In this case the results will either contain one band or two bands and the direct comparison is made between the reference sample and the evidence (NNIV). If there is no difference in the multi-locus genetic profile of the reference sample and the unknown, then the reference sample donor cannot be excluded as a source of the material. If there are a small number of discrepancies at single alleles across the genotype

Table 9.7. The results of typing bone NNIV presented as a forensic identification case, in which 'p_i' and 'p_j' represent the allele frequencies in a Croatian population sample (Schanfield *et al.*, 2002), 'formula' represents the formula used to calculate the HWE for the genotype, 'likelihood' is the HWE genotype frequency and 'combined' is the joint likelihood of the multi-locus genotype. 'p^2+' represents the correction of the homozygous frequency due to possible substructuring. See text for details.

Locus	NNIV	p_i	p_j	formula	likelihood	combined
D3S1358	15,17	0.267	0.168	$2^*p_i{}^*p_j$	0.09	0.09
VWA	19	0.098		p_i^2+	0.01	0.00094
FGA	21,22	0.17	0.185	$2^*p_i{}^*p_j$	0.063	5.92E-05
THOI	8,9	0.173	0.213	$2^*p_i{}^*p_j$	0.074	4.37E-06
TPOX	9,10	0.094	0.05	$2^*p_i{}^*p_j$	0.009	4.07E-08
CSFIPO	10,11	0.223	0.292	$2^*p_i{}^*p_j$	0.13	5.29E-09
D5S818	10,12	0.044	0.398	$2^*p_i{}^*p_j$	0.035	1.84E-10
DI3S3I7	11,12	0.379	0.238	$2^*p_i{}^*p_j$	0.18	3.32E-11
D7S820	8,10	0.178	0.223	$2^*p_i{}^*p_j$	0.079	2.63E-12
Amelogenin	X,Y					

the reference donor would be excluded, but it is likely that a relative is the donor such as a brother. In general a true exclusion appears as multiple mismatches including complete loci non-identity.

Once a sample cannot be excluded, how is the weight of the evidence determined? Just as anthropological geneticists study population variation (see Chapter 7), if a population is in HWE the allele frequencies from a population sample can be used to generate the HWE likelihood or frequency of specific genotypes. Assuming populations are in HWE with no significant substructuring ($F=0$) then the expected frequency of heterozygous genotypes will be the HWE expectation which in its most general form will be:

$$\text{HWE HT} = 2^*p_i{}^*p_j \tag{9.8}$$

where 'i' and 'j' represent the two different alleles detected at that locus. For single band patterns it is assumed that the individual is homozygous. The HWE expectation for homozygous individuals is:

$$\text{HWE HM} = p_i^2 \tag{9.9}$$

However, in the United States, because of the guidelines established by the DAB, most laboratories follow the guidelines of NRC (1996), which recommend making the value as conservative as possible. This is done by assuming that there is substructuring in the population. Substructuring as defined in Chapter 7 can include everything from inbreeding to stratification. Such that under substructuring of any kind there is an increase in homozygosity and a decrease in heterozygosity, leading to the well-known expected values for homozygotes and heterozygotes:

$$\text{Substructuring HM} = p_i^2 + F^*p_i{}^*(1 - p_i) \tag{9.10}$$

$$\text{Substructuring HT} = 2^*p_i{}^*p_j - 2^*F^*p_i{}^*(1 - p_i) \tag{9.11}$$

where 'F' represents F_{IT}, which is the combination of F_{IS} and F_{ST}. In practice as substructuring reduces heterozygosity, forensic scientists do not include the effect of substructuring on heterozygotes, only on homozygotes, maximizing the genotype frequencies. According to the NRC II guidelines the F recommended for large urban populations is 0.01 and that to be used for smaller potentially endogamous populations such as native American Indian tribes would be 0.03, unless specific significant F has been detected. Thus the values calculated in Table 9.7 reflect formulas (9.7) and (9.9). For the example the uncorrected value is 0.0096, while the corrected value is 0.0105 or a net increase of approximately 9%. The data used is from Croatia, where this is no indication of deviations from HWE or substructuring.

As each expected value is the expected percentage of the population that cannot be excluded, this population frequency is a direct parallel of the RMNE in parentage testing. To calculate the joint likelihood of a multi-locus genotype the results would be combined in the same manner:

$$\text{Combined likelihood of random match} = \prod LL \qquad (9.12)$$

where 'LL' is locus likelihood for each locus as determined by formulas (9.7) and (9.9), and $\prod$ is the product operator, indicating that the likelihood is the joint product of all of the individual likelihoods.

Differences between forensic parentage and identity information content

As seen in Tables 9.6 and 9.7 the information content or degree of individualization is markedly different between a parentage test and forensic identification. To ease the visualization of this comparison Table 9.8 compares the RMNE to the population's frequency on a locus by locus basis, as well as the cumulative RMNE and populations frequency. The average RMNE for the nine loci is 0.363, the average population frequency for the nine loci is 0.075, and the average ratio is 6.946, or on average the individualizing power of identification is about seven-fold greater than parentage testing. In this case using nine loci forensic identity testing is approximately 6 million times as informative as forensic parentage testing. The only way this difference can be made up is by adding additional loci.

Table 9.8. Comparison of paternity RMNE to the forensic population frequency for nine loci using Profiler Plus, in the identification of NNIV. Columns from left to right are the locus identification, results of testing NNIV, RMNE from parentage analysis, the combined RMNE, the forensic population frequency, the combined population frequency or likelihood of a match, the ratio of the RMNE/population frequency, and the combined ratio. On the average the RMNE is approximately seven times less individualizing than the population frequency with a range of 2.5 to 18 fold. The difference between the two is approximately six million fold. However, for paternity testing purposes even with the more stringent European standard NNIV would be determined to have come from the father of the child in question.

Locus	NNIV	RMNE	Product	Pop Frq	Product	RMNE/ Pop	Product Ratio
D3S1358	15,17	0.463	0.463	0.090	0.09	5.158	5
VWA	19	0.186	0.086	0.010	0.00094	17.772	91
FGA	21,22	0.583	0.0502	0.063	5.9E-05	9.264	849
THOI	8,9	0.513	0.0258	0.074	4.4E-06	6.958	5,911
TPOX	9,10	0.098	0.0025	0.009	4.1E-08	10.372	61,470
CSF1PO	10,11	0.396	0.001	0.130	5.3E-09	3.043	188,929
D5S818	10,12	0.086	0.0001	0.035	1.8E-10	2.457	542,888
D13S317	11,12	0.614	0.0001	0.180	3.3E-11	3.405	3,013,864
D7S820	8,10	0.324	1.68E-05	0.079	2.6E-12	4.085	6,370,539
Average		0.363		0.075		6.946	

Y-STR

Though temporally out of context, while STR loci were being identified for gene mapping purposes, the Y chromosome was also being searched for STR loci. The rise of Y chromosome STR loci actually came after mitochondrial DNA; however, the amplification and detection systems are identical to the nuclear DNA STR loci. The genetic variation associated with the Y chromosome is described in Chapter 7. From a forensic standpoint in mixtures with a large amount of female DNA, such that there is more than ten times female to male DNA, the male DNA will not be amplified. By using male specific primers it is possible to obtain male profiles as these are the only markers detected. This has a tremendous impact on sexual assault evidence which is largely female DNA, precluding detection of the male DNA of interest. Thus, Y specific STR loci would be very useful in detecting the male DNA. However, because they are inherited in haplotypes or haplogroups, the alleles are not indepen-dent and cannot be simply multiplied to obtain a haplotypes frequency. This requires the creation of Y STR haplotypes databases. One of the largest of these is the database set up by Peter de Knijff at the University of Leiden in the Netherlands (http://www.yhrd.org), which covers populations on a worldwide basis; however there are other databases. Reliagene, the first producer of a commercially successful Y STR kit, has a searchable database (http://www.reliagene .com) of US populations. A disadvantage of the Y STR loci is that all males in a paternal lineage will have the same Y STR haplogroup/ haplotypes; thus brothers, or cousins in the same male lineage could not be differentiated. Further, because you have to use haplogroups/ haplotypes the frequency is dependent on the size of the database such that the frequency of a new and novel haplogroup will be 1/database $+ 1$. Because of this Y haplogroups will never reach the level of individualization reached with nuclear STRs. However, Y haplogroups are very useful in paternity or inheritance cases when the alleged father is dead, as well as some forensic cases.

Just as with nuclear STRs, SWGDAM and the International user's group representing Europe and Asia settled on a minimum list of loci for testing. Religene's Y-PLEX™12 was the first commercially available kit to simultaneously analyse DYS19, DYS385a/b, DYS389I, DYS389II, DYS390, DYS391, DYS392, DYS393, DYS438 and DYS439, plus a sex determinant locus Amelogenin, which includes the 11 SWGDAM loci and the 9 Y-STR loci generating minimal haplotypes identified by the International Y-STR User Group. Subsequent to the creation of the Reliagene Y-plex the Promega Corporation created a similar 11 plex and Applied Biosystems created a 17 plex based on their new five colour technology used in the Identifiler nuclear STR test kit. As Y STR behave as nuclear STR match criteria are identical to those for nuclear STRs, which is the alleles must be the same to have a match.

The weight of the match is determined by the population frequency of the haplogroup.

Mitochondrial DNA

Another source of DNA is found in the mitochondria in the cytoplasm of cells. Unlike nuclear DNA, which only has two copies of each genetic region, mitochondrial DNA is involved in energy production within the cell and can have between 100 and 10,000 copies per cell. An extensive description of the variation in mtDNA is found in Chapter 7. Hairs are one of the primary forms of transfer evidence found in crime scene evidence. Unfortunately, microscopic analysis of hairs is not terribly reliable. If two hairs match (cannot be differentiated), the most powerful statement that can be made for two hairs that are not dissimilar is that they are 'consistent' with each other. With the advent of mtDNA it was possible to test hair shafts as there is little or no nuclear DNA in the hair shaft. It was determined that not all hair shafts that visually matched were genetically identical (Mark Wilson, FBI, personal communication). With the advent of multicolor (BigDye™) technology it is not difficult to do DNA sequencing if you are trained in the technology; however, most forensic laboratories do not do DNA sequencing. Thus, there was interest forensically to develop a simple test to do mtDNA on hairs and other forms of evidence. The successor to the Cetus Corporation, first Perkin Elmer and later Roche Molecular, developed a test similar to DQA1 and PM relying on the amplification of HV1 and HV2 segments and then hybridizing them to oligonucleotide probes or SSO probes that detected some of the hypervariable SNPs in HV1 and HV2. This product was of limited use in detecting sequence variation in HV1 and HV2; however, the power of discrimination ranged between 92% and 98%, depending on the population (Gabriel et al., 2003).

As with the Y haplotypes there are shortcomings found with mtDNA that do not occur in other typing systems. One of these is the presence of heteroplasmy. Heteroplasmy is the presence of two or more sequences in a single individual. There are several forms of heteroplasmy: (1) individuals showing more than one sequence in a single tissue; (2) individuals with one sequence in one tissue and a second sequence in a different tissue, and (3) individuals with heteroplasmic in one tissue and homoplasmic in another tissue (Budowle et al., 2002a). It is hypothesized that a mutation occurs in an individual and during the division of cytoplasm containing both DNA sequences are captured in the ova or that during differentiation the unequal separation of cytoplasm creates heteroplasmy. The frequency of heteroplasmy appears to vary by tissue such that it appears that heteroplasmy is more frequent in hair shafts than in bone marrow derived tissues. Thus, the tissue that is of most interest for forensic scientists may be heteroplasmic. However, there is disagreement on how much heteroplasmy is common in hairs (Budowle *et al.*, 2002 a,b).

The area of most common application of mtDNA is the identification of remains. It is not clear whether it is the small circular structure, the large number of copies or both, but the likelihood of recovering mtDNA from degraded remains is much higher than getting nuclear or Y STR results. Thus, this method has been used extensively for the repatriation of remains. The Armed Forces DNA Identification Laboratory (AFDIL) in Rockville, MD is known for the identification of Vietnam Era remains. The oldest remains successfully tested have been from Neandertal and early modern humans (Ovchinnikov *et al.*, 2000).

Example 3: Identification of the Romanov skeletons

Though not as old as Neandertal remains the identification of the missing family of the Tsar Nicholas was of great interest. There has been a great deal of speculation and many a romantic tale about the murder and disappearance of Tsar Nicholas II and his family and family retainers. After the fall of the Soviet Union remains were exhumed at Yekaterinburg thought to be that of the royal family. Using an array of tools, including facial reconstruction and STR analysis, the related individuals were identified (Gill *et al.*, 1994; Ivanov *et al.*, 1996a,b). To identify the male and female lineages mtDNA was used to test the Tsarina and the Tsar. There was no difficulty in identifying the Tsarina, a descendant of Queen Victoria; however, it turned out that the individual thought to be Tsar Nicholas II was heteroplasmic and the two linear descendants used as references were homoplasmic. The dilemma was resolved with the collection of samples from the Tsar's Brother Georgij. The mtDNA data on the Tsar and his family is presented in Figure 9.7.

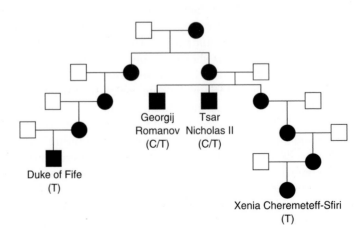

Fig 9.7 mtDNA data on Tsar Nicholas II and his family.

Ancient DNA and forensic evidence

Though not intuitively obvious there are shared attributes between testing of DNA evidence by forensic scientists and the study of ancient DNA by anthropological geneticists. When testing freshly collected samples neither group usually has a great deal of difficulty. However, both ancient DNA sources and forensic evidence share many traits. They both have been environmentally insulted to some extent by exposure to variation in temperature, humidity and ultraviolet light. All of these environmental components have the propensity to destroy DNA, such that at the extreme it may be possible to only test for mtDNA. The techniques to extract and test ancient DNA and forensic evidence are very similar in many cases. There is a shared need to make sure that inhibitors of the PCR reaction are removed, and the ever present concern that the samples do not become contaminated with exogenous DNA. Further, the development of technology such as whole genome amplification (see below) will benefit both the study of ancient DNA and forensic evidence.

Future of forensic DNA testing

With the acceptance of the CODIS loci there will be limited innovation in markers, as the number of individuals tested using the 13 CODIS STR loci will make it cost prohibitive to change. Thus, for the reasonable future the same STR loci will be used. However, clearly there are improvements that can be made. Slab gel electrophoresis was or is being replaced by capillary array electrophoresis which is amenable to automated sample loading, taking some of the labour away from the testing and reducing the size of the STR PCR product such that more degraded DNA can be amplified. This can be done by redesigning the primers to decrease the amount of flanking sequence. Recent work performed at the National Institute of Standards and Technology (NIST) has allowed for the development of mini-STR loci; however, these new mini-STR are not linked to the current CODIS loci, thus they will be limited to special applications (Coble and Butler, 2005).

Additional progress will be made by the use of nanotechnology to detect the markers already in CODIS. One such device which holds some promise is the micro fabrication created by the Whitehead Institute at MIT that uses 16 parallel channels and uses only 0.5 ul of PCR product (Goedecke et al., 2004). There has been much discussion in the field about automated devices that will include PCR as well as separation and detection strategies. These hold the promise of reducing time analysts spend on the technology, as well as decrease reagent usage (cost) and time.

Future improvements in mtDNA can come about by using SNPs to identify haplotypes rather than just use HV1 and HV2 sequencing. Investigators at AFDIL and NIST are developing panels of SNPs to help separate individuals with identical HV1 or HV2 sequences, due to the presence of the same haplogroup (Just *et al.*, 2004).

An area that will impact both ancient DNA studies and forensic testing is the development of procedures for whole genome amplification in which degraded or low level DNA can be randomly pre-amplified before looking for either SNP, STR or mtDNA. There are many different procedures out there and the research is still ongoing (Buchanan *et al.*, 2000; Tanabe *et al.*, 2003; Lasken and Egholm, 2003).

Another area that is changing in forensic DNA testing is the increased use of DNA testing in non-human evidence. Animal hairs, both feline and canine, are being investigated for evidential value (Menotti-Raymond *et al.*, 2003; Padar *et al.*, 2002; Eichmann *et al.*, 2004). Further, the use of DNA to characterize insects involved in either postmortem interval (Zenner *et al.*, 2004) or insects that feed off humans such as lice (Mumcuoglu *et al.*, 2004) are becoming more and more prevalent. Finally, two new areas that have developed is the use of DNA to characterize plants with controlled substances such as marijuana (Coyle *et al.*, 2003) and in the post 9/11 worry about controlled substances the use of DNA to characterize select agents (Jones *et al.*, 2005).

Thus it is evident that even as the nature of forensic science has changed in response to terrorism and other events the ability to do forensic DNA typing is rapidly developing to keep pace with the changes. Due to a lack of space many details on modern forensic testing have not been included; they can be found in Schanfield (2000 a,b,c,d). Additional reading on RFLP and mtDNA can be found in the *Encyclopedia of Forensic Sciences*, by Academic Press. Those interested in a more detailed exposition of the mathematics of parentage testing can look at Primorac *et al.* (2000).

References

American Association of Blood Banks, Annual Report Summary for American Association of Blood Banks, Annual Report Summary for Testing in 2002, Parentage Testing Program Unit. aabb.org.

Boerwinkle, E., Xiong, W., Fourest, E. and Chan, L. (1989). Rapid typing of tandemly repeated hypervariable loci by polymerase chain reaction: Application to the apolipoprotein B 3′ hypervariable region. *Proc. Nat. Acad. Sci.*, **86**, 212–16.

Budowle, B., Allard, M. W. and Wilson, M. (2002a). Characterization of heteroplasmy and hypervariable sites HV1: critique of D'Eustachio's interpretation. *Forensic Science International*, **130**, 68–70.

Budowle, B., Allard, M. W. and Wilson, M. (2002b). Critique of interpretation of high levels of heteroplasmy in the human mitochondrial DNA hypervariable region I from hair. *Forensic Science International*, **126**, 30–3.

Buchanan, A. V., Risch, G. M., Robichaux, M., Sherry, S. T., Batzer, M. A. and Weiss, K. M. (2000). Long DOP-PCR of rare archival anthropological samples *Human Biology*, **72**, 911–25.

Coble, M. and Butler, J. M. (2005). Characterization of new MiniSTR loci to analysis of degraded DNA. *Journal of Forensic Sciences*, **50**, 43–53.

Coyle, H. M., Palmbach, T., Juliano, N., Ladd, C. and Lee, H. C. (2003). An Overview of DNA Methods for the Identification and Individualization of Marijuana. *CMJ*, **44**, 315–21.

Daubert *et al.* v. Merrell Dow Pharmaceuticals, Inc (1993). No. 92–102 CERTIORARI TO THE UNITED STATES COURT OF APPEALS FOR THE NINTH CIRCUIT No. 92–102. Argued March 30, 1993. Decided June 28, 1993 U.S. Supreme Court.

Dykes, D. D., Fondell, J., Watkins, J. and Polesky, H. (1986). The use of biotinylated DNA probes for detecting single copy human restriction fragment length polymorphisms separated by electrophoresis. *Electrophoresis*, **7**, 278–82.

Dykes, D. D., Miller, S. A. and Schanfield, M. S. (1990a). Simultaneous DNA probing of paternity cases using non-Isotopic methods. In eds. Polesky, H. F. and Mayer, W. R., *Advance in Forensic Haemogenetics*, **3**, pp. 43–47.

Dykes, D. D., Miller, S. A., Schanfield, M. S., Langhoff, D. P. and Danilovs, J. A. (1990b). Two sources of error in size determinations of RFLP's using molecular weight standards. International Association of Forensic Sciences, August 1990, Adelaide, Australia.

Eichmann, C., Berger, B., Reinhold, M., Lutz, M. and Parson, W. (2004). Canine specific STR typing of saliva traces on dog bite wounds. *In J. Leg. Med.*, **118**, 337–42.

Flatal, R. A. (2005). DNA testing and the Frye standard. http://www.totse.com/en/law/justice_for_all/dnatest.html

Frye v. United States (1923). 54 App DC 46, 293 F. 1013, No. 3968, Court of Appeals of District of Columbia, submitted 7 November 1923, decided December 3, 1923.

Gabriel, M., Calloway, C. D., Reynolds, R. L. and Primorac, D. (2003). Identification of human remains by immobilized sequence specific oligonucleotides probe analysis of mtDNA hypervariable regions I and II. *Croatian Medical Journal*, **44**, 293–8.

Gill, P. (2001). Application of low copy number DNA profiling. *Croatian Medical Journal*, **42**, 229–32.

Gill, P., Ivanov, P. L., Kimpton, C., Piercy, R., Benson, P. *et al.* (1994). Identification of the remains of the Romanov family by DNA analysis: *Nature Genetics*, **6**, 130–5.

Goedecke, N., McKenna, B., El-Difrawy, S., Carey, L., Matsudaira, P. and Ehrlich, D. (2004). A high-performance multilane microdevice system designed for the DNA forensics laboratory. *Electrophoresis*, **25**, 1678–86.

Goltsov, A. A., Eisensmith, R. C., Konecki, D. S., Lichter-Konecki, U. and Woo, S. L. C. (1992). Association between mutations and a VNTR in the human Phenylalanine Hydroxylase gene. *American Journal of Human Genetics*, **51**, 627–36.

Helmuth, R., Fildes, N., Blake, E., Luce, M. C. Chimera, J., *et al.* (1990). HLA-DQα allele and genotype frequencies in various human populations, determined by using enzymatic amplification and oligonucleotides probes. *American Journal of Human Genetics*, **47**, 515–23.

Horn, G.T., Richards, B. and Klinger, K.W. (1989). Amplification of a highly polymorphic VNTR segment by the polymerase chain reaction. *Nuceleic Acids Research*, **17**, 2140.

Ivanov, P.L., Abramov, S.S., Gill, P., Sullivan, K.M., Kimpton, C.P., Ewett, I.W. and Plaksin, V.O. (1996a). Authentication of the skeletal remains of the last Russian Tsar and royal family. In eds. Schanfield, M.S., Al Khayat, A., Al Shamali, F.M. and Bergren, J.C., *International Methods of Forensic DNA Analysis: Proceedings of the First Forensic Experts Conference, Dubai, Emirates* 1994, pp. 21–35.

Ivanov, P.L., Wadhams, M.J., Roby, R.K., Holland, M.M., Weedn, V.W. and Parsons, T.J. (1996b). Mitochondrial DNA sequence heteroplasmy in the Grand Duke of Russia Georgij Romanov establishes the authenticity of the remains of Tsar Nicholas II. *Nature Genetics*, **12**, 417–20.

Jeffreys, A.J., Wilson, V. and Thein, S.L. (1985). Individual-specific 'fingerprints' of human DNA. *Nature*, **316**, 76.

Jones, S.W., Dobson, M.E., Francesconi, S., Schoske, R. and Crawford, R. (2005). DNA assays for the detection, identification and individualization of select agent microorganisms. *Croatian Medical Journal* (in press).

Just, R., Irwin, J., O'Callighan, J., Saunier, J.L., Coble, M.D. *et al.* (2004). Toward increased utility of mtDNA in forensic identification. *Forensic Science International*, **146** (suppl S), (S147–S149).

Kasai, K., Nakamura, Y. and White, R. (1990). Amplification of a variable number tandem repeats (VNTR) locus (pMCT118) by the polymerase chain reaction (PCR) and its application to forensic science. *Journal of Forensic Sciences*, **35**, 1196–2000.

Lasken, R.S. and Egholm, M. (2003). Whole genome amplification: Abundant supplies of DNA from precious samples or clinical specimens. *Trends in Biotechnology*, **21**, 531–5.

Latorra, D., Stern, C.M. and Schanfield, M.S. (1994). Characterization of human AFLP systems Apolipoprotein B, Phenylalanine Hydroxylase, and D1S80. *PCR Methods and Applications*, **3**, 351–8.

Ludwig, E.H., Friedl, W. and McCarthy, B.J. (1989). High-resolution analysis of a hypervariable region in the human Apolipoprotein B gene. *Am. J. Hum. Genet.*, **45**, 458–64.

McComb, J., Crawford, M.H., Osipova, L., Karaphet, T., Posukh, O. and Schanfield, M.S. (1996). DNA inter-population variation in Siberian indigenous population: The mountain Altai. *Amer. J. Hum. Biol.*, **8**, 599–607.

McComb, J., Blagitko, N., Comuzzie, A.G., Schanfield, M.S., Sukernik, R.I., Leonard, W.R. and Crawford, M.H. (1995). VNTR DNA variation in Siberian Indigenous populations. *Hum. Biol.*, **67**, 217–29.

Menotti-Raymond, M., David, V., Wachter, L., Yuhki, N. and O'Brien, S.J. (2003). Quantitative polymerase chain reaction-based assay for estimating DNA yield extracted from domestic cat specimens. *CMJ*, **44**, 327–33.

Mullis, K.B., Faloona, F.A., Scharf, S.J., Saiki, R.K., Horn, G.T. and Erlich, H.A. (1986). Specific enzymatic amplification of DNA *in vitro*: The polymerase chain reaction. *Cold Spring Harbor Symposium on Quantitative Biology*, **51**, 263–73.

Mumcuoglu, K., Gallili, N., Reshef, A., Brauner, P. and Grant, H. (2004). Use of human lice in forensic entomology. *J. Med. Entomol.*, **41**, 803–6.

NRC (1992). *DNA Technology in Forensic Sciences*. Washington, DC: National Academy.

NRC (1996). *The Evaluation of Forensic DNA Evidence.* Washington, DC: National Academy.

Ovchinnikov, I.V., Götherström, An., Romanova, G.P., Kharitonov, V.M., Lidén, K. and Goodwin, W. (2000). Molecular analysis of Neanderthal DNA from the northern Caucus. *Nature*, **404**, 490–3.

Padar, Z., Egyed, B., Kontadakis, K., Furedi, S., Woller, J., Zoldag, L. and Fekete, S. (2002). Canine STR analysis in forensic practice. Observation of a possible mutation in a dog hair. *Int. J. Leg. Med.*, **116**, 286–8.

Presley, L.A. (1999). The evolution of quality standards for forensic DNA analysis in the United States. *Profiles in DNA/September 1999.* http://www.promega.com/

Primorac, D., Andelinovic, S., Definis-Gojanovic, M., Drmic, I., Rezic, B., Baden, M.M., Kennedy, M.A., Schanfield, M.S., Skakel, S.B. and Lee, H.C. (1996). Identification of war victims from mass graves in Croatia, Bosnia and Herzegovina by the use of standard forensic methods and DNA typing. *J. Forensic Sciences*, **41**, 891–4.

Primorac, D., Schanfield, M.S. and Primorac, D. (2000). Application of forensic DNA testing in the legal system. *CMJ*, **41**, 32–46.

Schanfield, M. (2000a). Deoxyribonucleic Acid: Basic principles. In eds. Siegel, J.A., Saukko, P.J. and Knupfer, G.C., *Encyclopedia of Forensic Sciences.* London: Academic Press, pp. 479–85.

Schanfield, M. (2000b). Deoxyribonucleic Acid: Parentage Testing. In eds. Siegel, J.A., Saukko, P.J. and Knupfer, G.C., *Encyclopedia of Forensic Sciences.* London: Academic Press, pp. 504–15.

Schanfield, M. (2000c). Deoxyribonucleic Acid: Polymerase Chain Reaction. In eds. Siegel, J.A., Saukko, P.J. and Knupfer, G.C., *Encyclopedia of Forensic Sciences.* London: Academic Press, pp. 515–25.

Schanfield, M. (2000d). Deoxyribonucleic Acid: Polymerase Chain Reaction – Short Tandem Repeats. In eds. Siegel, J.A., Saukko, P.J. and Knupfer, G.C., *Encyclopedia of Forensic Sciences.* London: Academic Press, pp. 526–35.

Schanfield, M.S., Gabriel, M.N., Andelinovic, S., Reynolds, R., Ladd, C., Lee, H.C. and Primorac, D. (2002). Allele frequencies for the 13 CODIS STR loci in a sample of Southern Croatians. *J. For. Sci.*, **47**, 669–70.

Tanabe, C., Aoyagi, K., Sakiyama, T., Kohno, T., Yanagitani, N., *et al.* (2003). Evaluation of a whole-genome amplification method based on adaptor-ligation PCR of randomly sheared genomic DNA. *Genes Chromosomes & Cancer*, **38**, 68–176.

Von Beroldingen, C.H., Blake, E.T., Higuchi, R., Sensabaugh, G. and Ehrlich, H. (1989). Application of PCR to the analysis of biological evidence. In ed. Ehrlich, H.A., *PCR Technology: Principles and Applications for DNA Amplification.* New York: Stockton Press, pp. 209–23.

Walsh, P.S., Metzger, D.A. and Higuchi, R. (1991). Chelex® 100 as a medium for simple extraction of DNA for PCR based typing from forensic material. *BioTechniques*, **10**, 506–13.

Weber, J.L. and May, P.E. (1989). An abundant class of human DNA polymorphism which can be typed using the polymerase chain reaction. *American Journal of Human Genetics*, **44**, 388–96.

White, R., Leppert, M., Bishop, D.T., Barker, D., Berkowitz, J., Brown, C., Callahan, P., Holm, T. and Jerominski, L. (1985). Construction of linkage maps with DNA markers for human chromosomes. *Nature*, **313**, 192–8.

Weiner, A. (1976). Likelihood of parentage. In ed. Sussman, L., *Paternity Testing by Blood Grouping*, 2nd edn. Springfield, IL: Charles C. Thomas, pp. 124–31.

Wyman, A. and White, R. (1980). A highly polymorphic locus in human DNA. *Proc. Natl. Acad. Sci. (USA)*, **77**, 6754–8.

Zenner, R., Amendt, J., Schuett, S., Sauer, J., Krettek, R. and Povolny, D. (2004). Genetic identification of forensically important flesh flies (Diptera Sarcophagidae). *Int. J. Leg. Med.*, **118**, 245–7.

Chapter 10

Emerging Technologies: The Bright Future of Fluorescence

Eric J. Devor, Ph.D.

Senior Research Scientist, Molecular Genetics and Bioinformatics, Integrated DNA Technologies

Introduction

Oligonucleotide-driven, polymerase-catalyzed *in vitro* molecular reactions, specifically the polymerase chain reaction (PCR) and chain-termination DNA sequencing, have revolutionized our access to and understanding of genetics. Born less than three decades ago, these two techniques have together led to the phenomena of whole-genome sequencing, mass gene expression analyses, high-throughput drug discovery, and 'disease-of-the-week' mutation mapping – to name but a few. Major players in these advances range from the very small, like the bacterium *Thermus aquaticus*, that gave us thermal-stable DNA polymerase, to almost larger than life, like H. Gobind Khorana, under whose guidance the basic chemistries of oligonucleotide synthesis were developed. No less important is the smallest player of all, the fluorescent molecule. Appreciation for the potential of fluorescence as a tool in molecular biology pre-dates the advent of both chain-termination DNA sequencing and PCR, but it is only in the past few years that specific applications have begun to flower and pay huge dividends.

In this chapter I will present the basics of fluorescence relevant to molecular biology, including fluorescence resonance energy transfer (FRET). From there, the three applications in which fluorescence has made a significant contribution will be discussed. These are: chain-termination DNA sequencing, kinetic (real-time) PCR, and DNA microarrays. Finally, I will assess the role of fluorescence-aided

Correspondence: Dr. Eric J. Devor, Molecular Genetics and Bioinformatics, Integrated DNA Technologies, 1710 Commercial Park, Coralville, Iowa, USA 52241. Tel 319-626-8450, E-mail rdevor@idtdna.com molecular tools in Anthropological Genetics in the future as well as preview potential new fluorescence tools on the horizon.

Fluorescence and FRET

To begin, let us first distinguish *fluorescence* from *luminescence*. Luminescence is the production of light through excitation by means other than increasing temperature. These include chemical means (chemiluminescence), electrical discharges (electroluminescence), or crushing (triboluminescence). Fluorescence is a short-lived type of luminescence created by electromagnetic excitation. That is, fluorescence is generated when a substance absorbs light energy at a short (higher energy) wavelength and then emits light energy at a longer (lower energy) wavelength. The length of time between absorption and emission is usually relatively brief, often on the order of 10^{-9} to 10^{-8} seconds. The history of a single fluorescence event can be shown by means of a Jablonski diagram, named for the Ukranian born physicist Aleksander Jablonski (Fig. 10.1). As shown, in Stage 1 a photon of given energy $h\nu_{ex}$ is supplied from an outside source such as a laser or a lamp. The fluorescent molecule, lying in its *ground energy state* S_0, absorbs the energy creating an *excited electronic singlet state* S_1'. This excited state will last for a finite time, usually one to ten nanoseconds (sec^{-9}), during which time the fluorescent molecule (aka, *fluorophore*) undergoes conformational changes and can be subject to myriad potential interactions with its molecular environment. The first phase of Stage 2 is characterized by the fluorophore partially dissipating some of the absorbed energy, creating a *relaxed singlet excited state* S_1. It is from this state that the fluorophore will enter the second phase, the emission of energy, $h\nu_{em}$. Finally, in Stage 3, the fluorophore will return to its ground

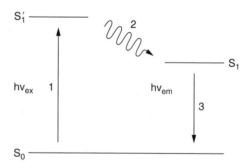

Fig 10.1   Jablonski diagram of a single fluorescence event. (1) Energy of the proper wavelength excites the fluorescent molecule to undergo a transition from its ground state, S_0, to an excited singlet state, S_1'. (2) The excited molecule enters a relaxed singlet state, S_1. (3) Energy in the form of emitted light is released by the fluorescent molecule from its relaxed singlet state, S_1, and the molecule returns to its ground state, S_0. The duration of the entire event is on the order of ten nanoseconds (10^{-8} seconds).

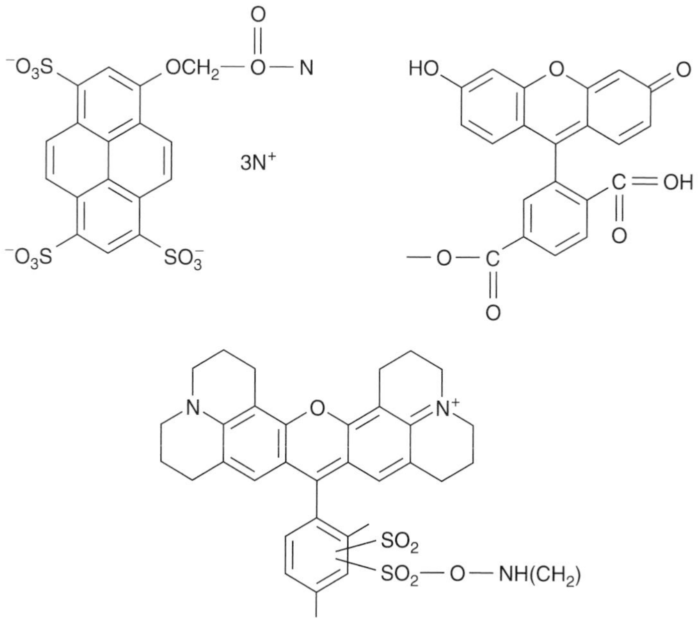

Fig 10.2 Molecular structure of three typical fluorescent molecules. Note that each molecule is a large, planar structure composed of rings with various, specific side groups attached. These molecules are, from left to right, Cascade Blue (peak absorbance 396 nanometers (nm); peak emission 410nm), fluorescein (peak absorbance 492nm; peak emission 520nm), and Rhodamine (peak absorbance 524nm; peak emission 550nm).

state, S_0. The term fluorescence comes from the mineral fluorspar (calcium fluoride) when Sir George G. Stokes observed in 1852 that fluorspar would give off visible light (fluoresce) when exposed to electromagnetic radiation in the ultraviolet wavelength. Stokes' studies of fluorescent substances led to the formulation of Stokes' Law, which states that the wavelength of fluorescent light is always greater than that of the exciting radiation.

As noted, molecules that display fluorescence are called fluorophores or fluorochromes. One group of fluorophores that is routinely used in molecular biology consists of planar, heterocyclic molecules, exemplified by fluorescein (aka FAM), Rhodamine, and Cascade Blue (Fig. 10.2). Each of these molecules has a characteristic absorbance spectrum and a characteristic emission spectrum. The specific wavelength at which one of these molecules will most efficiently absorb energy is called the peak absorbance and the wavelength at which it will most efficiently emit energy is called the peak emission. An example of these characteristic spectra is shown in Fig. 10.3 for the molecule TAMRA. The difference between peak absorbance and peak emission is known as the Stokes Shift, after Sir George Stokes. Peak absorbance and peak emission wavelengths for most of the fluorophores used in molecular applications are shown in Table 10.1.

Energy emitted from a fluorophore can also be in the form of heat dissipation. Molecules that dissipate absorbed energy as heat are a special class known as quenchers. Quenchers have the useful properties that they will absorb energy over a wide range of wavelengths and that they remain dark. As a result of these properties, quenchers have become very useful as energy acceptors

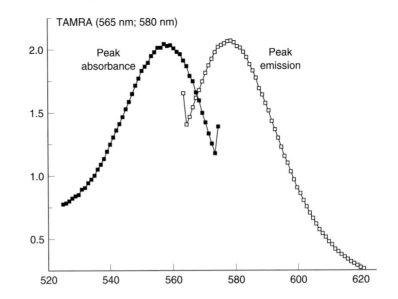

Fig 10.3 Absorbance and emission spectra for the fluorophore TAMRA. Peak absorbance and peak emission values are given. The shapes and ranges of these spectra are typical for fluorophores used in molecular biology. (Source: IDT TechVault Technical Report: Spectra of DNA-conjugated Fluorescent Dyes.)

in fluorescent resonance energy transfer (FRET) pairs. The FRET phenomenon involves the direct excitation of an acceptor fluorophore by a donor fluorophore following excitation of the donor by electromagnetic radiation in the proper wavelength. Acceptance of donor energy by a FRET acceptor requires that two criteria must simultaneously apply. One is compatibility and the other is proximity. Compatibility is precisely defined. A compatible acceptor is a molecule whose absorbance spectrum overlaps the emission spectrum of the donor molecule. If the absorbance spectrum of a molecule does not overlap the emission spectrum of the donor, the emitted energy will not be able to excite the potential acceptor. If the absorbance spectrum of the acceptor does overlap the emission spectrum of the donor, energy from the donor will excite the acceptor molecule provided that the proximity criterion is met.

Proximity is less precisely defined in operational terms. Proximity means that a compatible acceptor molecule is 'close enough' to the donor for the energy to excite it. In practical terms, it is assumed that the mechanism for excitation energy transfer between a compatible donor-acceptor fluorophore pair is the Förster mechanism in which the singlet energy transfer rate $\kappa(R)$ is,

$$\kappa(R) = \kappa_F(1/(1 + (R/R_F)^6)) \tag{1}$$

where R is the distance between the two molecules, R_F is the Förster radius and κ_F is the rate of transfer between donor and acceptor when the distance between them is small; i.e. $R/R_F \to 0$ (Förster, 1948). From (1) it can be seen that, when $R = R_F$, $\kappa(R) = \frac{1}{2}$. Thus, for convenience, we may define the Förster radius as the distance at which resonance energy transfer between compatible FRET pairs drops to 50%. What this means in molecular biology terms is that there is a maximum length of an oligonucleotide, with one member of a FRET pair tethered at each end, beyond which FRET will not be sufficiently

Table 10.1. Peak absorbance and peak emission wavelength, Stokes shift, and Extinction Coefficient, ε, for 43 Common Fluorophores.[&]

Dye	Ab(nM)	Em(nM)	SS(nM)	Extinction Coef[#]
Acridine	362	462	100	11,000
AMCA	353	442	89	19,000
BODIPY FL-Br2	531	545	14	75,000
BODIPY 530/550	534	545	10	77,000
BODIPY TMR	544	570	26	56,000
BODIPY 558/568	558	559	11	97,000
BODIPY 564/570	563	569	6	142,000
BODIPY 576/589	575	588	13	83,000
BODIPY 581/591	581	591	10	136,000
BODIPY TR	588	616	28	68,000
BODIPY 630/650[*]	625	640	15	101,000
BODIPY 650/665[*]	646	660	14	102,000
Cascade Blue	396	410	14	29,000
Cy2	489	506	17	150,000
Cy3[*]	552	570	18	150,000
Cy3.5	581	596	15	150,000
Cy5[*]	643	667	24	250,000
Cy5.5[*]	675	694	19	250,000
Cy7	743	767	24	250,000
Dabcyl[*]	453	none	0	32,000
Edans	335	493	158	5,900
Eosin	521	544	23	95,000
Erythrosin	529	553	24	90,000
Fluorescein[*]	492	520	28	78,000
6-Fam[*]	494	518	24	83,000
TET[*]	521	536	15	-
Joe[*]	520	548	28	71,000
HEX	535	556	21	-
LightCycler 640	625	640	15	110,000
LightCycler 705	685	705	20	-
NBD	465	535	70	22,000
Oregon Green 488[*]	492	517	25	88,000
Oregon Green 500	499	519	20	78,000
Oregon Green 514[*]	506	526	20	85,000
Rhodamine 6G	524	550	26	102,000
Rhodamine Green[*]	504	532	28	78,000
Rhodamine Red[*]	560	580	20	129,000
Rhodol Green	496	523	27	63,000
TAMRA[*]	565	580	15	91,000
ROX[*]	585	605	20	82,000
Texas Red[*]	595	615	20	80,000
NED	546	575	29	-
VIC	538	554	26	-

(Source: IDT Technical Bulletin: Fluorescence Excitation and Emission).

[*]Routinely offered by IDT.

[#]Energy capture efficiency.

[&]Figures are given for an activated NHS-ester with a linker arm.

efficient for reliable assays (see Kinetic PCR and Dual-labelled Probes below). In practice, this maximum length is greater than 60–70 nucleotides (nt) for many FRET pairs.

In terms of fluorescence assays using FRET pairs, consider the example of the classic FRET pair of FAM and TAMRA. Peak absorbance wavelength for FAM is 494 nanometers (nm) with a peak emission wavelength at 518nm. If FAM and TAMRA are tethered at the 5′ and 3′ ends respectively of a 35-mer oligonucleotide and this construct is excited at 494nm, so long as the oligonucleotide remains intact emission will be at 580nm and not at 518nm due to FAM transferring its energy to TAMRA. Once the oligonucleotide is disrupted by, say, an exonucleolytic reaction, excitation at 494nm will result in emission at 518nm. This is due to the fact that the pair is no longer tethered and, even though they are compatible, they are no longer proximate.

In recent years TAMRA has been replaced with one or another of the growing family of dark quencher molecules. Quenchers are chemically related to fluorophores but instead of emitting absorbed fluorescence resonance energy as light they have the useful property of transforming the light energy to heat. Heat dissipation of fluorescence energy means that replacing a fluorescent acceptor like TAMRA with a quencher such as Iowa Black FQ^{TM} will result in no measurable fluorescence so long as the tether remains intact. Such constructs can greatly simplify many fluorescence assays. As will be discussed below, dark fluorophore-quencher probes have become a standard in kinetic (real-time) PCR. A compilation of commonly used FRET pairs, including quencher pairings, is provided in Table 10.2.

Table 10.2. | Reporter/Dark Quencher pairs based upon the dynamic range of each quencher.

DABCYL	Iowa Black^{TM}-FQ	Iowa Black^{TM}-RQ
Oregon Green^{TM} 488-X	6-FAM^{TM}	Oregon Green^{TM} 514
6-FAM^{TM}	Rhodamine Green^{TM}-X	TET^{TM}
TET^{TM}	Oregon Green^{TM} 514	JOE
JOE	TET^{TM}	HEX^{TM}
HEX^{TM}	JOE	Cy3^{TM}
Cy3^{TM}	HEX^{TM}	Rhodamine Red^{TM}-X
(TAMRA^{TM})	Cy3^{TM}	ROX^{TM}
(ROX^{TM})	Rhodamine Red^{TM}-X	Texas Red^{TM}-X
(Texas Red$^®$)	ROX^{TM}	TAMRA^{TM}
	Texas Red^{TM}-X	Bodipy 630/650^{TM}-X
		Bodipy 650/665^{TM}-X
		Cy5^{TM}

Source: IDT TechVault Technical Report: Fluorescence and Fluorescence Applications.

Automated fluorescence DNA sequencing

Prior to the late 1970s no generally useful method for directly sequencing DNA was available. In 1977 two separate papers appeared that offered different ways of directly sequencing DNA. Allan Maxam and Walter Gilbert developed a method that took advantage of a two-step catalytic process involving piperidine and two chemicals that selectively attacked purines and pyrimidines (Maxam and Gilbert, 1977). Using radioactive labels the order of the bases in a DNA strand could be read with a fair degree of accuracy. To be sure it was no high throughput method and it involved the simultaneous use of radio-activity, polyacrylamide gels, and hydrazine, a neurotoxin. Fragments were resolved in polyacrylamide gels and then autoradiographed. Base calling was a manual task. In spite of the labour-intensive nature of Maxam-Gilbert chemical cleavage DNA sequencing, the revolution in understanding that it ushered in was almost instantaneous. Among the surprises that direct DNA sequencing had in store for geneticists was the discovery of introns (Breathnach *et al.*, 1977; Jeffries and Flavell, 1977; Breathnach *et al.*, 1978).

At the same time, Fred Sanger and colleagues announced an alternative DNA sequencing method based upon the use of dideoxyribose chain terminators (Sanger *et al.*, 1977). This method was enzymatic rather than chemical and made use of synthetic oligonucleotides to prime the reaction. In the presence of an appropriate primer sequence, a complementary copy of a DNA target strand was synthesized by a DNA polymerase. By carefully adjusting the relative proportions of deoxyribose nucleotides and dideoxyribose nucleotides, products of increasing lengths, each ending in a known nucleotide, would be produced. The initial reporting method involved radiolabelling the oligonucleotide primer, resolving the fragments on polyacrylamide gels, and autoradiography. Base calling remained a manual task but some of the inefficiency and one of the nasty chemicals had been removed. In 1986, Leroy Hood and colleagues reported on a variation of Sanger chain-termination DNA sequencing in which the primer-linked radioactive reporter was replaced with four primer-linked fluorescent reporters, each sequestered in a separate reaction containing only one of the four dideoxyribose chain terminators (Smith *et al.*, 1986). Fluorescence wavelength detection permitted the introduction of automated base-calling and the method was commercialized in 1987 by Applied Biosystems.

James M. Prober and his colleagues at DuPont took automated fluorescence DNA sequencing to the next level by developing 'a more elegant chemistry' (Dovichi and Zhang, 2000: 4465). Instead of fluorescence-labelled primers, they labelled the terminators themselves. The first 'dye set' was based upon succinylfluorescein. Slight shifts in the emission wavelengths of the dyes were achieved by changing the side groups. The dyes SF505, SF512, SF519 and SF526

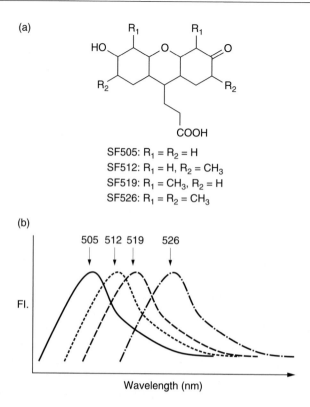

Fig 10.4 Chemical structures and emission spectra of the original succinylfluorescein dye terminators developed by DuPont in the 1980s. A. The central structure of the molecule is shown along with the specific side-group (R_1 and R_2) modifications. B. The peak emission wavelength shifts achieved by each side-group change are shown. The 7nm shifts obtained by each modification of the central structure are sufficient to uniquely identify each modification. (Figure adapted from Prober et al., 1987.)

were attached to dideoxy terminators ddG, ddA, ddC, and ddT respectively. The four dyes and their emission spectra are shown in Figure 10.4. All four dye labelled terminators could be excited by an argon ion laser at 488nm and each would produce a peak emission that could be distinguished by the detector. This detection system meant that the sequencing reaction could be carried out in a single tube with all four terminators present and fragment resolution would require only one gel lane (Prober et al., 1987). DuPont commercialized this technology for a brief period and then sold the licence to Applied Biosystems.

Applied Biosystems continued to refine both the terminator chemistries and the detection/base calling systems into the 1990s. Major refinements of the chemistry involved changing the dye labels on the terminators and improving fragment resolution. The fluorescent dyes were changed to a series of rhodamine derivatives; ddG was tagged with dichloroROX, ddA with dichloroR6G, ddC with dichloroR110, and ddT with dichloroTAMRA. Fragment resolution was improved by substituting deoxyInosine triphosphate (dITP) for dGTP and deoxyUridine triphosphate (dUTP) for dTTP. The former helped eliminate band compression on the gels and the latter helped with ddT incorporation in the sequencing reactions. Even though these improvements led to significant increases in DNA sequencing throughput, they were still acrylamide gel-based systems. In spite of the improvements in the reactions, detection and data interpretation, gel-based sequencing was still labour intensive and not well suited to a high throughput environment. In the early 1990s Harold Swerdlow and colleagues reported on the use of capillaries to obtain DNA sequences (Swerdlow *et al.*, 1990, 1991). Capillary electrophoresis was a well-established technique in analytical chemistry in the late 1980s. Capillaries are small, a 50μm inner diameter, and they dissipate heat very efficiently due to their high surface area to volume ratios. This means that a capillary-based system can be run with much higher voltages thus dramatically lowering the run times. Most importantly, capillary systems can be automated, a major limitation in gel-based systems. In 1993, B. L. Karger and colleagues reported on the use of a low viscosity separation matrix that could be pumped into capillaries at relatively low pressure (Ruiz-Martinez *et al.*, 1993). This matrix could replace cross-linked polyacrylamide and remove the final obstacle to the development of a truly automated DNA sequencing platform. With cross-linked polymers the capillary could not be reused. The low viscosity non-cross-linked polymer could be flushed out after a run and replaced for the next run without having to touch the capillary. Studies of thermal stability by Zhang *et al.* (1995) established that a non-cross-linked polymer would be stable at 60°C and would deliver high quality sequence data. Here, then, were all the elements required for the development of a fully automated, high throughput DNA sequencing platform.

DNA sequencing reactions can be carried out in a single reaction tube and be prepared for loading once the reaction reagents have been filtered out. The capillary system is set up to deliver new polymer to the capillary, load the sequencing reaction into the capillary, apply a constant electrical current through the capillary, and have the resolved fragments migrate past an optical window where a laser would excite the dye terminator, a detector would collect the fluorescence emission wavelengths, and software would interpret the emission wavelengths as nucleotides. At the present time such systems can deliver 500–1000 bases of high quality DNA sequence in a couple of hours.

The human genome
Once the separate paths of fluorescence dye-terminator DNA sequencing and capillary electrophoresis crossed, the era of high throughput sequencing began. Though it had been suggested in the early 1980s and actually initiated in 1988, The Human Genome Project did not settle into the massive, high throughput effort it eventually became until a true high throughput sequencing platform became a reality in the mid-1990s (see *Science* 291: 1145–1434, 2001 and *Nature* 409: 745–964, 2001). Today, the human genome has been sequenced and reasonably well annotated and many other genomes have been added to the list. These include mouse, rat, *E. coli*, *Arabidopsis thaliana*, rice, *C. elegans*, *Danio rario*, and dozens of microbial and viral genomes. Indeed, the contemporary high throughput DNA sequencing environment churns out completed genomes in ever-shorter time spans. A case

in point is the SARS corona virus. First discovered in 2002, the SARS genome was fully annotated by August of 2003 (Snijder *et al.*, 2003). Efforts are currently under way to sequence the chimpanzee and Rhesus macaque genomes and others will follow in time.

Mitochondrial genomes

Anderson *et al.* (1981) reported on the entire 16,569 base sequence of the human mitochondrial genome. While it would be a day's work today, this was a landmark effort in the sequencing environment of the early 1980s. Among the surprising features of this sequence was that the mitochondrial genome is as compact as any genome ever seen. Genes are packed in with little or no intergenic non-coding sequence and the genes themselves lack many of the traits normally expected in eukaryotic genes. Mitochondrial mRNAs lack non-translated leader and trailing sequences and more than half do not even have a stop codon. The mitochondrial genetic code is different from the eukaryotic code; UGA is read as tryptophan rather than as STOP; AGA and AGG, normally read as arginine, are read as STOPs; AUA is methionine and not isoleucine; and the ubiquitous AUG start codon is sometimes replaced by AUA or AUU in mitochondrial genes. Subsequent studies of other mtDNAs have shown that the mitochondrial genetic code is not even universal among mitochondria. Yeast mitochondrial genomes, for example, are much larger and have not reassigned the AUA, AGA, and AGG codons. Yeast has reassigned CTN as leucine rather than threonine.

The human mitchondrial D-loop

While the vast majority of the mitochondrial genome is under the scrutiny of selection because mutations in these areas are usually deleterious, there is a region in which there are no coding sequences and mutations are free to accumulate at will. This region is in the mitochondrial D-loop. The D-loop is the location of mitochondrial transcription promoters. MtDNA replication begins in the D-loop, resulting in the formation of a displacement loop with a newly synthesized heavy, or H, strand of about 700nt, known as 7S DNA (Anderson *et al.*, 1981). Both strands of the mtDNA are completely transcribed from the promoters in the D-loop. In addition to the promoter sequences, there are two small regions known as the hypervariable regions I and II (HVI and HVII). Mutation rates in HVI and HVII are especially high on average and there is evidence that the rates vary *within* the regions as well (Jazin *et al.*, 1998).

As a result of the high average mutation rates and the lack of coding or regulatory sequences in the hypervariable regions, they have become a tremendously valuable source of presumably neutral human genetic variation. In addition, since mtDNA is maternally inherited (sperm do not have mitochondria), there is no recombination between parental genomes. Thus, in every generation, you only have *one* mitochondrial ancestor whereas in nuclear DNA the number of ancestors increases by a factor of 2^n, where n is the generation number. This direct inheritance of mtDNA led to the idea that all humans alive today had a single common mitochondrial ancestor at some point in the dim past (cf. Cann *et al.*, 1987; Vigilant *et al.*, 1991; Relethford, 2001).

Of relevance here is that the high throughput sequencing environment in which we find ourselves today is a consequence of oligonucleotide-directed, DNA polymerase catalyzed, fluorescence dye terminator chemistries that have made direct DNA sequencing an everyday research tool. Anthropological Genetics laboratories routinely sequence the hypervariable D-loops of dozens, or, even hundreds, of members of the genus Primates. Sophisticated analytical tools are constantly being developed to place these sequences in their proper contexts and ever-increasing efficiency is dropping costs to a level where everyone can play.

Kinetic (Real-Time) PCR

The picture of PCR amplification that has been reproduced in dozens of textbooks and papers is the conventional view. It is important to note that this picture is, in fact, an over-simplification of the actual course of amplicon production in PCR. While the conservation of template molecules and the relatively plodding arithmetic production of anchored PCR products are accurately portrayed, the true shape of the amplicon curve is more complex. The actual trajectory of amplicon production is composed of three separate, sequential phases. The first phase is the exponential phase in which the DNA polymerase is madly churning out short, double-stranded amplicons at near-optimum capacity. However, after a time, amplification rate begins to slow and amplicon production enters a log-linear phase sometimes called a quasi-linear phase. Finally, the reaction changes amplification rate again and enters the final, plateau phase. The trajectory of amplicon production is shown in Figure 10.5.

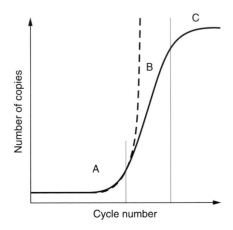

Fig 10.5 Amplicon production trajectory in a typical PCR reaction. A. Amplicon production follows a standard 2^n exponential course. B. The reaction transitions from an exponential trajectory to a log-linear trajectory. C. The reaction enters the plateau phase as the mass of double-stranded species in the tube begins to bind sufficient numbers of polymerase molecules to inhibit amplicon production (see Kainz, 2000).

A number of plausible explanations have been offered to explain the phenomenon of trajectory change in PCR that, theoretically, should not occur under ideal conditions. Among these are: (1) utilization of reagents, (2) thermal inactivation of the polymerase enzyme, (3) polymerase enzyme inhibition via increasing pyrophosphate concentration, (4) reductions in denaturation efficiency, and (5) exonuclease activity. None of these explanations has proved true. Reagent utilization has been eliminated by experiments in which each component was titrated without any apparent effect on trajectory changes. Thermal inactivation has also been eliminated by the observation of trajectory change in the presence of extremely thermal stable polymerases. Pyrophosphate build-up was an attractive idea because these 'waste products' of *in vitro* nucleotide addition; i.e. the reaction $dNTP + (dNMP)_n \rightarrow (dNMP)_{n+1} + pp_i$ wherein each polymerase-catalyzed nucleotide addition results in the irreversible addition of a pyrophosphate, pp_i, to the reaction environment, could potentially be inhibiting at late cycle number concentrations. However, this idea, too, failed under direct experimental scrutiny. Nor were reductions in denaturation efficiency or exonuclease activities able to pass muster in late cycle conditions. In the end, the answer came from a not-so-simple (except in hindsight) thermodynamic phenomenon.

Kainz (2000), who beautifully defined the plateau phase as 'attenuation in the rate of exponential product accumulation', showed that it is accumulation of the amplicons themselves that induces trajectory change. Once amplicons begin to be produced they exist as double-stranded species that are melted along with the target DNA and the anchored products during the denaturation step. At the end of the denaturation step the reaction is cooled down to the primer annealing temperature. Target DNA begins to re-anneal and, if it is genomic DNA, renaturation follows the familiar C_0t curve trajectory with single copy sequences renaturing last. So, too, will the anchored products begin to renature. However, it is the short, renaturing amplicons that begin to cause havoc because they present inviting targets for polymerases. Polymerase molecules will not be affected to any great degree by the relatively small numbers of double-stranded species presented by the renaturing target DNAs and anchored products. On the other hand, the mass increase of double-stranded species represented by the amplicons that have average T_Ms in the middle to high $80\,^{\circ}C$ ranges will begin to exert an effect on the polymerase population over time. Thus, showed Kainz, the production of amplicons is a self-limiting process.

A consequence of the non-exponential trajectory of the later part of a PCR reaction is that true quantitation of amplicon production is, at best, difficult. In the early 1990s Higuchi and colleagues showed that amplicon production could be monitored in real time by the addition of a fluorescent intercalating agent to the PCR reaction (Higuchi *et al.*, 1992, 1993). The agent, ethidium bromide, would intercalate, or bind to the minor groove of a DNA helix, and display

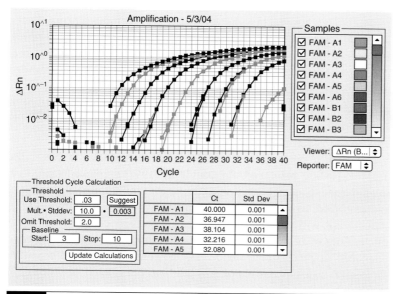

Fig 10.6 Screen shot of a typical Kinetic PCR experiment using a dual-labelled probe. The shape of the curves follows the trajectory shown in Figure 10.5. The threshold is set at the point at which the trajectory for each reaction transitions from exponential to log-linear. Note that each set of reactions ultimately plateaus at the same place regardless of where it enters the log linear phase. Under the conditions of these assays, the point at which the replicate curves cross the threshold is determined by the absolute amount of starting material. Thus, cycle thresholds are a measure of comparative starting concentration.

a significant increase in fluorescence over unbound molecules. Taking initial, pre-cycling fluorescence as the baseline, total fluorescence would increase in proportion to the increase in the number of double-stranded amplicons present in the reaction. As the amount of fluorescence in the reaction tube can be measured with great accuracy, the relative number of amplicons can be estimated with equal accuracy. In theory, fluorescence increase and amplicon production are the same. As shown in Figure 10.6, the fluorescence signal from a reaction containing the appropriate primers, probe, and template (R_n) has a trajectory that looks like the amplicon trajectory seen in Figure 10.5. The threshold is set just above the background fluorescence due to 'noise' and the amount of signal that is given by a No Template Control (NTC) reaction. The difference, ΔR_n, is the actual increase of fluorescence over the background due solely to the production of amplicons. The point at which the fluorescence signal first rises above the background/NTC level is called the threshold cycle, C_T. The numerical value assigned to the threshold cycle is a number interpolated between the fluorescence intensity measured at the whole cycle just below C_T and that measured at the whole cycle just above C_T. Fluorescence intensity is reported in relative fluorescence units (RFUs).

In theory, any fluorescence reporter that either anneals to or binds with an amplicon sequence will produce the fluorescence profile above. In practice, however, the use of intercalating agents such as ethidium bromide requires additional optimization of the PCR reaction because of the phenomenon of primer dimer (Vandesompele *et al.*, 2002). Primer dimers are also double-stranded and, as such, will permit intercalation just as easily as amplicons.

Fluorescent reporters in real-time PCR

The key to real-time monitoring of PCR amplicon production is the presence of a reporter molecule designed to specifically detect the amplicon being produced. Holland *et al.* (1991) employed the 5′→3′ exonuclease function of Taq polymerase in conjunction with a radiolabelled hybridizing oligonucleotide to detect amplicons. A polymerase with 5′→3′ exonuclease ability will cleave 5′ terminal nucleotides of an annealed DNA provided that there is a leading edge displaced single-stranded region. The polymerase will then continue to cleave nucleotides until the entire hybridized molecule is displaced. Recognizing this, Holland *et al.* designed hybridizing oligonucleotides complementary to one of the strands of the amplicon in which there was a 5′ displacement and a radioactive reporter at the 5′-most end. This construct would anneal to the amplicon and the polymerase would cleave the radioactive reporter releasing the label into solution. They tuned this system such that free label would be produced only if the desired amplicon was present in the reaction. As noted, Higuchi *et al.* (1992, 1993) substituted the double-stranded DNA intercalating agent ethidium bromide for the radioactive label, thus creating a more sensitive reporter for amplicon production. The drawback of the ethidium bromide system was that there was no guarantee that the amplicon being produced was really the desired product and the potentially confounding influence of primer dimer has already been mentioned.

The problem of specificity was solved by Livak *et al.* (1995) when they introduced the first amplicon-specific hybridizing probe having fluorescent molecules at both the 5′ and 3′ ends. The two fluorescent molecules were chosen such that they were compatible for FRET (fluorescence resonance energy transfer). These dual-labelled probes would permit 'real-time' monitoring of a specific amplicon in the same *in vitro* reaction as the PCR itself and could do so with greater sensitivity and safety than either radioactivity or ethidium bromide could provide. Since 1995 a wide range of fluorescent monitoring systems have been developed. Though they go by a number of different names, these systems fall into one or another of four basic categories. These are: (1) intercalating reporters, (2) hybridizing FRET reporters, (3) hydrolysis reporters, or (4) molecular beacons (Figure 10.7).

Intercalating reporters

Intercalating reporters, or intercalating dyes, are the only molecules that do not specifically hybridize to an amplicon. Rather, they are

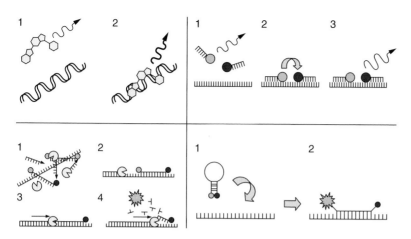

Fig 10.7 Mechanism of action of the four classes of fluorescent reporter. On the upper left, an intercalating agent binds to the minor groove of a double-stranded DNA molecule resulting in a large increase in fluorescence. On the upper right a pair of hybridizing FRET probes anneal to their target sites of an amplicon. Annealing places the fluorescent reporter molecules proximate to each other and fluorescence resonance energy transfer occurs. On the lower left a dual labelled hydrolysis reporter, with a fluorophore at the 5′ end and a dark quencher at the 3′ end, anneals to its amplicon target sequence and awaits the arrival of the polymerase. Polymerase exonucleolytic action hydrolyzes the oligonucleotide tether, releasing the fluorophore from the quencher. Finally, on the lower right, a molecular beacon in its closed form approaches its amplicon target, the bimolecular reaction becomes the preferred state for the probe, it anneals to the target sequence and the fluorophore is removed from the influence of the DABCYL quencher.

members of a class of fluorescent molecules that undergo a significant increase in fluorescent radiation when they bind in the minor groove of a double-stranded DNA molecule. The classic member of this group is, of course, ethidium bromide which has been used for many years to detect double-stranded DNA in gels. However, it is well known that ethidium bromide is a mutagen and both handling and disposal present a continuous set of problems for laboratories using it. In the late 1990s several reports surfaced on the use of another intercalator possessing much the same sensitivity of ethidium bromide without the drawbacks. This agent, called SYBR Green, exhibits low fluorescence in solution but increases in fluorescence energy when bound to double-stranded DNA (Wittwer et al., 1997; Morrison et al., 1998). When present in solution in an ongoing PCR reaction, fluorescence increases during the elongation step as more and more SYBR Green molecules bind to the amplicons. Fluorescence then drops off dramatically during the denaturing step of the next cycle as the SYBR Green molecules drop off the now single stranded DNA species. Fluorescence detection of SYBR Green takes place at the end of the extension step with excitation at 494nM and emission at 521nM.

The relative ease of use and disposal of SYBR Green has made it a popular reporter for real-time PCR. There are, however, many other minor groove binders that display properties similar to SYBR Green. Molecules such as oxazole yellow (YO) and thiazole orange (TO) are intercalating asymmetric cyanine dyes that also exhibit a large increase in fluorescence intensity upon DNA binding (Nygren *et al.*, 1998). Unlike SYBR Green, these dyes do not perform well in PCR conditions and they have not been used often. One dye that does behave well is DAPI (4′,6-diamidino-2-phenylindole) and it has been incorporated successfully as a conjugate with 5′ hydrolysis oligonucleotides (Kutyavin *et al.*, 2000). A new stand-alone intercalator is BEBO (4-[3-methyl-6-(benzothiazol-2-yl)-2,3-dihydro-(benzol-1,3-thiazole)-2-methylidene) has been reported on by Bengtsson *et al.* (2003). This dye is a derivative of the intercalating asymmetric cyanine dye 1-methyl-4-[3-methyl-2(3H)-benzothiazolidene)methyl]-pyridinium iodide (BO) in which a benzothiazole group has been added, giving the molecule the bent shape similar to minor groove binders like DAPI. Bengtsson *et al.* show that BEBO compares favourably in every respect with SYBR Green when used in real-time PCR assays. They note, however, that direct comparison with SYBR Green is limited as the structure, binding mode and stock solution concentration for SYBR Green is unavailable from Molecular Probes (cf. Wittwer *et al.*, 1997).

While there may be a plethora of intercalating fluorescent dyes available in the near future, one factor remains problematic. An intercalator is indifferent to the nature and source of double-stranded DNA species *in vitro*. Thus the problem of non-specific priming and primer dimer remains. For this reason systems that use intercalating dyes like SYBR Green suggest incorporation of a dissociation curve assay as the final step in a real-time PCR assay. As the final step in a real-time assay, the dissociation curve consists of a continuous monitoring of fluorescence starting at the primer annealing temperature, usually 60°C, and ending at 95°C. At 60°C all of the DNA in the reaction will be double stranded and fluorescence will be maximal. As the temperature increases all of the double-stranded species will dissociate beginning with primer dimer and ending with template DNA. As dissociation progresses fluorescence will decrease due to the release of bound dye into solution. Classically, there will be a sharp drop in fluorescence followed by slow approach to background levels. A theoretical dissociation profile is shown in Figure 10.8. The inflection point of the dissociation curve corresponds to the melting temperature of the amplicon. A better way to see this is to take the derivative of fluorescence intensity versus temperature increase (dRFU/dT). This is also shown in Figure 10.8.

Hybridizing FRET reporters

As noted above the issue of specificity was addressed with the use of probes that hybridized to the amplicons. Two variations on this theme, to be discussed later, involve a single reporter molecule.

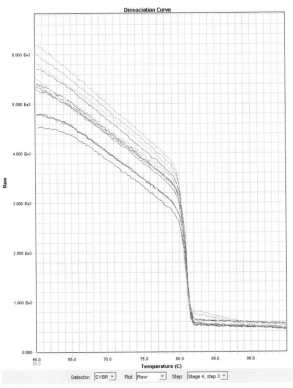

Detector: SYBR ▾   Plot: Raw ▾   Step: Stage 4, step 3 ▾

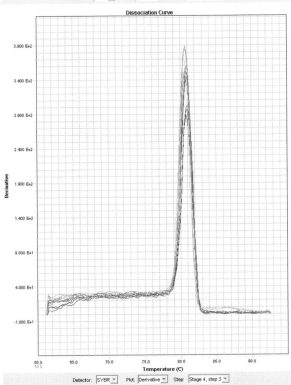

Detector: SYBR ▾   Plot: Derivative ▾   Step: Stage 4, step 3 ▾

Fig 10.8 A typical dissociation (melt) curve obtained from a SYBR Green kinetic amplification experiment. A. Once the final amplification cycle is completed, an overwhelming mass of SYBR Green molecules are bound to double-stranded amplicons. As the temperature increases the amplicons begin to melt and release the intercalated SYBR Green molecules. The inflection point of the melt curve is the point at which half of the amplicons are melted. B. Taking the derivative of the melt curve shows the inflection point as a peak corresponding to the T_M of the amplicon by definition. The different melt curves reflect differences in the amount of starting materials in the reactions.

Here, we will discuss a method involving the use of two hybridizing probes that are designed to anneal in tandem to the amplicons. One of the probes carries a fluorescent tag on its 3′ end while the other carries a compatible FRET acceptor tag on its 5′ end (Wittwer *et al.*, 1997). This design, which maximizes specificity of the assay, is shown in Figure 10.7. Transfer of fluorescent energy from the donor probe to the acceptor probe requires that both probes be hybridized to the amplicon target.

The standard composition of tandem hybridization probes is to place fluorescein at the 3′ end of one probe as the donor moiety and any one of several fluorescent molecules at the 5′ end of the adjacent probe to serve as an unambiguous acceptor moiety. Those familiar with LightCycler technology will recognize dyes such as LCRed 640 and LCRed 705 as popular choices. This scheme also permits multiplexing of tandem probe pairs using a common donor excitation wavelength and different acceptor emission wavelengths.

An excellent application of a dissociation curve end-point assay using tandem hybridizing probes is SNP genotyping. As an example, Bestmann *et al.* (2002) assayed a SNP in the human tumour necrosis factor beta (TNF-β) gene called A329G. This polymorphism is relatively common (40%−45%) and has been associated, along with several others, with clinical phenomena such as septic shock and various autoimmune disorders. Bestmann *et al.* used a FRET probe pair containing fluorescein (upstream) and LCRed640 (downstream) against a 200bp TNF-β amplicon. They designated the upstream probe as the 'anchor' and positioned it in an invariant sequence adjacent to the 'sensor' probe that was designed to anneal to the G allele. Their first derivative of the dissociation curve clearly differentiated a homozygous wild-type (AA) peak, a homozygous mutant (GG) peak, and a heterozygote (AG) double peak.

Hydrolysis reporters

Conventional PCR assays, whether they target DNA or, secondarily, mRNAs, employ a pair of priming oligonucleotides that serve as the substrate for a DNA polymerase to make new, defined DNA strands *in vitro*. Many DNA polymerases will have both a polymerase and a proof-reading protein domain. These domains function to generate the complementary copy of the DNA template and to check the copy for accuracy. In addition, however, some polymerases such as Taq and Tth also have a 5′→3′ exonuclease domain. The function of this domain is to clear the path for the polymerase and proof-reading domains. The best analogy I have come upon is that of a 'cow catcher' on a locomotive. Its job is to clear the tracks of any foreign objects. In the case of a DNA polymerase, any nucleotides remaining annealed to the target DNA strand are foreign objects that must be cleared so that the polymerase can perform its function.

A significant utility for the 5′→3′ exonuclease function of Taq polymerase was recognized by Holland *et al.* (1991). A polymerase with the 5′→3′ exonuclease domain will cleave the 5′ terminal nucleotides of an annealed DNA, provided that there is a leading edge-displaced, single-stranded region, and will continue to cleave nucleotides until the entire molecule is displaced. Recognizing this, Holland *et al.* designed hybridizing oligonucleotides with a 5′ displacement and a radioactive reporter at the 5′-most end. They found that Taq polymerase would digest the annealed oligonucleotide and release the 5′ label into solution. They further realized that such a system could be tuned to provide detection of specific PCR amplicon sequences without the requirement of agarose gel electrophoretic amplicon resolution. If free label was present, then the amplicon must be as well. Higuchi *et al.* (1992, 1993) suggested that the use of free Ethidium Bromide (EtBr) could replace radioactivity since EtBr would only bind to double-stranded DNAs. If an amplicon were being made *in vitro*, the amount of EtBr fluorescence would increase proportionately. The drawback to the EtBr method was that, even though the fluorescence would increase proportionately, there was no guarantee that the amplicon being monitored was the correct amplicon. This problem was solved by Livak *et al.* (1995) when they designed an amplicon-specific probe with FRET compatible fluorescent dyes on either end. This *dual-labelled probe* design would permit 'real-time' monitoring of a specific amplicon in the same *in vitro* reaction as the PCR and could do so with greater sensitivity and safety than either radioactivity or ethidium bromide could provide. They examined a number of designs in which a 5′ 6-FAM (fluorescein) fluorophore was coupled with a 3′ TAMRA at various positions of an oligonucleotide. One construct they examined was a 26-mer specific for β-actin in which 6-FAM was placed at the 5′ end and TAMRA was placed in position 2, 7, 14, 19, 22, or 26. The best results were realized when the TAMRA acceptor was placed at the 3′ end.

The mechanics of the Livak *et al.* assay were simple. During the course of the PCR amplification, one strand of the amplicon would serve as the target for the dual-labelled probe. As the primer annealed to the single-stranded target amplicon the probe would anneal as well. Once the polymerase began to process along the target strand it would encounter the annealed probe. The 5′→3′ exonuclease function of the polymerase would degrade the probe and release the 6-FAM reporter from its TAMRA tether (Fig. 10.7). If the reaction was excited by energy at 494nm any emission detected at 525nm was due to degraded probe. A tethered 6-FAM would transfer energy to TAMRA so that excitation at 494nm would yield emission at 573nm. This construct proved to be a sensitive monitor of the correct amplicon. In the Holland *et al.* (1991) paper, the DNA polymerase was symbolized by a character reminiscent of the video game 'Pac-Man'. Because the polymerase enzyme used in the Livak *et al.* (1995) assay was Taq polymerase, the whole assay became known as 'Taq-Man®' and the dual-labelled probes containing 6-FAM at one end and TAMRA at the other end became known as 'Taq-Man' probes.

Molecular beacons

At about the same time that the enzymatic dual-labelled probe system was being developed, Fred R. Kramer and colleagues in the Department of Genetics at the Public Health Research Institute in New York were perfecting a dual-labelled probe system that relied on DNA:DNA hybridization thermodynamics. In their method, a dark molecular construct incorporating the 'Universal Quencher' DABCYL was designed such that, in a particular temperature regime, the bi-molecular reaction involving the dual-labelled probe and its target sequence would be thermodynamically favoured. In order to achieve this, Kramer and colleagues created a dual-labelled probe structure that contained the target sequence complement plus a self-complementary non-target sequence added to the 5′ end and the 3′ end. The fluorescent reporter molecule and the DABCYL quencher were then affixed with the reporter at the end of the 5′ self-complementary sequence and the DABCYL quencher at the end of the 3′ self-complementary sequence. The sequence of the probe was, therefore: 5′-Reporter-Self Complement-Target Complement-Self Complement-Quencher-3′.

The target complement sequence was long enough to uniquely recognize the target in whatever background was present and the non-target self-complement was usually, 5–6 bases long. If the target complement was, say, 24 bases long, the entire probe would be 34 to 36 bases long depending on the length of the self-complement.

The motivation behind this design is that the phenomenon of annealing a nucleic acid sequence to its complement is a highly specific molecular recognition event. It was known that the proper mix of oligonucleotide sequence, length, and hybridization conditions can even lead to single nucleotide discrimination. The question was how to devise a dual-labelled probe that could take advantage of this. In 1996, Tyagi and Kramer reported on a solution to this question when they announced the development of the *Molecular Beacon*, the dual-labelled probe discussed above that would fluoresce upon hybridization to a specific target sequence. The key to their system was the non-target self-complementary sequence that would form a stem-loop structure. This stem-loop would bring the DABCYL quencher into contact with the fluorescent reporter and keep it dark. Within a defined range of temperatures, however, hybridization of the target complement sequence with its target would be favoured over closure of the self-complementary stem structure, the probe would open up, DABCYL would no longer be able to quench the reporter and it would be free to fluoresce (Figure 10.7). In their original paper, Tyagi and Kramer (1996) demonstrated that molecular beacon probes were in fact extraordinarily target-specific, ignoring targets that differed from the desired target by as little as a single nucleotide.

While molecular beacons are excellent fluorogenic probes for monitoring PCR amplifications in real-time, subsequent studies have shown them to also be quite versatile. Vet *et al.* (1999) reported on a set of four molecular beacons, each with a different fluorescent reporter that could simultaneously detect any combination of four pathogenic retroviruses in the same assay. Marras *et al.* (1999) employed a two probe system, in which the target complement sequence differed by a single base, to accurately genotype a single nucleotide polymorphism (SNP). Moreover, the design characteristics of molecular beacons were ideal for thermodynamic studies of probe state transitions (Bonnet *et al.*, 1999).

Monitoring PCR amplifications in real-time can be accomplished through the use of probes that fluoresce upon target sequence hybridization or probes that are designed to be enzymatically degraded. The two probe types present different sets of design considerations because they operate differently. Molecular beacons must be designed to have a T_M that lies in a fairly restricted range, about 5–7°C above the primer T_Ms, whereas dual labelled probes that are meant to be enzymatically degraded are more open ended. This difference in thermodynamic considerations is due to the fact that the molecular beacon must release from the target before they are degraded by the advancing polymerase but dual labelled probes are *supposed* to stay put so that the enzyme will degrade them. Thus, one is a reusable reagent and the other is a consumable reagent.

This thermodynamic difference also means that fluorescence is monitored at different *stages* of the PCR reaction and the reactions themselves will have different profiles. Molecular beacons and dual labelled probes will both anneal to their target sequences at about the same time as the primers anneal. However, since the molecular beacon is designed to release from the target prior to polymerase extension, fluorescence is monitored during the *annealing step* while dual labelled probes stay put, are degraded, and fluorescence is monitored during the *polymerase extension step*. Given this, a molecular beacon PCR profile will usually be a three-step cycle (i.e. 94°C, 58°C, and 72°C), but a dual labelled probe PCR profile is nearly always a two-step cycle with the primer annealing step and the polymerase extension step set at the same temperature (i.e. 94°C and 60°C).

One final point to be made about these assays is that probe annealing necessitates that a target must already exist. Thus, in both cases, the probes anneal to targets that exist *in the previous PCR cycle* and not in the current one. While this is, at best, a quibble, the concept of a real-time assay can be taken to mean that amplicons are being monitored as they are being made and not after they are made in the previous cycle. For this reason, and because real-time is often written RT and RT classically refers to reverse transcription, we prefer the term *kinetic PCR* as it has been used by Bustin (2000, 2002).

Kinetic PCR and ancient DNA

While the vast majority of kinetic PCR applications focus on differential gene expression, the nearly limitless potential for this technique was recently evidenced by Pruvost and Geigl (2004).

Ancient DNA analyses are necessarily subject to very stringent controls and cross checks. The very power of PCR becomes an enemy in the ancient DNA world against which elaborate precautions should, and are, being taken. Pruvost and Geigl have shown that kinetic PCR is a powerful ally in this struggle in that contaminants can be specifically identified and sorted out from the true signal. Moreover, this can be accomplished without loss of the precious ancient substrate. Kinetic PCR, with appropriate fluorescent reporters, uses far less substrate than conventional PCR and is much more specific.

Microarrays

Commonly used methods for assessing gene expression have been RT-PCR (reverse transcription), Western blots, and now kinetic PCR. These methods permit analysis of expression of one gene at a time. Beginning in the early 1990s, a method for fixing representative sequences from a large number of genes of interest to a solid support was developed such that expression of all of the genes could be simultaneously interrogated. At first, the solid support of choice was pure silica but rapid developments ensued in the preparation of derivatized glass supports which were more uniform in their ability to bind nucleic acid via one or another of a host of chemical linkers. With this technology it became possible to 'array' genes in precise patterns such that each column/row position in the array was a known sequence. From the outset, however, regardless of support/ attachment combination, it was clear that there were only two basic ways to produce these nucleic acid 'microarrays' or gene 'chips'. One was to synthesize the sequences right on the solid support and the other was to attach the sequences after they were synthesized (or purified). The former method is the technology employed by Affymetrix Corporation of Santa Clara, California. The latter method, generically referred to as a 'spotted array', is in use in hundreds of public and private sector laboratories worldwide.

Synthesis on the chip makes use of a method known as photolithography. Fodor *et al.* (1991) reported on the technique that relies on derivatizing synthesis units with a photolabile protecting group. In the original application the photolabile protecting group was nitroveratryloxycarbonyl (NVOC). The solid support is coated with linkers containing NVOC derivatized amino groups. Each step in the synthesis reaction begins with a deprotection step in which the amino group is rendered reactive by light. Subsequent synthesis units are themselves protected with NVOC photolabile groups until they are required for additional coupling. Translation of this technology to the manufacture of oligonucleotide arrays was a matter of derivatizing the hydroxyls on individual nucleotides and establishing a specific pattern of masks that would permit the synthesis of any desired oligonucleotide sequence in any desired pattern. Each round of synthesis would consist of four photodeprotections and four couplings, each with a different photoprotected nucleotide. Thus, for n-steps in a synthesis, there would be a total of 4^n different sequences that could be synthesized. This would translate as 1,073,741,824 *different* 15-mers for example. Using this method, with its subsequent refinements, synthesized oligonucleotide arrays containing hundreds of thousands of individual 'features' are routinely manufactured.

The alternative array synthesis technology was developed at nearly the same time as the gene chip technology. The difference was that the individual feature, whether it was an oligonucleotide or a cDNA, was synthesized or isolated and purified *prior* to attaching it to the dervatized solid support. Two separate attachment methodologies have emerged. One, the piezoelectric method, involves the same technology used in ink jet printing, while the other, classically referred to as 'spotting', involves an automated pipetting procedure (cf. Cheung *et al.*, 1999). Each of the array construction methods has advantages and disadvantages. General characteristics of the three array methods are presented in Table 10.3.

Regardless of the method employed to make the microarray, data acquisition from interrogating the array proceeds via a consistent process (Fig. 10.9). Each column by row feature on an array will consist of a finite number of copies of a specific DNA sequence. Standardization of copy number per feature across the array is an important step in array manufacture. Assuming that feature copy number consistency is achieved is essential to subsequent analyses and numerous checks are usually incorporated into the array. Once the array is assembled, interrogation begins with isolating and purifying nucleic acid from the source of interest. Often, the nucleic acid is RNA, either total or mRNA, and the source of choice is determined by the experiment. Frequently, cells in culture are the source of the nucleic acid that will be used to interrogate the array.

Taking an mRNA from a cell culture as an example, reverse transcription resulting in single-stranded cDNA is carried out. This cDNA is then processed by cleavage into small fragments and the

Table 10.3. Microarray technologies.

Characteristic	Photolithography	Piezoelectric	Microspotting
Combinatorial synthesis	Yes	Yes	No
Ink-jetting	No	Yes	No
Surface printing	No	No	Yes
Sample tracking	No	No	Yes
Feature density (cm^{-2})	~500,000	> 10,000	>6,000
Length restrictions	~25nt	none	none
Array elements	Oligos only	Oligos and cDNAs	Oligos and cDNAs

Adapted from: M. Schena (ed.) *DNA Microarrays: A Practical Approach*, 1999.

Fig 10.9 A typical process for interrogating a DNA microarray: (1) mRNA is purified from a cell source and a first-strand cDNA reverse transcription is carried out. (2) The cDNAs are then fragmented. (3) Fragments are tagged with a fluorescent reporter. (4) The tagged fragments are hybridized to the array. (5) The array is then read for the amount of fluorescence hybridized to each feature (target sequence). (6) Fluorescence is normalized against a set of controls (here, DI, A2, and F3). Normalized relative fluorescence of each experimental feature is assigned a colour corresponding to its relationship to the controls. Green is traditionally assigned to relative fluorescence lower than the norm of the array and red is traditionally assigned to relative fluorescence greater than the norm of the array. The photo is representative of colour-coded, normalized relative fluorescence on a typical spotted array.

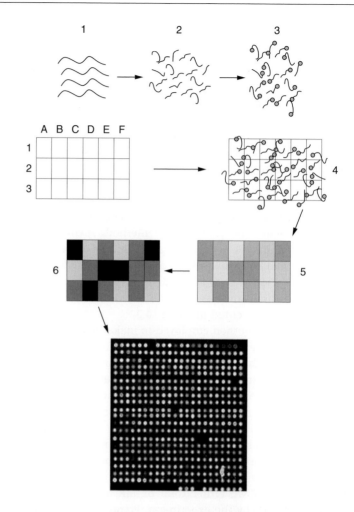

fragments are tagged with a fluorescent molecule, say, fluorescein. These tagged fragments are hybridized to the array and the assumption is made that the amount of fluorescence hybridized to any one feature will be an accurate representation of the relative amount of mRNA message initially present in the cells at the time the nucleic acid was harvested. Thus, each feature of the array will report a range of total fluorescence from zero to some finite maximum where that maximum is equivalent to the most abundant genetic message present in those cells. At this point, when the array is 'read' for total per feature fluorescence, the raw data consists of unprocessed fluorescence intensities. Processing of the raw fluores-cence data begins with assessing a set of control features that have been placed on the array in specific locations determined by the manufacturer. These control features serve two primary functions. The first is to assess the quality of the hybridization. Redundant controls scattered over the array will be expected to display the same amount of raw fluorescence intensity regardless of position if the hybridization is uniform over the array. The second function of the controls is to provide a base-line against which the other features can be compared. For this reason, controls in the form of 'housekeeping' genes are favoured. These are genes expressed in all cells regardless of type and condition. Simply put, some essential components are required for a cell to be a cell and it is the genes that encode these components that make good controls. Examples of housekeeping genes include glyceraldehydes-3-phosphate dehydrogenase (GAPDH), β-actin, glucose-6-phosphate dehydrogenase (G6PDH), β2-microglobulin (β2M), hypoxanthin-phosphoribosyl transferase (HPRT), and porphobilinogen deaminase (PBGD). Another control that has been used is 18S rRNA.

Once the quality of the array hybridization is established, the raw fluorescence intensities of the controls can be averaged and used as a base-line against which to compare the fluorescence intensities of the other features on the array. Relative fluorescence is, simply, the fluorescence intensity of an individual feature minus the base-line. On most arrays the majority of the features will not vary too much from the base-line while a few features will display a range of positive relative fluorescence intensities and a few features will display a range of negative relative fluorescence intensities. Since base-line fluorescence is a mean that will have a standard deviation, relative fluorescence intensity can be represented on a scale determined by the standard deviation around that mean. Each level of relative fluorescence at each individual feature can then be assigned a colour ranging, by convention, from bright green at the low end to bright red at the high end. This provides a rapid visual tool for assessing the overall array as well as for classifying the individual features.

Array-based comparative functional genomics

DNA microarrays permit analyses of gene expression and genetic variation at scales never before possible. Comparisons among cells from different tissues or within tissues at different stages of the cell cycle, comparisons among developmental stages, comparisons between normal cells and cells displaying various pathologies or cells responding to exposure to toxins, comparisons across species have all been carried out on platforms featuring thousands of gene sequences. For Anthropological Genetics, however, nowhere has the impact of arrays been more timely or relevant than in studies of comparative gene expression between humans and our closest biological relatives. Here, microarray results are beginning to reshape the discussion about the true differences between humans and non-human primates.

It has been axiomatic that humans and chimpanzees are nearly identical at the DNA sequence level, yet humans and chimpanzees are very different creatures both physically and behaviourally. Could just 2% or 3% of the entire genome account for this? The answer that is beginning to emerge from microarray studies is that the key lies not in the DNA sequences but rather in gene regulation. Microarray studies show that, regardless of the level or type of comparison, most genes do not differ substantially in their expression. Further, these studies show that, among those genes that do display significant variation in expression, there is a nearly even split between those that display higher expression levels (upregulation) and those that display lower expression level (downregulation). This pattern is observed in comparisons between humans and non-human primates for all tissues so far studied except the brain. Enard *et al.* (2002), using moderate-density arrays (~12,000 features), compared gene expression levels between humans, chimp, orangutan, and rhesus macaque for blood leukocytes, liver, and brain. By far the most pronounced differences in gene expression were seen between humans and chimpanzees for gene expression in the cerebral cortex. Enard *et al.* demonstrated that this difference is due to a 5.5-fold increase in the rate of change in gene expression levels in the lineage leading to *Homo sapiens*. Caceres *et al.* (2003) reported a similar finding. They identified 169 genes that exhibited differential expression in a human–chimpanzee cerebral cortex comparison. In contrast to liver and heart, where the ratios of upregulation to downregulation were nearly equal, most of the cortical differences involved upregulation and, of these, ~90% were genes that are more highly expressed in the human brain. Genome (DNA sequence) comparisons would not have predicted the striking results observed at the transcriptome (gene expression and regulation) level.

The future of fluorescence

It is always a challenge to predict the future of anything, but some predictions are easier than others. The adoption of fluorescence by molecular biologists and the meteoric rise of fluorescence-based applications over the past five years places predicting the future of this technology in the easier category. Fluorescence is extremely versatile, sensitive and safe. It is uniquely suited to the post-genome, high-throughput, parallel processing environment in which we now find ourselves. A case in point is the use of DNA microarrays for large-scale whole genome SNP (single nucleotide polymorphism) genotyping (Kennedy *et al.*, 2003). Another is verification of the transcriptome differences between humans and chimpanzees at the level of DNA methylation (Enard *et al.*, 2004). Variations on the basic theme of fluorescence-based reporting appear time and again in the form of unique or specialized assays. Moreover, new fluorophores such as BEBO (Bengtsson *et al.*, 2003), quenchers like Iowa Black FQ™ and Iowa Black RQ™ (www.idtdna.com), the unique electrostatic polythiophene biosensor recently reported by Dore *et al.* (2004), and primer/probe constructs like Scorpions™ (Whitcombe *et al.*, 1999) are constantly being developed. All of these advances and developments serve to open even more avenues for the use of fluorescence. Thus, it is safe to say that the future of fluorescence is bright indeed.

References

Anderson, S., Bankier, A. T., Barrell, B. G., de Bruijn, M. H. L., Couson, A. R., Drouin, J., Eperon, I. C., Nierlich, D. P., Roe, B. A, Sanger, F., Schreier, P. H, Smith, A. J. H., Staden, R. and Young, I. G. (1981). Sequence and organization of the human mitochondrial genome. *Nature*, **290**, 457–65.

Bengtsson, M., Karlsson, H. J., Westman, G. and Kubista, M. (2003). A new minor groove binding asymmetric cyanine reporter dye for real-time PCR. *Nucleic Acids Research*, **31**, e45.

Bestmann, L., Helmy, N., Garofalo, F., Demirtas, A., Vonderschmitt, D. and Maly, F. E. (2002). LightCycler PCR for the polymorphisms −308 and −238 in the TNF Alpha gene and for the TNFB1/B2 polymorphism in the LT Alpha gene. In *Rapid Cycle Real-Time PCR- Methods and Applications*, ed. W. Dietmaier, C. Wittwer and N. Sivasubramanian, Berlin: Springer.

Bonnet, G., Tyagi, S., Libchaber, A. and Kramer, F. R. (1999). Thermodynamic basis of the chemical specificity of structured DNA probes. *Proceedings of the National Academy of Sciences USA*, **96**, 6171–6.

Breathnach, R., Mandel, J. L. and Chambon, P. (1977). Ovalbumin gene is split in chicken DNA. *Nature*, **270**, 314–19.

Breathnach, R., Benoist, C., O'Hare, K., Gannon, F. and Chambon, P. (1978). Ovalbumin gene: evidence for a leader sequence in mRNA and DNA sequences at the exon-intron boundaries. *Proceedings of the National Academy of Sciences USA*, **75**, 4853–7.

Bustin, S. A. (2000). Absolute quantification of mRNA using real-time reverse transcription polymerase chain reaction. *Journal of Molecular Endocrinology*, **25**, 169–93.

Bustin, S. A. (2002) Quantification of mRNA using real-time reverse transcription PCR (RT-PCR): trends and problems. *Journal of Molecular Endocrinology*, **29**, 23–39.

Caceres, M., Lachuer, J., Zapala, M. A., Redmond, J. C., Kudo, L., Geschwind, D. H., Lockhart, D. J., Preuss, T. M. and Barlow, C. (2003). Elevated gene expression levels distinguish human from non-human primate brains. *Proceedings of the National Academy of Sciences USA*, **100**, 13030–5.

Cann, R. L., Stoneking, M. and Wilson, A. C. (1987). Mitochondrial DNA and human evolution. *Nature*, **325**, 31–6.

Cheung, V. G., Morley, M., Aguilar, F., Massimi, A., Kucherlapati, R. and Childs, G. (1999). Making and reading microarrays. *Nature Genetics Supplement*, **21**, 15–19.

Dore, K., Dubus, S., Ho, H.-A., Levesque, I., Brunette, M., Corbeil, G., Boissinot, M., Boivin, G., Bergeron, M. G., Boubreau, D. and Leclerc, M. (2004). Fluorescent polymeric transducer for the rapid, simple, and specific detection of nucleic acids at the zeptomole level. *Journal of the American Chemical Society*, **126**, 4240–4.

Dovichi, N. J. and Zhang, J. (2000). How capillary electrophoresis sequenced the human genome. *Angew Chemistry International Edition*, **39**, 4463–8.

Enard, W., Khaitovich, P., Klose, J., Zollner, S., Heissig, F., Giavalisco, P., Nieselt-Struwe, K., Muchmore, E., Varki, A., Ravid, R., Doxiadis, G.M., Bontrop, R. E. and Paabo, S. (2002). Intra- and interspecific variation in primate gene expression patterns. *Science*, **296**, 340–3.

Enard, W., Fassbender, A., Model, F., Adorjan, P., Paabo, S. and Olek, A. (2004). Differences in DNA methylation patterns between humans and chimpanzees. *Current Biology*, **14**, R148–R149.

Fodor, S.P.A., Read, J.L., Pirrung, M.C., Stryer, L., Lu, A.T. and Solas, D. (1991). Light-directed, spatially addressable parallel chemical synthesis. *Science*, **251**, 767–73.

Förster, T. (1948). Zwischenmolekulare energiewnaderung und fluoreszenz. *Annals of Physics*, **2**, 55–75.

Higuchi, R., Dollinger, G., Walsh, P.S. and Griffith, R. (1992). Simultaneous amplification and detection of specific DNA sequences. *Bio Technology*, **10**, 413–17.

Higuchi, R., Fockler, C., Dollinger, G. and Watson, R. (1993). Kinetic PCR analysis: Real-time monitoring of DNA amplification reactions. *Bio Technology*, **11**, 1026–30.

Holland, P.M., Abramson, R.D., Watson, R. and Gelfand, D.H. (1991). Detection of specific polymerase chain reaction product by utilizing the 5′ to 3′ exonuclease activity of *Thermus aquaticus* DNA polymerase. *Proceedings of the National Academy of Sciences USA*, **88**, 7276–80.

Jazin, E., Soodyall, H., Jalonen, P., Lindholm, E., Stoneking, M. and Gyllensten, U. (1998). Mitochondrial mutation rate revisited: hot spots and polymorphism. *Nature Genetics*, **18**, 109–10.

Jeffries, A.J. and Flavell, R.A. (1977). The rabbit β-globin gene contains a large insert in the coding sequence. *Cell*, **12**, 1097–1108.

Kainz, P. (2000). The PCR plateau phase — towards an understanding of its limitations. *Biochemistry and Biophysics Acta*, **1494**, 23–7.

Kennedy, G.C., Matsuzaki, H., Dong, S., Liu, W.-M., Huang, J., Liu, G., Su, X., Cao, M., Chen, W., Zhang, J., Liu, W., Yang, G., Di, X., Ryder, T., He, Z., Surti, U., Phillips, M.S., Boyce-Jacino, M.T., Fodor, S.P.A. and Jones, K.W. (2003). Large-scale genotyping of complex DNA. *Nature Biotechnology*, **21**, 1233–7.

Kutyavin, I.V., Afonia, I.A., Mills, A., Gorn, V.V., Lukhtanov, E.A., Belousov, E.S., Singer, M.J., Walburger, D.K., Lokhov, S.G., Gall, A.A., Dempcy, R., Reed, M.W., Meyer, R.B. and Hedgpeth, J. (2000). 3′-minor groove binder-DNA probes increase sequence specificity at PCR extension temperatures. *Nucleic Acids Research*, **28**, 655–61.

Livak, K., Flood, S., Marmaro, J., Giusti, W. and Deetz, K. (1995). Oligonucleotides with fluorescent dyes at opposite ends provide a quenched probe system useful for detecting PCR product and nucleic acid hybridization. *PCR Methods and Applications*, **4**, 357–62.

Marras, S.A.E., Kramer, F.R. and Tyagi, S. (1999). Multiplex detection of single-nucleotide variation using molecular beacons. *Genetic Analysis: Biomolecular Engineering*, **14**, 151–6.

Maxam, A.M. and Gilbert, W. (1977). A new method for sequencing DNA. *Proceedings of the National Academy of Sciences USA*, **74**, 560–4.

Morrison, T.B., Weis, J.J. and Wittwer, C.T. (1998). Quantification of low-copy transcripts by continuous SYBR green I monitoring during amplification. *Biotechniques*, **24**, 954–62.

Nygren, J., Svanik, N. and Kubista, M. (1998). The interactions between fluorescent dye thiazole orange and DNA. *Biopolymers*, **46**, 39–51.

Pruvost, M. and Geigl, E.-M. (2004). Real-time quantitative PCR to assess the authenticity of ancient DNA amplification. *Journal of Archaeological Science*, **31**, 1191–7.

Relethford, J.H. (2001). *Genetics and the Search for Modern Human Origins.* New York: Wiley-Liss.

Ruiz-Martinez, M. C., Berka, J., Belenkii, A., Foret, F., Miller, A. W. and Karger, B. L. (1993). DNA sequencing by capillary electrophoresis with replaceable linear polyacrylamide and laser-induced fluorescence detection. *Analytical Chemistry*, **65**, 2851–8.

Sanger, F., Micklen, S. and Coulson, A. R. (1977). DNA sequencing and chain-terminating inhibitors. *Proceedings of the National Academy of Sciences USA*, **74**, 5463–7.

Schena, M. (ed.) (1999). *DNA Microarrays: A Practical Approach*. Oxford: Oxford University Press.

Smith, L. M., Sanders, J. Z., Kaiser, R. J., Hughes, P., Dodd, C., Connell, C. R., Heiner, C., Kent, S. B. H. and Hood, L. E. (1986). Fluorescence detection in automated DNA sequence analysis. *Nature*, **321**, 674–9.

Snijder, E. J., Bredenbeek, P. J., Dobbe, J. C., Thiel, V., Ziebuhr, J., Poon, L. L. M., Guan, Y., Rozanov, M., Spaan, W. J. M. and Gorbalenya, A. E. (2003). Unique and conserved features of genome and proteome of SARS-coronavirus, an early split-off from the Coronavirus Group 2 lineage. *Journal of Molecular Biology*, **33**, 991–1004.

Swerdlow, H., Wu, S. L., Harke, H. and Dovichi, N. J. (1990). Capillary gel electrophoresis for DNA sequencing. Laser-induced fluorescence detection with the sheath flow cuvette. *Journal of Chromatography*, **516**, 61–7.

Swerdlow, H., Zhang, J. Z., Chen, D. Y., Harke, H. R., Grey, R., Wu, S. L., Dovichi, N. J. and Fuller, C. (1991). Three DNA sequencing methods using capillary gel electrophoresis and laser-induced fluorescence. *Analytical Chemistry*, **63**, 2835–41.

Tyagi, S. and Kramer, F. R. (1996). Molecular beacons: probes that fluoresce upon hybridization. *Nature Biotechnology*, **14**, 303–8.

Vandesompele, J., De Paepe, A. and Speleman, F. (2002). Elimination of primer-dimer artifacts and genomic coamplification using a two-step SYBR green I real-time RT-PCR. *Analytical Biochemistry*, **303**, 95–8.

Vet, J. A. M., Majithia, A. R., Marras, S. A. E., Tyagi, S., Dube, S., Poiesz, B. J. and Kramer, F. R. (1999). Multiplex detection of four pathogenic retroviruses using molecular beacons. *Proceedings of the National Academy of Sciences USA*, **96**, 6394–9.

Vigilant, L., Stoneking, M., Harpending, H., Hawkes, K. and Wilson, A. C. (1991). African populations and the evolution of human mitochondrial DNA. *Science*, **253**, 1503–7.

Whitcombe, D., Theaker, J., Guy, T., Brown, S. P. and Little, S. (1999). Detection of PCR products using self-probing amplicons and fluorescence. *Nature Biotechnology*, **17**, 804–7.

Wittwer, C. T., Herrmann, M. G., Moss, A. A. and Rasmussen, R. P. (1997). Continuous fluorescence monitoring of rapid cycle DNA amplification. *Biotechniques*, **22**, 130–8.

Zhang, J. Z., Fang, Y., Hou, J. Y., Ren, H. J., Jiang, R., Roos, P. and Dovichi, N. J. (1995). Use of non-cross-linked polyacrylamide for four-color DNA sequencing by capillary electrophoresis separation of fragments up to 640 bases in length in two hours. *Analytical Chemistry*, **67**, 4589–93.

Chapter II

Mapping Genes Influencing Human Quantitative Trait Variation

John Blangero, Jeff T. Williams, Laura Almasy and Sarah Williams-Blangero

Department of Genetics, Southwest Foundation for Biomedical Research, San Antonio, TX

Introduction

In the post-genomic era, the genetic analysis of common diseases will be one of the most critically important areas of biomedical science. Over the past two decades, it has become clear that many of the diseases that constitute the major public health burden in the United States – diseases such as diabetes, atherosclerosis, obesity, hypertension, depression, alcoholism, osteoporosis, and cancer – have a substantial genetic component. The genetic architecture of such diseases is complex, however, involving multiple genetic and environmental components and their interactions. The specific quantitative trait loci (QTLs) that are involved in the biological pathways of these diseases, and the individual effects of these QTLs in the general population, are still largely unknown. The stochastic complexity of the genotype-phenotype relationship of a common disease requires that statistical inference plays a prominent role in the dissection of the underlying genetic architecture. However, statistical genetic methods suitable for this immense task are still in their infancy. The genomic localization and identification of QTLs and characterization of their causal functional polymorphisms will require new advanced statistical genetic tools.

Over the past decade, we have been successful in developing the theoretical and empirical foundation requisite to a thorough

Correspondence address: John Blangero, Department of Genetics, Southwest Foundation for Biomedical Research, P.O. Box 760549, San Antonio, TX 78245-0549. Tel. 210-258-9634, Fax 210-258-9444, E-mail john@darwin.sfbr.org understanding of the strengths and weaknesses of variance component-based quantitative trait linkage methods. We have incorporated many of our statistical genetic developments into our freely available computer package, SOLAR (*Sequential Oligogenic Linkage Analysis Routines*) (Almasy and Blangero, 1998). SOLAR is now used by nearly 2,000 researchers around the world and its use is growing rapidly. In this chapter, we review our work on variance component methods by describing a unified framework for the analysis of QTLs, from detection and initial localization to fine mapping and the identification of functional variants in positional candidate genes.

Progress in statistical genetic methods to detect QTLs

In recent years substantial advances have been made in the techniques for finding QTLs influencing quantitative traits. The emphasis has been on quantitative traits related to risk of various diseases. High-throughput genotyping methods have revolutionized the search for complex disease loci and the resulting emphasis on linkage studies utilizing total genome scans represents the current state of the science. Additional molecular advances in high-throughput resequencing and SNP typing will soon be of considerable aid for identifying the functional mutations in positional candidate loci identified by linkage-based genome scans. These new genetic technologies have the potential to change radically our approaches to both drug discovery (Gelbert and Gregg, 1997) and genetic risk evaluation.

Paralleling the advances in molecular genetic technology has been substantial progress in the development of a statistical framework for detecting and localizing genes involved in disease susceptibility. There has also been an increasing emphasis on linkage analysis of quantitative disease-related phenotypes because many diseases have quantitative correlates that are directly related to risk (Blangero, 1995; Lander and Schork, 1994). Such quantitative characters have many benefits over their discrete counterparts for genetic analysis (Blangero, 1995; Blangero *et al.*, 2000; Duggirala *et al.*, 1997; Wijsman and Amos, 1997; Williams and Blangero, 2004). In psychiatric genetics, for example, the search for quantitative endophenotypes to examine in tandem with qualitative disease affection status is growing rapidly with the recognition that quantitative traits represent targets whose causal genes may be mapped more readily (Almasy and Blangero, 2001; Begleiter *et al.*, 1998; Cloninger *et al.*, 1998; Williams *et al.*, 1999a).

Quantitative trait linkage analysis in humans has been oriented largely towards applications utilizing only pairs of related individuals, such as sib-pairs (Amos *et al.*, 1989; Eaves *et al.*, 1996; Elston *et al.*, 2000; Fulker and Cherny, 1996; Gessler and Xu, 1996; Gu and Rao, 1997; Gu *et al.*, 1996; Haseman and Elston, 1972; Kruglyak and Lander, 1995; Olson, 1995; Olson and Wijsman, 1993; Risch and Zhang, 1995; Stoesz *et al.*, 1997; Tiwari and Elston, 1997; Wang *et al.*, 1998;

Wilson *et al.*, 1991; Xu *et al.*, 2000b). Recently, however, there has been a realization that pair-based methods offer substantially lower power to localize genes than do methods that utilize larger configurations of relatives (Alcais and Abel, 2000; Blangero *et al.*, 2000a; Sham *et al.*, 2000b; Todorov *et al.*, 1997; Williams and Blangero, 1999a,b; Williams *et al.*, 1997). A large number of papers have been published on quantitative trait linkage analysis methods that exploit all of the information in nuclear families (Amos, 1994; Amos *et al.*, 1996, 2000; de Andrade *et al.*, 1997, 1999; Elston *et al.*, 2000; Goldgar, 1990; Page *et al.*, 1998; Schork, 1993; Vogler *et al.*, 1997) and in extended kindreds (Almasy and Blangero, 1998; Almasy *et al.*, 1997, 1999b; Blangero and Almasy, 1997; Blangero *et al.*, 2000a, 2001; Comuzzie *et al.*, 1997; Duggirala *et al.*, 1997; Heath, 1997; Heath *et al.*, 1997; Pratt *et al.*, 2000; Williams and Blangero, 1999a,b; Williams *et al.*, 1997). Similar methods have been developed independently to search for QTLs in experimental animal populations, including both inbred and outbred populations (Hoeschele *et al.*, 1997; Jansen *et al.*, 1998; Jiang and Zeng, 1995; Korol *et al.*, 1995; Meuwissen and Goddard, 1997; Uimari and Hoeschele, 1997; Uimari *et al.*, 1996; Xu, 1998b; Xu and Atchley, 1995, 1996; Yi and Xu, 2000a,b; Zeng, 1993, 1994).

Given an adequate sampling design, it is possible to localize important genes involved in both pathologic and normal physiologic human variation using a genome scan strategy coupled with follow-up fine mapping using linkage disequilibrium/association-based methods. Subsequent to this primary localization of QTLs, it is now clear with the advent of high-throughput resequencing that the next phase of statistical genetic research must involve the development of methods to identify functionally relevant polymorphisms within positional candidate genes. Such studies are currently in their infancy and there are many important common diseases whose determinants have yet to be pursued using such modern techniques.

Finding QTLs for common diseases in human pedigrees

Common diseases present a particular challenge for genetic studies. Although some researchers have expressed considerable scepticism over the potential for linkage-based studies to find genes for common diseases, much of this scepticism is unwarranted. The primary reason for the failure of most genetic studies of common disease arises from a misapplication of study designs that are not tailored to highly prevalent diseases. We have shown theoretically and empirically that studies of large extended families will dramatically outperform the widely favoured alternative of affected sibpair designs when the disease is common (prevalence > 10%) (Blangero *et al.*, 2000a, 2003; Williams and Blangero, 2004). This is especially true when quantitative traits can be used to characterize the disease or be used as correlated risk factors to be jointly analysed with disease status (Williams *et al.*, 1999a,b). With the variance component approach, we now have the tools to design and analyse studies of appropriate size for mapping QTLs in human pedigrees. However, a great deal of work remains to be done to make variance component methods of wider utility and to understand better their strengths and limitations.

Variance component linkage methods

Basic theory of variance component linkage analysis

The variance component linkage method is based upon the classical quantitative genetic decomposition of a phenotype. Let the quantitative phenotype, y, be written as a linear function of the n quantitative trait loci which influence it:

$$y = \mu + \sum_{i=1}^{n} \gamma_i + e,$$

where μ is the grand mean, γ_i is the effect of the i-th QTL, and e represents a random environmental deviation. (Covariates influencing the mean effects can be subsumed in this model and are always simultaneously incorporated in our SOLAR software.) Assume γ_i and e are uncorrelated random variables with expectation zero so that the variance of y is $\sigma_y^2 = \sum_{i=1}^{n} \sigma_{\gamma i}^2 + \sigma_e^2$. We also allow for both additive and dominance effects by decomposing $\sigma_{\gamma i}^2$ as $\sigma_{\gamma i}^2 = \sigma_{ai}^2 + \sigma_{di}^2$, where σ_{ai}^2 is the additive genetic variance due to the i-th locus and σ_{di}^2 is the dominance variance.

For this simple random effects model, we can easily obtain the expected phenotypic covariance between the trait values of any pair of relatives as $\mathrm{Cov}(y_1, y_2) = E[(y_1 - \mu)(y_2 - \mu)] = \sum_{i=1}^{n}[(k_{1i}/2 + k_{2i})\sigma_{ai}^2 + k_{2i}\sigma_{di}^2]$, where k_1, k_2 are the k-coeffcients of Cotterman (1940), with k_{ji} being the i-th QTL-specific probability of the pair of relatives sharing j alleles identical-by-descent (IBD). Similarly, the expected phenotypic correlation between any pair of relatives is given by $\rho(y_1, y_2) = \sum_{i=1}^{n}[(k_{1i}/2 + k_{2i})h_{ai}^2 + k_{2i}d_i^2]$, where h_{ai}^2 is the proportion of the total phenotypic variance due to the additive genetic contribution of the i-th QTL and d_i^2 is the proportion due to the dominance effect. In the classical quantitative genetic variance component model, we do not have information on specific QTLs but use the expectation of the k probabilities over the genome to obtain the approximation $\mathrm{Cov}(y_1, y_2) \approx 2\phi\sigma_a^2 + \delta_7\sigma_d^2$, where $\sigma_a^2 = \sum_{i=1}^{n}\sigma_{ai}^2$ is the total additive genetic variance, $\sigma_d^2 = \sum_{i=1}^{n}\sigma_{di}^2$ is the total dominance genetic variance, $\phi = \frac{1}{2}E[(k_{1i}/2 + k_{2i})]$ is the expected kinship coefficient over the genome with $2\phi = r$ giving the expected coeffcient of relationship, and $\delta_7 = E[k_{2i}]$ is the expected probability of sharing two alleles IBD. As we are generally interested in the examination of one or a few QTLs at a time, we can exploit this approximation to reduce the number of parameters that need to be considered. For example, if we are focusing on the analysis of the i-th QTL, we can absorb the effects of the other QTLs in residual components of covariance. Employing these residual covariance terms, the expected phenotypic covariance between relatives is well approximated by $\text{Cov}(y_1, y_2) = \pi_i\sigma_{ai}^2 + K_{2i}\sigma_{di}^2 + 2\phi\sigma_g^2 + \delta_7\sigma_d^2$, where $\pi_i = (k_{1i}/2 + k_{2i})$ is the coefficient of relationship or the probability of a random allele being IBD at the i-th QTL, σ_g^2 now represents the residual additive genetic variance, and σ_d^2 now represents the residual dominance genetic variance. The π and k_2 coefficients and their expectations effectively structure the expected phenotypic covariances and are the basis for much of quantitative trait linkage analysis, such as the sibpair difference method of Haseman and Elston (1972). For any given chromosomal location, π and k_2 can be estimated from genetic marker data and information about the genetic map.

Given the simple model for phenotypic variation described above, it is possible to use data from pedigree structures of arbitrary complexity to make inferences regarding the localization and effect sizes of QTLs. For the simple additive model in which n QTLs and an unknown number of residual polygenes influence a trait, the covariance matrix for a pedigree can be written

$$\mathbf{\Omega} = \sum_{i=1}^{n} \hat{\prod}_i \sigma_{ai}^2 + 2\mathbf{\Phi}\sigma_g^2 + \mathbf{I}\sigma_e^2,$$

where $\hat{\prod}_i$ is the matrix whose elements $\hat{\pi}_{ijl}$ provide the predicted proportion of genes that individuals j and l share IBD at a QTL i that is linked to a genetic marker locus, Φ is the kinship matrix, and I is an identity matrix. The matrix $\hat{\prod}_i$ is a function of the estimated IBD matrix $\hat{\prod}_m$ for genetic marker and a matrix B of correlations between the proportions of genes IBD at the marker and at the QTL.

By assuming multivariate normality as a working model within pedigrees, the likelihood of any pedigree can easily be written and numerical procedures used to estimate the model parameters. For the covariance model above the ln-likelihood of a pedigree of t individuals with phenotypic vector y is

$$\text{In } L(\mu, \sigma_{ai}^2, \sigma_g^2, \sigma_e^2, \beta \,|\, \mathbf{y}, \mathbf{X}) = -\frac{t}{2}\text{In}(2\pi) - \frac{1}{2}\text{In}\,|\,\mathbf{\Omega}\,| - \frac{1}{2}\mathbf{\Delta'\Omega^{-1}\Delta},$$

where μ is the grand trait mean, $\Delta = (\mathbf{y} - \mu - \mathbf{X}\beta)$, $\mathbf{X}$ is a matrix of covariates, and β is the matrix of regression coeffcients associated with these covariates. Likelihood estimation will yield consistent parameter estimates even when the assumption of multivariate normality is violated (Amos, 1994; Beaty et al., 1985).

Using the variance component model, we can test the null hypothesis that the additive genetic variance due to the i-th QTL equals zero (i.e. no linkage) by comparing the likelihood of this restricted model with that of a model in which the variance due to the i-th QTL is estimated. The difference between the two $\log_{10}$-likelihoods produces a LOD score that is the equivalent of the classical LOD score of linkage analysis. Twice the difference in the $\log_e$-likelihoods of these two models yields a test statistic that is asymptotically distributed as a $\frac{1}{2}:\frac{1}{2}$ mixture of a χ_1^2 variable and a point mass at zero (Self and Liang, 1987).

Advantages of the variance component linkage method

Variance component linkage/disequilibrium methods have several advantages over classical quantitative trait linkage methods based on penetrance models. The simpler parameterization of variance component models leads to a more parsimonious and better estimated set of salient parameters. In variance component-based linkage analysis, the focal parameter is the QTL-specific additive genetic variance or QTL-specific heritability. For a simple one-locus model, the QTL-specific additive genetic variance can be written as $\sigma_q^2 = 2p_q(1 - p_q)\alpha^2$ where p_q is the allele frequency of the QTL polymorphic variant and α is half the displacement between the two homozygous genotypes. Thus, whereas a penetrance-based linkage analysis would need to specify or estimate both p_q and α, in variance component-based linkage analysis these two parameters are absorbed into σ_q^2. We have shown recently that little power is gained over the simpler variance component parameterization by using a penetrance-based model, even when the penetrance model can be specified exactly (Goring *et al.*, 2001). However, the power for variance component-based linkage analysis can greatly exceed that of penetrance-model based analysis when the penetrance model is misspecified, as is always the case when there are multiple QTLs or many functional variants at a QTL. Similarly, the computational requirements of fitting complex penetrance models based on finite mixtures of QTL genotypes can be excessive relative to the much less burdensome computations required for estimating variance components. The variance component approach can also accommodate pedigrees of any size and complexity while penetrance-based models rapidly become computationally intractable as pedigree size or complexity increases. Since it is now clear that large complex pedigrees have substantially more power per sampled individual than do smaller families (Blangero *et al.*, 2000a, 2001, 2003; Williams and Blangero, 1999a), the advantage of using variance component methods for localizing QTLs can be considerable.

Recent advances in variance component theory

There has been substantial progress over the past five years in the development of the variance component method for QTL analysis. Theoretical advances include the development of an analytical framework for power calculations (Sham *et al.*, 2000b; Williams and Blangero, 1999a,c, 2004), ascertainment correction (Blangero *et al.*, 2001; Burton *et al.*, 2000; Comuzzie and Williams, 1999; de Andrade *et al.*, 2000; Sham *et al.*, 2000a), assessment of the role of marker frequency misspecification (Borecki and Province, 1999), multivariate analysis (Almasy *et al.*, 1997; Amos *et al.*, 2001; Iturria and Blangero, 2000; Williams *et al.*, 1999a,b), longitudinal analysis (de Andrade *et al.*, 2002; Diego *et al.*, 2003; Soler and Blangero, 2003), the extension to dichotomous phenotypes (Burton *et al.*, 1999; Duggirala *et al.*, 1997; Williams *et al.*, 1999a,b; Williams and Blangero, 2004), models of age-of-onset and censored phenotypes (Duggirala *et al.*, 1999; Palmer *et al.*, 1999; Scurrah *et al.*, 2000), multiple locus models including epistasis (Blangero and Almasy, 1997; Blangero *et al.*, 2000, 1999; Cloninger *et al.*, 1998; Mitchell *et al.*, 1997), imprinting and parent-of-origin effects (Hanson *et al.*, 2001), X-linked inheritance (Ekstrom, 2004), mitochondrial effects (Czerwinski *et al.*, 2001), new estimation methods (Amos *et al.*, 2000; Blangero *et al.*, 2001; Iturria and Blangero, 2000; Wan *et al.*, 1998; Xu, 1998a), and considerations of robustness (Allison *et al.*, 1999; Blangero *et al.*, 2000a, 2001; Iturria *et al.*, 1999; Wang *et al.*, 1999).

Successful applications of variance component linkage analysis

The number of published papers utilizing variance component-based analysis is increasing rapidly. We published the first genome scan using the variance component method in 1996 in a study of variation in plasma glucose levels (Stern *et al.*, 1996), and there have since been numerous genome scans using the variance component method for a variety of quantitative traits related to common diseases. We have systematically searched the scientific literature to find such applications. The use of the variance component-based linkage method for analysing genome scan data is now well established for phenotypes related to obesity (Arya *et al.*, 2004; Cai *et al.*, 2004a; Chagnon *et al.*, 2000; Comuzzie *et al.*, 1997; Duggirala *et al.*, 1996; Hanson *et al.*, 1998; Hsueh *et al.*, 2001; Kissebah *et al.*, 2000; Mitchell *et al.*, 1999; Perusse *et al.*, 2001; Walder *et al.*, 2000), diabetes (Cai *et al.*, 2004b; Duggirala *et al.*, 1999; Martin *et al.*, 2002b; Mitchell *et al.*, 2000b; Stern *et al.*, 1996; Watanabe *et al.*, 2000), atherosclerosis (Almasy *et al.*, 1999a; Broeckel *et al.*, 2002; Duggirala *et al.*, 2000; Hixson and Blangero, 2000; Imperatore *et al.*, 2000; Mahaney *et al.*, 2003; North *et al.*, 2005; Rainwater *et al.*, 1999; Shearman *et al.*, 2000; Sonnenberg *et al.*, 2004; Yuan *et al.*, 2000), hypertension (Hsueh *et al.*, 2000; Levy *et al.*, 2000; North *et al.*, 2004; Pankow *et al.*, 2000; Rice *et al.*, 2000), thrombosis (Almasy and Blangero, 2004; Buil *et al.*, 2004; Soria *et al.*, 2000, 2002, 2003; Souto *et al.*, 2005), Alzheimer's disease (Ertekin-Taner *et al.*, 2000), depression (Visscher *et al.*, 1999), alcoholism endophenotypes (Begleiter *et al.*, 1998; Cloninger *et al.*, 1998; Williams *et al.*, 1999a), asthma (Mathias *et al.*, 2001; Xu *et al.*, 2000a), renal function (DeWan *et al.*, 2001), infectious disease (Williams-Blangero *et al.*, 2002), osteoporosis (Kammerer *et al.*, 2003; Mitchell *et al.*, 2000a) and oxygen uptake (Bouchard *et al.*, 2000).

Robustness of the variance component method

One of the main concerns about the utility of the variance component linkage method has been the robustness of the procedure when applied to traits with non-normal distributions. Given that the calculation of the usual pedigree likelihood explicitly assumes multivariate normality, it is important to understand when the resulting inferences are valid and to develop alternative procedures that do not depend upon the assumption of normality. We have addressed this issue in detail and examined the validity of the asymptotic distribution of the likelihood-ratio test or LOD score (Allison *et al.*, 1999; Blangero *et al.*, 2000a, 2001). Using computer simulations, we found that for traits with marked leptokurtosis (i.e. a standardized kurtosis coeffcient greater than 2), use of the asymptotic distribution is clearly contraindicated due to excessive Type I error (Allison *et al.*, 1999; Blangero *et al.*, 2001). However, most quantitative phenotypes related to common diseases do not show this magnitude of kurtosis when measured in randomly ascertained samples (Blangero *et al.*, 2001). To obviate the lack of robustness in the presence of extreme non-normality, we developed robust methods including the use of the multivariate-*t* distribution as an alternative to the normal distribution, and we showed that this method does lead to valid tests in the presence of non-normality (Blangero *et al.*, 2001).

Typically, simulation approaches for assessing robustness lack generality and fail to lead to exact inferences regarding the source of the deviation from asymptotic expectations under a misspecified model. Therefore, we also developed a comprehensive analytical theory to deal with the effect of non-normality on variance component-based linkage analysis (Blangero *et al.*, 2000a, 2001). Using misspecification theory, we derived analytical formulae relating the deviation from the expected asymptotic distribution of the LOD score to the kurtosis and total heritability of the quantitative trait. The usual LOD score, when multiplied by simple correction constant, yields a robust LOD score for any deviation from normality and for any pedigree structure. This correction effectively eliminates the problem of inflated Type I error due to misspecification of the underlying probability model in variance component-based linkage analysis. Figure 11.1 illustrates the relationship between the correction factor, kurtosis, and total trait heritability. Clearly, the deviation

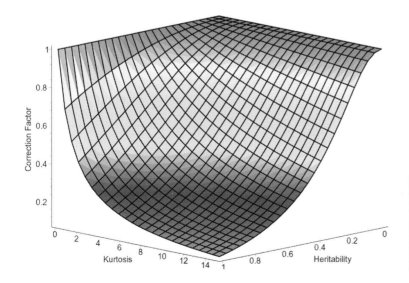

Fig 11.1 Relationship between the magnitude of the expected deviation of the LOD statistic (measured by the correction factor), the total trait heritability, and the trait kurtosis.

from the normal expectation is greatest for high kurtosis and high total heritability.

Power of the variance component method

A critical requirement for study design is the ability to compare the power of competing designs. We have therefore developed a body of analytical theory for evaluating the power of variance component linkage studies (Williams and Blangero, 1999a, 2004). Previously, considerations of power were limited to simulation-based assessments which were specific to each implementation. We derived a completely general linkage power function for arbitrary pedigrees and have obtained closed form equations for the expected LOD score for any relative pair, sibships, and nuclear families. We found that the expected LOD score (ELOD) is a function of the QTL-specific heritability (h_q^2), the total trait heritability (h_T^2), and the sample size and pedigree complexity. Importantly, we determined that the ELOD is proportional to h_q^4. Expressions were derived for the sample size required to achieve a given power in variance component linkage analysis of a quantitative trait in unascertained samples. Equations were developed relating sample size to trait heritability for sibpairs, sibtrios, nuclear families having two and three sibs, and arbitrary relative pairs. The effects of nonzero residual additive genetic variance and parental information were investigated, and a scale relationship for sample sizes with sibships and nuclear families was derived. For larger sampling structures such as extended pedigrees the inheritance space is randomly sampled and the relevant equations are solved numerically. Simulation results were used to confirm the validity of the theoretical results.

We have also used our analytical power theory to understand the consequences of breaking very large and complex pedigrees into smaller units. Traditionally, large pedigrees with many loops are broken into more manageable configurations to reduce computational burden. Using both analytical and simulation-based approaches to the evaluation of the ELOD, we investigated the loss of power that results when a large inbred pedigree (674 genotyped individuals in a single pedigree) is broken into smaller units (Dyer et al., 2001). Our results indicated that pedigree breaking and reduction of inbreeding loops can quickly and dramatically diminish the power to localize QTLs.

Comparison of variance component methods with alternative methods

It is important to compare statistical genetic methods for quantitative trait linkage analysis in order to help decide which method is most appropriate in a given situation. As a benchmark dataset, we used the simulated family data from Genetic Analysis Workshop 10 (GAW10), and different methods were used to screen all 200 replications of the GAW10 genome for evidence of linkage to a simulated quantitative trait (Williams and Blangero, 1999b; Williams *et al.*, 1997). The sibpair and variance component methods were each applied to datasets comprising single sibpairs and complete sibships, and for further comparison we also applied the variance component method to the nuclear family and extended pedigree datasets. For each analysis, the unbiasedness and efficiency of parameter estimation, the power to detect linkage, and the Type I error rate were estimated empirically. Sibpair and variance component methods exhibited comparable performance in terms of the unbiasedness of the estimate of QTL location and the Type I error rate. Within the single sibpair and sibship sampling units, the variance component approach gave consistently superior power and efficiency of parameter estimation. Within each method, the statistical performance was improved by the use of the larger and more informative sampling units.

We have also compared penetrance-based linkage analysis with the variance component method (Göring *et al.*, 2001). Using the computer programs PAP and SOLAR as representative software implementations, we conducted an empirical comparison of their power to map QTLs in extended, randomly ascertained pedigrees using simulated data. Two-point linkage analyses were conducted on several quantitative traits of varying genetic and environmental etiology and the LOD scores were compared. The two methods have similar power when the underlying QTL is diallelic. However, when the QTL is multiallelic, the variance component method exhibited much greater power. These findings suggest that the allelic architecture of a quantitative trait influences the power of these methods. Given that it is unlikely that a penetrance model will be correctly specified, and the high likelihood that QTLs will be multiallelic, we believe that the variance component method will generally outperform penetrance-based approaches to linkage analysis.

Oligogenic linkage models

Our approach to variance component-based linkage analysis has always allowed for an arbitrary number of QTLs to influence a phenotype, and for epistatic interactions among these loci (Blangero *et al.*, 2000a). To choose among potential oligogenic models we have developed a rational and comprehensive approach based on Bayesian model selection and model averaging (Blangero *et al.*, 1999; Martin *et al.*, 2001).

The traditional likelihood-based approach to hypothesis testing is not an optimal strategy for evaluating oligogenic models of inheritance. Under oligogenic inheritance the number of possible multilocus models can become very large, there may be several competing linkage models having similar likelihoods, and comparisons among non-nested models can be required to determine if a given multilocus model provides a significantly better fit to observed phenotypic variation than an alternative model. Our effcient Bayesian approach to oligogenic model selection makes use of existing model likelihoods, and explicitly allows model uncertainty to be incorporated into parameter estimation. This approach can be used to evaluate a large number of possible multilocus models and can be shown to be resistant to false positives. In fact, the Bayesian approach to oligogenic linkage analysis eliminates the problem of multiple testing in genome scans. This procedure was tested using data from Genetic Analysis Workshop 12 (GAW12) (Martin *et al.*, 2001). Using 20 replications of a genome scan for a trait influenced by two QTLs, the generating model was recovered with high accuracy and a greatly reduced rate of false positives.

Distribution of LOD scores in oligogenic linkage analysis

As part of our work on oligogenic models, we noticed that the conditional LOD score for a secondary locus may not exhibit its expected asymptotic distribution (Williams *et al.*, 2001). This occurs when the residual genetic variance bounds to zero when estimating the effect of the secondary QTL. Using computer simulation, we compared the observed LOD scores from oligogenic linkage analysis with the empirical LOD score distribution. We found that when the residual genetic variance bounds to zero during a conditional test for a secondary locus, the resulting asymptotic test distribution represents an unusual mixture of distributions including $\frac{3}{4}$ of a point mass at zero. The most important inference that we draw from this observation is that the absence of a residual heritability (conditional upon some number of QTLs being included in a genetic model) cannot be taken as evidence that all of the genetic variance has been accounted for.

Multivariate linkage analysis

While most published quantitative trait linkage analyses have examined one phenotype at a time, it is usually the case that multiple phenotypes are available. By jointly analysing related phenotypes, additional power to localize pleiotropic QTLs should result. Our group has published a number of papers regarding the development of multivariate variance component linkage methods (Iturria and Blangero, 2000; Williams *et al.*, 1999a,b). For the joint analysis of several phenotypes, finding the maximum likelihood estimates for a multivariate normal model can be a difficult computational task due to complex constraints among the model parameters. We developed an Expectation/Maximization (EM) algorithm for computing maximum likelihood estimates in a multi-phenotype variance component linkage model that readily accommodates these parameter constraints (Iturria and Blangero, 2000). Simulated data were used to demonstrate the potential increase in power to detect linkage that can be obtained if correlated phenotypes are analysed jointly rather than individually.

Linkage analysis of discrete traits and joint analysis of discrete and continuous traits

While we focus primarily on the genetic analysis of quantitative, continuously distributed traits, it is often the case in medical research that some of the most critical phenotypes, such as disease affection status, are dichotomous in nature. The variance component method for multipoint linkage analysis has been generalized to allow joint consideration of a discrete trait and a correlated continuous biological marker (e.g. a disease precursor or associated risk factor) in pedigrees of arbitrary size and complexity (Williams *et al.*, 1999b). This area of our research was in direct response to the needs of investigators in psychiatric genetics who wish to analyse disease affection status jointly with continuous endophenotypes. The discrete trait is modelled using a threshold model and an efficient method for calculating MVN integrals of high dimension was developed (Williams and Blangero, submitted). Formal likelihood-based tests were described for coincident linkage (i.e. linkage of the traits to distinct QTLs that happen to be linked) and pleiotropy (i.e. the same QTL influences both the discrete trait and the correlated continuous phenotype). The properties of the method were demonstrated by use of simulated data. The method has also been applied in a bivariate linkage analysis of alcoholism diagnoses and P300 amplitude of event-related brain potentials (Williams *et al.*, 1999a).

Robustness of variance component linkage analysis methods

Given that our working assumption of multivariate normality within pedigrees is often likely to be incorrect, it is useful to examine the effect of violations of non-normality on Type I error when using the variance component method. We have developed an analytical theory to predict the deviation of the distribution of the LOD score from its asymptotic expectation (Blangero *et al.*, 2000a, 2001). Using this theory, our results suggest that extended pedigrees are more robust than smaller configurations. The greater robustness of extended pedigrees is illustrated in Figure 11.2. For a trait with an expected kurtosis of 2, the expected LOD correction factor for an extended pedigree of size 50 is substantially closer to unity than for smaller family configurations. Thus, non-normality is less of a problem for large extended pedigrees than it is for a linkage study design using simpler units such as sibpairs.

The effect of non-normality on Type II errors and power has received less attention. We have used a combination of analytical approaches and computer simulations to examine this question. Specifically, we are interested to know the power of our robust alternatives, such as our robust LOD score correction method and the multivariate-t distribution method, under different distributions for a quantitative phenotype.

Preliminarily, we have compared the traditional method, based on the assumption of multivariate normality within pedigrees, to our

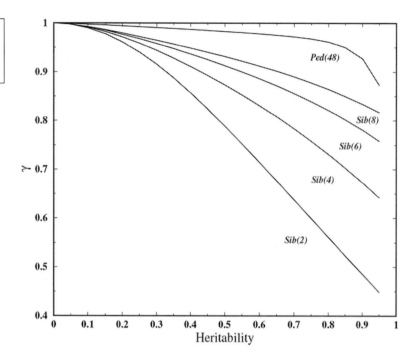

Fig 11.2 Expected LOD correction factors γ for different sampling designs. The total trait kurtosis was fixed at 2.

robust LOD score correction method, the multivariate-t distribution method, and the robust estimation method of Wang *et al.* (1999). The latter method is a Winsorization approach in which outlying values of the phenotype are truncated to a pre-specified value. By simulating traits with varying degrees of kurtosis and for which the source of kurtosis varies (i.e. is due to genes or environmental factors or both), we can assess the strengths and weaknesses of each of these methods. For most classical major gene models the kurtosis comes from the major locus, while for most typical robust statistical models the kurtosis is assumed to come from the environmental (or error) component.

To compare the various methods, we simulated a number of different traits with varying degrees of kurtosis due to either a genetic or environmental source. We simulated two thousand sibpairs with phenotypes exhibiting kurtosis from 0 to 4.3, a total heritability of 0.60, and a QTL-specific heritability of 0.30. Table 11.1 gives the results comparing four analytical methods: (1) the method assuming multivariate normality [MVN], (2) our robust LOD score correction method [cMVN], (3) our method assuming a multivariate-t distribution [MVT], and (4) the robust estimation method [RE] of Wang *et al.* (1999). For the case of no kurtosis, all of the methods yield equivalent LOD scores. When there is substantial kurtosis, however, the methods have different powers to detect linkage. For example, when the kurtosis is due to an environmental source, both the MVT and RE methods produce higher ELODs than the robust cMVN method. This is because the MVT and RE methods effectively

Table II.I. | ELODs for genetic and environmental sources of kurtosis. MVN: model assuming multivariate normality; cMVN: multivariate normal model with a robust LOD score correction; MVT: model assuming a multivariate-t distribution; RE: the robust estimation method of Wang *et al.* (1999).

				Kurtosis			
	None	Environmental			Genetic		
Method	0.00	0.68	1.58	4.28	0.68	1.58	4.28
MVN	6.77	6.88	6.98	7.23	6.81	6.83	6.69
cMVN	6.77	6.06	5.32	3.90	6.00	5.20	3.61
MVT	6.77	6.83	8.54	11.52	4.72	2.59	0.78
RE	6.73	7.22	8.04	10.97	6.04	4.95	1.96

downweight the tails of the distribution, such that the genetic signal is increased if those phenotypic values in the tail are in fact due to the environment. However, when the kurtosis is due to the major locus, the cMVN method is superior since downweighting the tails now reduces the genetic signal in the MVT and RE methods. Thus, these methods will perform differently depending upon the source of the kurtosis, which is generally unknown. However, by always performing both an MVT and cMVN analysis and comparing them, it may be possible to make inferences about the source of the kurtosis. Consideration should be limited to these two methods since they appear always to provide the correct Type I error levels while the MVN and RE do not.

Genotype × environment interaction

Genotype × environment (G×E) interaction is likely to be an important influence on continuous physiological variation. We have pursued a number of approaches for examining G×E interaction in quantitative genetic analysis (Blangero, 1993; Blangero *et al.*, 1990; Jaquish *et al.*, 1996, 1997) and in variance component-based linkage analysis (Towne *et al.*, 1999, 1997). Our previous work has concentrated on G×E interactions involving dichotomous environments, such as sex (Towne *et al.*, 1999, 1997) or smoking (Martin *et al.*, 2002a). We have recently begun to expand our work on G×E interaction methods for continuous environmental variables, such as age (Almasy *et al.*, 2001).

A simple way to model G×E interaction in a quantitative genetic analysis is to use a matrix of environmental differences between individuals to structure the additive genetic component of variance. The covariance is then $\mathbf{\Omega} = 2\mathbf{\Phi} \odot \Upsilon\sigma_g^2 + \mathbf{I}\sigma_e^2$, where Υ is a matrix of scaled similarities among individuals with respect to environment, Φ is a kinship matrix, $\mathbf{I}$ is an identity matrix, and σ_e^2 is a residual environment component. One approach we have used is a simple exponential decay (or Ornstein-Uhlenbeck random process) model for

Υ, i.e. $v_{ij} = \exp(-\lambda |x_i - x_j|)$, where x_i and x_j represent the environment values for two individuals i and j. We test for G×E interaction by testing whether λ is significantly different from zero in a likelihood-ratio test where the difference in likelihoods of the nested models is distributed as χ_1^2. The above model can be extended by separating out a specific additive genetic G×E term $\sigma_{G\times E}^2$ from the general additive genetic component such that $\Omega = 2\Phi\sigma_g^2 + 2\Phi \odot \Upsilon\sigma_{G\times E}^2 + I\sigma_e^2$.

Another approach to modelling G×E interaction is to make the additive genetic variance a function of an environmental variable x, i.e. $\sigma_g^2 = f(x)$. We employed the variance function $\sigma_g^2 = \exp(\alpha + \beta x)$ since it maintains positivity. If β is significantly different from 0, there is evidence for changing genetic variance with environment. We can also include in the model a correlation in genetic effects, ρG, at different levels of the environmental variable. This correlation is modelled as an exponential decay across environment value, i.e. for values x_i and x_j, $\rho G = \exp(-\lambda |x_i - x_j|)$. If ρG is significantly different from 1, this suggests that different genes influence the trait in different environments. Similarly, we can also allow the residual variance to change with environment. The complete quantitative genetic model for the covariance between individuals i and j is:

$$\Omega = 2\Phi\rho G \exp(\alpha + \beta[x_i - \bar{x}]) \exp(\alpha + \beta[x_j - \bar{x}])$$

$$+ I \exp(\alpha_e + \beta_e[x_i - \bar{x}]) \exp(\alpha_e + \beta_e[x_j - \bar{x}]).$$

Comparisons of these two approaches in simulated data from GAW12 suggest that, at least for linear changes in phenotype with environment, modelling genetic variance as a function of environment is a more powerful test of G×E interaction than using a matrix of environmental differences (Almasy et al., 2001). Additionally, the strategy of modelling variance as a function of environment is easily extended to G×E interaction at a QTL by making the QTL variance a function of environment, e.g. $\sigma_q^2 = \exp(\alpha_q + \beta_q x)$. In this case we are testing a specific QTL so we do not include a between-locus correlation. The covariance between individuals i and j is modelled as:

$$\Omega = \hat{\prod} \exp(\alpha_q + \beta_q[x_i - \bar{x}]) \exp(\alpha_q + \beta_q[x_j - \bar{x}])$$

$$+ 2\Phi\rho G \exp(\alpha + \beta[x_i - \bar{x}]) \exp(\alpha + \beta[x_j - \bar{x}])$$

$$+ I \exp(\alpha_e + \beta_e[x_i - \bar{x}]) \exp(\alpha_e + \beta_e[x_j - \bar{x}]).$$

For both analyses that model variance as a function of x, the distribution of the test statistic for a likelihood-ratio test is not immediately obvious — there are two additional parameters compared to the nested model (ρG and β, or αm and βm), but these two parameters are confounded. Preliminary analyses using a test with two degrees of freedom suggest that treating the parameters independently provides an overly conservative approximation (Almasy et al., 2001). This method has been extended and employed in an empirical example by Diego et al. (2003).

A QTL for parasite burden in the Jirels

A major concern in statistical human genetics has been the seemingly insurmountable computational burden associated with the analysis of very large pedigrees. This concern has been one of the reasons why studies of extended pedigrees are relatively uncommon in complex disease genetics. We believe that our variance component approach to linkage analysis can circumvent this problem. Using Markov Chain Monte Carlo methods, such as those implemented in the programme LOKI (Heath, 1997; Heath *et al.*, 1997), to generate the necessary IBD matrices, we have been able to use our SOLAR software to analyse pedigrees that we believe represent the upper range of size and complexity for most human studies. The following example demonstrates the potential of our approach to the analysis of large extended pedigrees. To our knowledge, this example cannot be handled by any other extant computer programme for variance component linkage analysis.

In one of our family-based studies, we have performed a genome scan in members of the isolated Jirel population of eastern Nepal to search for QTLs influencing risk of parasitic worm infection. A 10 cM genome scan was performed using short tandem repeat polymorphisms on 444 members of the sample chosen from a single branch of the large Jirel pedigree (Williams-Blangero *et al.*, 2002). This is an extremely complex and challenging pedigree for linkage analysis. By analysing the Jirel pedigree in its entirety, we eliminated the loss of power that results from dividing a pedigree into smaller units to reduce computation.

Our analyses of this data set have focused on *Ascaris* worm burden as measured by egg count per gram of faeces (EPG). Due to the extreme kurtosis of this measure, we applied a ln-transformation, $\ln(\text{EPG} + 1)$, but even after transformation this trait remains markedly non-normal (kurtosis 2.11). We therefore employed our robust LOD score calculation method that applies a scalar correction to the LOD score to recover the correct Type I error distribution irrespective of the trait distribution (Blangero *et al.*, 2000, 2001).

Figure 11.3 provides an example of a SOLAR string plot which shows the linkage results for all chromosomes. All *p*-values were obtained analytically and then validated empirically using intensive computer simulation. We found strong evidence for three distinct loci. Our largest signal is found near the *q*-terminus (13q32−q34) of chromosome 13: the maximum multipoint robust LOD score occurring at 101 cM was 4.30 (nominal *p*-value $= 4.3 \times 10^{-6}$, genome-wide *p*-value $= 0.0013$). This example indicates that it is clearly possible to localize genes that influence human quantitative variation to specific chromosomal regions. Additionally, it points out the value of such gene localization projects in large extended human pedigrees. Such extended human pedigrees are often easily available when working with genetic isolates such as the Jirels.

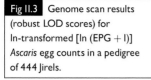

Fig II.3 Genome scan results (robust LOD scores) for ln-transformed [ln (EPG + 1)] *Ascaris* egg counts in a pedigree of 444 Jirels.

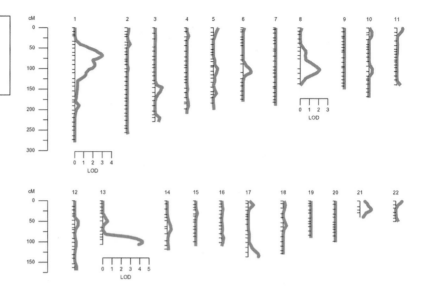

Conclusions

Anthropological genetics has contributed a great deal to both the methodological development and the practical application of the QTL mapping methods as applied to human populations. While most studies to date have focused on disease-related traits, we anticipate that in the future these methods will be employed routinely by biological anthropologists to assess the genetic basis of normal human variation. Only when we know the identity and architecture of these QTLs underlying all of human phenotypic variation, can we begin to consider adequately the evolutionary forces that helped to shape them.

Acknowledgements

This research was supported in part by National Institutes of Health grants MH59490, HL45522, GM31575, GM18897, AI44406, AI37091 and HL28972. We are grateful to Tom Dyer and Charles Peterson for their expert programming assistance. Interested individuals can obtain our computer package SOLAR by visiting our web site at http://www.sfbr.org/sfbr/public/software/solar/solar.html.

References

Alcais, A. and Abel, L. (2000). Linkage analysis of quantitative trait loci: sib pairs or sibships? *Hum Hered*, **50**, 251–6.

Allison, D.B., Neale, M.C., Zannolli, R., Schork, N.J., Amos, C.I. and Blangero, J. (1999). Testing the robustness of the likelihood-ratio test in a variance-component quantitative-trait loci-mapping procedure. *Am J Hum Genet*, **65**, 531–4.

Almasy, L. and Blangero, J. (1998). Multipoint quantitative trait linkage analysis in general pedigrees. *Am J Hum Genet*, **62**, 1198–1211.

Almasy, L. and Blangero, J. (2001). Endophenotypes as quantitative risk factors for psychiatric disease: Rationale and study design. *Am J Med Genet*, **105**, 42–4.

Almasy, L. and Blangero, J. (2004). Exploring positional candidate genes: linkage conditional on measured genotype. *Behav Genet*, **34**, 173–7.

Almasy, L., Dyer, T.D. and Blangero, J. (1997). Bivariate quantitative trait linkage analysis: pleiotropy versus co-incident linkages. *Genet Epidemiol*, **14**, 953–8.

Almasy, L., Hixson, J.E., Rainwater, D.L., Cole, S., Williams, J.T., Mahaney, M.C., VandeBerg, J.L., Stern, M.P., MacCluer, J.W. and Blangero, J. (1999a). Human pedigree-based quantitative-trait-locus mapping: localization of two genes influencing HDL-cholesterol metabolism. *Am J Hum Genet*, **64**, 1686–93.

Almasy, L., Williams, J.T., Dyer, T.D. and Blangero, J. (1999b). Quantitative trait locus detection using combined linkage/disequilibrium analysis. *Genet Epidemiol*, **17**, S31–S36.

Almasy, L., Towne, B., Peterson, C. and Blangero, J. (2001). Detecting genotype-by-age interaction. *Genet Epidemiol*, **21** (Suppl 1), S819–S824.

Almasy, L., Soria, J.M., Souto, J.C., Coll, I., Bacq, D., Faure, A., Mateo, J., Borrell, M., Munoz, X., Sala, N., Stone, W.H., Lathrop, M., Fontcuberta, J. and Blangero, J. (2003). A quantitative trait locus influencing free plasma protein S levels on human chromosome 1q: results from the Genetic Analysis of Idiopathic Thrombophilia (GAIT) project. *Arterioscler Thromb Vasc Biol*, **23**, 508–11.

Amos, C.I. (1994). Robust variance-components approach for assessing genetic linkage in pedigrees. *Am J Hum Genet*, **54**, 535–43.

Amos, C.I., Zhu, D.K. and Boerwinkle, E. (1996). Assessing genetic linkage and association with robust components of variance approaches. *Ann Hum Genet*, **60**, 143–60.

Amos, C., de Andrade, M. and Zhu, D. (2001). Comparison of multivariate tests for genetic linkage. *Hum Hered*, **51**, 133–44.

Amos, C.I., Elston, R.C., Wilson, A.F. and Bailey-Wilson, J.E. (1989). A more powerful robust sib-pair test of linkage for quantitative traits. *Genet Epidemiol*, **6**, 435–49.

Amos, C.I., Gu, X., Chen, J. and Davis, B.R. (2000). Least squares estimation of variance components for linkage. *Genet Epidemiol*, **19**, S1–S7.

Arya, R., Duggirala, R., Jenkinson, C.P., Almasy, L., Blangero, J., O'Connell, P. and Stern, M.P. (2004). Evidence of a novel quantitative-trait locus for obesity on chromosome 4p in Mexican Americans. *Am J Hum Genet*, **74**, 272–82.

Begleiter, H., Porjesz, B., Reich, T., Edenberg, H.J., Goate, A., Blangero, J., Almasy, L., Foroud, T., Van Eerdewegh, P., Polich, J., Rohrbaugh, J., Kuperman, S., Bauer, L.O., O'Connor, S.J., Chorlian, D.B., Li, T.K., Conneally, P.M., Hesselbrock, V., Rice, J.P., Schuckit, M.A., Cloninger, R., Nurnberger, J., Crowe, R. and Bloom, F.E. (1998). Quantitative trait loci analysis of human event-related brain potentials: P3 voltage. *Electroencephalogr Clin Neurophysiol*, **108**, 244–50.

Beaty, T.H., Self, S.G., Liang, K.Y., Connolly, M.A., Chase, G.A. and Kwiterovich, P.O. (1985). Use of robust variance components models to analyze triglyceride data in families. *Ann Hum Genet*, **49**, 315–28.

Blangero, J. (1993). Statistical genetic approaches to human adaptability. *Hum Biol*, **65**, 941–66.

Blangero, J. (1995). Genetic analysis of a common oligogenic trait with quantitative correlates: Summary of GAW9 results. *Genet Epidemiol*, **12**, 689–706.

Blangero, J. (2004). Localization and identification of human quantitative trait loci: King Harvest has surely come. *Curr Opin Genet Dev*, **14**, 233–40.

Blangero, J. and Almasy, L. (1997). Multipoint oligogenic linkage analysis of quantitative traits. *Genet Epidemiol*, **14**, 959–64.

Blangero, J., Williams, J.T. and Almasy, L. (2000a). Quantitative trait locus mapping using human pedigrees. *Hum Biol*, **72**, 35–62.

Blangero, J., Williams, J.T. and Almasy, L. (2000b). Robust LOD scores for variance component-based linkage analysis. *Genet Epidemiol*, **19** (Suppl 1), S8–S14.

Blangero, J., Williams, J.T. and Almasy, L. (2001). Variance component methods for detecting complex trait loci. *Adv Genet*, **42**, 151–81.

Blangero, J., Williams, J.T. and Almasy, L. (2003). Novel family-based approaches to genetic risk in thrombosis. *J Thromb Haemost*, **1**, 1391–97.

Blangero, J., Williams, J.T., Iturria, S.J. and Almasy, L. (1999). Oligogenic model selection using the Bayesian Information Criterion: linkage analysis of the P300 Cz event-related brain potential. *Genet Epidemiol*, **17**, S67–S72.

Blangero, J., MacCluer, J.W., Kammerer, C.M., Mott, G.E., Dyer, T.D. and McGill, H.C. Jr. (1990). Genetic analysis of apolipoprotein A-I in two dietary environments. *Am J Hum Genet*, **47**, 414–28.

Borecki, I.B. and Province, M.A. (1999). The impact of marker allele frequency misspecification in variance components quantitative trait locus analysis using sibship data. *Genet Epidemiol*, **17**, S73–S77.

Bouchard, C., Rankinen, T., Chagnon, Y.C., Rice, T., Perusse, L., Gagnon, J., Borecki, I., An, P., Leon, A.S., Skinner, J.S., Wilmore, J.H., Province, M. and Rao, D.C. (2000). Genomic scan for maximal oxygen uptake and its response to training in the HERITAGE Family Study. *J Appl Physiol*, **88**, 551–9.

Buil, A., Soria, J.M., Souto, J.C., Almasy, L., Lathrop, M., Blangero, J. and Fontcuberta, J. (2004). Protein C levels are regulated by a quantitative trait locus on chromosome 16: results from the Genetic Analysis of Idiopathic Thrombophilia (GAIT) Project. *Arterioscler Thromb Vasc Biol*, **24**, 1321–5.

Broeckel, U., Hengstenberg, C., Mayer, B., Holmer, S., Martin, L.J., Comuzzie, A.G., Blangero, J., Nurnberg, P., Reis, A., Riegger, G.A., Jacob, H.J. and Schunkert, H. (2002). A comprehensive linkage analysis for myocardial infarction and its related risk factors. *Nat Genet*, **30**, 210–14.

Burton, P.R., Palmer, L.J., Jacobs, K., Keen, K.J., Olson, J.M. and Elston, R.C. (2000). Ascertainment adjustment: where does it take us? *Am J Hum Genet*, **67**, 1505–14.

Burton, P.R., Tiller, K.J., Gurrin, L.C., Cookson, W.O., Musk, A.W. and Palmer, L.J. (1999). Genetic variance components analysis for binary phenotypes using generalized linear mixed models (GLMMs) and Gibbs sampling. *Genet Epidemiol*, **17**, 118–40.

Cai, G., Cole, S.A., Freeland-Graves, J.H., MacCluer, J.W., Blangero, J. and Comuzzie, A.G. (2004a). Principal component for metabolic syndrome risk maps to chromosome 4p in Mexican Americans: the San Antonio Family Heart Study. *Hum Biol*, **76**, 651–65.

Cai, G., Cole, S.A., Freeland-Graves, J.H., MacCluer, J.W., Blangero, J. and Comuzzie, A.G. (2004b). Genome-wide scans reveal quantitative trait Loci on 8p and 13q related to insulin action and glucose metabolism: the San Antonio Family Heart Study. *Diabetes*, **53**, 1369–74.

Chagnon, Y.C., Borecki, I.B., Perusse, L., Roy, S., Lacaille, M., Chagnon, M., Ho-Kim, M.A., Rice, T., Province, M.A., Rao, D.C. and Bouchard, C. (2000). Genome-wide search for genes related to the fat-free body mass in the Quebec family study. *Metabolism*, **49**, 203–7.

Cloninger, C.R., Van Eerdewegh, P., Goate, A., Edenberg, H.J., Blangero, J., Hesselbrock, V., Reich, T., Nurnberger, Jr J., Schuckit, M., Porjesz, B., Crowe, R., Rice, J.P., Foroud, T., Przybeck, T.R., Almasy, L., Bucholz, K., Wu, W., Shears, S., Carr, K., Crose, C., Willig, C., Zhao, J., Tischfield, J.A., Li, T.K., Conneally, P.M. and Begleiter, H. (1998). Anxiety proneness linked to epistatic loci in genome scan of human personality traits. *Am J Med Genet*, **81**, 313–17.

Comuzzie, A.G., Hixson, J.E., Almasy, L., Mitchell, B.D., Mahaney, M.C., Dyer, T.D., Stern, M.P., MacCluer, J.W. and Blangero, J. (1997). A major quantitative trait locus determining serum leptin levels and fat mass is located on human chromosome 2. *Nat Genet*, **15**, 273–5.

Comuzzie, A.G. and Williams, J.T. (1999). Correcting for ascertainment bias in the COGA data set. *Genet Epidemiol*, **17**, S109–S114.

Cotterman, C.W. (1940). A calculus for statistico-genetics. Unpublished Ph.D. dissertation, Ohio State University, Columbus, OH.

Czerwinski, S.A., Williams, J.T., Demerath, E.W., Towne, B., Siervogel, R.M. and Blangero, J. (2001). Does accounting for mitochondrial genetic variation improve the fit of genetic models? *Genet Epidemiol*, **21** (Suppl 1), S779–S782.

de Andrade, M. and Amos, C.I. (2000). Ascertainment issues in variance components models. *Genet Epidemiol*, **19**, 333–44.

de Andrade, M., Amos, C.I. and Thiel, T.J. (1999). Methods to estimate genetic components of variance for quantitative traits in family studies. *Genet Epidemiol*, **17**, 64–76.

de Andrade, M., Thiel, T.J., Yu, L. and Amos, C.I. (1997). Assessing linkage on chromosome 5 using components of variance approach: univariate versus multivariate. *Genet Epidemiol*, **14**, 773–8.

de Andrade, M., Gueguen, R., Visvikis, S., Sass, C., Siest, G. and Amos, C.I. (2002). Ex-tension of variance components approach to incorporate temporal trends and longitudinal pedigree data analysis. *Genet Epidemiol*, **22**(3), 221–32.

Dempster, A.P., Laird, N.M. and Rubin, D.B. (1977). Maximum likelihood from incomplete data via the *EM* algorithm (with discussion). *J Roy Stat Soc Ser B*, **39**, 1–38.

De Wan, A.T., Arnett, D.K., Atwood, L.D., Province, M.A., Lewis, C.E., Hunt, S.C. and Eckfeldt, J. (2001). A genome scan for renal function among hypertensives: the HyperGEN study. *Am J Hum Genet*, **68**, 136–44.

Diego, V.P., Almasy, L., Dyer, T.D., Soler, J.M. and Blangero, J. (2003). Strategy and model building in the fourth dimension: a null model for genotype × age interaction as a Gaussian stationary stochastic process. *BMC Genet*, **4** (Suppl 1), S34.

Duggirala, R., Blangero, J., Almasy, L., Dyer, T.D., Williams, K.L., Leach, R.J., O'Connell, P. and Stern, M.P. (1999). Linkage of type 2 diabetes mellitus and of age at onset to a genetic location on chromosome 10q in Mexican Americans. *Am J Hum Genet*, **64**, 1127–40.

Duggirala, R., Blangero, J., Almasy, L., Dyer, T.D., Williams, K.L., Leach, R.J., O'Connell, P. and Stern, M.P. (2000). A major susceptibility locus influencing plasma triglyceride concentrations is located on chromosome 15q in Mexican Americans. *Am J Hum Genet*, **66**, 1237–45.

Duggirala, R., Stern, M.P., Mitchell, B.D., Reinhart, L.J., Shipman, P.A., Uresandi, O.C., Chung, W.K., Leibel, R.L., Hales, C.N., O'Connell, P. and Blangero, J. (1996). Quantitative variation in obesity-related traits and insulin precursors linked to the OB gene region on human chromosome 7. *Am J Hum Genet*, **59**, 694–703.

Duggirala, R., Williams, J.T., Williams-Blangero, S. and Blangero, J. (1997). A variance component approach to dichotomous trait linkage analysis using a threshold model. *Genet Epidemiol*, **14**, 987–92.

Dupuis, J., Brown, P.O. and Siegmund, D. (1995). Statistical methods for linkage analysis of complex traits from high-resolution maps of identity by descent. *Genetics*, **140**, 843–56.

Dyer, T.D., Blangero, J., Williams, J.T., Göring, H.H.H. and Mahaney, M.C. (2001). The effect of pedigree complexity on quantitative trait linkage analysis. *Genet Epidemiol*, **21** (Suppl 1), S236–S243.

Eaves, L.J., Neale, M.C. and Maes, H. (1996). Multivariate multipoint linkage analysis of quantitative trait loci. *Behav Genet*, **26**, 519–25.

Ekstrom, C.T. (2004). Multipoint linkage analysis of quantitative traits on sex-chromosomes. *Genet Epidemiol*, **26**, 218–30.

Elston, R.C., Buxbaum, S., Jacobs, K.B. and Olson, J.M. (2000). Haseman and Elston revisited. *Genet Epidemiol*, **19**, 1–17.

Ertekin-Taner, N., Graff-Radford, N., Younkin, L.H., Eckman, C., Baker, M., Adamson, J., Ronald, J., Blangero, J., Hutton, M. and Younkin, S.G. (2000). Linkage of plasma Aβ42 to a quantitative locus on chromosome 10 in late-onset Alzheimer's disease pedigrees. *Science*, **290**(5500), 2303–4.

Fulker, D.W. and Cherny, S.S. (1996). An improved multipoint sib-pair analysis of quantitative traits. *Behav Genet*, **26**, 527–32.

Fulker, D.W., Cherny, S.S., Sham, P.C. and Hewitt, J.K. (1999). Combined linkage and association sib-pair analysis for quantitative traits. *Am J Hum Genet*, **64**(1), 259–67.

Gelbert, L.M. and Gregg, R.E. (1997). Will genetics really revolutionize the drug discovery process? *Curr Opin Biotechnol*, **8**, 669–74.

Gessler, D.D. and Xu, S. (1996). Using the expectation or the distribution of the identity by descent for mapping quantitative trait loci under the random model. *Am J Hum Genet*, **59**, 1382–90.

Goldgar, D.E. (1990). Multipoint analysis of human quantitative genetic variation. *Am J Hum Genet*, **47**, 957–67.

Göring, H.H.H., Williams, J.T. and Blangero, J. (2001). Linkage analysis of quantitative traits in randomly ascertained pedigrees: comparison of penetrance-based and variance component analysis. *Genet Epidemiol*, **21** (Suppl 1), S783–S788.

Gu, C. and Rao, D.C. (1997). A linkage strategy for detection of human quantitative-trait loci. I. Generalized relative risk ratios and power of sib pairs with extreme trait values. *Am J Hum Genet*, **61**, 210–210.

Gu, C., Todorov, A. and Rao, D. C. (1996). Combining extremely concordant sibpairs with extremely discordant sibpairs provides a cost effective way to linkage analysis of quantitative trait loci. *Genet Epidemiol*, **13**, 513–33.

Hanson, R. L., Kobes, S., Lindsay, R. S. and Knowler, W. C. (2001). Assessment of parent-of-origin effects in linkage analysis of quantitative traits. *Am J Hum Genet*, **68**, 951–62.

Hanson, R. L., Ehm, M. G., Pettitt, D. J., Prochazka, M., Thompson, D. B., Timberlake, D., Foroud, T., Kobes, S., Baier, L., Burns, D. K., Almasy, L., Blangero, J., Garvey, W. T., Bennett, P. H. and Knowler, W. C. (1998). An autosomal genomic scan for loci linked to type II diabetes mellitus and body-mass index in Pima Indians. *Am J Hum Genet*, **63**, 1130–8.

Haseman, J. K. and Elston, R. C. (1972). The investigation of linkage between a quantitative trait and a marker locus. *Behav Genet*, **2**, 3–19.

Heath, S. C. (1997). Markov chain Monte Carlo segregation and linkage analysis for oligogenic models. *Am J Hum Genet*, **61**, 748–60.

Heath, S. C., Snow, G. L., Thompson, E. A., Tseng, C. and Wijsman, E. M. (1997). MCMC segregation and linkage analysis. *Genet Epidemiol*, **14**, 1011–16.

Hixson, J. E. and Blangero, J. (2000). Genomic searches for genes that influence atherosclerosis and its risk factors. *Ann N Y Acad Sci*, **902**, 1–7.

Hoeschele, I., Uimari, P., Grignola, F. E., Zhang, Q. and Gage, K. M. (1997). Advances in statistical methods to map quantitative trait loci in outbred populations. *Genetics*, **147**, 1445–57.

Hopper, J. L. and Mathews, J. D. (1982). Extensions to multivariate normal models for pedigree analysis. *Ann Hum Genet*, **46**, 373–83.

Hsueh, W. C., Mitchell, B. D., Schneider, J. L., St Jean, P. L., Pollin, T. I., Ehm, M. G., Wagner, M. J., Burns, D. K., Sakul, H., Bell, C. J. and Shuldiner, A. R. (2001). Genome-wide scan of obesity in the old order amish. *J Clin Endocrinol Metab*, **86**, 1199–1205.

Hsueh, W. C., Mitchell, B. D., Schneider, J. L., Wagner, M. J., Bell, C. J., Nanthakumar, E. and Shuldiner, A. R. (2000). QTL influencing blood pressure maps to the region of PPH1 on chromosome 2q31–34 in Old Order Amish. *Circulation*, **101**, 2810–16.

Imperatore, G., Knowler, W. C., Pettitt, D. J., Kobes, S., Fuller, J. H., Bennett, P. H. and Hanson, R. L. (2000). A locus influencing total serum cholesterol on chromosome 19p: results from an autosomal genomic scan of serum lipid concentrations in Pima Indians. *Arterioscler Thromb Vasc Biol*, **20**, 2651–6.

Iturria, S. J. and Blangero, J. (2000). An EM algorithm for obtaining maximum likelihood estimates in the multi-phenotype variance components linkage model. *Ann Hum Genet*, **64**, 349–62.

Iturria, S. J., Williams, J. T., Almasy, L., Dyer, T. D. and Blangero, J. (1999). An empirical test of the significance of an observed quantitative trait locus effect that preserves additive genetic variation. *Genet Epidemiol*, **17** (Suppl 1), S169–S173.

Jansen, R. C., Johnson, D. L. and Van Arendonk, J. A. (1998). A mixture model approach to the mapping of quantitative trait loci in complex populations with an application to multiple cattle families. *Genetics*, **148**, 391–9.

Jaquish, C. E., Blangero, J., Haffner, S. M., Stern, M. P. and MacCluer, J. W. (1996). Quantitative genetics of dehydroepiandrosterone sulfate and its relation to possible cardiovascular disease risk factors in Mexican Americans. *Hum Hered*, **46**, 301–9.

Jaquish, C. E., Leland, M. M., Dyer, T., Towne, B. and Blangero, J. (1997). Ontogenetic changes in genetic regulation of fetal morphometrics in baboons (*Papio hamadryas subspp.*). *Hum Biol*, **69**, 831–48.

Jiang, C. and Zeng, Z. B. (1995). Multiple trait analysis of genetic mapping for quantitative trait loci. *Genetics*, **140**, 1111–27.

Kammerer, C. M., Schneider, J. L., Cole, S. A., Hixson, J. E., Samollow, P. B., O'Connell, J. R., Perez, R., Dyer, T. D., Almasy, L., Blangero, J., Bauer, R. L. and Mitchell, B. D. (2003). Quantitative trait loci on chromosomes 2p, 4p, and 13q influence bone mineral density of the forearm and hip in Mexican Americans. *J Bone Miner Res*, **18**, 2245–52.

Kissebah, A. H., Sonnenberg, G. E., Myklebust, J., Goldstein, M., Broman, K., James, R. G., Marks, J. A., Krakower, G. R., Jacob, H. J., Weber, J., Martin, L., Blangero, J. and Comuzzie, A. G. (2000). Quantitative trait loci on chromosomes 3 and 17 influence phenotypes of the metabolic syndrome. *Proc Natl Acad Sci USA*, **97**, 14478–83.

Korol, A. B., Ronin, Y. I. and Kirzhner, V. M. (1995). Interval mapping of quantitative trait loci employing correlated trait complexes. *Genetics*, **140**, 1137–47.

Kruglyak, L. and Lander, E. S. (1995). Complete multipoint sib-pair analysis of qualitative and quantitative traits. *Am J Hum Genet*, **57**, 439–54.

Kruglyak, L., Daly, M. J., Reeve-Daly, M. P. and Lander, E. S. (1996). Parametric and nonparametric linkage analysis: a unified multipoint approach. *Am J Hum Genet*, **58**, 1347–63.

Lander, E. S. and Schork, N. J. (1994). Genetic dissection of complex traits. *Science*, **265**, 2037–48.

Lange, K. (1978). Central limit theorems for pedigrees. *J Math Bio*, **6**, 59–66.

Lange, K. (1997). *Mathematical and Statistical Methods for Genetic Analysis.* New York: Springer-Verlag.

Lange, K. and Boehnke, M. (1983). Extensions to pedigree analysis. IV. Covariance components models for multivariate traits. *Am J Med Genet*, **14**, 513–24.

Lange, K., Little, R. J. A. and Taylor, J. M. G. (1989). Robust statistical modeling using the *t* distribution. *JASA*, **84**, 881–96.

Lange, K., Weeks, D. and Boehnke, M. (1988). Programs for pedigree analysis: Mendel, Fisher, and dGene. *Genet Epidemiol*, **5**, 471–2.

Levy, D., DeStefano, A. L., Larson, M. G., O'Donnell, C. J., Lifton, R. P., Gavras, H., Cupples, L. A. and Myers, R. H. (2000). Evidence for a gene influencing blood pressure on chromosome 17. Genome scan linkage results for longitudinal blood pressure phenotypes in subjects from the Framingham Heart Study. *Hypertension*, **36**, 477–83.

Lewis, C. M. and Kort, E. N. (1997). Multilocus quantitative trait analysis using the multipoint identity-by-descent method. *Genet Epidemiol*, **14**, 839–44.

Mahaney, M. C., Almasy, L., Rainwater, D. L., VandeBerg, J. L., Cole, S. A., Hixson, J. E., Blangero, J. and MacCluer, J. W. (2003). A quantitative trait locus on chromosome 16q influences variation in plasma HDL-C levels in Mexican Americans. *Arterioscler Thromb Vasc Biol*, **23**, 339–45.

Martin, L. J., Comuzzie, A. G., North, K. E., Williams, J. T. and Blangero, J. (2001). The utility of Bayesian model averaging for detecting known oligogenic effects. *Genet Epidemiol*, **21** (Suppl 1), S789–S793.

Martin, L.J., Cole, S.A., Hixson, J.E., Blangero, J. and Comuzzie, A.G. (2002a). Genotype by smoking interaction for leptin levels in the San Antonio Family Heart Study. *Genet Epidemiol*, **22**(2), 105–15.

Martin, L.J., Comuzzie, A.G., Dupont, S., Vionnet, N., Dina, C., Gallina, S., Houari, M., Blangero, J. and Froguel, P. (2002b). A quantitative trait locus influencing type 2 diabetes susceptibility maps to a region on 5q in an extended French family. *Diabetes*, **51**, 3568–72.

Mathias, R.A., Freidhoff, L.R., Blumenthal, M.N., Meyers, D.A., Lester, L., King, R., Xu, J.F., Solway, J., Barnes, K.C., Pierce, J., Stine, O.C., Togias, A., Oetting, W., Marshik, P.L., Hetmanski, J.B., Huang, S.K., Ehrlich, E., Dunston, G.M., Malveaux, F., Banks-Schlegel, S., Cox, N.J., Bleecker, E., Ober, C., Beaty, T.H. and Rich, S.S. (2001). Genome-wide linkage analyses of total serum IgE using variance components analysis in asthmatic families. *Genet Epidemiol*, **20**, 340–55.

Meuwissen, T.H. and Goddard, M.E. (1997). Estimation of effects of quantitative trait loci in large complex pedigrees. *Genetics*, **146**, 409–16.

Mitchell, B.D., Ghosh, S., Schneider, J.L., Birznieks, G. and Blangero, J. (1997). Power of variance component linkage analysis to detect epistasis. *Genet Epidemiol*, **14**, 1017–22.

Mitchell, B.D., Cole, S.A., Comuzzie, A.G., Almasy, L., Blangero, J., MacCluer, J.W. and Hixson, J.E. (1999). A quantitative trait locus influencing BMI maps to the region of the beta-3 adrenergic receptor. *Diabetes*, **48**, 1863–7.

Mitchell, B.D., Cole, S.A., Bauer, R.L., Iturria, S.J., Rodriguez, E.A., Blangero, J., MacCluer, J.W. and Hixson, J.E. (2000a). Genes influencing variation in serum osteocalcin concentrations are linked to markers on chromosomes 16q and 20q. *J Clin Endocrinol Metab*, **85**, 1362–6.

Mitchell, B.D., Cole, S.A., Hsueh, W.C., Comuzzie, A.G., Blangero, J., MacCluer, J.W. and Hixson, J.E. (2000b). Linkage of serum insulin concentrations to chromosome 3p in Mexican Americans. *Diabetes*, **49**, 513–6.

North, K.E., Rose, K.M., Borecki, I.B., Oberman, A., Hunt, S.C., Miller, M.B., Blangero, J., Almasy, L. and Pankow, J.S. (2004). Evidence for a gene on chromosome 13 influencing postural systolic blood pressure change and body mass index. *Hypertension*, **43**, 780–4.

North, K.E., Miller, M.B., Coon, H., Martin, L.J., Peacock, J.M., Arnett, D., Zhang, B., Province, M., Oberman, A., Blangero, J., Almasy, L., Ellison, R.C. and Heiss, G. (2005). Evidence for a gene influencing fasting LDL cholesterol and triglyceride levels on chromosome 21q. *Atherosclerosis*, **179**, 119–25.

Olson, J.M. (1995). Robust multipoint linkage analysis: an extension of the Haseman-Elston method. *Genet Epidemiol*, **12**, 177–93.

Olson, J.M. and Wijsman, E.M. (1993). Linkage between quantitative trait and marker loci: methods using all relative pairs. *Genet Epidemiol*, **10**, 87–102.

Page, G.P., Amos, C.I. and Boerwinkle, E. (1998). The quantitative LOD score: test statistic and sample size for exclusion and linkage of quantitative traits in human sibships. *Am J Hum Genet*, **62**, 962–8.

Palmer, L.J., Tiller, K.J. and Burton, P.R. (1999). Genome-wide linkage analysis using genetic variance components of alcohol dependency-associated censored and continuous traits. *Genet Epidemiol*, **17**, S283–S288.

Pankow, J.S., Rose, K.M., Oberman, A., Hunt, S.C., Atwood, L.D., Djousse, L., Province, M.A. and Rao, D.C. (2000). Possible locus on chromosome 18q influencing postural systolic blood pressure changes. *Hypertension*, **36**, 471–6.

Perusse, L., Rice, T., Chagnon, Y. C., Despres, J. P., Lemieux, S., Roy, S., Lacaille, M., Ho-Kim, M. A., Chagnon, M., Province, M. A., Rao, D. C. and Bouchard, C. (2001). A genome-wide scan for abdominal fat assessed by computed tomography in the Quebec Family Study. *Diabetes*, **50**, 614–21.

Pratt, S. C., Daly, M. J. and Kruglyak, L. (2000). Exact multipoint quantitative-trait linkage analysis in pedigrees by variance components. *Am J Hum Genet*, **66**, 1153–7.

Rainwater, D. L., Almasy, L., Blangero, J., Cole, S. A., VandeBerg, J. L., MacCluer, J. W. and Hixson, J. E. (1999). A genome search identifies major quantitative trait loci on human chromosomes 3 and 4 that influence cholesterol concentrations in small LDL particles. *Arterioscler Thromb Vasc Biol*, **19**, 777–83.

Rainwater, D. L., Mahaney, M. C., VandeBerg, J. L., Brush, G., Almasy, L., Blangero, J., Dyke, B., Hixson, J. E., Cole, S. A. and MacCluer, J. W. (2004). A quantitative trait locus influences coordinated variation in measures of ApoB-containing lipoproteins. *Atherosclerosis*, **176**, 379–86.

Rice, T., Rankinen, T., Province, M. A., Chagnon, Y. C., Perusse, L., Borecki, I. B., Bouchard, C. and Rao, D. C. (2000). Genome-wide linkage analysis of systolic and diastolic blood pressure: the Quebec Family Study. *Circulation*, **102**, 1956–63.

Risch, N. and Zhang, H. (1995). Extreme discordant sib pairs for mapping quantitative trait loci in humans. *Science*, **268**, 1584–9.

Schork, N. J. (1993) Extended multipoint identity-by-descent analysis of human quantitative traits: efficiency, power and modeling considerations. *Am J Hum Genet*, **53**, 1306–19.

Scurrah, K. J., Palmer, L. J. and Burton, P. R. (2000). Variance components analysis for pedigree-based censored survival data using generalized linear mixed models (GLMMs) and Gibbs sampling in BUGS. *Genet Epidemiol*, **19**, 127–48.

Self, S. G. and Liang, K. Y. (1987). Asymptotic properties of maximum likelihood estimators and likelihood ratio tests under nonstandard conditions. *J Am Stat Assoc*, **82**, 605–10.

Sham, P. C., Zhao, J. H., Cherny, S. S. and Hewitt, J. K. (2000a). Variance-components QTL linkage analysis of selected and non-normal samples: conditioning on trait values. *Genet Epidemiol*, **19**, S22–S28.

Sham, P. C., Cherny, S. S., Purcell, S. and Hewitt, J. K. (2000b). Power of linkage versus association analysis of quantitative traits, by use of variance-components models, for sibship data. *Am J Hum Genet*, **66**, 1616–30.

Shearman, A. M., Ordovas, J. M., Cupples, L. A., Schaefer, E. J., Harmon, M. D., Shao, Y., Keen, J. D., DeStefano, A. L., Joost, O., Wilson, P. W., Housman, D. E. and Myers, R. H. (2000). Evidence for a gene influencing the TG/HDL-C ratio on chromosome 7q32.3-qter: a genome-wide scan in the Framingham study. *Hum Mol Genet*, **9**, 1315–20.

Soler, J. M. and Blangero, J. (2003). Longitudinal familial analysis of blood pressure involving parametric (co)variance functions. *BMC Genet*, **4** (Suppl 1), S87.

Sonnenberg, G. E., Krakower, G. R., Martin, L. J., Olivier, M., Kwitek, A. E., Comuzzie, A. G., Blangero, J. and Kissebah, A. H. (2004). Genetic determinants of obesity-related lipid traits. *J Lipid Res*, **45**, 610–15.

Soria, J.M., Almasy, L., Souto, J.C., Tirado, I., Borell, M., Mateo, J., Slifer, S., Stone, W., Blangero, J. and Fontcuberta, J. (2000). Linkage analysis demonstrates that the prothrombin G20210A mutation jointly influences plasma prothrombin levels and risk of thrombosis. *Blood*, **95**, 2780–5.

Soria, J.M., Almasy, L., Souto, J.C., Bacq, D., Buil, A., Faure, A., Martinez-Marchan, E., Mateo, J., Borrell, M., Stone, W., Lathrop, M., Fontcuberta, J. and Blangero, J. (2002). A quantitative-trait locus in the human factor XII gene influences both plasma factor XII levels and susceptibility to thrombotic disease. *Am J Hum Genet*, **70**, 567–74.

Soria, J.M., Almasy, L., Souto, J.C., Buil, A., Martinez-Sanchez, E., Mateo, J., Borrell, M., Stone, W.H., Lathrop, M., Fontcuberta, J. and Blangero, J. (2003). A new locus on chromosome 18 that influences normal variation in activated protein C resistance phenotype and factor VIII activity and its relation to thrombosis susceptibility. *Blood*, **101**, 163–7.

Souto, J.C., Blanco-Vaca, F., Soria, J.M., Buil, A., Almasy, L., Ordonez-Llanos, J., Martin-Campos, J.M., Lathrop, M., Stone, W., Blangero, J. and Fontcuberta, J. (2005). A genomewide exploration suggests a new candidate gene at chromosome 11q23 as the major determinant of plasma homocysteine levels: results from the GAIT project. *Am J Hum Genet*, **76**, 925–33.

Stern, M., Duggirala, R., Mitchell, B., Reinhart, J.L., Shivakumar, S., Shipman, P.A., Uresandi, O.C., Benavides, E., Blangero, J. and O'Connell, P. (1996). Evidence for linkage of regions on chromosomes 6 and 11 to plasma glucose concentrations in Mexican Americans. *Genome Res*, **6**, 724–34.

Stoesz, M.R., Cohen, J.C., Mooser, V., Marcovina, S. and Guerra, R. (1997). Extension of the Haseman-Elston method to multiple alleles and multiple loci: theory and practice for candidate genes. *Ann Hum Genet*, **61**, 263–74.

Terwilliger, J.D. (2001). On the resolution and feasibility of genome scanning approaches. *Adv Genet*, **42**, 351–91.

Tiwari, H.K. and Elston, R.C. (1997). Linkage of multilocus components of variance to polymorphic markers. *Ann Hum Genet*, **61**, 253–61.

Todorov, A.A., Province, M.A., Borecki, I.B. and Rao, D.C. (1997). Trade-off between sibship size and sampling scheme for detecting quantitative trait loci. *Hum Hered*, **47**, 1–5.

Towne, B., Siervogel, R.M. and Blangero, J. (1997). Effects of genotype-by-sex interaction on quantitative trait linkage analysis. *Genet Epidemiol*, **14**, 1053–8.

Towne, B., Almasy, L., Siervogel, R.M. and Blangero, J. (1999). Effects of genotype × sex interaction on linkage analysis of visual event-related evoked potentials. *Genet Epidemiol*, **17**, S355–S360.

Towne, B., Parks, J.S., Brown, M.R., Siervogel, R.M., Roche, A.F. and Blangero, J. (2000). Longitudinal quantitative genetic analysis of childhood skeletal maturation. *Genet Epidemiol*, **19**, 275.

Uimari, P. and Hoeschele, I. (1997). Mapping-linked quantitative trait loci using Bayesian analysis and Markov chain Monte Carlo algorithms. *Genetics*, **146**, 735–43.

Uimari, P., Thaller, G. and Hoeschele, I. (1996). The use of multiple markers in a Bayesian method for mapping quantitative trait loci. *Genetics*, **143**, 1831–42.

Visscher, P.M., Haley, C.S., Heath, S.C., Muir, W.J. and Blackwood, D.H. (1999). Detecting QTLs for uni- and bipolar disorder using a variance component method. *Psychiatr Genet*, **9**, 75–84.

Vogler, G.P., Tang, W., Nelson, T.L., Hofer, S.M., Grant, J.D., Tarantino, L.M. and Fernandez, J.R. (1997). A multivariate model for the analysis of sibship covariance structure using marker information and multiple quantitative traits. *Genet Epidemiol*, **14**, 921–6.

Walder, K., Hanson, R.L., Kobes, S., Knowler, W.C. and Ravussin, E. (2000). An autosomal genomic scan for loci linked to plasma leptin concentration in Pima Indians. *Int J Obes Relat Metab Disord*, **24**, 559–65.

Wan, Y., De Andrade, M., Yu, L., Cohen, J. and Amos, C.I. (1998). Genetic linkage analysis using lognormal variance components. *Ann Hum Genet*, **62**, 521–30.

Wang, J., Guerra, R. and Cohen, J. (1998). Statistically robust approaches for sib-pair linkage analysis. *Ann Hum Genet*, **62**, 349–59.

Wang, J., Guerra, R. and Cohen, J. (1999). A statistically robust variance-components approach for quantitative trait linkage analysis. *Ann Hum Genet*, **63**, 249–62.

Watanabe, R.M., Ghosh, S., Langefeld, C.D., Valle, T.T., Hauser, E.R., Magnuson, V.L., Mohlke, K.L., Silander, K., Ally, D.S., Chines, P., Blaschak-Harvan, J., Douglas, J.A., Duren, W.L., Epstein, M.P., Fingerlin, T.E., Kaleta, H.S., Lange, E.M., Li, C., McEachin, R.C., Stringham, H.M., Trager, E., White, P.P., Balow, Jr J., Birznieks, G., Chang, J. and Eldridge, W. (2000). The Finland-United States investigation of non-insulin-dependent diabetes mellitus genetics (FUSION) study. II. An autosomal genome scan for diabetes-related quantitative-trait loci. *Am J Hum Genet*, **67**, 1186–1200.

Wijsman, E.M. and Amos, C.I. (1997). Genetic analysis of simulated oligogenic traits in nuclear families and extended pedigrees: summary of GAW10 contributions. *Genet Epidemiol*, **14**, 719–35.

Williams, J.T. and Blangero, J. (1999a). Power of variance component linkage analysis to detect quantitative trait loci. *Ann Hum Genet*, **63**, 545–63.

Williams, J.T. and Blangero, J. (1999b). Comparison of variance components and sibpair-based approaches to quantitative trait linkage analysis in unselected samples. *Genet Epidemiol*, **16**, 113–34.

Williams, J.T. and Blangero, J. (1999c). Asymptotic power of likelihood-ratio tests for detecting quantitative trait loci using the COGA data. *Genet Epidemiol*, **17**, S397–S402.

Williams, J.T. and Blangero, J. (2004). Power of variance component linkage analysis-II. Discrete traits. *Ann Hum Genet*, **68**, 620–32.

Williams, J.T., Duggirala, R. and Blangero, J. (1997). Statistical properties of a variance-components method for quantitative trait linkage analysis in nuclear families and extended pedigrees. *Genet Epidemiol*, **14**, 1065–70.

Williams, J.T., Begleiter, H., Porjesz, B., Edenberg, H.J., Foroud, T., Reich, T., Goate, A., Van Eerdewegh, P., Almasy, L. and Blangero, J. (1999a). Joint multipoint linkage analysis of multivariate qualitative and quantitative traits. II. Alcoholism and event-related potentials. *Am J Hum Genet*, **65**, 1148–60.

Williams, J.T., Van Eerdewegh, P., Almasy, L. and Blangero, J. (1999b). Joint multipoint linkage analysis of multivariate qualitative and quantitative traits. I. Likelihood formulation and simulation results. *Am J Hum Genet*, **65**, 1134–47.

Williams, J.T., North, K.E., Martin, L.J., Comuzzie, A.G., Göring, H.H.H. and Blangero, J. (2001). Distribution of LOD scores in oligogenic linkage analysis. *Genet Epidemiol*, **21** (Suppl 1), S805–S810.

Williams-Blangero, S., VandeBerg, J. L., Subedi, J., Aivaliotis, M. J., Rai, D. R., Upadhayay, R. P., Jha, B. and Blangero, J. (2002). Genes on chromosomes 1 and 13 have significant effects on Ascaris infection. *Proc Natl Acad Sci USA*, **99**, 5533–8.

Wilson, A. F., Elston, R. C., Tran, L. D. and Siervogel, R. M. (1991). Use of the robust sib-pair method to screen for single-locus, multiple-locus, and pleiotropic effects: application to traits related to hypertension. *Am J Hum Genet*, **48**, 862–72.

Xu, S. (1998a). Iteratively reweighted least squares mapping of quantitative trait loci. *Behav Genet*, **28**, 341–55.

Xu, S. (1998b). Mapping quantitative trait loci using multiple families of line crosses. *Genetics*, **148**, 517–24.

Xu, S. and Atchley, W. R. (1995). A random model approach to interval mapping of quantitative trait loci. *Genetics*, **141**, 1189–97.

Xu, S. and Atchley, W. R. (1996). Mapping quantitative trait loci for complex binary diseases using line crosses. *Genetics*, **143**, 1417–24.

Xu, J., Postma, D. S., Howard, T. D., Koppelman, G. H., Zheng, S. L., Stine, O. C., Bleecker, E. R. and Meyers, D. A. (2000a). Major genes regulating total serum immunoglobulin E levels in families with asthma. *Am J Hum Genet*, **67**, 1163–73.

Xu, X., Weiss, S., Xu, X. and Wei, L. J. (2000b). A unified Haseman-Elston method for testing linkage with quantitative traits. *Am J Hum Genet*, **67**, 1025–8.

Yi, N. and Xu, S. (2000a). Bayesian mapping of quantitative trait loci under the identity-by-descent-based variance component model. *Genetics*, **156**, 411–22.

Yi, N. and Xu, S. (2000b). Bayesian mapping of quantitative trait loci for complex binary traits. *Genetics*, **155**, 1391–403.

Yuan, B., Neuman, R., Duan, S. H., Weber, J. L., Kwok, P. Y., Saccone, N. L., Wu, J. S., Liu, K. Y. and Schonfeld, G. (2000). Linkage of a gene for familial hypobetalipoproteinemia to chromosome 3p21.1–22. *Am J Hum Genet*, **66**, 1699–704.

Zeng, Z. B. (1993). Theoretical basis for separation of multiple linked gene effects in mapping quantitative trait loci. *Proc Natl Acad Sci USA*, **90**, 10972–6.

Zeng, Z. B. (1994). Precision mapping of quantitative trait loci. *Genetics*, **136**, 1457–68.

PART 4

The Human Diaspora

Michael H. Crawford

Chapter 12

Human Origins Within and Out of Africa

Sarah A. Tishkoff and Mary Katherine Gonder

Department of Biology, University of Maryland, College Park, MD

Abstract

Comparative studies of ethnically diverse human populations are important for testing historical hypotheses relating to the origin and dispersal of modern humans. In this chapter, we summarize the competing theories about how, when and where modern humans originated. We describe levels and patterns of genetic diversity across modern human populations and review the genetic evidence concerning modern human origins. We also discuss genetic signatures of population migrations within and out of Africa by contrasting and comparing these genetic signatures with global patterns of genetic diversity. Finally, we discuss implications of molecular data for reconstructing the demographic histories of African and non-African populations.

Introduction

The origin and dispersal of modern humans across the globe remains a topic of considerable interest and debate. While this topic has historically been within the realm of paleoanthropology based on fossil and archeological data, this topic is now being addressed in the fields of genetics and molecular biology. Genetic data, primarily gene frequency variation at polymorphisms, have been used to examine population similarities since the study of ABO blood type frequencies by Hirszfeld and Hirszfeld in 1919 (Hirszfeld and Hirszfeld, 1919).

Address for Correspondence: Dr Sarah A. Tishkoff, Department of Biology, Building #144, University of Maryland, College Park, MD 20742. Tel: (301) 405-6038. Fax: (301) 314-9358. E-mail: Tishkoff@umd.edu

In the subsequent decades, hundreds of populations have been studied for blood group loci, serum protein polymorphisms, and various enzyme electrophoretic polymorphisms (Cavalli-Sforza *et al.*, 1994). The advent of molecular biology techniques, including Restriction Fragment Length Polymorphisms (RFLPs) and Polymerase Chain Reaction (PCR) in the 1980s, mtDNA sequence variation in the 1980s, and nuclear sequence variation, Single Nucleotide Polymorphisms (SNPs), Short Tandem Repeat Polymorphisms (STRPs), and Alu polymorphisms in the 1990s until present, have made it feasible to do much more detailed, high-throughput studies of the distribution of molecular variation in globally diverse human populations.

The current draft of the human genome sequence covers 99% of the euchromatic genome and provides a blueprint for the structure of the human genome and location of the $\sim 20,000 - 25,000$ genes in our genome (Consortium, 2004). A major focus is now on identifying and characterizing differences in genetic diversity among individuals and populations (Collins *et al.*, 2003). Knowledge of the distribution of genetic variation in globally diverse populations will enable us to understand what makes each of us a morphologically and behaviourally unique individual, why some people are at a greater risk for disease or respond differently to drugs, and what differentiates us from our closest relatives, the great apes. Additionally, this data, together with new methods of statistical analysis, will enable us to test hypotheses of modern human origins and to reconstruct historic migration events within and out of Africa. In this chapter, we review the genetic evidence concerning modern human origins and human demographic history. We also discuss the genetic signatures of population migrations within and out of Africa by contrasting and comparing these genetic signatures with global patterns of genetic diversity.

Models of modern human origins

Patterns of genetic variation observed in modern human populations have been influenced by the demographic history of our ancestors. The origin and demographic history of modern humans is a topic that has not been resolved. The Multiregional Origin model (MRO) of human origins suggests that there was no single geographic origin for all modern humans (Figure 12.1a). Instead, proponents of the Multiregional Origin model argue that after the radiation of *H. erectus* from Africa into Europe and Asia 0.8 to 1.8 million years ago (Mya), there has been a continuous transition among regional populations from *H. erectus* to *H. sapiens* (Wolpoff *et al.*, 2000; Wolpoff *et al.*, 2001). This model is primarily supported by the observation of regional continuity of certain morphological traits in the fossil record, which suggests that they must have evolved over very long periods of time in the regions where they are found today (e.g. shovel shaped incisors

(a)

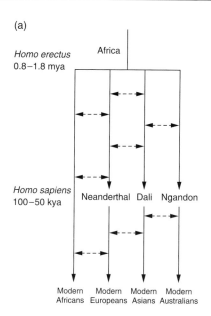

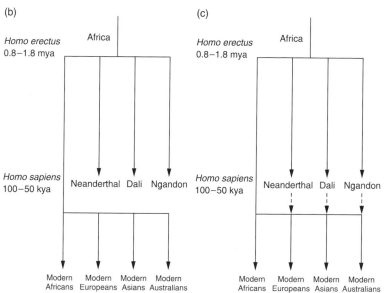

Fig 12.1 Models of modern human evolution. (a) Multiregional Origin (MRO) model, (b) Recent African Origin (RAO) model, (c) Hybridization/Assimilation model.

in *H. erectus* fossils from Asia and in modern Asians, and the robust cheekbones observed in *H. erectus* fossils from S.E. Asia and in modern Australian aborigines). It has been proposed that the 'parallel evolution' from *H. erectus* to *H. sapiens* among geographically dispersed populations could have been achieved through considerable amounts of gene flow between populations. This model predicts that genetic lineages in geographically diverse populations will have ancient coalescence times and that gene flow has been extensive over large geographic regions and through long tracts of time. Thus, the MRO model would imply a large effective population size in order to sustain gene flow amongst geographically diverse populations.

The Recent African Origin (RAO) model (also commonly referred to as the 'Out of Africa' model) suggests that all non-African populations descend from an anatomically modern *H. sapiens* ancestor that evolved in Africa approximately ~200 thousand years ago (Kya) and then spread and diversified throughout the rest of the world, supplanting any archaic *Homo* populations still present outside of Africa, such as Neanderthals (Stringer and Andrews, 1988; Stringer, 2002) (Figure 12.1b). The 'RAO' model is supported by the fossil record, in which the earliest anatomically modern human fossils are from Africa and the Middle East. In the Levant, the earliest dated AMHS were found at Qafzeh and Skhul at ~130–90 ka (Stringer and Andrews, 1988; Lahr, 1996; Stringer, 2002). However, recent research indicates the presence of early or near anatomically modern *Homo sapiens* (AMHS) in Ethiopia at ~195–154 ka (Clark *et al.*, 2003; Haile-Selassie *et al.*, 2004; McDougall *et al.*, 2005). The fossil data correlate well with archaeological data, dating the earliest evidence for modern human behaviour to ~90–70 Kya in Eastern and Southern Africa (McBrearty and Brooks, 2000; Henshilwood *et al.*, 2002; Henshilwood and Marean, 2003). After the initial appearance of AMHS in Ethiopia and the Levant ~180–90 ka, they do not reappear in that region, or in Europe, until ~50–40 ka, although there is evidence for modern humans in Australia by ~60 ka (Tattersall, 2002). Based on these data, multiple migrations have been proposed from NE Africa across the globe; an early route through the Horn/Arabia towards South Asia ~100–60 ka and a later one through North Africa/Middle East towards Eurasia ~70–40 ka (Stringer and Andrews, 1988; Bar-Yosef, 1994; Lahr and Foley, 1994; Ambrose, 1998).

A modified version of the RAO model, the 'Weak Garden of Eden Hypothesis' (WGE) suggests that populations may have remained small and subdivided for some time after the initial migration of modern humans out of Africa, followed by recent and rapid population expansion within the past 50,000 years (Harpending *et al.*, 1993). The RAO and WGE models predict that all genetic lineages in human populations stem from a recent common African ancestor. If modern humans originated in Africa and a subset of this population migrated out of Africa, then we would expect to observe in non-Africans a subset of the genetic diversity present in modern African populations. Levels of diversity in the non-African populations will depend on the severity of a bottleneck during migration out of Africa. The WGE model predicts high levels of genetic subdivision across major regions of the world (e.g. Africa, Asia and Europe), although not as high as would be expected under a multiregional origin model. The RAO model does not require worldwide gene flow as a mechanism for the emergence of modern humans.

Lastly, there have been some intermediate models proposed which are sometimes referred to as 'assimilation' or 'hybridization' models (Figure 12.1c). These models suggest that gene flow between the different regional populations of early humans might not have been equal through time and space (Stringer and Andrews, 1988; Stringer, 2002). Hybridization models allow for some gene flow between anatomically modern humans migrating from Africa and archaic populations outside of Africa. Thus, the evolution of modern humans could have been due to a blending of modern characters derived from African populations with local characteristics in archaic Eurasian populations such as Neanderthals. This model predicts a varying contribution of genes from archaic African and non-African populations to the modern gene pool (Tishkoff and Verrelli, 2003).

Levels and patterns of genetic variation in humans

Recent studies suggest that humans and chimpanzees differ by less than 1.2% at the nucleotide level (Chen and Li, 2001; Chen *et al.*, 2001; Wildman *et al.*, 2003; Watanabe *et al.*, 2004; Consortium, 2005), prompting some researchers to propose including chimpanzees as part of our genus, *Homo* (Uddin *et al.*, 2004). The intra-specific histories of the great apes and humans, however, have been considerably different. Chimpanzees, for example, possess two to six times greater genetic diversity at the nucleotide level than humans (Kaessmann *et al.*, 1999b; Stone *et al.*, 2002; Kitano *et al.*, 2003; Fischer *et al.*, 2004). Human chromosomes differ on average at only one out of every 500– 1,000 nucleotides (Chakravarti, 1999; Przeworski *et al.*, 2000; Jorde *et al.*, 2001), making us approximately 99.6% identical at the nucleotide level. This similarity corresponds to approximately 10 million DNA variants that can occur in many combinations. Additionally, recent analyses of the human genome indicate that humans also vary for insertion/deletion and inversion polymorphisms, often several thousand nucleotides in size (She *et al.*, 2004; Salem *et al.*, 2003; Boissinot *et al.*, 2004; Otieno *et al.*, 2004; Weber *et al.*, 2002). Because they are difficult to detect, we do not know how frequently large-size insertion/deletion and inversion polymorphisms occur (She *et al.*, 2004). Although the genetic differences between humans are very small, it is still more than enough variation to ensure individual uniqueness at the DNA level.

Sequencing of the human genome, and recent advances in identifying and genotyping genetic variation at hundreds of loci in hundreds of individuals, is providing a more detailed understanding of global patterns of genetic variation and of modern human origins. As discussed in detail in subsequent sections of this chapter, studies of mtDNA, Y chromosome, X chromosome and autosomal loci consistently show higher levels of genetic variation in African populations compared to non-African populations (Table 12.1). Exceptions to this pattern are RFLP and SNP studies which yield higher variability in Europeans, but are biased due to ascertainment in non-African populations (Tishkoff and Verrelli, 2003; Kidd *et al.*, 2004). Additionally, studies of autosomal and X-chromosomal

Table 12.1. Genetic diversity and population subdivision estimates for Africa, Europe, and Asia (Adapted from Tishkoff and Verrelli, 2003).

Locus	Marker system	n^a Africa	n^a Europe	n^a Asia	H^b Africa	H^b Europe	H^b Asia	F_{ST}^c Africa	F_{ST}^c Europe	F_{ST}^c Asia	Refs
MtDNA	HVRI	21	15	24	2.08^d	1.08^d	1.75^d	N/A	N/A	N/A	(Vigilant et al., 1991)
MtDNA	HVRI	72	120	63	0.022^e	0.009^e	0.015^e	0.088	0.045	0.032	(Jorde et al., 2000)
MtDNA	HVRII	72	120	63	0.030^e	0.010^e	0.011^e	0.092	0.013	0.017	(Jorde et al., 2000)
NRY	6 STRPs	72	120	63	0.576	0.498	0.472	0.026	0.602	0.092	(Jorde et al., 2000)
NRY	43 biallelic	360	507	1415	0.841	0.852	0.904	0.220	0.128	0.271	(Hammer et al., 2001)
Autosome	60 STRPs	72	120	63	0.76	0.732	0.697	N/A	N/A	N/A	(Relethford and Jorde, 1999)
Autosome	45 STRPs	216	246	302	0.792	0.773	0.677	0.03	0.01	0.02	(Calafell et al., 1998)
Autosome	30 RFLPs	72	120	63	0.293	0.401	0.350	0.027	0.013	0.017	(Jorde et al., 2000)
Autosome	100 Alus	152	118	75	0.349	0.297	0.557	0.042	0.024	0.010	(Watkins et al., 2003)
Autosome	13 Alus	72	120	63	0.276	0.243	0.233	0.017	0.022	0.009	(Jorde et al., 2000)
Autosome	8 Alus	176	334	359	0.402	0.396	0.377	0.088	0.011	0.058	(Stoneking et al., 1997)
PLAT	STRP1/Alu	924	352	386	0.903	0.739	0.772	3.18^f	2.31^f	2.09^f	(Tishkoff et al., 2000)
PLAT	STRP2/Alu	1030	410	422	0.909	0.800	0.767	3.28^f	1.22^f	1.66^f	(Tishkoff et al., 2000)
CD4	STRP/Alu	806	658	600	0.850	0.690	0.520	N/A	N/A	N/A	(Tishkoff et al., 1996)
DMD	STRP/Alu/RFLP	828	438	454	0.870	0.800	0.770	N/A	N/A	N/A	(Tishkoff et al., 1998a)
1q24	nucleotide	20	21	20	0.076^e	0.045^e	0.047^e	N/A	N/A	N/A	(Yu et al., 2001)
22q11	nucleotide	20	20	20	0.085^e	0.077^e	0.075^e	N/A	N/A	N/A	(Zhivotovsky et al., 2000)

[a] Number of chromosomes.

[b] Diversity reflects heterozygosity values in most cases ($1-\Sigma p_i^2$), where p_i represents the frequency of the ith allele.

[c] Values shown were estimated using F_{ST} or the equivalent measures of G_{ST} or R_{ST}.

[d] Average number of pairwise differences (Li, 1997).

[e] Gene diversity ($n/(n-1) \, \Sigma \, x_i x_j d_{ij}$), where n is the number of sequences, x_i and x_j are the population frequencies of the ith and jth type of DNA sequence, and d_{ij} is the proportion of nucleotides that differ between the ith and jth sequence.

[f] Subdivision estimated using DLR, a likelihood ratio statistic (Tishkoff et al., 2000).

haplotype variation, as well as mtDNA variation, indicate that Africans have the largest number of population-specific alleles and that non-African populations harbour a subset of the genetic diversity that is present in Africa, as expected under a RAO model if there were a genetic bottleneck at the time of migration of modern humans out of Africa (Tishkoff and Williams, 2002; Tishkoff and Verrelli, 2003; Tishkoff and Kidd, 2004).

Although levels of genetic diversity in humans are low relative to chimpanzees, we would like to know how the variation that does exist is distributed among human populations. Isolation by distance is the norm in human populations because humans do not mate at random; individuals living in the same geographic region and sharing a language are more likely to mate with each other than with individuals from more distant regions (Cavalli-Storza, 1997; Tishkoff and Kidd, 2004). Therefore, due to the process of genetic drift, populations have differentiated over time. The classic measure of population subdivision (or genetic differentiation) is Sewall Wright's F_{ST} (Wright, 1951; Wright, 1969). F_{ST} partitions variance into within- and among-population components. An F_{ST} value of zero indicates no differentiation between populations whereas a value of one indicates that they share no genetic variation. Under a model of random genetic drift, F_{ST} increases with time populations are separated at rates inversely related to Ne (the effective population size). At equilibrium between gene flow among populations and genetic drift within populations, the value of F_{ST} will depend on the number of migrants (Nm) exchanged between populations each generation. Thus, a low F_{ST} value could reflect either recent common ancestry and/or high levels of migration (Tishkoff and Kidd, 2004).

Estimates of F_{ST} (or equivalent measures) within and between major geographic regions (African/Europe/Asia) typically range from 0.11−0.16 for protein polymorphisms, blood groups, RFLPs, and autosomal STRPs, indicating that only 11%−16% of observed variation is due to differences among populations (Nei and Livshits, 1989; Bowcock and Cavalli-Sforza, 1991; Barbujani et al., 1997; Jorde et al., 1997; Jorde et al., 2000; Romualdi et al., 2002). F_{ST} estimates based on variation in mtDNA ($F_{ST} = 0.24 -0.38$) (Seielstad et al,. 1998; Jorde et al., 2000; Wilder et al., 2004a) or the Y-chromosome ($F_{ST} = 0.23$ to 0.64) (Hammer et al., 1997; Poloni et al., 1997; Seielstad et al., 1998; Jorde et al., 2000; Hammer et al., 2001; Romualdi et al., 2002; Kayser et al., 2003; Wilder et al., 2004a) are higher than estimates from autosomal DNA. This observation may be due to the smaller effective size for mtDNA and the Y-chromosome (one-quarter that of autosomes), which results in more genetic drift, and/or due to selection acting on mtDNA and/or Y-chromosome DNA (Jorde et al., 1998; Mountain, 1998; Seielstad et al., 1998; Hammer et al., 2001; Tishkoff and Verrelli, 2003). Seielstad et al. (1998) proposed that the lower F_{ST} estimates for mtDNA relative to the Y-chromosome indicated a higher female migration rate on a global scale. However, Wilder et al. (2004a) used an unbiased method for comparing levels of variation in mtDNA and the

Y-chromosome for the same set of individuals in the same set of populations and observed similar Φ_{ST} values (a statistic similar to F_{ST}) for mtDNA and Y-chromosome sequences (0.382 and 0.334, respectively). Thus, they conclude that there is no evidence for a higher migration rate of females compared to males on a global level (Wilder et al., 2004a).

These F_{ST} values indicate that the majority of genetic variability in human populations is accounted for by variation within populations rather than by variation among populations, consistent with a recent common ancestry and/or gene flow between populations. Although the amount of genetic diversity between populations is relatively small compared to the amount of genetic diversity within populations, populations usually cluster by major geographic region based on genetic distance. For example, a recent study of 377 autosomal microsatellite loci in 1,056 globally diverse individuals from 52 populations indicates that within-population differences among individuals account for 93–95% of the observed genetic variation whereas differences among major groups constitute only 3–5% of the observed variation (Rosenberg et al., 2002). The neighbour-joining tree shown in Figure 12.2 illustrates the clustering patterns of genetic distance estimates based on 353 of the 377 microsatellites originally reported in Rosenberg et al., 2002. The tree was constructed from genetic distance estimates of R_{ST}, which is a genetic distance measure that is closely related to F_{ST} but differs in that it is calibrated

Fig 12.2 Unrooted neighbour-joining phylogram of genetic distances based on 353 autosomal microsatellites (Rosenberg et al. (2002)).

to account for the special mutational properties of microsatellites (Slatkin, 1995). The branching patterns shown in this tree indicate that populations cluster by major geographic region (i.e. Africa, Europe, Asia, Oceania, New World). Populations from geographically intermediate regions (e.g. the Middle East and Central Asia) are placed in intermediate positions in the tree, indicating a continuous gradient of variation between geographic regions and/or admixture in these regions. The most likely explanation of this pattern is genetic drift resulting from isolation by distance and founding of major geographic regions following an initial expansion out of Africa (Tishkoff and Kidd, 2004).

Studies of population subdivision within major geographic regions have observed high levels of population subdivision in sub-Saharan Africa relative to Europe or Asia (Tishkoff *et al.*, 2000; Tishkoff and Williams, 2002; Tishkoff and Kidd, 2004). Phylogenetic analyses of haplotypes on the Y-chromosome (Underhill *et al.*, 2000; Hammer *et al.*, 2001; Underhill *et al.*, 2001; Semino *et al.*, 2002), mtDNA (Chen *et al.*, 1995; Watson *et al.*, 1997; Chen *et al.*, 2000), autosomes (Alonso and Armour, 2001; Yu *et al.*, 2001) and X chromosome (Labuda *et al.*, 2000; Zietkiewicz *et al.*, 2003) indicate that African populations have highly divergent lineages that suggest a long history of population subdivision. Several studies of mtDNA and Y-chromosome variation have observed higher levels of subdivision in African populations compared to other regions as measured by F_{ST} (Jorde *et al.*, 2000; Hammer *et al.*, 2001). Additionally, recent studies using autosomal RFLPs (Jorde *et al.*, 2000), STRPs (Jorde *et al.*, 2000; Rosenberg *et al.*, 2002), or Alu elements (Batzer and Deininger, 2002; Stoneking *et al.*, 1997; Watkins *et al.*, 2001; Watkins *et al.*, 2003) as markers, as well as studies which combine analysis of these markers as haplotypes (Tishkoff *et al.*, 1996, 1998a, 2000; Labuda *et al.*, 2000), in a broad range of African populations, have observed higher F_{ST} values for African populations compared to other geographic regions, indicating higher levels of subdivision (Table 12.1). Thus, the pattern of genetic diversity in Africa indicates that African populations have maintained a large and subdivided population structure throughout much of their evolutionary history (Figure 12.3). Historic subdivision among African populations is likely due to ethnic and linguistic barriers, as well as a number of geographic, ecological, and climatic factors (including periods of glaciation and warming) that could have contributed to population expansions, contractions, fragmentations, and extinctions during recent human evolution in Africa (Lahr and Foley, 1994; Tishkoff and Williams, 2002; Tishkoff and Verrelli, 2003).

Nucleotide studies of modern human origins

Many researchers active in the field of modern human origins have progressed past testing the relatively simple 'MRO' versus

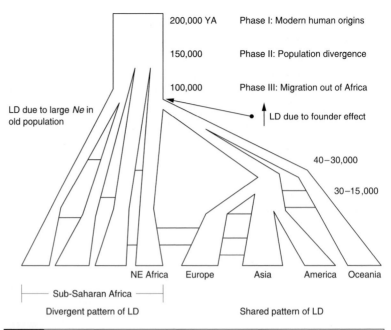

200,000 YA   Phase I: Modern human origins

150,000   Phase II: Population divergence

100,000   Phase III: Migration out of Africa

LD due to large *Ne* in old population

LD due to founder effect

40–30,000

30–15,000

NE Africa   Europe   Asia   America   Oceania

Sub-Saharan Africa

Divergent pattern of LD   Shared pattern of LD

Fig 12.3 Model of human demographic history. Fossil and archeological data suggest that modern humans may have originated within the past 200 Ky in Africa. East African and other sub-Saharan African populations may have diverged early in African history. A small subset of this east African population may have migrated out of Africa within the past 100–50 Ky, rapidly expanding throughout a broad geographical region. Ancestral African populations have maintained a large and subdivided population structure, resulting in fewer sites being in linkage disequilibrium (LD) compared to non-African populations. Non-African populations experienced a population bottleneck that resulted in reduced genetic variation and more LD in the genome. Horizontal lines indicate geneflow. (Adapted from Mountain, 1998; Tishkoff and Williams, 2002; and Tishkoff and Verrelli, 2003.)

'RAO' models. Instead, studies in molecular anthropology and paleoanthropology have become increasingly focused on what particular patterns and processes of diversification were involved when modern humans left Africa ~100–50 Kya (Jobling *et al.*, 2004) and testing more complex demographic scenarios that consider fluctuations in population size, migration, and admixture over time and space (Wall and Przeworski, 2000). A number of statistical methods have been developed to analyse patterns of nucleotide variation obtained by DNA sequencing in order to distinguish the effects of demographic history from gene-specific factors such as recombination, gene conversion, and natural selection on levels and patterns of genetic variation (Tishkoff and Verrelli, 2003). Below, we summarize results of recent nucleotide studies of mtDNA, the Y chromosome, the X chromosome, and the autosomes and implications for inferring human demographic history and for testing hypotheses of modern human origins. Overall these genetic data support the demographic model of human evolutionary history shown in Figure 12.3 that is discussed in detail below.

Mitochondrial DNA

Most of the early genetic evidence on modern human origins came from analyses of mitochondrial (mt) DNA. The human mitochondrial genome is 16,569 bp in length and is 0.00006% the size of the total human genome. The majority of the mtDNA genome consists of DNA that codes for 13 polypeptides, all of which are involved in the process of oxidative phosphorylation within the mitochondrial organelle. A non-coding region ~1200 bp in size (the D loop) consists of two ~350 bp hypervariable regions (HVRI and HVRII) that are frequently used in evolutionary genetics studies. The mtDNA genome is maternally inherited and does not undergo recombination, making its effective population size one-fourth that of nuclear autosomes. Because mtDNA is inherited from mother to offspring, it is possible to trace the ancestry of mtDNA lineages back to a single female ancestor, reflecting the evolutionary history of the female lineage. Another advantage of mtDNA is that mutations accumulate at a high rate (average of $2-3 \times 10^{-7}$ per nucleotide per generation for the coding region and as high as 0.043 mutations per nucleotide per generation in the hypervariable regions of the D-loop), making the molecule amenable to analysis in evolutionary studies based on relatively small sequences and allowing reconstruction of evolutionary events on a recent time scale (Meyer *et al.*, 1999; Tishkoff and Verrelli, 2003). However, the high mutation rate of mtDNA can also be problematic for phylogenetic analyses due to a high incidence of homoplasy (e.g. convergent mutation events), as discussed below.

One of the earliest and most widely publicized studies of mtDNA from a global panel of humans concluded that all human mtDNA lineages can be traced back through maternal lineages to an ancestral mtDNA lineage present in an African population ~200 Kya (Cann *et al.*, 1987). Subsequent studies of HVRI and HVRII, assorted genes in the mtDNA coding region, and complete mtDNA genome sequences from global human panels have largely supported this conclusion (Table 12.1). In general, these studies have shown that African mtDNA lineages cluster together nearest to the root of the human phylogenetic tree, and that the oldest lineages are specific to Africans. Both African and non-African mtDNA lineages are found throughout the more recent branches of the human phylogenetic tree. The most parsimonious explanation for the observed pattern of mtDNA variation is that a subset of African individuals migrated from Africa into the rest of the world, leaving additional variation behind. Thus, non-Africans carry a subset of the African mtDNA variation, with additional mutations that occurred after the migration from Africa (Vigilant *et al.*, 1991; Chen *et al.*, 1995; Watson *et al.*, 1996, 1997; Quintana-Murci *et al.*, 1999; Jorde *et al.*, 2000; Tishkoff and Verrelli, 2003; Salas *et al.*, 2004).

Although the majority of the mtDNA genetic data support a RAO model of modern human evolution, variations in the RAO model, which take into account gene flow between the anatomically modern humans migrating from Africa and certain archaic populations outside of Africa, are difficult to rule out based on current data (Nordborg, 1998; Wall, 2000; Templeton, 2002; Tishkoff and Verrelli, 2003). This is because low levels of gene flow between Neanderthal and AMHS may be difficult to detect. Recent morphological surveys suggest, however, that Neanderthals form a distinct group that should be recognized as a separate species (Schillaci and Froehlich, 2001; Harvati, 2003; Harvati et al., 2004). Analysis of nucleotide sequences of the HVRI and HVRII of the mtDNA D loop from eight Neanderthals also suggests that Neanderthals should be recognized as a separate species (Krings et al., 1997, 1999, 2000; Ovchinnikov et al., 2000; Schmitz et al., 2002; Serre et al., 2004). The Neanderthal mtDNA is highly differentiated from modern human mtDNA; HVRI and HVRII mtDNA sequences of Neanderthals and modern humans differ by approximately 32–36 nucleotide substitutions (Krings et al., 1997, 2000). The Neanderthal sequences are no closer to Europeans (the geographic region from which the samples were obtained) than to Africans or Asians. Furthermore, sequence analysis of the mtDNA D loop amplified from two anatomically modern Cro-Magnon specimens from Italy dated to ~25 Kya, indicates a high level of sequence similarity compared to modern humans (particularly those originating from Europe) and a high level of sequence divergence when compared to several Neanderthal sequences (Caramelli et al., 2003). These results suggest that Neanderthals did not contribute significantly to the modern human gene pool (Krings et al., 2000; Caramelli et al., 2003). The estimated TMRCA of human and Neanderthal mtDNA lineages is ~300–850 Kya, considerably older than the TMRCA of modern humans (Krings et al., 1997, 1999, 2000; Ovchinnikov et al., 2000). Despite the ancient TMRCA between Neanderthal and human mtDNA, Neanderthals and modern humans have similar levels of nucleotide diversity; pairwise sequence diversity of HVRI and HVRII is 1.7%–3.7% and 1.8%–3.4% for Neanderthals and humans, respectively (Krings et al., 2000; Jobling et al., 2004). This observation suggests that Neanderthals and humans both had small historic population sizes (Krings et al., 2000).

Although the limited studies of Neaderthal HVRI and HVRII mtDNA sequences add support to a RAO model of modern human origins, they do not rule out the possibility of limited gene flow between Neanderthals and AMHS (Serre et al., 2004). Unfortunately, it is technically not feasible to obtain DNA from large numbers of Neanderthal individuals (Paabo et al., 2004). One could look for evidence of Neanderthal lineages in the modern human genome. However, identification of Neanderthal DNA in the modern human gene pool would depend on the length of time and extent of gene flow between the two groups and is unlikely to be identified based on analysis of one or a few genetic loci because the Neanderthal genetic contribution to any particular locus may be lost due to genetic drift. Rather, the identification of archaic *Homo* contributions to the modern *H. sapiens* gene pool would require analysis of tens or hundreds of loci (Nordborg, 1998; Wall, 2000; Nordborg, 2001). As we proceed into the genomics era and large-scale comparative sequence analyses become more readily available, it may be possible to definitively distinguish amongst these hypotheses in the future (Tishkoff and Verrelli, 2003).

Although studies of mtDNA have added to our knowledge of human evolutionary history, there are some features of mtDNA that are problematic for evolutionary genetics studies and warrant caution in interpreting studies of human mtDNA variation. Most studies of human mtDNA variation are based on analyses of small segments of the non-coding hypervariable region (HVRI and HVRII). The HVRI and HVRII regions are each ~350 bp in size, representing only ~5% of the mitochondrial genome. The hypervariable region mutates rapidly and is subject to homoplasy (Meyer *et al.*, 1999; Stoneking, 2000). Several equally likely gene trees can often be inferred from HVRI and HVRII sequence data, particularly when the numbers of taxa analysed is large (Maddison *et al.*, 1992). Mutations are not randomly distributed across the length of the HVRI and HVRII sequences, making rate heterogeneity an important issue in calculating divergence dates from these loci (Excoffier and Yang, 1999; Meyer *et al.*, 1999; Stoneking, 2000). In addition, nuclear mitochondrial pseudogenes (NUMTs) occur frequently among humans. These NUMTs contain regions that are homologous to large regions of the mtDNA genome (Mourier *et al.*, 2001; Tourmen *et al.*, 2002), some of which have been integrated into the nuclear genome subsequent to the appearance of modern humans (Mishmar *et al.*, 2004). NUMTs can be accidentally amplified via PCR using primers that should only target mtDNA, particularly when using low-quantity or badly degraded DNA (Bensasson *et al.*, 2001; Thalmann *et al.*, 2004). Because NUMTs are subject to recombination and possibly natural selection, and because they evolve at a different rate from mtDNA, accidental amplification of NUMTs can confound phylogenetic and population genetic analyses that assume a lack of recombination and a neutral rate of evolution (Mishmar *et al.*, 2004; Ricchetti *et al.*, 2004; Thalmann *et al.*, 2004).

The utility of the mtDNA genome is further limited in that it has been associated with human disease and could be subject to effects of selection (either positive or negative). Recent analyses of complete mtDNA genomes among global humans, for example, have shown that regional variation in human mtDNA has been subject to the effects of purifying selection (Mishmar *et al.*, 2003; Ruiz-Pesini *et al.*, 2004; Kivisild *et al.*, 2006) Due to the effects of genetic hitchhiking, any mutations linked to a mutation under selection will increase or decrease in frequency along with the site under selection (Kaplan *et al.*, 1989; Chen *et al.*, 2000; Bamshad *et al.*, 2003; Schlotterer, 2003). Because all loci in the mtDNA genome are linked, all sites are subject to the effects of hitchhiking. Thus, a selective sweep at any site on the mtDNA genome will cause a reduction of neutral variation at linked sites. Therefore, the mtDNA genome is subject to the stochastic processes of genetic drift as well as to natural selection which can alter the shape of the haplotype phylogenies, patterns of variation, and estimates of TMRCA (Excoffier, 1990; Di Rienzo and Wilson, 1991; Maddison et al., 1992; Marjoram and Donnelly, 1994; Mishmar et al., 2003; Ruiz-Pesini et al., 2004).

Y chromosome

The Y chromosome is a ~60Mb haploid chromosome that is paternally inherited in humans, which makes it ideal for tracing the male lineage of our species (Jobling and Tyler-Smith, 2003). The majority of the Y chromosome does not undergo recombination (Non-Recombining Y, NRY or, more recently, male specific region, MSY), except in a 3Mb region that recombines with the X chromosome at the tips of the Y chromosome (e.g. the pseudoautosomal region [PAR]). However, analysis of the complete sequence of the Y chromosome indicates that the NRY is prone to gene conversion events in areas enriched with repetitive genomic elements, which can result in shuffling of genetic material across loci on the Y chromosome (Rozen et al., 2003; Skaletsky et al., 2003). The NRY contains ~25 genes, many of which play a role in spermatogenesis (Jobling and Tyler-Smith, 2003). Most of the human evolutionary studies discussed in this review are based on analyses of the NRY.

Nucleotide studies of the Y chromosome indicate lower levels of genetic diversity compared to mtDNA, the X chromosome, or autosomal sequences (Hammer, 1995; Jorde et al., 2000; Thomson et al., 2000), with most NRY sequences differing by one nucleotide every 3–4 kb (Stumpf and Goldstein, 2001). One reason that relatively low levels of genetic variation have been observed across Y chromosomes is that the effective population size of the Y chromosome is one-quarter that of the autosomes and one-third that of the X chromosome. The reduced rate of recombination in most of the Y chromosome, combined with relatively low effective population size, makes the Y chromosome more subject to genetic drift than the autosomes. Thus, genetic analyses of the Y chromosome observe high levels of geographically structured genetic drift compared to analyses of other loci. Despite relatively low levels of genetic diversity, an abundant number of polymorphisms have recently been identified on the Y-chromosome and analysed for variation across ethnically diverse human populations (Jobling and Tyler-Smith, 2003). These include studies of SNPs (Hammer et al., 1998; Underhill et al., 2000; Hammer et al., 2001; Underhill et al., 2001), Alu polymorphisms (Hammer, 1994), microsatellites (Santos et al., 1996; Hammer et al., 1997; Zerjal et al., 1997, 2002; Pritchard et al., 1999; Kayser et al., 2001, 2004; Zhivotovsky et al., 2004) and combinations of these markers (de Knijff, 2000; Forster *et al.*, 2000; Malaspina *et al.*, 2000; Thomas *et al.*, 2000).

Consistent with studies of mtDNA variation, studies of genetic variation in the NRY region of the Y chromosome have been used to trace haplotype lineages back to a single male ancestral Y chromosome that existed in Africa ~65 K–200 Kya (Hammer, 1995; Hammer *et al.*, 1998; Pritchard *et al.*, 1999; Thomson *et al.*, 2000; Underhill *et al.*, 2001) (Table 12.2). Haplotype analyses of Y-chromosome markers indicate that both African and non-African populations have many population-specific haplotypes and that haplotypes in each region are distinct. Phylogenetic trees constructed from Y-chromosome haplotype data indicate that the two most ancestral Y-chromosome lineages, from which all non-African lineages are derived, are African specific, consistent with an RAO model of modern human origins (Underhill *et al.*, 2000; Hammer *et al.*, 2001; Underhill *et al.*, 2001). The estimated time of emergence of an ancestral Y-chromosome lineage out of Africa, 44 Kya, is later than the estimation based on mtDNA (Underhill *et al.*, 2000). However, patterns of Y-chromosome haplotype diversity also suggest the possibility of reverse migration of Y-chromosome lineages back to Africa (Hammer *et al.*, 1998; Hammer *et al.*, 2001; Cruciani *et al.*, 2002; Luis *et al.*, 2004).

TMRCA estimates of human mtDNA and Y chromosomes differ. The human mtDNA TMRCA is estimated to range from ~171–238 Kya (Ingman *et al.*, 2000; Tang *et al.*, 2002). In contrast, TMRCA estimates for the NRY range from ~46–109 Kya (Pritchard *et al.*, 1999; Thomson *et al.*, 2000; Hammer and Zegura, 2002; Tang *et al.*, 2002; Wilder *et al.*, 2004b). The TMRCA of a selectively neutral locus is influenced primarily by its effective population size (*Ne*). The observed disparity between estimates of TMRCA for mtDNA and the NRY is unexpected because they are assumed to have equal *Ne* values and are, therefore, also expected to have similar TMRCAs. This observation raises the possibility that the more recent TMRCA of the Y chromosome could be due to natural selection or sex-specific demographic processes (Wilder *et al.*, 2004b). The Y chromosome is unique because of its significantly reduced gene density compared to mtDNA or autosomes (Quintana-Murci and Fellous, 2001) and for its functional specialization in accumulating genes that specifically benefit male fitness (Lahn *et al.*, 2001). It has been hypothesized that the gene density of the Y chromosome is low because the chromosome is haploid and most of it does not recombine, making it prone to genetic decay (via Muller's ratchet) (Charlesworth, 1978; Lahn *et al.*, 2001; Marshall Graves, 2002). Like the mtDNA genome, the Y chromosome has genes associated with human disease, in addition to genes associated with male fertility, and could be subject to effects of selection (either positive or negative) and genetic hitchhiking (Makova and Li, 2002). However, there is little evidence thus far to suggest that natural selection has had a significant impact on the geographic partitioning of contemporary Y-chromosome haplotype lineages (Jobling and Tyler-Smith, 2003; Wilder *et al.*, 2004a; Wilder *et al.*, 2004b).

Table 12.2. | Genetic diversity and parameter estimates for different genomic regions (Adapted from Tisinkoff and Verrelli, 2003).

Locus	Region	n^a	Length[b]	MRCA[c]	S[d]	π^e	D[f]	N_e^g	References
Whole Genome	mtDNA	53	16500	170	657	0.370	−1.50	10.9	(Ingman et al., 2000)
Whole Genome	mtDNA	179	15500	240	971	NE	NE	NE	(Tang et al., 2002)
HVR-I/HVR-2	mtDNA	>5000	350	150–350	NE	NE	NE	10.0	(Penny et al., 1995; Rogers and Jorde, 1995)
YAP	Y	24	2600	188	NE	NE	NE	10.0	(Hammer, 1995)
9 di-allelic	Y	1924	N/A	150	NE	NE	NE	10.0	(Hammer et al., 1998)
8 STRPs	Y	674	N/A	45–95	NE	NE	NE	NE	(Pritchard et al., 1999)
Yp11.3	Y	205	676	N/A	1	0.001	−0.95	14.0	(Jaruzelska et al., 1999)
SMCY	Y	53	40000	41–68	47	0.007	−2.31	8.9	(Shen et al., 2000; Thomson et al., 2000)
DBY	Y	70	9000	39–100	12	0.105	−2.04	NE	(Shen et al., 2000; Thomson et al., 2000)
DFFRY	Y	70	15000	40–65	17	0.098	−1.79	NE	(Shen et al., 2000; Thomson et al., 2000)
ALAS2	Xp11.21	41	4697	656	7	0.015	−1.527	NE	(Nachman et al., 2004)
MAO-A	Xp11.4	56	18800	N/A	41	0.050	0.33	8.4	(Gilad et al., 2002)
PDHA1	Xp22.1	35	4200	1860	25	0.178	0.78	18.2	(Harris and Hey, 1999)
ZFX	Xp22.13	335	1100	700–1100	10	0.082	−0.95	21.7	(Jaruzelska et al., 1999)
AXPL	Xp22.3	41	4638	840	19	0.080	−0.541	NE	(Hammer et al., 2004)
AMELX	Xp22.3	41	4638	1178	17	0.061	−0.601	NE	(Hammer et al., 2004)
MSN	Xq11.2	41	4581	605	9	0.035	−0.662	NE	(Nachman et al., 2004)
Intergenic	Xq13.3	70	10200	535	33	0.033	−1.61	16.0	(Kaessmann et al., 1999a)
DMD intron 7	Xq21	41	2390	210	12	0.034	−1.79	3.5	(Nachman and Crowell, 2000b)
DMD intron 44	Xq21	41	3000	1560	19	0.141	−0.16	26.0	(Nachman and Crowell, 2000b)
PLP	Xq22.2	10	769	864	2	0.095	0.12	24.4	(Nachman et al., 1998)
GK	Xq22.3	10	1861	500	1	0.019	0.02	4.9	(Nachman et al., 1998)
HPRT	Xq26.1	10	2485	777	4	0.038	−1.25	6.5	(Nachman et al., 1998)
TNFS5	Xq25.2	41	5239	1340	16	0.035	−1.634	NE	(Hammer et al., 2004)
RRM2P4	Xq27.3	41	2385	1714	13	0.119	−0.206	NE	(Garrigan et al., 2005b; Hammer et al., 2004)
FIX intron 4	Xq27.2	36	3700	282	6	0.014	−1.71	8.4	(Harris and Hey, 2001)

Locus	Region	n^a	Length^b	MRCA^c	S^d	π^e	D^f	$N_e{}^g$	References
G6PD	Xq28	216	5200	620	23	0.073	−0.95	15.0	(Verrelli et al., 2002)
OPN1LW	Xq28	236	5500	NE	74	0.240	−1.05	NE	(Verrelli and Tishkoff, 2004)
Duffy (FY)	1q21	82	1931	NE	22	0.128	0.19	17.1	(Hamblin and Di Rienzo, 2000; Hamblin et al., 2002)
Intergenic	1q24	122	8991	1376	52	0.058	−1.22	12.6	(Yu et al., 2001)
CCR5 5'	3p21	224	1123	2100	13	0.240	0.67	30.0	(Bamshad et al., 2002)
LPL	8p22	142	9734	1200	79	0.200	0.91	16.9	(Clark et al., 1998)
b-globin	11p15.5	349	2670	800	28	0.180	1.16	11.6	(Harding et al., 1997)
EDN	14q24	67	1200	1150	9	0.063	−1.28	NE	(Zhang and Rosenberg, 2000)
ECP	14q31	54	1200	1090	7	0.110	0.04	NE	(Zhang and Rosenberg, 2000)
CYP1A2 5'	15q22	226	3669	NE	18	0.047	−1.15	10.2	(Wooding et al., 2002)
MS205 intron	16q13.3	100	1700	1040	75	0.089	−1.60	16.8	(Alonso and Armour, 2001)
MCR1	16q24.3	356	951	1000	16	0.114	−1.06	14.3	(Harding et al., 2000)
MCR1 5'	16q24.3	108	6700	1520	72	0.141	−1.04	12.0	(Makova et al., 2001)
ACE	17q23	22	24070	1113	78	0.093	−0.32	10.2	(Rieder et al., 1999)
FUT2	19p13.3	355	1000	NE	21	0.410	0.73	NE	(Koda et al., 2001)
FUT6	19p13.3	486	1000	NE	18	0.200	−0.43	NE	(Koda et al., 2001)
APOE	19q13.2	192	5500	311	22	0.050	−0.62	7.0	(Fullerton et al., 2000)
Intergenic	22q11.2	128	9834	1288	75	0.089	0.09	12.2	(Zhao et al., 2000)

NE indicates that statistics or values were not estimated.

^{a}Number of chromosomes.

^{b}Number of base pairs (bp) in analysis.

^{c}Estimate of Time to Most Recent Common Ancestor $\times 10^3$ years.

^{d}Number of polymorphisms in sample.

^{e}Estimate of nucleotide diversity from average pairwise differences among individuals (Li, 1997).

^{f}Tajima's D, see text (Tajima, 1989).

^{g}Estimate of effective population size $\times 10^3$.

In a recent study, Wilder *et al.* (2004b) proposed a model in which the human effective population size is skewed toward an excess of females due to sex-biased demographic processes. They observe that the human mating system, which tends to be polygynous, would increase the variance in reproductive success among males, thereby lowering their *Ne* relative to females. Thus, the more recent TMRCA of the Y-chromosome lineages is likely due to sex-specific demographic processes that tend to favour a larger female *Ne*.

X chromosome

The X chromosome is roughly three times larger in size than the Y chromosome and is unique among the nuclear chromosomes because females have two copies, whereas males only have one (e.g. they are hemizygotes). Because it is possible to unambiguously determine haplotype phase in males (e.g. the arrangement of alleles along a chromosome), there have been a relatively large number of studies of nucleotide variation in intergenic and coding regions of the X chromosome compared to the autosomes (Table 12.2). These studies have been used to investigate correlations between recombination rate and levels of nucleotide diversity (Nachman *et al.*, 1998; Nachman and Crowell, 2000b; Nachman and Crowell, 2000a), patterns of linkage disequilibrium and haplotype diversity (Tishkoff *et al.*, 2001; Verrelli *et al.*, 2002; Verrelli and Tishkoff, 2004; Nachman *et al.*, 2004; Saunders *et al.*, 2002), and in regions of unusually low recombination (i.e. Xq13.3) (Kaessmann *et al.*, 1999a; Yu *et al.*, 2002; Nachman *et al.*, 2004), to trace back ancestral haplotype lineages to reconstruct population histories (Table 12.2). The findings of these studies have varied considerably amongst each other due to several factors. Because the X chromosome is hemizygous in males, its effective population size is three-quarters that of the autosomes but three times larger than the effective population size of mtDNA or the Y chromosome. Generally speaking, the larger effective population size of the X chromosome makes TMRCA estimates larger for the X chromosome compared to mtDNA and the Y chromosome, but smaller compared to the autosomes, with some exceptions (Table 12.2; Kaessmann *et al.*, 1999a; Jobling *et al.*, 2004). Owing to the smaller effective population size of the X chromosome, it may be more sensitive to the effects of genetic drift and to selective sweeps than the autosomes (Hammer *et al.*, 2004).

Most studies of X chromosome variation suggest that the roots of population trees are composed of African populations and/or that Africans have the most divergent lineages, as observed for both mtDNA and Y chromosome lineages. In addition, some X chromosome studies indicate that Africans have the largest number of population-specific alleles and that non-African populations harbour a subset of the genetic diversity that is present in Africa, as expected if there was a genetic bottleneck at the time of migration of modern humans out of Africa (Kaessmann *et al.*, 1999a; Takahata *et al.*, 2001; Tishkoff *et al.*, 2001; Verrelli *et al.*, 2002; Verrelli and Tishkoff, 2004). Patterns of variation across the X chromosome, however, differ considerably in the magnitude of that reduction in genetic diversity (Hammer *et al.*, 2004; Nachman *et al.*, 2004).

TMRCA estimates vary widely in studies of genetic variation at the X chromosome (Table 12.2), from 210 Kya (DMD intron 7) to 1.7 Mya (RRM2P4) which may be attributable to a variety of factors, including historic natural selection which can either result in a loss of genetic variation and more recent TMRCA estimates (e.g. positive selection) or an increase in genetic variation and more ancient TMRCA estimates (e.g. balancing selection). Several X chromosome studies have shown that there was considerable population subdivision in the ancestral human population, which may have preceded the migration of modern humans out of Africa within the past ~100 Kya (Harris and Hey, 1999; Yu *et al.*, 2002; Zietkiewicz *et al.*, 2003; Garrigan *et al.*, 2005a,b). For example, there are fixed nucleotide differences at *PDHA1* in African and non-African populations which suggest the onset of population subdivision began 179 ± 79.6 Kya, predating the earliest modern human fossils outside of Africa (Harris and Hey, 1999). A recent survey of genetic diversity at the *dys44* locus in 1343 X chromosomes from 33 populations of diverse geographic origin also revealed that at least one haplotype in the ancestral lineages at the root of the population tree for *dys44* originated outside of Africa, prior to population expansion out of Africa (Zietkiewicz *et al.*, 2003). Zietkiewicz *et al.* propose that this lineage could have left Africa before migration of modern humans out of Africa (~160 Kya). Yu *et al.* (2002) analysed an ~10 kb non-coding region of the X chromosome and identified a lineage specific to Eurasians that arose > 140 Kya. Recently, Garrigan *et al.* (2005b) reported an ancient RRM2P4 lineage that forms the basal branches of the RRM2P4 gene tree and is nearly exclusive to Asians. The TMRCA of this ancient lineage is nearly 2 my. Garrigan *et al.* (2005b) suggest that this ancient lineage is a remnant of introgressive hybridization between expanding anatomically modern humans from Africa and archaic populations in Eurasia.

It is not surprising that some ancient lineages occurring outside Africa predate the human population expansion out of Africa. It is well known that coalescence times of genes often precede population divergence times (Rosenberg and Nordborg, 2002; Knowles, 2004). Takahata *et al.* (2001) analysed DNA sequence data from ten X-chromosomal regions, five autosomal regions, and one Y-chromosomal region, in addition to mitochondrial DNA. They were able to distinguish ancestral haplotypes for ten sequences, and of these, nine occurred in Africa and only one occurred in Asia during the

Pleistocene. They demonstrate by computer simulation that it would be highly unlikely to observe restriction of nine of ten ancestral haplotypes in a single geographic region (Africa) under an MRO model with high levels of gene flow, unless the effective size of the African populations is ~10 times larger than outside of Africa, which is not supported by current data (Takahata *et al.*, 2001). Thus, this study demonstrates that the observation of a small number of lineages with ancient coalescent times outside of Africa is still consistent with a RAO model. However, large-scale sequence analyses in a broader region of the genome across a geographically diverse set of populations will be necessary to test the hypothesis of ancient genetic lineages originating outside of Africa, possibly due to admixture of AMHS with archaic humans in that region.

Autosomes

Until recently, little was known about the extent of variation in single-copy DNA sequences on the autosomes. A growing number of population studies of autosomal nucleotide variation, as well as haplotype analyses using SNPs, In/dels, Alu, and or STR polymorphisms, have been published in the past ten years (Table 12.2) (summarized in Tishkoff and Verrelli, 2003). These studies have included autosomal intergenic regions (Zhao *et al.*, 2000; Alonso and Armour, 2001; Yu *et al.*, 2001) and segments of autosomal genes (Table 12.2). Similar to mtDNA and the X and Y chromosomes, these studies generally suggest that modern humans originated in Africa, owing to the high nucleotide and haplotype diversity observed among African populations and the subset of that diversity observed among non-African populations. TMRCA estimates for these loci tend to be large, ranging from approximately 311 Kya (*APOE*) (Fullerton *et al.*, 2000) to 2.1 Mya (*CCR5*) (Bamshad *et al.*, 2002), as expected for autosomal loci with an effective size larger than the mtDNA, NRY, or X chromosome.

Many of these autosomal loci, as well as X-chromosome loci, were first identified as being important in medical conditions, such as hemoglobinopathies (*β-globin*) (Harding *et al.*, 1997), cardiovascular disease (*LPL*) (Clark *et al.*, 1998) and neurological diseases (*PDHA1*) (Harris and Hey, 1999). Other loci are important in immune response (e.g. G6PD; CCR5) (Tishkoff *et al.*, 2001; Sabeti *et al.*, 2002; Bamshad *et al.* 2002; Saunders *et al.*, 2002) or skin pigmentation (e.g. *MCR1*) (Harding *et al.*, 2000). Thus, these loci may be subject to natural selection that can potentially influence genetic signatures of migration, population bottlenecks, and population expansions (Kaessmann and Paabo, 2002; Tishkoff and Verrelli, 2003). For that reason, analysis of potentially thousands of unlinked loci in the autosomal genome will be important for reconstructing human demographic and evolutionary history.

Global patterns of linkage disequilibrium

Analysis of the non-random association of genetic markers along a chromosome (i.e. linkage disequilibrium (LD)) can be particularly informative for inferring human demographic history and the time of migration and population differentiation events. This is because levels and patterns of LD depend on a number of demographic factors including population size, population structure, founder effect, and admixture, as well as gene-specific factors such as selection and rates of mutation and recombination. For example, Tishkoff *et al.* (1996) studied the association of a highly variable STRP and an insertion/deletion polymorphism located 10 kb away. Haplotypes composed of both slow-evolving markers (SNPs, polymorphic *Alu* elements, in/dels) and fast-evolving (STRP or minisatellite) markers are particularly informative because of the ability to define haplogroups based on the bi-allelic markers and to estimate divergence dates based on the amount of STRP or minisatellite diversity and LD associated with a defined SNP, *Alu*, or in/del haplogroup (Tishkoff *et al.*, 1996, 1998a, 1998b, 2000; Scozzari *et al.*, 1999; Mountain *et al.*, 2002; Jobling and Tyler-Smith, 2003; Hey *et al.*, 2004). Tishkoff *et al.* (1996) found that within Africa, the deletion allele was associated with a broad range of STRP alleles. By contrast, in populations from outside of Africa, the deletion allele was strongly associated with just a single STRP allele (that containing six repeats). Based on an estimated STRP mutation rate and an estimate of the rate of decay of association between the STRP and the in/del polymorphism, the upper bound for emergence of the deletion chromosome out of Africa (which presumably corresponded with migration out of Africa) was ~150,000 ya (Tishkoff *et al.*, 1996; Tishkoff *et al.*, 1998b). Thus, this study supports an RAO model of human origins.

A number of subsequent studies using large number of SNPs have supported these initial findings, demonstrating lower levels of LD in Africa compared to other regions of the world (Frisse *et al.*, 2001; Reich *et al.*, 2001; Gabriel *et al.*, 2002). The distinct pattern of LD in African and non-African populations reflects their distinct demographic histories (Figure 12.3; Tishkoff and Williams, 2002; Tishkoff and Verrelli, 2003). Ancestral African populations should have lower LD because there has been more time for recombination and mutations to cause LD to decay and because they have maintained a larger effective population size compared to non-Africans (Tishkoff *et al.*, 1996; Tishkoff *et al.*, 1998a; Tishkoff *et al.*, 2000; Wall, 2001; Tishkoff and Williams, 2002). A larger, more subdivided population structure in Africa could result in divergent patterns of association amongst ethnically distinct African populations. The divergent pattern of LD in non-African populations relative to African populations is likely the result of a founding event during expansion of modern humans out of Africa within the past 100,000 years (Tishkoff and Williams, 2002). LD that has been established after a founding event is likely to be maintained in rapidly expanding non-African populations.

Estimates of effective population size

Estimates of within population variation, which can be estimated from allele frequency or nucleotide diversity data, reflect the history of the size of a population. The effective population size Ne can be estimated based on observed levels of autosomal genetic diversity as $Ne = \theta/4\mu$, where μ is the mutation rate and θ is the nucleotide diversity population parameter. For the X chromosome, $Ne = \theta/3\mu$ and for the haploid mtDNA and Y chromosome, $Ne = \theta/2\mu$. Estimates of Ne are most strongly influenced by population sizes when they are at their smallest and it may take many generations to recover from a bottleneck event. Thus, estimates of Ne in modern populations reflect the size of the human population prior to population expansion. Studies of nuclear sequence diversity in humans consistently estimate an effective population size of ~10,000 and studies of mtDNA diversity estimate a haploid effective population size of ~5,000 (Table 12.2). By contrast, an estimate of Ne in chimpanzees based on autosomal nucleotide diversity is ~35,000, indicating that humans have gone through a bottleneck at some time since their divergence from chimpanzee ~5–6 Mya (Kaessmann et al., 1999b; Won and Hey, 2005). Estimates of Ne in humans are usually based on pooled data from African and non-African populations, which are biased towards larger samples from non-Africans (Tishkoff and Verrelli, 2003). This pooling could result in a biased estimate of Ne, particularly if non-African populations went through a strong bottleneck event during migration out of Africa and lost much of their genetic diversity. Studies that have analysed data from African populations separately from non-African populations indicate a larger effective population size in Africans (Ne as high as 20 K; Harris and Hey, 1999; Jaruzelska et al., 1999; Sherry et al., 1997; Stoneking et al., 1997; Calafell et al., 1998; Harpending et al., 1998; Relethford and Jorde, 1999; Hammer et al., 1997, 1998; Pritchard et al., 1999; Scozzari et al., 1999). The larger effective size of Africans relative to non-Africans likely reflects a bottleneck at the time of migration of humans out of Africa, resulting in a loss of genetic diversity in non-African populations.

It has been argued that an effective population size of 10–20 K, which would be equivalent to a census size of ~ 20–60 K, would not be large enough to sustain the high levels of gene flow required under a multiregional origin model (Harpending and Rogers, 2000). However, it should be noted that if the ancestral human population was highly subdivided, the census size could have been larger, although still not large enough to sustain high levels of gene flow across major geographic regions (Harris and Hey, 1999; Relethford and Jorde, 1999).

Although there is still uncertainty about the long-term effective population size of humans, it is clear that after an initial speciation event from *Homo erectus* to modern *H. sapiens*, humans spread across a broad geographic region and rapidly increased in population size within the past 50–10 K years. An even more rapid population expansion likely occurred after the development and spread of agriculture within the past 10 K years (Cavalli-Sforza *et al.*, 1994). Slatkin and Hudson used coalescent theory to demonstrate that gene trees inferred for populations that have undergone an expansion have a more starlike pattern than do the gene trees of populations of constant size (Slatkin and Hudson, 1991). The starlike pattern is reflected in a unimodal distribution of sequence differences between pairs of individuals (Rogers and Harpending, 1992). In a rapidly expanding population, the pairwise sequence difference, or mismatch distribution, is usually smooth and wave shaped, whereas it has a 'ragged' shape in a small population of constant size (Rogers and Harpending, 1992; Rogers and Jorde, 1995; Harpending and Rogers, 2000). Based on the shape and position of the wave it is possible to make inferences about the timing and magnitude of past population expansion events (Rogers and Jorde, 1995; Harpending and Rogers, 2000). However, this type of distribution can also be caused by an episode of positive selection and it can often be difficult to differentiate selective from demographic factors (Slatkin and Hudson, 1991). Analyses of mtDNA mismatch distributions in globally diverse populations indicate approximate dates for population expansion in Asia of 23 Kya, in Europe of 52 Kya, and in Africa of 99 Kya (Harpending *et al.*, 1993; Sherry *et al.*, 1994; Rogers, 1995; Jorde *et al.*, 1998; Harpending and Rogers, 2000; Jorde *et al.*, 2001). These results are supported by studies of the mismatch distribution of highly variable microsatellites, indicating African populations expanded in size earlier than non-African populations (Shriver *et al.*, 1997; Di Rienzo *et al.*, 1998; Kimmel *et al.*, 1998; Reich and Goldstein, 1998; Relethford and Jorde, 1999; Harpending and Rogers, 2000). Y-chromosome variation also shows an excess of rare frequency haplotypes, as is expected under a model of rapid population growth (Pritchard *et al.*, 1999; Thomson *et al.*, 2000; Underhill *et al.*, 2000; Underhill *et al.*, 2001) and studies of Y-chromosome microsatellites indicate a signature of population expansion between 16 and 126 Kya across global populations (Pritchard *et al.*, 1999).

The patterns of sequence variation of autosomal and X-chromosome genes do not give such a clear-cut indication of rapid population growth. Approximately 30% of the currently available nuclear DNA data indicate an excess of intermediate frequency haplotypes, which is not consistent with a model of rapid population growth (Hey, 1997; Harpending and Rogers, 2000; Przeworski *et al.*, 2000; Wall and Przeworski, 2000; Fay and Wu, 2001; Tishkoff and Verrelli 2003) (Table 12.2). When African samples were analysed separately from non-African samples, more loci showed evidence for population growth (Wall and Przeworski, 2000). Simulations of the expected distribution of variation under models of a population bottleneck or population subdivision, followed by rapid population expansion, show that these models could explain the pattern for some, but not all loci (Wall and Przeworski, 2000). These studies suggest that the pattern of variation observed in many nuclear genes could be the result of historic balancing selection acting at some of these loci. Many of the genes that do not show evidence of population growth have likely been historic targets of selection, such as β-globin (HBB) (Harding et al., 1997), Duffy (FY) (Hamblin and Di Rienzo, 2000, 2002), and pyruvate dehydrogenase E1 alpha subunit (PDAH1) (Harris and Hey, 1999). By contrast, most of the nucleotide studies of pseudo-genes, introns, or non-coding regions show evidence for population expansion (Table 12.2). Wakeley et al. studied several SNP datasets and found that evidence for population expansion could be detected only when one accounts for population substructure, as well as the ascertainment bias resulting from identifying SNPs in Europeans (Wakeley et al., 2001).

However, several other analyses of nuclear genetic diversity have supported the hypothesis of population growth following a bottle-neck event. A study of 500,000 nuclear SNPs obtained from the public SNP database (from which ethnicity is obscured but biased towards European samples), found evidence for a population collapse approximately 40 Kya (corresponding to the initial appearance of anatomically modern humans in Europe) followed by a signature of modest recovery of population size (Marth et al., 2003). Additionally, a study of 3,899 SNPs from 313 genes observed an excess of rare-frequency alleles at 281 of the genes (Stephens et al., 2001), as expected under a model of historic population expansion. Estimates of historic population expansion based on two nuclear diversity studies range from 120 to 190 Kya for Xq13.3 (Wooding and Rogers, 2000; Kaessmann et al., 2001) to 100−140 Kya for region 16p13.3 in Eurasian populations (Wooding and Rogers, 2000). A much more extensive analysis of nuclear sequence variability from multiple loci sampled from ethnically diverse populations is necessary to resolve these conflicting patterns of human demographic history.

Migration patterns out of Africa

The migration of modern humans out of Africa was the most significant event influencing patterns of genetic variation in modern humans. The primary split between African and non-African popula-tions occurred sometime between 44 to 200 Kya (Goldstein et al., 1995; Tishkoff et al., 1996; Stoneking et al., 1997; Tishkoff et al., 1998a; Harris and Hey, 1999; Ingman et al., 2000; Underhill et al., 2000). Multiple migrations out of Africa via Ethiopia and Arabia towards South Asia (~100−60 Kya) or via North Africa and the Midde East towards Euarasia (~ 70−40 Kya) have been proposed (Stringer and

Andrews, 1988; Bar-Yosef, 1994; Lahr and Foley, 1994; Lahr and Foley, 1998). However, a key question is whether there were migrations from one or more African populations. An analysis of nuclear autosomal haplotype variability at three loci, cluster differentiation 4 (CD4), myotonic dystrophy (DM), and tissue plasminogen activator (PLAT), in the same set of 33–42 globally diverse populations, found that all non-African populations share a similar pattern of haplotype variability and have a subset of the haplotypic variability that is present in Ethiopian and Somalian populations, which is in itself a subset of the variability that is present in other sub-Saharan African populations (Figure 12.3) (Tishkoff et al., 1996; Tishkoff et al., 1998a; Tishkoff et al., 2000). Distinct patterns of haplotype variability exist across sub-Saharan African populations. We would, therefore, expect to see divergent patterns of haplotype variability in non-African populations, if migration had occurred from multiple source populations out of Africa (Tishkoff et al., 1996; Tishkoff et al., 1998a; Tishkoff et al., 2000). If there were multiple migrations out of Africa, they would have had to originate from the same source population in Africa, or from different genetic lineages that subsequently went extinct in modern populations. However, if the latter case were true, we would not expect to observe a similar pattern of haplotype loss due to genetic drift across all non-African populations.

Analyses of mtDNA (Watson et al., 1996; Watson et al., 1997; Quintana-Murci et al., 1999; Chen et al., 2000; Ingman et al., 2000; Kivisild et al., 2004) and Y-chromosome diversity (Underhill et al., 2000; Hammer et al., 2001; Underhill et al., 2001; Semino et al., 2002) support a single East African source of migration out of Africa. However, the possibility still remains that there could have been an earlier migration event from Africa, across southeastern Asia, and into Australo-Melanesia (Melton et al., 1998; Kivisild et al., 2003; Thangaraj et al., 2003; Cruciani et al., 2004; Luis et al., 2004). If such an early migration event did occur, it is not clear whether it originated from a population that was genetically differentiated from the population(s) giving rise to subsequent migrations across Eurasia. Both source populations may have originated in northeast Africa from a single common ancestral population. Furthermore, if this earlier migration event did occur, it is likely that the gene pool of modern populations in Australo-Melanesia, which overall are most genetically similar to other non-African populations (Cavalli-Sforza et al., 1994; Ingman and Gyllensten, 2003), reflect admixture between early and later migrants into the region (Tishkoff and Verrelli, 2003).

Migration patterns within Africa

Human population history within Africa has been particularly rich and complex, as evidenced by the tremendous cultural, linguistic and genetic diversity, with more than 2,000 distinct ethnic groups and languages across continental Africa (http://www.ethnologue.org/). As the ancestral homeland of modern humans, understanding patterns of genetic diversity amongst the multitude of ethnically diverse African populations will shed light on many questions concerning human evolutionary history and the genetic basis of phenotypic variation. Despite the important contributions studies of African populations can make, these populations have been under-represented in human genetic studies compared to non-African populations (Tishkoff and Williams, 2002). The majority of studies that have included African populations have focused on analysis of mtDNA variation. These studies have revealed at least four major mitochondrial lineages, L0, L1, L2, and L3, which have an estimated coalescence date of 126–165 Kya, which sets an upper bound for the time of common ancestry of African populations (Chen *et al.*, 1995, 2000; Watson *et al.*, 1996; Watson *et al.*, 1997; Salas *et al.*, 2002; Kivisild *et al.*, 2004; Mishmar *et al.*, 2004; Salas *et al.*, 2004). The L0 lineage is most ancient and is common in South African !Kung and is found less frequently in East Africa (Salas *et al.*, 2002). L1 lineages are also common among groups in southeastern and eastern Africa, as well as among Biaka Pygmies from Central African Republic (Chen *et al.*, 2000; Salas *et al.*, 2002). The L2 and L3 lineages diverged from the L1 lineage approximately 60–103 Kya (Chen *et al.*, 1995, 2000; Watson *et al.*, 1997). The L2 lineage is common in Mbuti Pygmies from the Democratic Republic of the Congo and in West African Bantu-speaking populations, whereas the L3 lineage is dispersed widely throughout East Africa and is rare in West and South Africa (Chen *et al.*, 1995, 2000; Watson *et al.*, 1997). These lineages may reflect historic migration events in Africa within the past 30–77 K years (Watson *et al.*, 1997). Phylogenetic analysis indicates that the L3 haplogroup is the likely precursor of modern European and Asian mtDNA haplotypes (Chen *et al.*, 2000) and a subset of this lineage (M) which is prevalent in Ethiopians, Sudanese, and Asians may have expanded out of Africa approximately 60–50 Kya, founding modern Eurasian populations (Watson *et al.*, 1997; Quintana-Murci *et al.*, 1999; Kivisild *et al.*, 2004). The latter observation adds strength to the proposal that the dispersal of modern humans out of Africa may have occurred via Ethiopia (Tishkoff *et al.*, 1996; Tishkoff *et al.*, 1998a; Tishkoff *et al.*, 2000; Kivisild *et al.*, 2004). The M lineages have been found at low frequency in West African populations that trace their ancestry to the Sudan, indicating historic long-range migration events across Africa (Rosa *et al.*, 2004).

Several other long-range migration events have shaped the genetic landscape in Africa. Analyses of mtDNA data and the Y chromosome supports studies of classical polymorphisms as well as archeological data indicating that Khoisan-speaking populations (those whose language contains clicks, which includes the !Kung San) may have originated in Eastern Africa and migrated into southern Africa > 20–10 Kya (Cavalli-Sforza *et al.*, 1994; Cavalli-Sforza, 1997; Scozzari *et al.*, 1999). Analyses of Y-chromosome haplotype variation have identified that the most ancestral Y-chromosome haplotype is present at moderate to high frequency in East African Sudanese and Ethiopians, as well as in southern African !Kung San and is absent in all other African populations (Passarino *et al.*, 1998; Scozzari *et al.*, 1999; Underhill *et al.*, 2000; Hammer *et al.*, 2001; Underhill *et al.*, 2001; Luis *et al.*, 2004). Analysis of the mtDNA D loop and complete mtDNA genome sequences in click speakers of Tanzania (i.e. the Hadza and the Sandawe) and the Khoisan speakers of southern Africa (i.e. !Kung San) indicate that the Sandawe (and groups with which they may have recently admixed) harbour the L0d mtDNA lineage that is nearly exclusively found among southern African Khoisan speakers (Salas *et al.*, 2002; Gonder *et al.*, in review; Tishoff *et al.*, in prep). The L0d lineages found in Tanzania have a more recent TMRCA estimate (30.6 $\pm$ 17.8 Kya) than the L0d lineages of the sourthern African Khosian speakers (90.4 $\pm$ 18.9 Kya) (Gonder *et al.*, in review). These data indicate that there is a unique, but ancient, genetic connection between the click-speaking Sandawe of Tanzania and the southern African Khoisan speakers (Tishkoff *et al.*, in prep; Gonder *et al.*, in review). Neither Hadza nor the Sandawe share a close relationship with other hunter-gatherer groups, such as the Bantu-speaking central African pygmies.

Archeological, linguistic, and genetic data indicate several other important long-range migration events in Africa within the past several thousand years that have shaped the pattern of genetic variation in modern African populations. The most significant of these long-range migration events is the expansion of Bantu-speaking agricultural populations from a postulated homeland in Cameroon into the Great Lakes region of eastern Africa and then into southern Africa within the past 3 K years (Soodyall *et al.*, 1996). Signatures of this migration event have been observed for mtDNA data (Soodyall *et al.*, 1996) as well as Y-chromosome data (Poloni *et al.*, 1997; Passarino *et al.*, 1998; Hammer *et al.*, 2001). There have also been recent migration events of non-Africans, particularly those of Semitic origin, along the eastern coast of Africa who have admixed with indigenous African populations. The Lemba population of Zimbabwe speak a Bantu language and claim Jewish ancestry (Thomas *et al.*, 2000). Two analyses of Y-chromosome variation have identified both Bantu and Semitic (which could include Arabic and/or Jewish) contributions to the Lemba Y-chromosome gene pool (Spurdle and Jenkins, 1996; Thomas *et al.*, 2000). High resolution microsatellite haplotype analysis identified a unique haplotype found at high frequency in one of the Lemba clans that is only found in the Cohen clan of Jewish priests, adding strength to the claim by the Lemba, based on oral history, that they are of Jewish descent. By contrast, studies of the Beta-Israel, an Ethiopian Jewish population, have not shown clear indication of Jewish ancestry (Hammer *et al.*, 2000; Ritte *et al.*, 1993). There have also been a number of recent studies of mtDNA and Y-chromosome variation in African and Middle Eastern populations, indicating high levels of gene flow in both directions (particularly between northeast

Ethiopia and Yemen) during the late Pleistocene and Holocene (Richards *et al.*, 2003; Cruciani *et al.*, 2004; Kivisild *et al.*, 2004; Luis *et al.*, 2004; Semino *et al.*, 2004). As we begin to include more African populations in human genetics studies, we will have a better understanding of their rich and complex population histories, the origin of major language families, and the origin of the human species. These studies will also shed light on the complex patterns of haplotype variation and linkage disequilibrium within and between African populations, which could have important implications for the design of more effective biomedical studies in people of recent African descent.

Conclusions

Genetic analyses of autosomal, X-chromosome, Y-chromosome, and mtDNA generally support an RAO model of modern humans. Africans have the highest levels of genetic diversity, both within and between populations, and non-Africans have a subset of the African diversity. The migration event most likely originated from East Africa between 100 and 50 Kya and was accompanied by a moderate to severe bottleneck event. The pattern of genetic differentiation in global human populations reflects isolation by distance and population differentiation due to genetic drift. However, there have been a number of short-range and long-range migration events both within and out of Africa that have shaped patterns of genetic variation in modern populations. Although most data support an RAO model, some ancient lineages have been found in Asia, raising the possibility of low levels of admixture between modern and archaic humans outside of Africa.

As novel high-throughput sequencing and SNP genotyping technology is developed, we will be able to move away from studies of human genetic variation that have focused on one or a few loci, often in differing sets of populations. We will soon be able to accumulate comparative data from thousands of loci across thousands of ethnically diverse individuals. These comparative data will facilitate accurate testing of hypotheses about human evolutionary history and reconstruction of human demographic history, including population expansions and bottlenecks, population sub-structure, and timing and patterns of migration events. The chimpanzee genome has been sequenced and sequencing of several other primate genomes has begun (Ruvolo, 2004; Consortium, T.C.S.A., 2005; Dennis, 2005). Comparative and functional genomic analyses of humans and their closest ancestors can help in understanding differences in genome architecture between humans and other primates and with the identification of regions that play an important role in gene expression and gene regulation. Lastly, an important area to further explore is the genetic basis of adaptation.

This will require a better understanding of genotype/phenotype relationships as well as development of better tests of selection so that we can determine the roles that selection and demographic history have played in shaping the human and primate genomes. From this, we will have a better understanding of the evolutionary processes that have led towards the emergence of anatomically and behaviourally modern humans.

Acknowledgements

We thank Floyd Reed for assistance with construction of Figure 12.2, the NJ tree of genetic distances from autosomal microsatellite data. Funded by NSF grant BCS-9905396, and Burroughs Wellcome Fund and David and Lucile Packard Career Awards to SAT.

References

Alonso, S. and Armour, J.A. (2001). A highly variable segment of human subterminal 16p reveals a history of population growth for modern humans outstide Africa. *Proc Natl Acad Sci USA*, **98**, 864–9.

Ambrose, S.H. (1998). Late Pleistocene human population bottlenecks, volcanic winter, and differentiation of modern humans. *J Hum Evol*, **34**, 623–51.

Bamshad, M.J., Mummidi, S., Gonzalez, E., Ahuja, S.S., Dunn, D.M., Watkins, W.S., Wooding, S., Stone, A.C., Jorde, L.B., Weiss, R.B. and Ahuja, S.K. (2002). A strong signature of balancing selection in the 5′ cis-regulatory region of CCR5. *Proc Natl Acad Sci USA*, **99**, 10539–44.

Bamshad, M.J., Wooding, S., Watkins, W.S., Ostler, C.T., Batzer, M.A. and Jorde, L.B. (2003). Human population genetic structure and inference of group membership. *Am J Hum Genet*, **72**, 578–89.

Bar-Yosef, O. (1994). The contribution of southwest Asia to the study of modern human origins. In *Origins of Anotomically Modern Humans,* eds. Nitaecki, M.H. and Nitecki, D.V. New York: Plenum Press, pp. 22–36.

Barbujani, G., Magagni, A., Minch, E. and Cavalli-Sforza, L.L. (1997). An apportionment of human DNA diversity. *Proc Natl Acad Sci USA*, **94**, 4516–19.

Batzer, M.A. and Deininger, P.L. (2002). Alu repeats and human genomic diversity. *Nat Rev Genet*, **3**, 370–9.

Bensasson, D., Zhang, D., Hartl, D.L. and Hewitt, G.M. (2001). Mitochondrial pseudogenes: evolution's misplaced witnesses. *Trends Ecol Evol*, **16**, 314–21.

Boissinot, S., Entezam, A., Young, L., Munson, P.J. and Furano, A.V. (2004). The insertional history of an active family of L1 retrotransposons in humans. *Genome Res*, **14**, 1221–31.

Bowcock, A. and Cavalli-Sforza, L. (1991). The study of variation in the human genome. *Genomics*, **11**, 491–8.

Calafell, F., Shuster, A., Speed, W.C., Kidd, J.R. and Kidd, K.K. (1998). Short tandem repeat polymorphism evolution in humans. *Eur J Hum Genet*, **6**, 38–49.

Cann, R.L., Stoneking, M. and Wilson, A.C. (1987). Mitochondrial DNA and human evolution. *Nature*, **325**, 31–6.

Caramelli, D., Lalueza-Fox, C., Vernesi, C., Lari, M., Casoli, A., Mallegni, F., Chiarelli, B., Dupanloup, I., Bertranpetit, J., Barbujani, G. and Bertorelle, G. (2003). Evidence for a genetic discontinuity between Neandertals and 24,000-year-old anatomically modern Europeans. *Proc Natl Acad Sci USA*, **100**, 6593–7.

Cavalli-Sforza, L. L. (1997). Genes, peoples, and languages. *Proc Natl Acad Sci USA*, **94**, 7719–24.

Cavalli-Sforza, L. L., Piazza, A. and Menozzi, P. (1994). *History and Geography of Human Genes*. Princeton: Princeton Univeristy Press.

Chakravarti, A. (1999). Population genetics—making sense out of sequence. *Nat Genet*, **21**, 56–60.

Charlesworth, B. (1978). Model for evolution of Y chromosomes and dosage compensation. *Proc Natl Acad Sci USA*, **75**, 5618–22.

Chen, F. C. and Li, W. H. (2001). Genomic divergences between humans and other hominoids and the effective population size of the common ancestor of humans and chimpanzees. *Am J Hum Genet*, **68**, 444–56.

Chen, F. C., Vallender, E. J., Wang, H., Tzeng, C. S. and Li, W. H. (2001). Genomic divergence between human and chimpanzee estimated from large-scale alignments of genomic sequences. *J Hered*, **92**, 481–9.

Chen, Y. S., Olckers, A., Schurr, T. G., Kogelnik, A. M., Huoponen, K. and Wallace, D. C. (2000). mtDNA variation in the South African Kung and Khwe-and their genetic relationships to other African populations. *Am J Hum Genet*, **66**, 1362–83.

Chen, Y. S., Torroni, A., Excoffier, L., Santachiara-Benerecetti, A. S. and Wallace, D. C. (1995). Analysis of mtDNA variation in African populations reveals the most ancient of all human continent-specific haplogroups. *Am J Hum Genet*, **57**, 133–49.

Clark, A. G., Weiss, K. M., Nickerson, D. A., Taylor, S. L., Buchanan, A., Stengard, J., Salomaa, V., Vartiainen, E., Perola, M., Boerwinkle, E. and Sing, C. F. (1998). Haplotype structure and population genetic inferences from nucleotide-sequence variation in human lipoprotein lipase. *Am J Hum Genet*, **63**, 595–612.

Clark, J. D., Beyene, Y., Wolde-Gabriel, G., Hart, W. K., Renne, P. R., Gilbert, H., Defleur, A., Suwa, G., Katoh, S., Ludwig, K. R., Boisserie, J. R., Asfaw, B. and White, T. D. (2003). Stratigraphic, chronological and behavioural contexts of Pleistocene Homo sapiens from Middle Awash, Ethiopia. *Nature*, **423**, 747–52.

Collins, F. S., Green, E. D., Guttmacher, A. E. and Guyer, M. S. (2003). A vision for the future of genomics research. *Nature*, **422**, 835–47.

Consortium, I. H. G. S. (2004). Finishing the euchromatic sequence of the human genome. *Nature*, **431**, 931–45.

Consortium, T. C. S. A. (2005). Initial sequence of the chimpanzee genome and comparison with the human genome. *Nature*, **437**, 69–87.

Cruciani, F., La Fratta, R., Santolamazza, P., Sellitto, D., Pascone, R., Moral, P., Watson, E., Guida, V., Colomb, E. B., Zaharova, B., Lavinha, J., Vona, G., Aman, R., Cali, F., Akar, N., Richards, M., Torroni, A., Novelletto, A. and Scozzari, R. (2004). Phylogeographic analysis of haplogroup E3b (E-M215) y chromosomes reveals multiple migratory events within and out of Africa. *Am J Hum Genet*, **74**, 1014–22.

Cruciani, F., Santolamazza, P., Shen, P., Macaulay, V., Moral, P., Olckers, A., Modiano, D., Holmes, S., Destro-Bisol, G., Coia, V., Wallace, D. C., Oefner, P. J., Torroni, A., Cavalli-Sforza, L. L., Scozzari, R. and Underhill, P. A. (2002).

A back migration from Asia to sub-Saharan Africa is supported by high-resolution analysis of human Y-chromosome haplotypes. *Am J Hum Genet*, **70**, 1197−214.

de Knijff, P. (2000). Messages through bottlenecks: on the combined use of slow and fast evolving polymorphic markers on the human Y chromosome. *Am J Hum Genet*, **67**, 1055−61.

Dennis, C. (2005). Chimp genome: branching out. *Nature*, **437**, 17−19.

Di Rienzo, A. and Wilson, A.C. (1991). Branching pattern in the evolutionary tree for human mitochondrial DNA. *Proc Natl Acad Sci USA*, **88**, 1597−601.

Di Rienzo, A., Donnelly, P., Toomajian, C., Sisk, B., Hill, A., Petzl-Erler, M.L., Haines, G.K. and Barch, D.H. (1998). Heterogeneity of microsatellite mutations within and between loci, and implications for human demographic histories. *Genetics*, **148**, 1269−84.

Excoffier, L. (1990). Evolution of human mitochondrial DNA: evidence for departure from a pure neutral model of populations at equilibrium. *J Mol Evol*, **30**, 125−39.

Excoffier, L. and Yang, Z. (1999). Substitution rate variation among sites in mitochondrial hypervariable region I of humans and chimpanzees. *Mol Biol Evol*, **16**, 1357−68.

Fay, J.C. and Wu, C.I. (2001). The neutral theory in the genomic era. *Current Opinions in Genetics and Development*, **11**, 642−46.

Fischer, A., Wiebe, V., Paabo, S. and Przeworski, M. (2004). Evidence for a complex demographic history of chimpanzees. *Mol Biol Evol*, **21**, 799−808.

Forster, P., Rohl, A., Lunnemann, P., Brinkmann, C., Zerjal, T., Tyler-Smith, C. and Brinkmann, B. (2000). A short tandem repeat-based phylogeny for the human Y chromosome. *Am J Hum Genet*, **67**, 182−96.

Frisse, L., Hudson, R.R., Bartoszewicz, A., Wall, J.D., Donfack, J. and Di Rienzo, A. (2001). Gene conversion and different population histories may explain the contrast between polymorphism and linkage disequilibrium levels. *Am J Hum Genet*, **69**, 831−43.

Fullerton, S.M., Clark, A.G., Weiss, K.M., Nickerson, D.A., Taylor, S.L., Stengard, J.H., Salomaa, V., Vartiainen, E., Perola, M., Boerwinkle, E. and Sing, C.F. (2000). Apolipoprotein E variation at the sequence haplotype level: implications for the origin and maintenance of a major human polymorphism. *Am J Hum Genet*, **67**, 881−900.

Gabriel, S.B., Schaffner, S.F., Nguyen, H., Moore, J.M. and Roy, J. *et al.* (2002). The structure of haplotype blocks in the human genome. *Science*, **296**, 2225−29.

Garrigan, D., Mobasher, Z., Kingan, S.B, Wilder, J.A. and Hammer, M.F. (2005a). Deep haplotype divergence and long-range linkage disequilibrium at xp21.1 provide evidence that humans descend from a structured ancestral population. *Genetics*, **170**, 1849−56.

Garrigan, D., Mobasher, Z., Severson, T., Wilder, J.A. and Hammer, M.F. (2005b). Evidence for Archaic Asian Ancestry on the Human X Chromosome. *Mol Biol Evol*, **22**, 189−92.

Gilad, Y., Rosenberg, S., Przeworski, M., Lancet, D. and Skorecki, K. (2002). Evidence for positive selection and population structure at the human MAO-A gene. *Proc Natl Acad Sci USA*, **99**, 862−7.

Goldstein, D.B., Ruiz Linares, A., Cavalli-Sforza, L.L. and Feldman, M.W. (1995). Genetic absolute dating based on microsatellites and the origin of modern humans. *Proc Natl Acad Sci USA*, **92**, 6723−7.

Haile-Selassie, Y., Asfaw, B. and White, T.D. (2004). Hominid cranial remains from upper Pleistocene deposits at Aduma, Middle Awash, Ethiopia. *Am J Phys Anthropol*, **123**, 1–10.

Hamblin, M.T. and Di Rienzo, A. (2000). Detection of the signature of natural selection in humans: evidence from the Duffy blood group locus. *Am J Hum Genet*, **66**, 1669–79.

Hamblin, M.T., Thompson, E.E. and Di Rienzo, A. (2002). Complex signatures of natural selection at the Duffy blood group locus. *Am J Hum Genet*, **70**, 369–83.

Hammer, M.F. (1994). A recent insertion of an alu element on the Y chromosome is a useful marker for human population studies. *Mol Biol Evol*, **11**, 749–61.

Hammer, M.F. (1995). A recent common ancestry for human Y chromosomes. *Nature*, **378**, 376–8.

Hammer, M.F., Garrigan, D., Wood, E., Wilder, J.A., Mobasher, Z., Bigham, A., Krenz, J.G. and Nachman, M.W. (2004). Heterogeneous patterns of variation among multiple human x-linked loci: the possible role of diversity-reducing selection in non-Africans. *Genetics*, **167**, 1841–53.

Hammer, M.F., Karafet, T., Rasanayagam, A., Wood, E.T., Altheide, T.K., Jenkins, T., Griffiths, R.C., Templeton, A.R. and Zegura, S.L. (1998). Out of Africa and back again: nested cladistic analysis of human Y chromosome variation. *Mol Biol Evol*, **15**, 427–41.

Hammer, M.F., Karafet, T.M., Redd, A.J., Jarjanazi, H., Santachiara-Benerecetti, S., Soodyall, H. and Zegura, S.L. (2001). Hierarchical patterns of global human Y-chromosome diversity. *Mol Biol Evol*, **18**, 1189–203.

Hammer, M.F., Spurdle, A.B., Karafet, T., Bonner, M.R., Wood, E.T., Novelletto, A., Malaspina, P., Mitchell, R.J., Horai, S., Jenkins, T. and Zegura, S.L. (1997). The geographic distribution of human Y chromosome variation. *Genetics*, **145**, 787–805.

Hammer, M.F., Redd, A.J., Wood, E.T., Bonner, M.R., Jarjanazi, H., Karafet, T., Santachiara-Benerecetti, S., Oppenheim, A., Jobling, M.A., Jenkins, T., Ostrer, H. and Bonne-Tamir, B. (2000). Jewish and Middle Eastern non-Jewish populations share a common pool of Y-chromosome biallelic haplotypes. *Proc Natl Acad Sci USA*, **97**, 6769–74.

Hammer, M.F. and Zegura, S.L. (2002). The human Y chromosome haplogroup tree: Nomenclature and phylogeography. *Annual Review of Anthropology*, **31**, 303–21.

Harding, R.M., Fullerton, S.M., Griffiths, R.C., Bond, J., Cox, M.J., Schneider, J.A., Moulin, D.S. and Clegg, J.B. (1997). Archaic African and Asian lineages in the genetic ancestry of modern humans. *Am J Hum Genet*, **60**, 772–89.

Harding, R.M., Healy, E., Ray, A.J., Ellis, N.S., Flanagan, N., Todd, C., Dixon, C., Sajantila, A., Jackson, I.J., Birch-Machin, M.A. and Rees, J.L. (2000). Evidence for variable selective pressures at MC1R. *Am J Hum Genet*, **66**, 1351–61.

Harpending, H. and Rogers, A. (2000). Genetic perspectives on human origins and differentiation. *Annu Rev Genomics Hum Genet*, **1**, 361–85.

Harpending, H.C., Sherry, S.T., Rogers, A.R. and Stoneking, M. (1993). The genetic structure of ancient human populations. *Current Anthropology*, **34**, 483–96.

Harpending, H.C., Batzer, M.A., Gurven, M., Jorde, L.B., Rogers, A.R. and Sherry, S.T. (1998). Genetic traces of ancient demography. *Proc Natl Acad Sci USA*, **95**, 1961–7.

Harris, E. E. and Hey, J. (1999). X chromosome evidence for ancient human histories. *Proc Natl Acad Sci USA*, **96**, 3320–4.

Harris, E. E. and Hey, J. (2001). Human populations show reduced DNA sequence variation at the factor IX locus. *Curr Biol*, **11**, 774–8.

Harvati, K. (2003). Quantitative analysis of Neanderthal temporal bone morphology using three-dimensional geometric morphometrics. *Am J Phys Anthropol*, **120**, 323–38.

Harvati, K., Frost, S. R. and McNulty, K. P. (2004). Neanderthal taxonomy reconsidered: implications of 3D primate models of intra- and interspecific differences. *Proc Natl Acad Sci USA*, **101**, 1147–52.

Henshilwood, C. S., d'Errico, F., Yates, R., Jacobs, Z., Tribolo, C., Duller, G. A., Mercier, N., Sealy, J. C., Valladas, H., Watts, I. and Wintle, A. G. (2002). Emergence of modern human behavior: Middle Stone Age engravings from South Africa. *Science*, **295**, 1278–80.

Henshilwood, C. S. and Marean, C. W. (2003). The origin of modern human behavior. *Curr Anthropol*, **44**, 627–51.

Hey, J. (1997). Mitochondrial and nuclear genes present conflicting portraits of human origins. *Molecular Biology and Evolution*, **14**, 166–72.

Hey, J., Won, Y. J., Sivasundar, A., Nielsen, R. and Markert, J. A. (2004). Using nuclear haplotypes with microsatellites to study gene flow between recently separated Cichlid species. *Mol Ecol*, **13**, 909–19.

Hirszfeld, L. and Hirszfeld, H. (1919). Serological differences between the blood of different races. *Lancet*, **2**, 675–9.

Ingman, M. and Gyllensten, U. (2003). Mitochondrial genome variation and evolutionary history of Australian and New Guinean aborigines. *Genome Res*, **13**, 1600–6.

Ingman, M., Kaessmann, H., Paabo, S. and Gyllensten, U. (2000). Mitochondrial genome variation and the origin of modern humans. *Nature*, **408**, 708–13.

Jaruzelska, J., Zietkiewicz, E., Batzer, M., Cole, D. E., Moisan, J. P., Scozzari, R., Tavare, S. and Labuda, D. (1999). Spatial and temporal distribution of the neutral polymorphisms in the last ZFX intron: analysis of the haplotype structure and genealogy. *Genetics*, **152**, 1091–101.

Jobling, M. A. and Tyler-Smith, C. (2003). The human Y chromosome: an evolutionary marker comes of age. *Nat Rev Genet*, **4**, 598–612.

Jobling, M. A., Hurles, M. E. and Tyler-Smith, C. (2004). *Human Evolutionary Genetics: Origins, Peoples and Disease*. New York: Gartlan.

Jorde, L. B., Bamshad, M. and Rogers, A. R. (1998). Using mitochondrial and nuclear DNA markers to reconstruct human evolution. *Bioessays*, **20**, 126–36.

Jorde, L. B., Watkins, W. S. and Bamshad, M. J. (2001). Population genomics: a bridge from evolutionary history to genetic medicine. *Hum Mol Genet*, **10**, 2199–207.

Jorde, L. B., Rogers, A. R., Bamshad, M., Watkins, W. S., Krakowiak, P., Sung, S., Kere, J. and Harpending, H. C. (1997). Microsatellite diversity and the demographic history of modern humans. *Proc Natl Acad Sci USA*, **94**, 3100–3.

Jorde, L. B., Watkins, W. S., Bamshad, M. J., Dixon, M. E., Ricker, C. E., Seielstad, M. T. and Batzer, M. A. (2000). The distribution of human genetic diversity: a comparison of mitochondrial, autosomal, and Y-chromosome data. *Am J Hum Genet*, **66**, 979–88.

Kaessmann, H. and Paabo, S. (2002). The genetical history of humans and the great apes. *J Intern Med*, **251**, 1–18.

Kaessmann, H., Heissig, F., von Haeseler, A. and Paabo, S. (1999a). DNA sequence variation in a non-coding region of low recombination on the human X chromosome. *Nat Genet*, **22**, 78–81.

Kaessmann, H., Wiebe, V. and Paabo, S. (1999b). Extensive nuclear DNA sequence diversity among chimpanzees. *Science*, **286**, 1159–62.

Kaessmann, H., Wiebe, V., Weiss, G. and Paabo, S. (2001). Great ape DNA sequences reveal a reduced diversity and an expansion in humans. *Nat Genet*, **27**, 155–6.

Kaplan, N. L., Hudson, R. R. and Langley, C. H. (1989). The 'hitchhiking effect' revisited. *Genetics*, **123**, 887–99.

Kayser, M., Brauer, S., Schadlich, H., Prinz, M., Batzer, M. A., Zimmerman, P. A., Boatin, B. A. and Stoneking, M. (2003). Y chromosome STR haplotypes and the genetic structure of U.S. populations of African, European, and Hispanic ancestry. *Genome Res*, **13**, 624–34.

Kayser, M., Krawczak, M., Excoffier, L., Dieltjes, P., Corach, D., Pascali, V., Gehrig, C., Bernini, L. F., Jespersen, J., Bakker, E., Roewer, L. and de Knijff, P. (2001). An extensive analysis of Y-chromosomal microsatellite haplotypes in globally dispersed human populations. *Am J Hum Genet*, **68**, 990–1018.

Kayser, M., Kittler, R., Erler, A., Hedman, M., Lee, A. C., Mohyuddin, A., Mehdi, S. Q., Rosser, Z., Stoneking, M., Jobling, M. A., Sajantila, A. and Tyler-Smith, C. (2004). A comprehensive survey of human Y-chromosomal microsatellites. *Am J Hum Genet*, **74**, 1183–97.

Kidd, K. K., Pakstis, A. J., Speed, W. C. and Kidd, J. R. (2004). Understanding human DNA sequence variation. *J Hered*, **95**, 406–20.

Kimmel, M., Chakraborty, R., King, J. P., Bamshad, M., Watkins, W. S. and Jorde, L. B. (1998). Signatures of population expansion in microsatellite repeat data. *Genetics*, **148**, 1921–30.

Kitano, T., Schwarz, C., Nickel, B. and Paabo, S. (2003). Gene diversity patterns at 10 X-chromosomal loci in humans and chimpanzees. *Mol Biol Evol*, **20**, 1281–9.

Kivisild, T., Reidla, M., Metspalu, E., Rosa, A., Brehm, A., Pennarun, E., Parik, J., Geberhiwot, T., Usanga, E. and Villems, R. (2004). Ethiopian mitochondrial DNA heritage: tracking gene flow across and around the gate of tears. *Am J Hum Genet*, **75**, 752–70.

Kivisild, T., Rootsi, S., Metspalu, M., Mastana, S., Kaldma, K., Parik, J., Metspalu, E., Adojaan, M., Tolk, H. V., Stepanov, V., Golge, M., Usanga, E., Papiha, S. S., Cinnioglu, C., King, R., Cavalli-Sforza, L., Underhill, P. A. and Villems, R. (2003). The genetic heritage of the earliest settlers persists both in Indian tribal and caste populations. *Am J Hum Genet*, **72**, 313–32.

Kivisild, T., Shen, P., Wall, D. P., Do, B., Sung, R., Davis, K., Passarino, G., Underhill, P. A., Scharfe, C., Torroni, A., Scozzari, R., Modiano, D., Coppa, A., de Knijff, P., Feldman, M., Cavalli-Sforza, L. L. and Oefner, P. J. (2006). The role of selection in the evolution of human mitochondrial genomes. *Genetics*, **172**, 373–87.

Knowles, L. L. (2004). The burgeoning field of statistical phylogeography. *J Evol Biol*, **17**, 1–10.

Koda, Y., Tachida, H., Pang, H., Liu, Y., Soejima, M., Ghaderi, A. A., Takenaka, O. and Kimura, H. (2001). Contrasting patterns of polymorphisms at the ABO-secretor gene (FUT2) and plasma alpha(1,3)fucosyltransferase gene (FUT6) in human populations. *Genetics*, **158**, 747–56.

Krings, M., Geisert, H., Schmitz, R.W., Krainitzki, H. and Paabo, S. (1999). DNA sequence of the mitochondrial hypervariable region II from the neandertal type specimen. *Proc Natl Acad Sci USA*, **96**, 5581–5.

Krings, M., Stone, A., Schmitz, R.W., Krainitzki, H., Stoneking, M. and Paabo, S. (1997). Neandertal DNA sequences and the origin of modern humans. *Cell*, **90**, 19–30.

Krings, M., Capelli, C., Tschentscher, F., Geisert, H., Meyer, S., von Haeseler, A., Grosschmidt, K., Possnert, G., Paunovic, M. and Paabo, S. (2000). A view of Neandertal genetic diversity. *Nat Genet*, **26**, 144–6.

Labuda, D., Zietkiewicz, E. and Yotova, V. (2000). Archaic lineages in the history of modern humans. *Genetics*, **156**, 799–808.

Lahn, B.T., Pearson, N.M. and Jegalian, K. (2001). The human Y chromosome, in the light of evolution. *Nat Rev Genet*, **2**, 207–16.

Lahr, M.M. (1996). *The Evolution of Modern Human Diversity*. Cambridge: Cambridge University Press.

Lahr, M.M. and Foley, R.A. (1994). Multiple dispersals and modern human origins. *Evolutionary Anthropology*, **3**, 48–60.

Lahr, M.M. and Foley, R.A. (1998). Towards a theory of modern human origins: geography, demography, and diversity in recent human evolution. *Am J Phys Anthropol*, Suppl **27**, 137–76.

Li, W.H. (1997). *Molecular Evolution*. Sunderland, MA: Sinauer Associates.

Luis, J.R., Rowold, D.J., Regueiro, M., Caeiro, B., Cinnioglu, C., Roseman, C., Underhill, P.A., Cavalli-Sforza, L.L. and Herrera, R.J. (2004). The Levant versus the Horn of Africa: evidence for bidirectional corridors of human migrations. *Am J Hum Genet*, **74**, 532–44.

Maddison, D.R., Ruvolo, M. and Swofford, D.L. (1992). Geographic origins of human mitochondrial DNA: phylogenetic evidence from control region sequences. *Systematic Biology*, **41**, 111–24.

Makova, K.D. and Li, W.H. (2002). Strong male-driven evolution of DNA sequences in humans and apes. *Nature*, **416**, 624–6.

Makova, K.D., Ramsay, M., Jenkins, T. and Li, W.H. (2001). Human DNA sequence variation in a 6.6-kb region containing the melanocortin 1 receptor promoter. *Genetics*, **158**, 1253–68.

Malaspina, P., Cruciani, F., Santolamazza, P., Torroni, A., Pangrazio, A., Akar, N., Bakalli, V., Brdicka, R., Jaruzelska, J., Kozlov, A., Malyarchuk, B., Mehdi, S.Q., Michalodimitrakis, E., Varesi, L., Memmi, M.M., Vona, G., Villems, R., Parik, J., Romano, V., Stefan, M., Stenico, M., Terrenato, L., Novelletto, A. and Scozzari, R. (2000). Patterns of male-specific inter-population divergence in Europe, West Asia and North Africa. *Ann Hum Genet*, **64**, 395–412.

Marjoram, P. and Donnelly, P. (1994). Pairwise comparisons of mitochondrial DNA sequences in subdivided populations and implications of early human evolution. *Genetics*, **136**, 673–83.

Marshall Graves, J.A. (2002). The rise and fall of SRY. *Trends Genet*, **18**, 259–64.

Marth, G., Schuler, G., Yeh, R., Davenport, R., Agarwala, R., Church, D., Wheelan, S., Baker, J., Ward, M., Kholodov, M., Phan, L., Czabarka, E., Murvai, J., Cutler, D., Wooding, S., Rogers, A., Chakravarti, A., Harpending, H.C., Kwok, P.Y. and Sherry, S.T. (2003). Sequence variations in the public human genome data reflect a bottlenecked population history. *Proc Natl Acad Sci USA*, **100**, 376–81.

McBrearty, S. and Brooks, A. (2000). The revolution that wasn't: a new interpretation of the origin of modern human behaviour. *Journal of Human Evolution*, **39**, 453–563.

McDougall, I., Brown, F.H. and Fleagle, J.G. (2005). Stratigraphic placement and age of modern humans from kibish, Ethiopia. *Nature*, **433**, 733–6.

Melton, T., Clifford, S., Martinson, J., Batzer, M. and Stoneking, M. (1998). Genetic evidence for the proto-Austronesian homeland in Asia: mtDNA and nuclear DNA variation in Taiwanese aboriginal tribes. *Am J Hum Genet*, **63**, 1807–23.

Meyer, S., Weiss, G. and von Haesler, A. (1999). Pattern of nucleotide substitution and rate heterogeneity in the hypervariable regions I and II of human mtDNA. *Genetics*, **152**, 1103–10.

Mishmar, D., Ruiz-Pesini, E., Brandon, M. and Wallace, D.C. (2004). Mitochondrial DNA-like sequences in the nucleus (NUMTs): insights into our African origins and the mechanism of foreign DNA integration. *Hum Mutat*, **23**, 125–33.

Mishmar, D., Ruiz-Pesini, E., Golik, P., Macaulay, V., Clark, A.G., Hosseini, S., Brandon, M., Easley, K., Chen, E., Brown, M.D., Sukernik, R.I., Olckers, A. and Wallace, D.C. (2003). Natural selection shaped regional mtDNA variation in humans. *Proc Natl Acad Sci USA*, **100**, 171–6.

Mountain, J.L. (1998). Molecular evolution and modern human origins. *Evolutionary Anthropology*, **7**, 21–37.

Mountain, J.L., Knight, A., Jobin, M., Gignoux, C., Miller, A., Lin, A.A. and Underhill, P.A. (2002). SNPSTRs: empirically derived, rapidly typed, autosomal haplotypes for inference of population history and mutational processes. *Genome Res*, **12**, 1766–72.

Mourier, T., Hansen, A.J., Willerslev, E. and Arctander, P. (2001). The Human Genome Project reveals a continuous transfer of large mitochondrial fragments to the nucleus. *Mol Biol Evol*, **18**, 1833–7.

Nachman, M.W., Bauer, V.L., Crowell, S.L. and Aquadro, C.F. (1998). DNA variability and recombination rates at X-linked loci in humans. *Genetics*, **150**, 1133–41.

Nachman, M.W. and Crowell, S.L. (2000a). Contrasting evolutionary histories of two introns of the duchenne muscular dystrophy gene, Dmd, in humans. *Genetics*, **155**, 1855–64.

Nachman, M.W. and Crowell, S.L. (2000b). Estimate of the mutation rate per nucleotide in humans. *Genetics*, **156**, 297–304.

Nachman, M.W., D'Agostino, S.L., Tillquist, C.R., Mobasher, Z. and Hammer, M.F. (2004). Nucleotide variation at Msn and Alas2, two genes flanking the centromere of the X chromosome in humans. *Genetics*, **167**, 423–37.

Nei, M. and Livshits, G. (1989). Genetic relationships of Europeans, Asians and Africans and the origin of modern Homo sapiens. *Hum Hered*, **39**, 276–81.

Nordborg, M. (1998). On the probability of Neanderthal ancestry. *Am J Hum Genet*, **63**, 1237–40.

Nordborg, M. (2001). 'Coalescent Theory.' In *Handbook of Statistical Genetics*, eds. Balding, D.J., Bishop, M.J. and Cannings, C. Chichester, UK: Wiley and Sons. pp. 197–212.

Otieno, A.C., Carter, A.B., Hedges, D.J., Walker, J.A., Ray, D.A., Garber, R.K., Anders, B.A., Stoilova, N., Laborde, M.E., Fowlkes, J.D., Huang, C.H., Perodeau, B. and Batzer, M.A. (2004). Analysis of the human Alu Ya-lineage. *J Mol Biol*, **342**, 109–18.

Ovchinnikov, I. V., Gotherstrom, A., Romanova, G. P., Kharitonov, V. M., Liden, K. and Goodwin, W. (2000). Molecular analysis of Neanderthal DNA from the northern Caucasus. *Nature*, **404**, 490–3.

Paabo, S., Poinar, H., Serre, D., Jaenicke-Despres, V., Hebler, J., Rohland, N., Kuch, M., Krause, J., Vigilant, L. and Hofreiter, M. (2004). Genetic analyses from ancient DNA. *Annu Rev Genet*, **38**, 645–79.

Passarino, G., Semino, O., Quintana-Murci, L., Excoffier, L., Hammer, M. and Santachiara-Benerecetti, A. S. (1998). Different genetic components in the Ethiopian population, identified by mtDNA and Y-chromosome polymorphisms. *Am J Hum Genet*, **62**, 420–34.

Penny, D., Steel, M., Waddell, P. J. and Hendy, M. D. (1995). Improved analyses of human mtDNA sequences support a recent African origin for Homo sapiens. *Mol Biol Evol*, **12**, 863–82.

Poloni, E. S., Semino, O., Passarino, G., Santachiara-Benerecetti, A. S., Dupanloup, I., Langaney, A. and Excoffier, L. (1997). Human genetic affinities for Y-chromosome P49a,f/TaqI haplotypes show strong correspondence with linguistics. *Am J Hum Genet*, **61**, 1015–35.

Pritchard, J. K., Seielstad, M. T., Perez-Lezaun, A. and Feldman, M. W. (1999). Population growth of human Y chromosomes: a study of Y chromosome microsatellites. *Molecular Biology and Evolution*, **16**, 1791–98.

Przeworski, M., Hudson, R. R. and Di Rienzo, A. (2000). Adjusting the focus on human variation. *Trends Genet*, **16**, 296–302.

Quintana-Murci, L. and Fellous, M. (2001). The human Y chromosome: The biological role of a 'Functional Wasteland'. *J Biomed Biotechnol*, **1**, 18–24.

Quintana-Murci, L., Semino, O., Bandelt, H. J., Passarino, G., McElreavey, K. and Santachiara-Benerecetti, A. S. (1999). Genetic evidence of an early exit of Homo sapiens from Africa through eastern Africa. *Nat Genet*, **23**, 437–41.

Reich, D. E., Cargill, M., Bolk, S., Ireland, J., Sabeti, P. C., Richter, D. J., Lavery, T., Kouyoumjian, R., Farhadian, S. F., Ward, R. and Lander, E. S. (2001). Linkage disequilibrium in the human genome. *Nature*, **411**, 199–204.

Reich, D. E. and Goldstein, D. B. (1998). Genetic evidence for a Paleolithic human population expansion in Africa. *Proc Natl Acad Sci USA*, **95**, 8119–23.

Relethford, J. H. and Jorde, L. B. (1999). Genetic evidence for larger African population size during recent human evolution. *Am J Phys Anthropol*, **108**, 251–60.

Ricchetti, M., Tekaia, F. and Dujon, B. (2004). Continued colonization of the human genome by mitochondrial DNA. *PLoS Biol*, **2**, E273.

Richards, M., Rengo, C., Cruciani, F., Gratrix, F., Wilson, J. F., Scozzari, R., Macaulay, V. and Torroni, A. (2003). Extensive female-mediated gene flow from sub-Saharan Africa into near eastern Arab populations. *Am J Hum Genet*, **72**, 1058–64.

Rieder, M. J., Taylor, S. L., Clark, A. G. and Nickerson, D. A. (1999). Sequence variation in the human angiotensin converting enzyme. *Nat Genet*, **22**, 59–62.

Ritte, U., Neufeld, E., Broit, M., Shavit, D. and Motro, U. (1993). The differences among Jewish communities – maternal and paternal contributions. *J Mol Evol.*, **37**, 435–40.

Rogers, A. R. (1995). Genetic evidence for a Pleistocene population explosion. *Evolution*, **49**, 608–15.

Rogers, A. R. and Harpending, H. (1992). Population growth makes waves in the distribution of pairwise genetic differences. *Molecular Biology and Evolution*, **9**, 552–69.

Rogers, A. R. and Jorde, L. B. (1995). Genetic evidence on modern human origins. *Hum Biol*, **67**, 1–36.

Romualdi, C., Balding, D., Nasidze, I. S., Risch, G., Robichaux, M., Sherry, S. T., Stoneking, M., Batzer, M. A. and Barbujani, G. (2002). Patterns of human diversity, within and among continents, inferred from biallelic DNA polymorphisms. *Genome Res*, **12**, 602–12.

Rosa, A., Brehm, A., Kivisild, T., Metspalu, E. and Villems, R. (2004). MtDNA profile of West Africa Guineans: towards a better understanding of the Senegambia region. *Ann Hum Genet*, **68**, 340–52.

Rosenberg, N. A. and Nordborg, M. (2002). Genealogical trees, coalescent theory and the analysis of genetic polymorphisms. *Nat Rev Genet*, **3**, 380–90.

Rosenberg, N. A., Pritchard, J. K., Weber, J. L., Cann, H. M., Kidd, K. K., Zhivotovsky, L. A. and Feldman, M. W. (2002). Genetic structure of human populations. *Science*, **298**, 2381–5.

Rozen, S., Skaletsky, H., Marszalek, J. D., Minx, P. J., Cordum, H. S., Waterston, R. H., Wilson, R. K. and Page, D. C. (2003). Abundant gene conversion between arms of palindromes in human and ape Y chromosomes. *Nature*, **423**, 873–6.

Ruiz-Pesini, E., Mishmar, D., Brandon, M., Procaccio, V. and Wallace, D. C. (2004). Effects of purifying and adaptive selection on regional variation in human mtDNA. *Science*, **303**, 223–6.

Ruvolo, M. (2004). Comparative primate genomics: the year of the chimpanzee. *Curr Opin Genet Dev*, **14**, 650–6.

Sabeti, P. C., Reich, D. E., Higgins, J. M., Levine, H. Z., Richter, D. J., Schaffner, S. F., Gabriel, S. B., Platko, J. V., Patterson, N. J., McDonald, G. J., Ackerman, H. C., Campbell, S. J., Altshuler, D., Cooper, R., Kwiatkowski, D., Ward, R. and Lander, E. S. (2002). Detecting recent positive selection in the human genome from haplotype structure. *Nature*, **419**, 832–7.

Salas, A., Richards, M., De la Fe, T., Lareu, M. V., Sobrino, B., Sanchez-Diz, P., Macaulay, V. and Carracedo, A. (2002). The making of the African mtDNA landscape. *Am J Hum Genet*, **71**, 1082–111.

Salas, A., Richards, M., Lareu, M. V., Scozzari, R., Coppa, A., Torroni, A., Macaulay, V. and Carracedo, A. (2004). The African diaspora: mitochondrial DNA and the Atlantic slave trade. *Am J Hum Genet*, **74**, 454–65.

Salem, A. H., Kilroy, G. E., Watkins, W. S., Jorde, L. B. and Batzer, M. A. (2003). Recently integrated Alu elements and human genomic diversity. *Mol Biol Evol*, **20**, 1349–61.

Santos, F. R., Bianchi, N. O. and Pena, S. D. (1996). Worldwide distribution of human Y-chromosome haplotypes. *Genome Res*, **6**, 601–11.

Saunders, M. A., Hammer, M. F. and Nachman, M. W. (2002). Nucleotide variability at G6pd and the signature of malarial selection in humans. *Genetics*, **162**, 1849–61.

Schillaci, M. A. and Froehlich, J. W. (2001). Nonhuman primate hybridization and the taxonomic status of Neanderthals. *Am J Phys Anthropol*, **115**, 157–66.

Schlotterer, C. (2003). Hitchhiking mapping – functional genomics from the population genetics perspective. *Trends Genet*, **19**, 32–8.

Schmitz, R. W., Serre, D., Bonani, G., Feine, S., Hillgruber, F., Krainitzki, H., Paabo, S. and Smith, F. H. (2002). The Neandertal type site revisited: interdisciplinary investigations of skeletal remains from the Neander Valley, Germany. *Proc Natl Acad Sci USA*, **99**, 13342–7.

Scozzari, R., Cruciani, F., Santolamazza, P., Malaspina, P., Torroni, A., Sellitto, D., Arredi, B., Destro-Bisol, G., De Stefano, G., Rickards, O., Martinez-Labarga, C., Modiano, D., Biondi, G., Moral, P., Olckers, A., Wallace, D. C. and Novelletto, A. (1999). Combined use of biallelic and microsatellite Y-chromosome polymorphisms to infer affinities among African populations. *Am J Hum Genet*, **65**, 829–46.

Seielstad, M. T., Minch, E. and Cavalli-Sforza, L. L. (1998). Genetic evidence for a higher female migration rate in humans. *Nat Genet*, **20**, 278–80.

Semino, O., Magri, C., Benuzzi, G., Lin, A. A., Al-Zahery, N., Battaglia, V., Maccioni, L., Triantaphyllidis, C., Shen, P., Oefner, P. J., Zhivotovsky, L. A., King, R., Torroni, A., Cavalli-Sforza, L. L., Underhill, P. A. and Santachiara-Benerecetti, A. S. (2004). Origin, diffusion, and differentiation of Y-chromosome haplogroups E and J: inferences on the neolithization of Europe and later migratory events in the Mediterranean area. *Am J Hum Genet*, **74**, 1023–34.

Semino, O., Santachiara-Benerecetti, A. S., Falaschi, F., Cavalli-Sforza, L. L. and Underhill, P. A. (2002). Ethiopians and Khoisan share the deepest clades of the human Y-chromosome phylogeny. *Am J Hum Genet*, **70**, 265–8.

Serre, D., Langaney, A., Chech, M., Teschler-Nicola, M., Paunovic, M., Mennecier, P., Hofreiter, M., Possnert, G. G. and Paabo, S. (2004). No evidence of Neandertal mtDNA contribution to early modern humans. *PLoS Biol*, **2**, E57.

She, X., Jiang, Z., Clark, R. A., Liu, G., Cheng, Z., Tuzun, E., Church, D. M., Sutton, G., Halpern, A. L. and Eichler, E. E. (2004). Shotgun sequence assembly and recent segmental duplications within the human genome. *Nature*, **431**, 927–30.

Shen, P., Wang, F., Underhill, P. A., Franco, C., Yang, W. H., Roxas, A., Sung, R., Lin, A. A., Hyman, R. W., Vollrath, D., Davis, R. W., Cavalli-Sforza, L. L. and Oefner, P. J. (2000). Population genetic implications from sequence variation in four Y chromosome genes. *Proc Natl Acad Sci USA*, **97**, 7354–9.

Sherry, S. T., Harpending, H. C., Batzer, M. A. and Stoneking, M. (1997). Alu evolution in human populations: using the coalescent to estimate effective population size. *Genetics*, **147**, 1977–82.

Sherry, S. T., Rogers, A. R., Harpending, H., Soodyall, H., Jenkins, T. and Stoneking, M. (1994). Mismatch distributions of mtDNA reveal recent human population expansions. *Human Biology*, **66**, 761–75.

Shriver, M. D., Jin, L., Ferrell, R. E. and Deka, R. (1997). Microsatellite data support an early population expansion in Africa. *Genome Res*, **7**, 586–91.

Skaletsky, H., Kuroda-Kawaguchi, T., Minx, P. J., Cordum, H. S., Hillier, L., Brown, L. G., Repping, S., Pyntikova, T., Ali, J., Bieri, T., Chinwalla, A., Delehaunty, A., Delehaunty, K., Du, H., Fewell, G., Fulton, L., Fulton, R., Graves, T., Hou, S. F., Latrielle, P., Leonard, S., Mardis, E., Maupin, R., McPherson, J., Miner, T., Nash, W., Nguyen, C., Ozersky, P., Pepin, K., Rock, S., Rohlfing, T., Scott, K., Schultz, B., Strong, C., Tin-Wollam, A., Yang, S. P., Waterston, R. H., Wilson, R. K., Rozen, S. and Page, D. C. (2003). The male-specific region of the human Y chromosome is a mosaic of discrete sequence classes. *Nature*, **423**, 825–37.

Slatkin, M. (1995). A measure of population subdivision based on microsatellite allele frequencies. *Genetics*, 457–62.

Slatkin, M. and Hudson, R. R. (1991). Pairwise comparisons of mitochondrial DNA sequences in stable and exponentially growing populations. *Genetics*, **129**, 555–62.

Soodyall, H., Vigilant, L., Hill, A.V., Stoneking, M. and Jenkins, T. (1996). mtDNA control-region sequence variation suggests multiple independent origins of an 'Asian-specific' 9-bp deletion in sub-Saharan Africans. *Am J Hum Genet*, **58**, 595–608.

Spurdle, A.B. and Jenkins, T. (1996). The origins of the Lemba 'Black Jews' of southern Africa: evidence from p12F2 and other Y-chromosome markers. *Am J Hum Genet*, **59**, 1126–33.

Stephens, J.C., Schneider, J.A., Tanguay, D.A., Choi, J., Acharya, T., Stanley, S.E., Jiang, R., Messer, C.J., Chew, A., Han, J.H., Duan, J., Carr, J.L., Lee, M.S., Koshy, B., Kumar, A.M., Zhang, G., Newell, W.R., Windemuth, A., Xu, C., Kalbfleisch, T.S., Shaner, S.L., Arnold, K., Schulz, V., Drysdale, C.M., Nandabalan, K., Judson, R.S., Ruano, G. and Vovis, G.F. (2001). Haplotype variation and linkage disequilibrium in 313 human genes. *Science*, **293**, 489–93.

Stone, A.C., Griffiths, R.C., Zegura, S.L. and Hammer, M.F. (2002). High levels of Y-chromosome nucleotide diversity in the genus Pan. *Proc Natl Acad Sci USA*, **99**, 43–8.

Stoneking, M. (2000). Hypervariable sites in the mtDNA control region are mutational hotspots. *American Journal of Human Genetics*, **67**, 1029–32.

Stoneking, M., Fontius, J.J., Clifford, S.L., Soodyall, H., Arcot, S.S., Saha, N., Jenkins, T., Tahir, M.A., Deininger, P.L. and Batzer, M.A. (1997). Alu insertion polymorphisms and human evolution: evidence for a larger population size in Africa. *Genome Res*, **7**, 1061–71.

Stringer, C. (2002). Modern human origins: progress and prospects. *Philos Trans R Soc Lond B Biol Sci*, **357**, 563–79.

Stringer, C.B. and Andrews, P. (1988). Genetic and fossil evidence for the origin of modern humans. *Science*, **239**, 1263–8.

Stumpf, M.P. and Goldstein, D.B. (2001). Genealogical and evolutionary inference with the human Y chromosome. *Science*, **291**, 1738–42.

Tajima, F. (1989). Statistical method for testing the neutral mutation hypothesis by DNA polymorphism. *Genetics*, **123**, 585–95.

Takahata, N., Lee, S.H. and Satta, Y. (2001). Testing multiregionality of modern human origins. *Mol Biol Evol*, **18**, 172–83.

Tang, H., Siegmund, D.O., Shen, P., Oefner, P.J. and Feldman, M.W. (2002). Frequentist estimation of coalescence times from nucleotide sequence data using a tree-based partition. *Genetics*, **161**, 447–59.

Tattersall, I. (2002). The case for saltational events in human evolution. In *The Speciation of Modern Homo sapiens*, ed. Crow, T.J. Oxford: Oxford University Press.

Templeton, A. (2002). Out of Africa again and again. *Nature*, **416**, 45–51.

Thalmann, O., Hebler, J., Poinar, H.N., Paabo, S. and Vigilant, L. (2004). Unreliable mtDNA data due to nuclear insertions: a cautionary tale from analysis of humans and other great apes. *Mol Ecol*, **13**, 321–35.

Thangaraj, K., Singh, L., Reddy, A.G., Rao, V.R., Sehgal, S.C., Underhill, P.A., Pierson, M., Frame, I.G. and Hagelberg, E. (2003). Genetic affinities of the Andaman Islanders, a vanishing human population. *Curr Biol*, **13**, 86–93.

Thomas, M.G., Parfitt, T., Weiss, D.A., Skorecki, K., Wilson, J.F., le Roux, M., Bradman, N. and Goldstein, D.B. (2000). Y chromosomes traveling south: the cohen modal haplotype and the origins of the Lemba – the 'Black Jews of Southern Africa'. *Am J Hum Genet*, **66**, 674–86.

Thomson, R., Pritchard, J. K., Shen, P., Oefner, P. J. and Feldman, M. W. (2000). Recent common ancestry of human Y chromosomes: evidence from DNA sequence data. *Proc Natl Acad Sci USA*, **97**, 7360–5.

Tishkoff, S. A. and Williams, S. M. (2002). Genetic analysis of African populations: human evolution and complex disease. *Nat Rev Genet*, **3**, 611–21.

Tishkoff, S. A. and Verrelli, B. C. (2003). Patterns of human genetic diversity: implications for human evolutionary history and disease. *Annu Rev Genomics Hum Genet*, **4**, 293–340.

Tishkoff, S. A. and Kidd, K. K. (2004). Implications of biogeography of human populations for 'race' and medicine. *Nature Genetics*, Suppl **1**, S21–7.

Tishkoff, S. A., Goldman, A., Calafell, F., Speed, W. C., Deinard, A. S., Bonne-Tamir, B., Kidd, J. R., Pakstis, A. J., Jenkins, T. and Kidd, K. K. (1998a). A global haplotype analysis of the myotonic dystrophy locus: implications for the evolution of modern humans and for the origin of myotonic dystrophy mutations. *Am J Hum Genet*, **62**, 1389–402.

Tishkoff, S. A., Kidd, K. K. and Clark, A. G. (1998b). Inferences of modern human origins from variation in CD4 haplotypes. In *Proceedings of the Trinational Workshop on Molecluar Evolution,* ed. Haessler, A. V. Durham, NC: Duke University Publishing Group, pp. 181–98.

Tishkoff, S. A., Dietzsch, E., Speed, W., Pakstis, A. J., Kidd, J. R., Cheung, K., Bonne-Tamir, B., Santachiara-Benerecetti, A. S., Moral, P. and Krings, M. (1996). Global patterns of linkage disequilibrium at the CD4 locus and modern human origins. *Science*, **271**, 1380–7.

Tishkoff, S. A., Pakstis, A. J., Stoneking, M., Kidd, J. R., Destro-Bisol, G., Sanjantila, A., Lu, R. B., Deinard, A. S., Sirugo, G., Jenkins, T., Kidd, K. K. and Clark, A. G. (2000). Short tandem-repeat polymorphism/alu haplotype variation at the PLAT locus: implications for modern human origins. *Am J Hum Genet*, **67**, 901–25.

Tishkoff, S. A., Varkonyi, R., Cahinhinan, N., Abbes, S., Argyropoulos, G., Destro-Bisol, G., Drousiotou, A., Dangerfield, B., Lefranc, G., Loiselet, J., Piro, A., Stoneking, M., Tagarelli, A., Tagarelli, G., Touma, E. H., Williams, S. M. and Clark, A. G. (2001). Haplotype diversity and linkage disequilibrium at human G6PD: recent origin of alleles that confer malarial resistance. *Science*, **293**, 455–62.

Tourmen, Y., Baris, O., Dessen, P., Jacques, C., Malthiery, Y. and Reynier, P. (2002). Structure and chromosomal distribution of human mitochondrial pseudogenes. *Genomics*, **80**, 71–7.

Uddin, M., Wildman, D. E., Liu, G., Xu, W., Johnson, R. M., Hof, P. R., Kapatos, G., Grossman, L. I. and Goodman, M. (2004). Sister grouping of chimpanzees and humans as revealed by genome-wide phylogenetic analysis of brain gene expression profiles. *Proc Natl Acad Sci USA*, **101**, 2957–62.

Underhill, P. A., Passarino, G., Lin, A. A., Shen, P., Mirazon Lahr, M., Foley, R. A., Oefner, P. J. and Cavalli-Sforza, L. L. (2001). The phylogeography of Y chromosome binary haplotypes and the origins of modern human populations. *Ann Hum Genet*, **65**, 43–62.

Underhill, P. A., Shen, P., Lin, A. A., Jin, L., Passarino, G., Yang, W. H., Kauffman, E., Bonne-Tamir, B., Bertranpetit, J., Francalacci, P., Ibrahim, M., Jenkins, T., Kidd, J. R., Mehdi, S. Q., Seielstad, M. T., Wells, R. S., Piazza, A.,

Davis, R. W., Feldman, M. W., Cavalli-Sforza, L. L. and Oefner, P. J. (2000). Y chromosome sequence variation and the history of human populations. *Nat Genet*, **26**, 358–61.

Verrelli, B. C., McDonald, J. H., Argyropoulos, G., Destro-Bisol, G., Froment, A., Drousiotou, A., Lefranc, G., Helal, A. N., Loiselet, J. and Tishkoff, S. A. (2002). Evidence for balancing selection from nucleotide sequence analyses of human G6PD. *Am J Hum Genet*, **71**, 1112–28.

Verrelli, B. C. and Tishkoff, S. A. (2004). Signatures of selection and gene conversion associated with human color vision variation. *Am J Hum Genet*, **75**, 363–75.

Vigilant, L., Stoneking, M., Harpending, H., Hawkes, K. and Wilson, A. C. (1991). African populations and the evolution of human mitochondrial DNA. *Science*, **253**, 1503–7.

Wakeley, J., Nielsen, R., Liu-Cordero, S. N. and Ardlie, K. (2001). The discovery of single-nucleotide polymorphisms – and inferences about human demographic history. *Am J Hum Genet*, **69**, 1332–47.

Wall, J. D. (2000). Detecting ancient admixture in humans using sequence polymorphism data. *Genetics*, **154**, 1271–9.

Wall, J. D. (2001). Insights from linked single nucleotide polymorphisms: What we can learn from linkage disequilibrium. *Current Opinions in Genetics and Development*, **11**, 647–51.

Wall, J. D. and Przeworski, M. (2000). When did the human population size start increasing? *Genetics*, **155**, 1865–74.

Watanabe, H., Fujiyama, A., Hattori, M., Taylor, T. D., Toyoda, A. *et al.* (2004). DNA sequence and comparative analysis of chimpanzee chromosome 22. *Nature*, **429**, 382–8.

Watkins, W. S., Ricker, C. E., Bamshad, M. J., Carroll, M. L., Nguyen, S. V., Batzer, M. A., Harpending, H. C., Rogers, A. R. and Jorde, L. B. (2001). Patterns of ancestral human diversity: an analysis of Alu-insertion and restriction-site polymorphisms. *Am J Hum Genet*, **68**, 738–52.

Watkins, W. S., Rogers, A. R., Ostler, C. T., Wooding, S., Bamshad, M. J., Brassington, A. M., Carroll, M. L., Nguyen, S. V., Walker, J. A., Prasad, B. V., Reddy, P. G., Das, P. K., Batzer, M. A. and Jorde, L. B. (2003). Genetic variation among world populations: inferences from 100 Alu insertion polymorphisms. *Genome Res*, **13**, 1607–18.

Watson, E., Bauer, K., Aman, R., Weiss, G., von Haeseler, A. and Paabo, S. (1996). mtDNA sequence diversity in Africa. *Am J Hum Genet*, **59**, 437–44.

Watson, E., Forster, P., Richards, M. and Bandelt, H. J. (1997). Mitochondrial footprints of human expansions in Africa. *Am J Hum Genet*, **61**, 691–704.

Weber, J. L., David, D., Heil, J., Fan, Y., Zhao, C. and Marth, G. (2002). Human diallelic insertion/deletion polymorphisms. *Am J Hum Genet*. **71**, 854–62.

Wilder, J. A., Kingan, S. B., Mobasher, Z., Pilkington, M. M. and Hammer, M. F. (2004a). Global patterns of human mitochondrial DNA and Y-chromosome structure are not influenced by higher migration rates of females versus males. *Nat Genet*, **36**, 1122–5.

Wilder, J. A., Mobasher, Z. and Hammer, M. F. (2004b). Genetic evidence for unequal effective population sizes of human females and males. *Mol Biol Evol*, **21**, 2047–57.

Wildman, D. E., Uddin, M., Liu, G., Grossman, L. I. and Goodman, M. (2003). Implications of natural selection in shaping 99.4% nonsynonymous DNA identity between humans and chimpanzees: enlarging genus Homo. *Proc Natl Acad Sci USA*, **100**, 7181–8.

Wolpoff, M.H., Hawks, J. and Caspari, R. (2000). Multiregional, not multiple origins. *Am J Phys Anthropol*, **112**, 129–36.

Wolpoff, M.H., Hawks, J., Frayer, D.W. and Hunley, K. (2001). Modern human ancestry at the peripheries: a test of the replacement theory. *Science*, **291**, 293–7.

Won, Y.J. and Hey, J. (2005). Divergence Population Genetics of Chimpanzees. *Mol Biol Evol*, **22**, 297–307.

Wooding, S. and Rogers, A. (2000). A Pleistocene population X-plosion? *Hum Biol*, **72**, 693–5.

Wooding, S.P., Watkins, W.S., Bamshad, M.J., Dunn, D.M., Weiss, R.B. and Jorde, L.B. (2002). DNA sequence variation in a 3.7-kb noncoding sequence 5′ of the CYP1A2 gene: implications for human population history and natural selection. *Am J Hum Genet*, **71**, 528–42.

Wright, S. (1951). The genetical structure of populations. *Annals of Eugenics*, **1**, 323–34.

Wright, S. (1969). *Evolution and the Genetics of Populations 2. The Theory of Gene Frequencies*. Chicago: University of Chicago Press.

Yu, N., Fu, Y.X. and Li, W.H. (2002). DNA polymorphism in a worldwide sample of human X chromosomes. *Mol Biol Evol*, **19**, 2131–41.

Yu, N., Zhao, Z., Fu, Y.X., Sambuughin, N., Ramsay, M., Jenkins, T., Leskinen, E., Patthy, L., Jorde, L.B., Kuromori, T. and Li, W.H. (2001). Global patterns of human DNA sequence variation in a 10-kb region on chromosome 1. *Mol Biol Evol*, **18**, 214–22.

Zerjal, T., Dashnyam, B., Pandya, A., Kayser, M., Roewer, L., Santos, F.R., Schiefenhovel, W., Fretwell, N., Jobling, M.A., Harihara, S., Shimizu, K., Semjidmaa, D., Sajantila, A., Salo, P., Crawford, M.H., Ginter, E.K., Evgrafov, O.V. and Tyler-Smith, C. (1997). Genetic relationships of Asians and Northern Europeans, revealed by Y-chromosomal DNA analysis. *Am J Hum Genet*, **60**, 1174–83.

Zerjal, T., Wells, R.S., Yuldasheva, N., Ruzibakiev, R. and Tyler-Smith, C. (2002). A genetic landscape reshaped by recent events: Y-chromosomal insights into central Asia. *Am J Hum Genet*, **71**, 466–82.

Zhang, J. and Rosenberg, H.F. (2000). Sequence variation at two eosinophil-associated ribonuclease loci in humans. *Genetics*, **156**, 1949–58.

Zhao, Z., Jin, L., Fu, Y.X., Ramsay, M., Jenkins, T., Leskinen, E., Pamilo, P., Trexler, M., Patthy, L., Jorde, L.B., Ramos-Onsins, S., Yu, N. and Li, W.H. (2000). Worldwide DNA sequence variation in a 10-kilobase noncoding region on human chromosome 22. *Proc Natl Acad Sci USA*, **97**, 11354–8.

Zhivotovsky, L.A., Bennett, L., Bowcock, A.M. and Feldman, M.W. (2000). Human population expansion and microsatellite variation. *Mol Biol Evol*, **17**, 757–67.

Zhivotovsky, L.A., Underhill, P.A., Cinnioglu, C., Kayser, M., Morar, B., Kivisild, T., Scozzari, R., Cruciani, F., Destro-Bisol, G., Spedini, G., Chambers, G.K., Herrera, R.J., Yong, K.K., Gresham, D., Tournev, I., Feldman, M.W. and Kalaydjieva, L. (2004). The effective mutation rate at Y chromosome short tandem repeats, with application to human population-divergence time. *Am J Hum Genet*, **74**, 50–61.

Zietkiewicz, E., Yotova, V., Gehl, D., Wambach, T., Arrieta, I., Batzer, M., Cole, D.E., Hechtman, P., Kaplan, F., Modiano, D., Moisan, J.P., Michalski, R. and Labuda, D. (2003). Haplotypes in the dystrophin DNA segment point to a mosaic origin of modern human diversity. *Am J Hum Genet*, **73**, 994–1015.

Chapter 13

The Peopling of Europe

Barbara Arredi

Istituto di Medicina Legale, Università Cattolica del Sacro Cuore di Roma, Rome, Italy (Present affiliation Dept of Histology, Microbiology and Medical Biotechnologies, University of Padua, Italy)

Estella S. Poloni

Dept of Anthropology, University of Geneva, Geneva, Switzerland

Chris Tyler-Smith

The Wellcome Trust Sanger Institute, Hinxton, UK

Summary

Although hominins were present in Europe as early as ~780 thousand years ago, there is broad agreement that these archaic humans, including Neanderthals, contributed little to the contemporary European gene pool. In contrast, there is vigorous debate about the relative contributions of humans who entered in the Upper Paleolithic and Neolithic. Here, we argue that the Y-chromosomal diversity pattern is likely to have a largely Neolithic or later origin. In addition to the genome-wide influences resulting from migration, admixture and drift, the effects of positive selection are detectable around some genes, such as lactase. Studies of species associated with humans, e.g. cattle, are providing additional insights.

Address for correspondence: Chris Tyler-Smith, The Wellcome Trust Sanger Institute, Wellcome Trust Genome Campus, Hinxton, Cambs. CB10 1SA, UK. Tel: [+44] (0) 1223 495376; Fax: [+44] (0) 1223 494919; E-mail: cts@sanger.ac.uk

Why study Europe?

Introductory remarks: why Europe?

The western edge of the Asian continent has a special status in studies of history, prehistory, anthropology and many other fields, more because of the origin of the people conducting these studies than the properties of the area. It is conventionally considered as a continent in its own right, Europe, although the geographical justification for this is difficult to see and a clear definition of parts of its eastern boundary, such as between the Urals and Bosphorus, is not easy to establish. Furthermore, much of the terminology we use and attempt to apply to the rest of the world, 'Stone Age', 'Bronze Age', etc. derives from studies of Europe. Now that we can take a more global view of some of these matters, does Europe still provide a legitimate unit of enquiry? It seems as valid to study Europe as, say, Northern or Eastern Asia, and the intensive studies of its archaeology, history and linguistics as well as genetics endow it with a particular interest. From this viewpoint, we will discuss the peopling of Europe. This chapter will not be a comprehensive account of the arrivals and subsequent fates of all the people who could claim to be Europeans. Instead, we will organize it around some of the key issues that have attracted attention and debate: the contribution of Lower or Middle Paleolithic, Upper Paleolithic and Neolithic migrants to modern Europeans, the importance of selection in shaping the gene pool, and the insights provided by studies of associated species. Within each section, we will concentrate on a few examples that illustrate general principles. Since we will focus on the modern European gene pool, we must first decide who the Europeans are.

Who are the Europeans?

Peopling is a dynamic and continuous process, and both long-distance and large-scale movements of people have been particularly common in recent times. This complicates finding representative 'Europeans', since the ancestors of a person encountered in Europe may have arrived at any date during historical or prehistorical times. For some purposes, such as forensic and some anthropological investigations, all can provide good subjects. In this chapter, however, we will be concerned largely with prehistoric events and will therefore concentrate on long-term inhabitants, often interpreted as those who can trace all four grandparents back to the same local region.

With this restriction, there are still significant issues to consider about sampling.

> Should individuals be chosen at random with respect to family relationships, or should an effort be made to select 'unrelated' individuals, which usually means 'unrelated for the last two or three generations'? The latter strategy is commonly used, but is it the best?

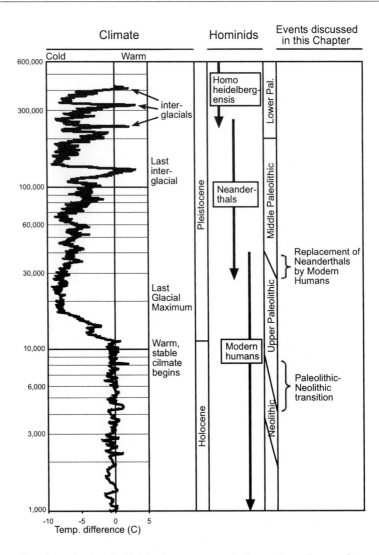

Fig 13.1 Chronology of the peopling of Europe. Note the logarithmic timescale, and that several dates are debated. Climate data are from the Vostok ice core (Petit *et al.*, 1999).

Should single individuals or populations be the units of investigation?

If populations are sampled, how should they be identified? Geography, language, ethnicity and politics can all provide criteria, but how should these be combined?

In practice, casual or hybrid schemes are usually adopted. A typical sample of Europeans might aim to include roughly equivalent numbers of self-identified individuals from each politically defined country, with information on language, and then add additional populations of particular interest to the investigators, for example the Basques because of their language or the Saami because of their distinct lifestyle. Some might aim to include Jews or Roma as distinct populations because of their separate histories. As genetic technologies improve, larger-scale surveys will become possible and sampling strategy may become less important, but at present sampling limitations must be borne in mind throughout this chapter.

Sources of information

'The peopling of Europe' is a multidisciplinary subject *par excellence*. In this section, we will consider some of the relevant sources of information, emphasizing their strengths and limitations.

Climate

Human evolution has been dominated by the climate, and this has fluctuated enormously during the last million years, with cold spells every 125 KY or so. Records of past temperatures with a resolution of just a few years can be obtained from sources such as ice cores (Figure 13.1). Humans expanded their range out of Africa (and probably their numbers as well) during warm periods and contracted or went extinct during cold ones. Recent expansions include a temporary exodus of anatomically (but not behaviourally) modern humans during the last interglacial period reaching Israel ~100 KYA and, most importantly, the colonization of much of the world when the climate improved ~50 KYA. The subsequent deterioration of the climate, culminating in the Last Glacial Maximum ~18–21 KYA when Northern Europe was covered by an ice sheet, rendered much of the area uninhabitable and allowed human survival only in the south. Any geographical patterns of variation established before this time would have been largely erased. As the climate warmed and stabilized, reaching its present form after ~11 KYA, people re-expanded from refugia and farming economies (with their demographic consequences) became effective, owing their success more to a suitable climate than any improved ability of the people.

Paleontology and archaeology

Fossils and archaeological remains provide clear evidence of human presence and can often be dated with high precision. If people left no descendants, such sources may provide the only evidence for their presence. For example, they show that *Homo* (classified as *heidelbergensis* by some, *antecessor* by others) was present in Spain ~780 KYA (Bermudez de Castro *et al.*, 1997) and became quite widespread at later dates. Neanderthals, probably descendants of *heidelbergensis*, were present between ~250 and ~27 KYA, while anatomically modern humans only appeared a little before 40 KYA (Figure 13.1), although recent recalibration of radiocarbon dates from this period suggest that the period of overlap may have been shorter than previously believed (Mellars, 2006). Archaeological remains are far more numerous than human fossils and allow discrimination between cultures, although the identity of their creators is not always clear. They can provide information about the geographical and temporal extent of cultures, and show, for example, that the Neolithic transition took place first in the southeast and spread along the Mediterranean coast and into Central Europe before extending over the rest of Europe. Paleontology and archaeology thus set the agenda for discussion of the peopling of Europe. Although we emphasize genetic data in this chapter, the questions discussed in the third and fourth sections would not even be posed without paleontology and archaeology.

Language

We do not know when language developed, but few linguists think that existing language families can be traced back more than ~10 KY. Europe is now dominated by a single language family: Indo-European. On a global scale, such uniformity is found in some areas (spread zones), but not in others (mosaic zones). One possible explanation for these patterns is the association of spread zone languages with the expansion of farming, a subject that is discussed further in a later section. This hypothesis receives some support, at least for the Indo-European languages, from an analysis based on vocabulary that favours an origin of Indo-European in Anatolia 7.8–9.8 KYA (Gray and Atkinson, 2003). However, the non-Indo-European languages found in Europe require additional explanations. Distinct languages belonging to the Uralic family are spoken both in Scandinavia (e.g. Finnish) and Hungary (Hungarian); Uralic languages are widespread in northern Asia and probably originated there, but what events led to their spread to separate parts of Europe? Basque forms an isolate, not clearly related to any other language, but what does this imply about the people who speak it?

Genetics

Genetic studies of Europe began with the analysis of blood groups and other 'classical' immunological and protein markers, an area admirably summarized by Cavalli-Sforza et al. (1994) and still relevant to discussions of the peopling of Europe (see later section headed 'The controversy about European Paleolithic and Neolithic contributions'). Current genetic studies are based on the analysis of DNA and can involve individual unlinked markers or, more informatively, closely linked markers forming haplotypes (Chapter 6). Large numbers of unlinked microsatellites have allowed clustering of individuals from around the world into groups on genetic grounds alone using the program STRUCTURE (Pritchard et al., 2000), and these were suggested to correspond to major geographical regions (Rosenberg et al., 2002), although other interpretations of the same data have emphasized that the pattern of variation, even between continents, forms a continuous gradient rather than falling into discrete categories (Serre and Pääbo, 2004). Europe was 'the most difficult region in which to detect population structure' with the highest-likelihood results detecting no sub-structure at all within Europe, although traditional outlying populations like the Basques and Sardinians could be distinguished from other Europeans when the programme was required (suboptimally) to identify more groups. The extent of linkage disequilibrium identified by SNP typing can be informative: it extends to around 60 kb on average in Europeans, compared with <5 kb in Africans, indicating, in one interpretation, a bottleneck 27–53 KYA in Europe (Reich *et al.*, 2001). In non-recombining regions (mtDNA and the Y chromosome), detailed phylogenies can be constructed and the geographical distribution of the separate branches is highly informative: the field of phylo-geography (Jobling and Tyler-Smith, 2003; Forster, 2004). Together, these approaches show that there is no specific genetic character-istic that identifies 'Europeans'. European populations are generally distinct from those to the south, North Africans, but less so from those to the east, Near Easterners or Western Central Asians, in keeping with the idea that worldwide genetic patterns are character-ized by gradients rather than discontinuities (Serre and Pääbo, 2004). Differentiation within Europe is low but can be detected by some approaches.

History

Historical records are confined to the last few millennia. They provide crucial information about events in the peopling of Europe that fall into this time frame, such as the colonization of Iceland ~1,000 YA from Scandinavia and the British Isles, and explanations for some surprises such as the Hungarian language, which results from the Magyar conquest ~1,100 YA. As mentioned at the beginning of this chapter, many of the people now living in Europe, and the languages they speak, have very recent origins and we can use historical information to exclude them when we consider ancient events. In addition, knowledge of the historical extent of diseases such as malaria can be important for understanding modern genetic patterns.

Did Neanderthals contribute to the modern European gene pool?

The discovery of Neanderthal fossils in the nineteenth century revealed the existence of a strikingly different, more robust form of humans, and raised the question of their relationship to modern Europeans. Were they completely separate, did they evolve into modern humans, did they interbreed with them? An understanding of the chronology of the fossil sequence in Israel revealed that anatomically modern humans were present in the warm period around 90–100 KYA (Valladas *et al.*, 1988) but were then replaced by Neanderthals between 50 and 60 KYA (Valladas *et al.*, 1987) when the climate was cooler. An earlier date for anatomically modern humans than Neanderthals excludes the simple possibility that Neanderthals evolved into modern humans. However, the relation-ship and extent of breeding (if any) between the two forms has been controversial. In Europe, both types of hominin were present together for perhaps 10 KY between about 40 and 30 KYA (Figure 13.1), so they could have encountered one another. Indeed, some archaeologists ascribe the development of an Upper Paleolithic culture by Neanderthals (the Châtelperonian) to the influence of the contemporary modern human Upper Paleolithic culture (the Aurignacian). The recent history of meetings between different human populations shows that mating usually occurs whenever there is an opportunity; if this was the case in early modern human-Neanderthal interactions, were offspring produced and if so, did they survive and reproduce, leading to gene flow between the populations? Paleontological evidence can, in principle, address this question. A fossil skeleton of a 4-year-old child from Portugal has been interpreted as showing morphological features of both Neanderthals and modern humans, thus representing a possible hybrid between them (Duarte *et al.*, 1999). This skeleton is dated to 24.5 KYA, but the last known evidence for Neanderthal survival is only ~27 KYA, implying that, if it was indeed a hybrid, a population with intermediate characteristics must have survived for several millennia. The morphological interpretation has, however, been questioned: other paleontologists view the child simply as a chunky modern human (Tattersall and Schwartz, 1999), so the evidence from paleontology is inconclusive. Attention has therefore focused on evidence from ancient DNA from Neanderthal fossils.

Neanderthal mtDNA

DNA was extracted from the Neanderthal type specimen (~40 KY old, known as 'Feldhofer' after the location in Germany where it was found in 1856) and short fragments of the mtDNA control region were amplified (Krings *et al.*, 1997). The enormous technical difficulties associated with authenticating ancient DNA sequences are discussed in Chapter 8. High standards, including replication in an independent laboratory, were used in this work and the results are widely accepted as authentic; compelling evidence comes from the distinct, yet related, nature of the sequences found. Subsequent analyses of Neanderthal fossils from the Caucasus (Mezmaiskaya, ~29 KY old, Ovchinnikov *et al.*, 2000) and Croatia (Vindija, ~42 KY old, Krings *et al.*, 2000) as well as a second individual from Feldhofer (also ~40 KY old, Schmitz *et al.*, 2002) revealed closely related mtDNA sequences. The mean pairwise difference between these four Neanderthal sequences was 1.7%, similar to the average value of $1.8 \pm 0.6\%$ obtained from comparisons between four randomly chosen modern human sequences, and contrasting with the ~6–7% difference between Neanderthals and modern humans (Schmitz *et al.*, 2002).

The known Neanderthal mtDNA sequences are thus distinct from present-day humans, but could some Neanderthals have carried sequences like those of modern humans, or could some anatomically modern humans living at the time of the Neanderthals have had Neanderthal-like sequences that have subsequently been lost by drift? The first question seems difficult to answer because modern human sequences can readily be amplified from almost any bones, whatever the species; such sequences would be considered contaminants, even if they were endogenous. Nevertheless, an attempt to address both questions has been reported recently (Serre *et al.*, 2004): 24 Neanderthal and 40 early modern human fossils were tested for their amino acid preservation, and only those showing the high amino acid concentration and low aspartic acid racemization indicative of good DNA preservation (four Neanderthals and five modern humans) were examined further. All nine fossils yielded modern human sequences, as did cave bear remains tested in parallel, as expected from the contamination usually present in such specimens, and this finding was therefore considered uninformative. However, attempts to amplify Neanderthal-specific fragments were more enlightening: all four Neanderthal, but none of the modern human, fossils produced such fragments. Both the positive and the negative aspects of these findings are significant: all the Neanderthal fossils provided sequences identical (in the short region examined) to the known Neanderthal mtDNAs, implying that Neanderthal mtDNA diversity did not include modern human mtDNA variants; similarly, early modern human fossils expected to allow amplification of Neanderthal mtDNA if they contained it because of their good preservation showed no such product, implying that early modern humans did not carry Neanderthal mtDNAs. Neanderthal and modern human mtDNAs therefore appear to be distinct.

What Neanderthal DNA contribution could be present in modern Europeans?

If we accept that Neanderthal and modern mtDNAs are distinct, does this mean that Neanderthals made no genetic contribution whatever to modern Europeans? This question was considered by Nordborg (1998) when the first Neanderthal sequence became available, and he concluded that the possibility of humans and Neanderthals constituting a single randomly mating population could be excluded. Under the more plausible scenario of two distinct populations merging, the absence of Neanderthal mtDNAs in modern populations was rather uninformative about the possibility of other loci being present since Neanderthal mtDNAs could have been present in the population >30 KYA, but, like most of the mtDNA lineages present then, subsequently lost by drift. Under standard genetic models, present-day mtDNAs would coalesce to around five lineages 25 KYA, so the early modern sequences analysed by Serre *et al.* (2004) almost double the number of early sequences sampled and these authors estimate that if the population size had been constant, the Neanderthal mtDNA contribution could have been as large as 25% and still been lost from present-day humans, although if demographic expansion had started when colonization of Europe began, the maximum possible contribution would be smaller. More extensive modelling of a modern human range expansion into a Europe populated by Neanderthals has led recently to a paper with the strongly worded title 'Modern humans did not admix with Neanderthals during their range expansion into Europe' (Currat and Excoffier, 2004). In the authors' model, even a low degree (0.1%) of Neanderthal admixture into an expanding modern human population would leave a visible signal, and the observed absence of Neanderthal mtDNA is compatible with a maximum of 120 admixture events.

Despite this, other recent work has led to the claim that small segments of Neanderthal autosomal DNA may have been incorporated into the modern genome. A segment of 17q21 containing *MAPT* (Microtubule-Associated Protein Tau) and other genes exists as a 900 kb inversion polymorphism, with the two orientations designated H1 and H2. H2 is common ($\sim$20%) only in Europe, where it appears to confer a selective advantage (Stefansson *et al.*, 2005). H1 and H2 have diverged substantially, with the common ancestor placed at $\sim$3 MYA. Hardy *et al.*, (2005) have argued that the H2 lineage entered the modern human population from Neanderthals, but Stefansson *et al.*, (2005) favour an African origin from a species such as *H. erectus* or *heidelbergensis* on the basis of the greater diversity of the rare African H2 lineages. Thus it remains uncertain whether or not any Neanderthal contribution to modern humans has been found, but it seems clear that such a contribution must have been small.

What were the relative contributions of the Paleolithic and Neolithic immigrants?

If we accept that the contribution of archaic humans to the current European gene pool was small, three early demographic events can be proposed as possible major influences: the entry of anatomically modern humans during the Upper Paleolithic, their retreat to small southern refugia at the time of the Last Glacial Maximum and subsequent re-expansion, and the entry of farmers with the Neolithic transition (Barbujani and Goldstein, 2004). All are likely to have had some influence, and additional events like gene flow from the Near East between the Glacial Maximum and the Neolithic may also have been important. The Neolithic transition, however, has attracted the most attention and debate, and we focus on it here. It is generally accepted that agriculture started in a region of Western Asia called the Fertile Crescent (Pinhasi *et al.*, 2005) and spread from there in several directions, reaching Europe, North Africa and the Indian subcontinent. The spread of languages may have accompanied it (Diamond and Bellwood, 2003; Renfrew, 2003). Two alternative models for the spread of the agriculture have been proposed: the demic diffusion model (Ammerman and Cavalli-Sforza, 1984), according to which the spread of the technology was accompanied by the movement of people, and the cultural diffusion model, in which the spread of agriculture was a purely cultural phenomenon and did not involve the movement of people. Of course, such models are drastic oversimplifications of reality and the 'pure' forms would find little support: the key question is the level of gene flow between the indigenous population and the incoming farmers.

Before plunging into the debate on Europe, it is worth considering what kinds of evidence might be most useful for distinguishing between the extreme models, and what is known about the transition in other parts of the world. Short of a time machine, the most direct evidence could in principle come from ancient DNA: if we could examine a snapshot of the genetic variation throughout Europe every millennium, we should be able to deduce a lot about its spread. Unfortunately, the difficulty of distinguishing between endogenous sequences and contaminants precludes this approach. If we limit ourselves to studies of modern DNA, an alternative would be to compare populations that have recently adopted agriculture with both those that have used it for a long time and those that have never practised it. This approach cannot be taken in Europe because of the omnipresence of agriculture, but has been applied to India. Here, one study found that the mtDNA and Y chromosomes of recent agriculturalists were more similar to those of traditional agriculturalists than to traditional hunter-gatherers (Cordaux et al., 2004); if the classifications used are accepted, this finding would emphasize the importance of demic diffusion. In North Africa, the Neolithic transition has been studied less, but the time depth of Y-chromosomal lineages has been examined: early work favoured a Paleolithic origin for the common lineages (Bosch et al., 2001) but a more recent study by the present authors and others favoured a Neolithic origin (Arredi et al., 2004). The implications of this work for Europe are discussed further below. Thus a number of lines of evidence support a model of the Neolithic transition outside Europe involving substantial demographic change.

The controversy about European Paleolithic and Neolithic contributions

We now consider the Neolithic transition in Europe. The first genetic analyses began when only classical markers were available. The information provided by each marker was small, but ways of combining information from many of them were developed so that the results could be presented in the form of a synthetic map for each principal component (Cavalli-Sforza et al., 1994). The first principal component based on gene frequencies (representing 28% of the variance) showed one pole in the southeast and the other in the northwest, paralleling the archaeological dates of the first appearance of the Neolithic in Europe. Moreover, maps of several individual allele frequencies (including autosomal DNA, mtDNA and Y-chromosomal markers) show clines that are similarly oriented (Sokal et al., 1991; Barbujani et al., 1994; Rosser et al., 2000; Simoni *et al.,* 2000). Genetic clines can arise in several ways, including adaptation to environment (unlikely in this case since the same patterns are observed for independent genetic systems), continuous gene flow between initially differing groups (Figure 13.2), and population expansion into an empty environment or involving little admixture with pre-existing populations (Figure 13.2). Thus, although some authors have attributed the clines observed in Europe to the demic diffusion model, the patterns themselves provide no information about the timescale over which they were established and are compatible with more than one scenario (Barbujani and Bertorelle, 2001). Furthermore, recent simulations that take into account both demography and the geographical structure of Europe suggest that clines are more likely to be observed when markers known in advance to be high-frequency are used, and reinforce the conclusion that clines are compatible with both scenarios (Currat and Excoffier, 2005).

At this point, all may appear straightforward: despite the possibility of alternative interpretations of clines, the available information could be taken as supporting a demic diffusion model.

Time

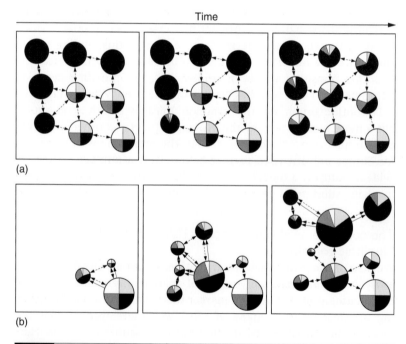

(a)

(b)

Fig 13.2 Two ways of generating a cline. a. Mixing of two populations that differ in allele frequencies (left panel, compare top left and bottom right). Gene flow between the populations (dotted arrows) creates a hybrid zone. b. Successive expansions of limited numbers of people into empty territory (solid arrows). Allele frequencies change because of drift, and one allele can eventually be fixed (right panel, top left); gene flow takes place between populations (dotted arrows). The resulting clines look similar, but make different predictions about the location of the highest diversity associated with the black allele: a, top left or b, bottom right.

This changed with the analysis of mtDNA and the Y chromosome (see discussion in Jobling *et al.*, 2004, section 10.5). Although each system represents a single locus, many lineages can be distinguished and the time to the most recent ancestor (TMRCA) estimated for any set of haplotypes. Most (~80%) mtDNA lineages were suggested to have entered Europe before the Neolithic, with the majority falling into the period between the end of the ice age and the beginning of the Neolithic (Richards *et al.*, 1996; Richards *et al.*, 2000). Extensive datasets have documented Y-chromosomal variation in Europe (Rosser *et al.*, 2000; Semino *et al.*, 2000; Roewer *et al.*, 2005), and one study has been interpreted as indicating a Paleolithic origin for over 70% of the chromosomes (Semino *et al.*, 2000), but the method used for dating lineages was poorly chosen and these calculations need to be re-evaluated. Even if lineages are securely dated, lineage ages cannot be equated with dates of migrations: if humans colonize Mars, their mtDNAs may have a Paleolithic MRCA, but we might be unwise to infer from this a Paleolithic Martian colonization (Barbujani *et al.*, 1998). Alternative interpretations of the same, and additional, datasets have arrived at different conclusions (Chikhi *et al.*, 2002; Dupanloup *et al.*, 2004). Here, the question was treated as one of estimating admixture between source populations: Neolithic (represented by modern Near Easterners), Paleolithic (represented by modern Basques) and, in some analyses, North Africans and North-West Asians. Contributions from North Africa and North-West Asia were low (<2% and <11%, respectively), while that from Near Easterners was high (54% on average, ranging from 22% in England to 96% in Finland). However, this approach suffers from a serious disadvantage: despite the linguistic uniqueness of the Basques, the assumption that they can be used to represent Paleolithic gene frequencies, albeit with allowance for drift, seems to us to have no sound basis (see also Jobling *et al.*, 2004, Box 10.3). If so, the conclusions of this analysis are uninformative about Paleolithic versus Neolithic origins, but it does illustrate the model-dependent nature of such conclusions.

A comparison of Southern Europe and North Africa

From the above results, it can be seen that the extent to which the European genetic landscape has been influenced by Neolithic demic diffusion remains unclear. In the rest of this section, we depart from the reviewers' usual practice of attempting to provide a balanced overview of the field to explore a possibility suggested by the parallels between Y-chromosomal variation in Southern Europe and North Africa (Arredi *et al.*, 2004). The similarities, both qualitative and quantitative, are very striking. Both regions show a predominant pattern of differentiation oriented east-west (Figure 13.3), with rather similar levels of genetic structure (Table 13.1). In each region, the pattern of variation is clinal (Figure 13.3a,b) and extends into the Middle East (Figure 13.3c,d), and the genetic diversity

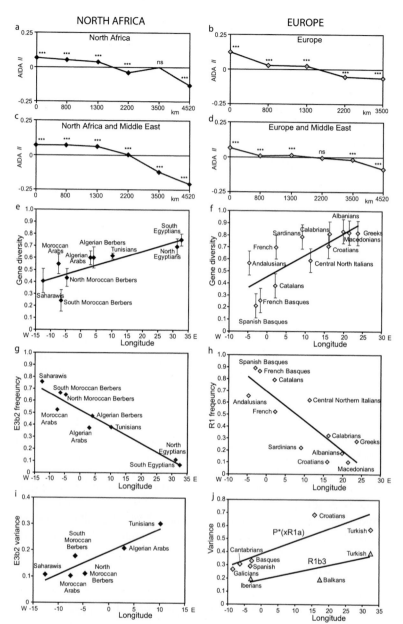

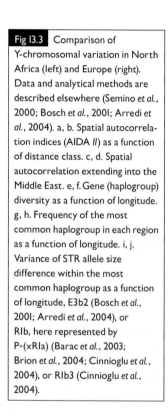

Fig 13.3 Comparison of Y-chromosomal variation in North Africa (left) and Europe (right). Data and analytical methods are described elsewhere (Semino et al., 2000; Bosch et al., 2001; Arredi et al., 2004). a, b. Spatial autocorrelation indices (AIDA II) as a function of distance class. c, d. Spatial autocorrelation extending into the Middle East. e, f. Gene (haplogroup) diversity as a function of longitude. g, h. Frequency of the most common haplogroup in each region as a function of longitude. i, j. Variance of STR allele size difference within the most common haplogroup as a function of longitude, E3b2 (Bosch et al., 2001; Arredi et al., 2004), or R1b, here represented by P*(xR1a) (Barac et al., 2003; Brion et al., 2004; Cinnioglu et al., 2004), or R1b3 (Cinnioglu et al., 2004).

within populations decreases towards the west (Figure 13.3e,f), largely because a single haplogroup predominates. In Europe this is R1b, while in North Africa it is E3b2, and these both become more frequent towards the west (Figure 13.3g, h). In addition, genetic distances between the two regions increase from east to west, so that genetic differentiation develops in two opposite directions: eastern samples from both regions have small genetic distances from the Middle East, whereas the largest differentiation is observed between the western parts of North Africa (Morocco) and Southern Europe (Spain) (Figure 13.4).

Table 13.1. | Comparison of Y-chromosomal variation in North Africa and Southern Europe.

	North Africa	Southern Europe
East-West clinal pattern	Yes	Yes
Predominant Y-chromosomal haplogroup	E3b2 (42%)	RIb (40%)
Proportion of genetic variation due to differentiation among populations (Φ_{ST})	9.9%, $P<0.0001$	13.1%, $P<0.0001$
Correlation coefficient between genetic and geographic distances (r)	0.46, $P<0.01$	0.37, $P<0.01$

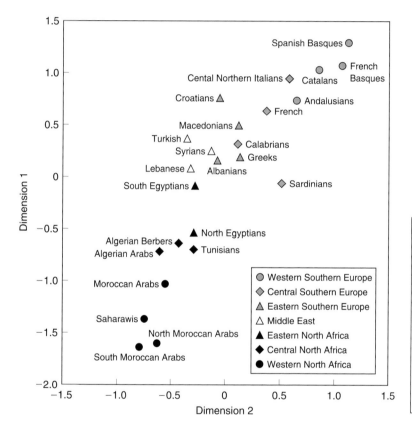

Fig 13.4 Y-chromosomal genetic distances between North African, Middle Eastern and Southern European populations. Population pairwise genetic distances were calculated as described previously (Arredi et al., 2004) from published data on haplogroup frequencies (Semino et al., 2000; Bosch et al., 2001; Arredi et al., 2004) and are presented as a multidimensional scaling plot.

How could these parallels arise? We can suggest three possible answers: (1) chance (convergent evolution), (2) common processes in the Paleolithic, or (3) common processes in the Neolithic. The correspondences seem too great to be explained by convergence. Little is known about the initial settlement of North Africa by modern humans, but, as in Europe, the genetic patterns established by the first colonists are likely to have been greatly modified by climatic changes at the time of the Last Glacial Maximum, and there is no evidence that subsequent re-expansions followed similar paths. In contrast, a Neolithic expansion originating in the Fertile

Crescent, and split into two branches by the geographical barrier of the Mediterranean Sea, would explain all of the parallels in a parsimonious fashion.

A Holocene origin for the North African Y-chromosomal variation was favoured because of the pattern of STR variation within the commonest haplogroups: TMRCAs were recent and variation in E3b2 decreased towards the west, where the frequency was highest. In Europe, a recent TMRCA and lower STR variation in the west would similarly be expected if R1b were of Neolithic origin, while expansion from an Iberian refuge after the Last Glacial Maximum would be an earlier event that should result in higher R1b diversity in the west. We have therefore examined published data on the variation of this haplogroup in order to try to distinguish between these two possibilities. No ideal dataset is available, but variances could be calculated for a major subset of the R1b chromosomes, R1b3 — 92% of R1b in Turkey — or for P*(xR1a), which includes R1b as its major component — 83% in Turkey; see Figure 13.5 for phylogenetic details. Variances of both sets of chromosomes decrease towards the west (Figure 13.3i,j). In addition, TMRCAs are recent: point estimates are mostly ~10–15 KYA for haplogroup P, ~9–13 KYA for R1, and ~5–8 KYA for R1b3 (Table 13.2). These findings must be viewed with caution: additional data are required, and TMRCA estimates are highly dependent on the mutation rate used (Ho and Larson, 2006) and the other assumptions made, including the effective population size (Table 13.2), but they appear to support a Neolithic timing for the expansion of R1b3 in Europe rather than a Paleolithic one. Since R1b is the most common haplogroup in Europe (Rosser et al., 2000;

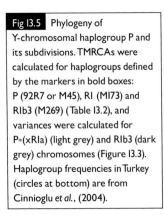

Fig 13.5 Phylogeny of Y-chromosomal haplogroup P and its subdivisions. TMRCAs were calculated for haplogroups defined by the markers in bold boxes: P (92R7 or M45), R1 (M173) and R1b3 (M269) (Table 13.2), and variances were calculated for P*(xR1a) (light grey) and R1b3 (dark grey) chromosomes (Figure 13.3). Haplogroup frequencies in Turkey (circles at bottom) are from Cinnioglu et al., (2004).

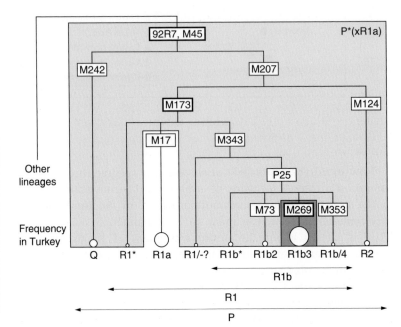

Table 13.2. Estimates of TMRCAs (Times to the Most Recent Common Ancestor) for the commonest North African and Southern European Y-chromosomal lineages.

Region	Lineage	TMRCA (95% CI) (KY)		Data source
		Nprior gamma (1,0.001)	Nprior gamma (3,0.003)	
Kayser-Dupuy observed mutation rate				
North Africa	E3b	8.3 (5.2−12.4)		(Bosch et al., 2001; Arredi et al., 2004)
	E3b2	4.2 (2.8−6.0)		(Bosch et al., 2001; Arredi et al., 2004)
Turkey	P	10.2 (7.5−13.9)	10.0 (7.1−14.1)	(Cinnioglu et al., 2004)
	RI	9.2 (6.9−12.4)	9.1 (6.6−12.5)	(Cinnioglu et al., 2004)
	RIb3	5.4 (4.0−7.7)	5.6 (4.0−7.9)	(Cinnioglu et al., 2004)
Iberia	P	10.7 (7.3−19.1)	7.3 (4.9−11.0)	(Brion et al., 2004)
Zhivotovsky 'effective' mutation rate				
North Africa	E3b	14.3 (9.3−19.2)		(Bosch et al., 2001; Arredi et al., 2004)
	E3b2	6.9 (5.9−8.2)		(Bosch et al., 2001; Arredi et al., 2004)
Turkey	P	12.7 (8.4−19.2)	15.0 (10.3−21.8)	(Cinnioglu et al., 2004)
	RI	11.7 (8.0−17.2)	13.4 (9.5−19.8)	(Cinnioglu et al., 2004)
	RIb3	7.0 (4.9−11.1)	8.2 (5.8−11.6)	(Cinnioglu et al., 2004)
Iberia	P	15.7 (8.7−29.5)	15.9 (9.3−26.2)	(Brion et al., 2004)

North African TMRCAs are from Arredi et al., (2004). Other TMRCAs were calculated with BATWING (Wilson and Balding, 1998) using mutation rate estimates derived either from the father-son measurements of Kayser et al., (2000) and Dupuy et al., (2004) (upper section), or the lower 'effective' rate calculated by Zhivotovsky et al., (2004) (lower section); other parameters were from Arredi et al., (2004), except that an alternative value for the effective population size (Nprior) was also used (Macpherson et al., 2004, gamma (3,0.003)).

Semino et al., 2000), and is proposed as the earliest entrant (Semino et al., 2000), a Neolithic or later origin for most Y-chromosomal diversity seems likely, including that in populations considered 'Paleolithic', such as the Basques. These conclusions apply only to the Y chromosome − different loci, of course, have different histories − and have only considered Southern Europe. More complicated explanations are possible: for example, the available time estimates cannot distinguish an early Neolithic expansion from a late Mesolithic one; and any search for a single major explanation for a genetic pattern in Europe may be too simple. Nevertheless, the hypothesis presented here of such a major Neolithic influence on Y diversity extends the range of models that should be considered for the peopling of Europe.

In conclusion, a wide range of opposing models can thus still be used to interpret the demographic impact of the Neolithic transition in Europe. More data from well-studied loci like the

Y chromosome and independent genetic systems are needed: some, like mitochondrial DNA, show a similar geographical pattern (Capelli *et al.*, 2006), while others such as the HLA class II loci DRB1 and DQB1 show different patterns (Abdennaji Guenounou *et al.*, 2006). It would also be particularly useful to have a better theoretical understanding of the genetic patterns that could be expected under different, and more realistic, evolutionary models: an area that is advancing rapidly (Currat and Excoffier 2005; Klopfstein *et al.*, 2006).

To what extent has selection shaped the European gene pool?

The previous sections have assumed, explicitly or implicitly, that the markers studied were neutral (or nearly so) so that their patterns of variation could be explained by processes such as drift and migration. There is debate about the extent to which non-neutral processes such as natural and sexual selection have influenced patterns of human genetic diversity. This section discusses examples of how selection seems to have influenced the European gene pool.

Pigmentation

One of the most obvious forms of phenotypic variation between humans is in their skin colour. Europe stands out on maps showing the worldwide distribution of skin colour because of the light shade (Barsh, 2003). Although it seems widely accepted this is due to selection, there has been extensive debate about whether natural selection or sexual selection has been the dominant force (Aoki, 2002). In Africa, dark skin is advantageous because it protects against sunburn and folate degradation. It is near-universal among modern Africans, although there is variation between sub-Saharan and North Africa, and it is often assumed that the people who left Africa ~50,000 years ago would also have been dark-skinned. If so, why did those who reached Europe develop light skin? The argument from natural selection proposes that when there is less sunlight, people with dark skin will synthesize insufficient vitamin D and develop rickets, a disease in which the bones become soft and deformed, leading to decreased fertility and survival. Although rickets has indeed been reported in modern dark-skinned children living in Europe, this may be a product of an indoor lifestyle; hunter-gatherers or farmers spending more time exposed to the sun should have synthesized sufficient vitamin D irrespective of their skin pigmentation (Aoki, 2002). The alternative explanation for light skin is that it results from sexual selection, a view that can be traced back to Darwin's *The Descent of Man*. If light-skinned mates were preferred, this phenotype would have increased in frequency and could have come to predominate in the population.

Little is known about the genetics of such complex phenotypes, but one gene related to normal variation in skin and hair pigmentation has been studied extensively: the melanocortin 1 receptor, *MC1R*. Variation in this gene appears to have a small but noticeable effect on skin colour and ultraviolet sensitivity, and a more obvious one on hair colour (Rees, 2003). In Africans, the amino acid sequence is strongly conserved: one survey of 106 African chromosomes detected only synonymous variants (i.e. ones that did not change the amino acid sequence), but in 356 European chromosomes, nine non-synonymous variants were found, together making up almost half of the total: a striking contrast to the usual finding of the greatest genetic diversity in Africa (Harding *et al.*, 2000). Three amino acid changes are strongly associated with pale skin and red hair, with most such individuals being homozygotes or compound heterozygotes for these variants (Rees, 2003), a frequency high enough to interest forensic scientists wishing to determine the phenotype of the donor of a DNA sample of unknown origin. The enhanced diversity in Europe fitted a model of neutral evolution, explained by loss of functional constraint 20–40 KYA, and no evidence for positive selection could be found in these studies (Harding *et al.*, 2000; Rees, 2003). Despite the lack of evidence for selection at MC1R, whether natural or sexual, it is possible that larger datasets or more sensitive tests might detect it; alternatively, selection may have acted primarily on other genes influencing the same phenotypes. Indeed, recent findings show evidence for positive selection at the *SLC24A5* and *AIM1* (=*MATP*) genes in Europeans (Lamason *et al.*, 2005; Soejima *et al.*, 2006).

Infectious diseases

In some countries at the present time, one in five children die before their fifth birthday and most of these deaths are due to infectious disease (World Health Organization Report on Infectious Diseases, http://www.who.int/infectious-disease-report/pages/textonly.html). If the figures in prehistoric Europe were similar, infectious diseases would have been powerful selective agents that could have influenced the gene pool. The best-studied example is malaria, a disease of warm climates that was endemic in much of Europe within historical times, and still accounts for 0.7–2.7 million deaths/year worldwide.

Partial resistance to malaria can be conferred by a number of genetic mechanisms involving, among others, the globin genes and glucose-6-phosphate dehydrogenase (*G6PD*). Evidence for the protective role of G6PD deficiency variants comes from their geographical distribution (which matches the historical extent of malaria), epidemiological and biochemical studies. Protection against malaria is associated with a corresponding disadvantage: acute haemolytic anaemia after infection by other agents or after exposure to foods such as broad beans ('favism'). One variant designated 'Med' (Ser188Phe) is found in the Mediterranean countries, the Middle East and Asia and may reach a frequency of 2–10%; other protective variants predominate in other regions, such as A- in Africa. Med chromosomes showed greatly reduced variability at closely-related microsatellites, suggesting that this allele (and adjacent sequences) arose or expanded recently, 3.3 (95% CI: 1.6–6.6) KYA according to one calculation (Tishkoff et al., 2001). Eight Med chromosomes from Calabria (Italy) were associated with a different haplotype that included red-green colour blindness due to a mutation 300 kb away, indicating a rare recombination event or local second origin for the Med mutation. G6PD thus provides an example of a rapid change to the European genome in response to selective pressure.

Diet

The European diet is always likely to have differed from that of our African ancestors because of the different flora and fauna of Europe, and it changed substantially after the transition to a Neolithic economy. The clearest example of the way that diet has influenced the European genome comes from studies of the lactase gene LCT. Milk only became available to adults as a food source after animals were domesticated and contains the disaccharide lactose, which must be digested to galactose and glucose in order to be utilized. All babies produce the enzyme lactase which catalyzes this reaction, but, on a global scale, most adults do not and thus cannot use fresh milk as a food source. In some parts of the world, however, lactase activity persists throughout adult life. This 'lactase persistence' is found at highest frequency in Europe, especially in Swedes and Danes (Figure 13.6) and is also prevalent in northern India and some Arabian and African populations (Swallow, 2003). Lactase persistence tends to be more frequent among milk-drinking populations (e.g. Tutsis) than in neighbouring populations where this is uncommon (e.g. Hutus). Persistence is inherited as a simple dominant trait and an analysis of its segregation in Finnish families identified a C/T variant ~14 kb upstream of the lactase gene (lying within an intron of an unrelated gene, MCM6) associated with the condition: all non-persistent individuals examined carried the CC genotype, while all persistent individuals had CT or TT genotypes (Enattah et al., 2002). Interestingly, although the T allele is present in India (Swallow, 2003) and North Africa (Myles et al., 2005) at a frequency that could account for the persistence there, it cannot account for persistence in most sub-Saharan African populations (Mulcare et al., 2004), suggesting that lactase persistence has more than one origin.

If the hypothesis that the T allele has risen in frequency as an adaptation to post-Neolithic milk-drinking in Europe is correct, such a recent episode of positive selection should have left a clear

(a)

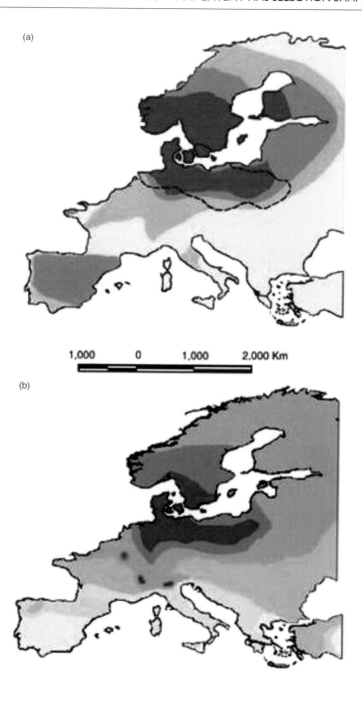

1,000   0   1,000   2,000 Km

(b)

Fig 13.6 a. Frequency of lactase persistence in European populations. Dark shading = high frequency; the dotted line indicates the extent of early Neolithic cattle-raising as judged by the archaeological remains of the Funnel Beaker Culture (~6,000–4,800 years ago). b. Genetic diversity of milk protein genes, presented as a synthetic map of the first principal component of the variation. From Beja-Pereira et al., (2003), with the authors' permission.

signature in the surrounding sequences. Relative extended haplotype homozygosity (REHH) has been used to test for such an effect (Bersaglieri et al., 2004). As an allele (or tightly linked set of alleles; a 'core haplotype') increases in frequency by genetic drift, it carries adjacent sequences with it, but recombination events soon whittle down the length of these. If, however, it increases more rapidly than expected by drift because of positive selection, it will be embedded in a longer haplotype than expected. The REHH measure shows that the short core haplotype containing the T allele is indeed commonly embedded in a stretch of homozygosity >800 kb in length, far longer than the stretches of homozygosity associated with non-persistence haplotypes or other regions of the genome. Estimates of the time at which the rapid rise began suggested 2,200–21,000 years ago for a European-American sample and 1,600–3,200 years ago for a Scandinavian one. These estimates require many assumptions and are therefore not very reliable, but would imply that the selective event was recent. The selective advantage of the persistence allele was estimated at 0.014–0.15 (European-Americans) or 0.09–0.19 (Scandinavians), comparable to that for G6PD deficiency (0.02–0.05) or sickle-cell trait (0.05–0.18) in regions where malaria is endemic. Thus diet appears to have exerted very strong local selection pressure.

There are substantial geographical differences in genes involved in drug response, a topic of considerable medical importance. For example, the major drug-metabolizing gene *CYP2D6* is highly polymorphic and its alleles are classified as functional, showing reduced function, or non-functional. Europeans show markedly higher levels of functional and non-functional variants, and fewer reduced ones (Bradford, 2002). These enzymes did not evolve to confound pharmacologists: the variation is likely to be the result of adaptations to diet, disease or other local environmental factors, and an understanding of these factors could provide additional insights into the peopling of Europe.

What insights can the genetics of associated species provide?

Humans share their environment with other organisms, so many species can provide useful insights into events like the ice ages that have far-ranging consequences. For example, during the Last Glacial Maximum, temperate species were confined to refugia in Iberia, Italy and the Balkans and their routes of postglacial spread can be traced (Taberlet *et al.*, 1998). Humans shared the same refugia and their expansions would have been heavily influenced by animal and plant availability. Some of the most important insights, however, come from the organisms most closely associated with humans: domesticates, pests and diseases. We will consider here two examples that provide information particularly relevant to Europe.

Helicobacter pylori
H. pylori is a bacterium that lives in the stomach and duodenum, where it can cause ulcers, and is usually acquired from the mother or another family member. It therefore provides an independent genome whose evolution should track that of humans, but has about

50-fold greater diversity. The most extensive study so far analysed eight regions in 370 strains, clustering them on the basis of the genetic information with the program STRUCTURE and then comparing the resulting genetic clusters with the known origins of the individuals (Falush *et al.*, 2003). Five ancestral *H. pylori* populations were identified: two in Africa, one in East Asia and two in Europe (AE1 and AE2), and the proportion of each of these in modern human populations was determined. Both AE1 and AE2 were present in all the European populations examined, but they were distributed differently. AE1 was most frequent in Finland and Estonia, while AE2 was most frequent in Spain, Israel and Sudan, suggesting different sources. The authors compared the *H. pylori* results with the human classical marker patterns (Cavalli-Sforza *et al.*, 1994) and associated AE2 with the migration of Neolithic farmers and AE1 with that of Uralic speakers. While these associations are possible, it would be more interesting to see additional populations sampled and an analysis of the data that tested, rather than assumed, particular hypotheses about human movements.

Cattle milk protein genes

Cattle were domesticated in the Fertile Crescent around 10,000 years ago and show their highest genetic diversity there with most markers, apparently spreading to other parts of Europe with farming (Bruford *et al.*, 2003). A large-scale study concentrating on non-synonymous variants in six milk protein genes in ~20,000 cattle, however, revealed a different pattern (Beja-Pereira *et al.*, 2003). With these loci, the highest diversity was found in North Central Europe (Figure 13.6). It therefore appears that artificial selection for milk yield and altered milk protein composition occurred in this region, and the early Neolithic people identified as the Funnel Beaker Culture, who lived here ~6,000–4,800 years ago, could have been responsible. The very striking correspondence between the archaeological remains, cattle milk protein diversity and human lactase persistence provides a good example of gene-culture coevolution.

The genetics of associated species has been studied in much less detail than that of humans. It provides one of the most promising areas for future work.

Concluding remarks

There is now general agreement about some features of the peopling of Europe, but not about others. Archaic humans contributed little or nothing to the modern gene pool but opinions differ sharply about the relative contributions of the Upper Paleolithic and Neolithic immigrants. Genome-wide patterns were established by a combination of migration, admixture and drift; superimposed on these, natural and sexual selection for phenotypes relevant to local survival and reproduction in Europe have had detectable effects on individual genes within the European gene pool.

How is the field likely to develop in the future? Technologies for genotyping and re-sequencing are improving rapidly, driven by the medical interest in identifying susceptibility genes for complex disorders. The Genographic Project aims to collect 100,000 samples from indigenous populations worldwide for anthropological genetic research and the HapMap project (The International HapMap Consortium, 2003) will reveal how haplotype structures in Europe compare with those in Africa and Asia, and is already identifying long regions of low recombination on autosomes (McVean *et al.*, 2004) which can be used for phylogeographic studies like mtDNA and the Y chromosome (Jobling and Tyler-Smith, 2003, Box 6). Genome scans for regions undergoing positive or balancing selection are beginning to reveal the extent of these forces (Voight *et al.*, 2006). Such an abundance of data will stimulate new theoretical approaches leading to better models and predictions. As re-sequencing costs approach the goal of $1,000/whole human genome, we will have an unprecedented wealth of genetic data and optimists can hope that a convincing consensus on the peopling of Europe will emerge.

Acknowledgements

We thank María Brion and Toomas Kivisild for providing data on Y-chromosomal variation in European populations, and both Toomas Kivisild and Martin Richards for particularly helpful comments on the manuscript. CTS is supported by The Wellcome Trust. BA was supported by a Ph.D. fellowship from the Università Cattolica di Roma. First draft: September 2004, updated August 2005, final additions July 2006.

References

Abdennaji Guenounou, B., Loueslati, B. Y., Buhler, S., Hmida, S., Ennafaa, H., Khodjet-Elkhil, H., Moojat, N., Dridi, A., Boukef, K., Ben Ammar Elgaaied, A. and Sanchez-Mazas, A. (2006). HLA class II genetic diversity in southern Tunisia and the Mediterranean area. *Int J Immunogenet*, **33**, 93–103.

Ammerman, A. J. and Cavalli-Sforza, L. L. (1984). *The Neolithic Transition and the Genetics of Populations in Europe*. Princeton, NJ: Princeton University Press

Aoki, K. (2002). Sexual selection as a cause of human skin colour variation: Darwin's hypothesis revisited. *Ann Hum Biol*, **29**, 589–608.

Arredi, B., Poloni, E. S., Paracchini, S., Zerjal, T., Fathallah, D. M., Makrelouf, M., Pascali, V. L., Novelletto, A. and Tyler-Smith, C. (2004). A predominantly Neolithic origin for Y-chromosomal DNA variation in North Africa. *Am J Hum Genet*, **75**, 338–45.

Barac, L., Pericic, M., Klaric, I. M., Rootsi, S., Janicijevic, B., Kivisild, T., Parik, J., Rudan, I., Villems, R. and Rudan, P. (2003). Y chromosomal heritage of Croatian population and its island isolates. *Eur J Hum Genet*, **11**, 535–42.

Barbujani, G. and Bertorelle, G. (2001). Genetics and the population history of Europe. *Proc Natl Acad Sci USA*, **98**, 22–5.

Barbujani, G., Bertorelle, G. and Chikhi, L. (1998). Evidence for Paleolithic and Neolithic gene flow in Europe. *Am J Hum Genet*, **62**, 488–92.

Barbujani, G. and Goldstein, D. B. (2004). Africans and Asians abroad: genetic diversity in Europe. *Annu Rev Genomics Hum Genet*, **5**, 119–50.

Barbujani, G., Pilastro, A., De Domenico, S. and Renfrew, C. (1994). Genetic variation in North Africa and Eurasia: Neolithic demic diffusion vs. Paleolithic colonisation. *Am J Phys Anthropol*, **95**, 137–54.

Barsh, G. S. (2003). What controls variation in human skin color? *PLoS Biol*, **1**, E27.

Beja-Pereira, A., Luikart, G., England, P. R., Bradley, D. G., Jann, O. C., Bertorelle, G., Chamberlain, A. T., Nunes, T. P., Metodiev, S., Ferrand, N. and Erhardt, G. (2003). Gene-culture coevolution between cattle milk protein genes and human lactase genes. *Nat Genet*, **35**, 311–13.

Bermudez de Castro, J. M., Arsuaga, J. L., Carbonell, E., Rosas, A., Martinez, I. and Mosquera, M. (1997). A hominid from the lower Pleistocene of Atapuerca, Spain: possible ancestor to Neandertals and modern humans. *Science*, **276**, 1392–5.

Bersaglieri, T., Sabeti, P. C., Patterson, N., Vanderploeg, T., Schaffner, S. F., Drake, J. A., Rhodes, M., Reich, D. E. and Hirschhorn, J. N. (2004). Genetic signatures of strong recent positive selection at the lactase gene. *Am J Hum Genet*, **74**, 1111–20.

Bosch, E., Calafell, F., Comas, D., Oefner, P. J., Underhill, P. A. and Bertranpetit, J. (2001). High-resolution analysis of human Y-chromosome variation shows a sharp discontinuity and limited gene flow between northwestern Africa and the Iberian Peninsula. *Am J Hum Genet*, **68**, 1019–29.

Bradford, L. D. (2002). CYP2D6 allele frequency in European Caucasians, Asians, Africans and their descendants. *Pharmacogenomics*, **3**, 229–43.

Brion, M., Quintans, B., Zarrabeitia, M., Gonzalez-Neira, A., Salas, A., Lareu, V., Tyler-Smith, C. and Carracedo, A. (2004). Micro-geographical differentiation in Northern Iberia revealed by Y-chromosomal DNA analysis. *Gene*, **329**, 17–25.

Bruford, M. W., Bradley, D. G. and Luikart, G. (2003). DNA markers reveal the complexity of livestock domestication. *Nat Rev Genet*, **4**, 900–10.

Capelli, C., Redhead, N., Romano, V., Cali, F., Lefranc, G., Delague, V., Megarbane, A., Felice, A. E., Pascali, V. L., Neophytou, P. I., Poulli, Z., Novelletto, A., Malaspina, P., Terrenato, L., Berebbi, A., Fellous, M., Thomas, M. G. and Goldstein, D. B. (2006). Population structure in the Mediterranean basin: a Y chromosome perspective. *Ann Hum Genet*, **70**, 207–25.

Cavalli-Sforza, L. L., Menozzi, P. and Piazza, A. (1994). *The History and Geography of Human Genes*. Princeton, NJ: Princeton University Press.

Chikhi, L., Nichols, R. A., Barbujani, G. and Beaumont, M. A. (2002). Y genetic data support the Neolithic demic diffusion model. *Proc Natl Acad Sci USA*, **99**, 11008–13.

Cinnioglu, C., King, R., Kivisild, T., Kalfoglu, E., Atasoy, S., Cavalleri, G. L., Lillie, A. S., Roseman, C. C., Lin, A. A., Prince, K., Oefner, P. J., Shen, P., Semino, O., Cavalli-Sforza, L. L. and Underhill, P. A. (2004). Excavating Y-chromosome haplotype strata in Anatolia. *Hum Genet*, **114**, 127–48.

Cordaux, R., Deepa, E., Vishwanathan, H. and Stoneking, M. (2004). Genetic evidence for the demic diffusion of agriculture to India. *Science*, **304**, 1125.

Currat, M. and Excoffier, L. (2004). Modern humans did not admix with Neanderthals during their range expansion into Europe. *PLoS Biol*, **2**, e421.

Currat, M. and Excoffier, L. (2005). The effect of the Neolithic expansion on European molecular diversity. *Proc R Soc B*, **272**, 679–88.

Diamond, J. and Bellwood, P. (2003). Farmers and their languages: the first expansions. *Science*, **300**, 597–603.

Duarte, C., Maurício, J., Pettitt, P. B., Souto, P., Trinkaus, E., van der Plicht, H. and Zilhão, J. (1999). The early Upper Paleolithic human skeleton from the Abrigo do Lagar Velho (Portugal) and modern human emergence in Iberia. *Proc Natl Acad Sci USA*, **96**, 7604–9.

Dupanloup, I., Bertorelle, G., Chikhi, L. and Barbujani, G. (2004). Estimating the impact of prehistoric admixture on the genome of Europeans. *Mol Biol Evol*, **21**, 1361–72.

Dupuy, B. M., Stenersen, M., Egeland, T. and Olaisen, B. (2004). Y-chromosomal microsatellite mutation rates: differences in mutation rate between and within loci. *Hum Mutat*, **23**, 117–24.

Enattah, N. S., Sahi, T., Savilahti, E., Terwilliger, J. D., Peltonen, L. and Jarvela, I. (2002). Identification of a variant associated with adult-type hypolactasia. *Nat Genet*, **30**, 233–7.

Falush, D., Wirth, T., Linz, B., Pritchard, J. K., Stephens, M., Kidd, M., Blaser, M. J., Graham, D. Y., Vacher, S., Perez-Perez, G. I., Yamaoka, Y., Megraud, F., Otto, K., Reichard, U., Katzowitsch, E., Wang, X., Achtman, M. and Suerbaum, S. (2003). Traces of human migrations in *Helicobacter pylori* populations. *Science*, **299**, 1582–5.

Forster, P. (2004). Ice Ages and the mitochondrial DNA chronology of human dispersals: a review. *Philos Trans R Soc Lond B Biol Sci*, **359**, 255–64.

Gray, R. D. and Atkinson, Q. D. (2003). Language-tree divergence times support the Anatolian theory of Indo-European origin. *Nature*, **426**, 435–9.

Harding, R. M., Healy, E., Ray, A. J., Ellis, N. S., Flanagan, N., Todd, C., Dixon, C., Sajantila, A., Jackson, I. J., Birch-Machin, M. A. and Rees, J. L. (2000). Evidence for variable selective pressures at MC1R. *Am J Hum Genet*, **66**, 1351–61.

Hardy, J., Pittman, A., Myers, A., Gwinn-Hardy, K., Fung, H. C., de Silva, R., Hutton, M. and Duckworth, J. (2005). Evidence suggesting that *Homo neanderthalensis* contributed the H2 *MAPT* haplotype to *Homo sapiens*. *Biochem Soc Trans*, **33**, 582–5.

Ho, S. Y. and Larson, G. (2006). Molecular clocks: when times are a-changin'. *Trends Genet*, **22**, 79–83.

Jobling, M. A., Hurles, M. E. and Tyler-Smith, C. (2004). *Human Evolutionary Genetics*. New York and Abingdon: Garland Science.

Jobling, M. A. and Tyler-Smith, C. (2003). The human Y chromosome: an evolutionary marker comes of age. *Nat Rev Genet*, **4**, 598–612.

Kayser, M., Roewer, L., Hedman, M., Henke, L., Henke, J., Brauer, S., Kruger, C., Krawczak, M., Nagy, M., Dobosz, T., Szibor, R., de Knijff, P., Stoneking, M. and Sajantila, A. (2000). Characteristics and frequency of germline mutations at microsatellite loci from the human Y chromosome, as revealed by direct observation in father/son pairs. *Am J Hum Genet*, **66**, 1580–8.

Klopfstein, S., Currat, M. and Excoffier, L. (2006). The fate of mutations surfing on the wave of a range expansion. *Mol Biol Evol*, **23**, 482–90.

Krings, M., Capelli, C., Tschentscher, F., Geisert, H., Meyer, S., von Haeseler, A., Grossschmidt, K., Possnert, G., Paunovic, M. and Pääbo, S. (2000). A view of Neandertal genetic diversity. *Nat Genet*, **26**, 144–6.

Krings, M., Stone, A., Schmitz, R.W., Krainitzki, H., Stoneking, M. and Pääbo, S. (1997). Neandertal DNA sequences and the origin of modern humans. *Cell*, **90**, 19–30.

Lamason, R. L., Mohideen, M. A., Mest, J. R., Wong, A. C., Norton, H. L., Aros, M. C., Jurynec, M. J., Mao, X., Humphreville, V. R., Humbert, J. E., Sinha, S., Moore, J. L., Jagadeeswaran, P., Zhao, W., Ning, G., Makalowska, I., McKeigue, P. M., O'Donnell, D., Kittles, R., Parra, E. J., Mangini, N. J., Grunwald, D. J., Shriver, M. D., Canfield, V. A. and Cheng, K. C. (2005). SLC24A5, a putative cation exchanger, affects pigmentation in zebrafish and humans. *Science*, **310**, 1782–6.

Macpherson, J.M., Ramachandran, S., Diamond, L. and Feldman, M.W. (2004). Demographic estimates from Y chromosome microsatellite polymorphisms: analysis of a worldwide sample. *Hum Genomics*, **1**, 345–54.

McVean, G.A., Myers, S.R., Hunt, S., Deloukas, P., Bentley, D.R. and Donnelly, P. (2004). The fine-scale structure of recombination rate variation in the human genome. *Science*, **304**, 581–4.

Mellars, P. (2006). A new radiocarbon revolution and the dispersal of modern humans in Eurasia. *Nature*, **439**, 931–5.

Mulcare, C.A., Weale, M.E., Jones, A.L., Connell, B., Zeitlyn, D., Tarekegn, A., Swallow, D.M., Bradman, N. and Thomas, M.G. (2004). The T allele of a single-nucleotide polymorphism 13.9 kb upstream of the lactase gene *(LCT)* *(C-13.9kbT)* does not predict or cause the lactase-persistence phenotype in Africans. *Am J Hum Genet*, **74**, 1102–10.

Myles, S., Bouzekri, N., Haverfield, E., Cherkaoui, M., Dugoujon, J.M. and Ward, R. (2005). Genetic evidence in support of a shared Eurasian-North African dairying origin. *Hum Genet*, **117**, 34–42.

Nordborg, M. (1998). On the probability of Neanderthal ancestry. *Am J Hum Genet*, **63**, 1237–40.

Ovchinnikov, I.V., Götherström, A., Romanova, G.P., Kharitonov, V.M., Liden, K. and Goodwin, W. (2000). Molecular analysis of Neanderthal DNA from the northern Caucasus. *Nature*, **404**, 490–3.

Petit, J.R., Jouzel, J., Raynaud, D., Barkov, N.I., Barnola, J.-M., Basile I., Bender, M., Chappellaz, J., Davis, M., Delayque, G., M. Delmotte, M., Kotlyakov, V.M., Legrand, M., Lipenkov, V.Y., Lorius, C., Pepin, L., Ritz, C., Saltzman, E. and Stievenard, M. (1999). Climate and atmospheric history of the past 420,000 years from the Vostok ice core, Antarctica. *Nature*, **399**, 429–36.

Pinhasi, R., Fort, J. and Ammerman, A.J. (2005). Tracing the origin and spread of agriculture in Europe. *PLoS Biol*, **3**, e410.

Pritchard, J.K., Stephens, M. and Donnelly, P. (2000). Inference of population structure using multilocus genotype data. *Genetics*, **155**, 945–59.

Rees, J.L. (2003). Genetics of hair and skin color. *Annu Rev Genet*, **37**, 67–90.

Reich, D.E., Cargill, M., Bolk, S., Ireland, J., Sabeti, P.C., Richter, D.J., Lavery, T., Kouyoumjian, R., Farhadian, S.F., Ward, R. and Lander, E.S. (2001). Linkage disequilibrium in the human genome. *Nature*, **411**, 199–204.

Renfrew, C. (2003). 'The emerging synthesis': the archaeogenetics of farming/language dispersals and other spread zones. In *Examining the Farming/Language Dispersal Hypothesis*, eds. P. Bellwood and C. Renfrew. McDonald Institute for Archaeological Research, Cambridge, UK, pp. 3–16.

Richards, M., Corte-Real, H., Forster, P., Macaulay, V., Wilkinson-Herbots, H., Demaine, A., Papiha, S., Hedges, R., Bandelt, H.J. and Sykes, B. (1996). Paleolithic and Neolithic lineages in the European mitochondrial gene pool. *Am J Hum Genet*, **59**, 185–203.

Richards, M., Macaulay, V., Hickey, E., Vega, E., Sykes, B., Guida, V., Rengo, C., Sellitto, D., Cruciani, F., Kivisild, T., Villems, R., Thomas, M., Rychkov, S., Rychkov, O., Rychkov, Y., Golge, M., Dimitrov, D., Hill, E., Bradley, D., Romano, V., Cali, F., Vona, G., Demaine, A., Papiha, S., Triantaphyllidis, C., Stefanescu, G., Hatina, J., Belledi, M., Di Rienzo, A., Novelletto, A., Oppenheim, A., Norby, S., Al-Zaheri, N., Santachiara-Benerecetti, S., Scozari, R., Torroni, A. and Bandelt, H.J. (2000). Tracing European founder lineages in the Near Eastern mtDNA pool. *Am J Hum Genet*, **67**, 1251–76.

Roewer, L., Croucher, P.J., Willuweit, S., Lu, T.T., Kayser, M., Lessig, R., de Knijff, P., Jobling, M.A., Tyler-Smith, C. and Krawczak, M. (2005). Signature of recent historical events in the European Y-chromosomal STR haplotype distribution. *Hum Genet*, **116**, 279–91.

Rosenberg, N.A., Pritchard, J.K., Weber, J.L., Cann, H.M., Kidd, K.K., Zhivotovsky, L.A. and Feldman, M.W. (2002). Genetic structure of human populations. *Science*, **298**, 2381–5.

Rosser, Z.H., Zerjal, T., Hurles, M.E., Adojaan, M., Alavantic, D., Amorim, A., Amos, W., Armenteros, M., Arroyo, E., Barbujani, G., Beckman, G., Beckman, L., Bertranpetit, J., Bosch, E., Bradley, D.G., Brede, G., Cooper, G., Corte-Real, H.B., de Knijff, P., Decorte, R., Dubrova, Y.E., Evgrafov, O., Gilissen, A., Glisic, S., Golge, M., Hill, E.W., Jeziorowska, A., Kalaydjieva, L., Kayser, M., Kivisild, T., Kravchenko, S.A., Krumina, A., Kucinskas, V., Lavinha, J., Livshits, L.A., Malaspina, P., Maria, S., McElreavey, K., Meitinger, T.A., Mikelsaar, A.V., Mitchell, R.J., Nafa, K., Nicholson, J., Norby, S., Pandya, A., Parik, J., Patsalis, P.C., Pereira, L., Peterlin, B., Pielberg, G., Prata, M.J., Previdere, C., Roewer, L., Rootsi, S., Rubinsztein, D.C., Saillard, J., Santos, F.R., Stefanescu, G., Sykes, B.C., Tolun, A., Villems, R., Tyler-Smith, C. and Jobling, M.A. (2000). Y-chromosomal diversity in Europe is clinal and influenced primarily by geography, rather than by language. *Am J Hum Genet*, **67**, 1526–43.

Schmitz, R.W., Serre, D., Bonani, G., Feine, S., Hillgruber, F., Krainitzki, H., Paabo, S. and Smith, F.H. (2002). The Neandertal type site revisited: interdisciplinary investigations of skeletal remains from the Neander Valley, Germany. *Proc Natl Acad Sci USA*, **99**, 13342–7.

Semino, O., Passarino, G., Oefner, P.J., Lin, A.A., Arbuzova, S., Beckman, L.E., De Benedictis, G., Francalacci, P., Kouvatsi, A., Limborska, S., Marcikiae, M., Mika, A., Mika, B., Primorac, D., Santachiara-Benerecetti, A.S.,

Cavalli-Sforza, L. L. and Underhill, P. A. (2000). The genetic legacy of Paleolithic *Homo sapiens sapiens* in extant Europeans: a Y chromosome perspective. *Science*, **290**, 1155–9.

Serre, D., Langaney, A., Chech, M., Teschler-Nicola, M., Paunovic, M., Mennecier, P., Hofreiter, M., Possnert, G. G. and Pääbo, S. (2004). No evidence of Neandertal mtDNA contribution to early modern humans. *PLoS Biol*, **2**, 313–17.

Serre, D. and Pääbo, S. (2004). Evidence for gradients of human genetic diversity within and among continents. *Genome Res*, **14**, 1679–85.

Simoni, L., Calafell, F., Pettener, D., Bertranpetit, J. and Barbujani, G. (2000). Geographic patterns of mtDNA diversity in Europe. *Am J Hum Genet*, **66**, 262–78.

Soejima, M., Tachida, H., Ishida, T., Sano, A. and Koda, Y. (2006). Evidence for recent positive selection at the human *AIM1* locus in a European population. *Mol Biol Evol*, **23**, 179–88.

Sokal, R. R., Oden, N. L. and Wilson, C. (1991). Genetic evidence for the spread of agriculture in Europe by demic diffusion. *Nature*, **351**, 143–5.

Stefansson, H., Helgason, A., Thorleifsson, G., Steinthorsdottir, V., Masson, G., Barnard, J., Baker, A., Jonasdottir, A., Ingason, A., Gudnadottir, V. G., Desnica, N., Hicks, A., Gylfason, A., Gudbjartsson, D. F., Jonsdottir, G. M., Sainz, J., Agnarsson, K., Birgisdottir, B., Ghosh, S., Olafsdottir, A., Cazier, J. B., Kristjansson, K., Frigge, M. L., Thorgeirsson, T. E., Gulcher, J. R., Kong, A. and Stefansson, K. (2005). A common inversion under selection in Europeans. *Nat Genet*, **37**, 129–37.

Swallow, D. M. (2003). Genetics of lactase persistence and lactose intolerance. *Annu Rev Genet*, **37**, 197–219.

Taberlet, P., Fumagalli, L., Wust-Saucy, A. G. and Cosson, J. F. (1998). Comparative phylogeography and postglacial colonization routes in Europe. *Mol Ecol*, **7**, 453–64.

Tattersall, I. and Schwartz, J. H. (1999). Hominids and hybrids: the place of Neanderthals in human evolution. *Proc Natl Acad Sci USA*, **96**, 7117–19.

The International HapMap Consortium (2003). The International HapMap Project. *Nature*, **426**, 789–96.

Tishkoff, S. A., Varkonyi, R., Cahinhinan, N., Abbes, S., Argyropoulos, G., Destro-Bisol, G., Drousiotou, A., Dangerfield, B., Lefranc, G., Loiselet, J., Piro, A., Stoneking, M., Tagarelli, A., Tagarelli, G., Touma, E. H., Williams, S. M. and Clark, A. G. (2001). Haplotype diversity and linkage disequilibrium at human G6PD: recent origin of alleles that confer malarial resistance. *Science*, **293**, 455–62.

Valladas, H., Joron, J. L., Valladas, G., Arensburg, B., Bar-Yosef, O., Belfer-Cohen, A., Goldberg, P., Laville, H., Meignen, L., Rak, Y., Tchernov, E., Tillier, A. M. and Vandermeersch, B. (1987). Thermoluminescence dates for the Neanderthal burial site at Kebara in Israel. *Nature*, **330**, 159–60.

Valladas, H., Reyss, J. L., Joron, J. L., Valladas, G., Bar-Yosef, O. and Vandermeersch, B. (1988). Thermoluminescence dating of Mousterian 'proto-Cro-Magnon' remains from Israel and the origin of modern man. *Nature*, **331**, 614–16.

Voight, B. F., Kudaravalli, S., Wen, X. and Pritchard, J. K. (2006). A map of recent positive selection in the human genome. *PLoS Biol*, **4**, e72.

Wilson, I. J. and Balding, D. J. (1998). Genealogical inference from microsatellite data. *Genetics*, **150**, 499–510.

Zhivotovsky, L. A., Underhill, P. A., Cinnioglu, C., Kayser, M., Morar, B., Kivisild, T., Scozzari, R., Cruciani, F., Destro-Bisol, G., Spedini, G., Chambers, G. K., Herrera, R. J., Yong, K. K., Gresham, D., Tournev, I., Feldman, M. W. and Kalaydjieva, L. (2004). The effective mutation rate at Y chromosome short tandem repeats, with application to human population-divergence time. *Am J Hum Genet*, **74**, 50–61.

Chapter 14

The Peopling of Oceania

Elizabeth Matisoo-Smith

Dept of Anthropology, University of Auckland, New Zealand

The peopling of the Pacific is particularly fascinating as it involves one of the earliest migrations of modern humans, the settlement of Australia and New Guinea, and the last major colonization event, the settlement of Polynesia. It required crossing vast distances of open ocean in small double-hulled canoes, suggesting highly developed sailing and navigational skills. The deep chronology, combined with the variations in Pacific environments, make the Pacific an ideal region to study colonization, adaptation, human genetics and human biology. This is not because islands represent laboratory conditions (Vayda and Rappaport, 1963), but because Pacific islands represent a range of environments and, perhaps, unique historical conditions that allow us to study the many factors that affect human variation.

The general geographic setting has significant implications for the settlement history and human biology in any region, but this is dramatically so in the Pacific. Most people are familiar with the original biogeographic dissection of the Pacific into Melanesia, Polynesia and Micronesia. While these crudely descriptive terms define the three regions as the 'black islands', 'many islands' and 'small islands', this familiar system of classification makes no sense in terms of biology, language and culture. So, prehistorians like Roger Green and others (Pawley and Green, 1973; Green, 1991; Kirch, 2000) have suggested an alternative system identifying two significant regions in the Pacific, Near Oceania and Remote Oceania, shown in Figure 14.1. This classification is based primarily on colonization history and, therefore, has biological, cultural and linguistic significance.

When we are considering the prehistoric settlement of the Pacific, we must also be aware of the fact that the landscape has changed significantly over the periods of human colonization. Due to sea level

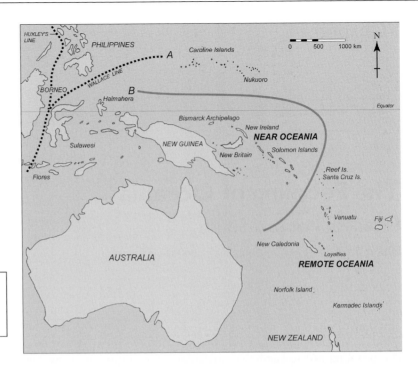

Fig 14.1 Map of the Western Pacific showing the Wallace line (A) and that delineating Near and Remote Oceania (B).

changes over the millennia, many of the low islands and atolls present today were under water until relatively recently, and many of the islands in Southeast Asia and Near Oceania were, in the past, joined either to each other or to larger continental landmasses nearby. These landscape changes will, of course, have had a significant impact on the colonization history and potentially the biology of peoples in the region.

When people first entered the Asian/Pacific region, sea levels were significantly lower than they are today, exposing two major continental shelves: Sunda, the greater Asian landmass, and Sahul, the southern continent made up of present-day Australia, New Guinea and Tasmania. The region in between, Wallacea, is identified by the biogeographical boundaries of Wallace's and Huxley's lines and is an ancient water barrier, as is seen by the differences between the Asian and Australian flora and fauna (White and O'Connell, 1982).

Archaeological evidence for Pacific settlement

The settlement of Near Oceania began at least 42,000 and possibly (though questionably) as much as 60,000 years ago when modern humans moved into the Sahul region from Sunda (O'Connell and Allen, 2004). Due to the deep channels in the Wallacea region, which separates Sunda and Sahul, people could not have walked across a land bridge as they did in the settlement of the Americas. The settlement of Sahul involved the crossing of water gaps of at least 60 km, even at the lowest sea levels; therefore, the settlers must have had some form of water craft (Chappell, 1993).

The dates for first human occupation of the Bismarck Archipelago are possibly as early as those of the New Guinea mainland (Leavesley *et al.*, 2002), but certainly by 29,000 BP people had moved throughout Sahul and out to the islands of the Bismarcks and the greater Solomon Islands. Analysis of food remains from the early sites in the region suggest that those living on the large landmass of Sahul had a broad based terrestrial economy, while out on the islands there was a more marine focus, with reef fish and shellfish remains making up a large component of the middens (Allen, 2000; O'Connor and Veth, 2000; Bellwood *et al.*, 1998). A significant factor in the settlement of Near Oceania was that all of the islands were, for the most part, intervisible – people never had to leave sight of one island before they caught sight of another. This appears to be a significant factor in limiting the extension of human settlement at this initial period of time. Long-distance exchange systems, moving obsidian at least intermittently through the Bismarck Archipelago, were developed 20,000 years ago with evidence of increasing interaction in the Holocene. By 13,000 BP people had settled the small island of Manus in the Admiralty group, which required sailing beyond the safety of island intervisibility. It has been suggested that the region from Wallacea through to the Solomon Islands was a 'voyaging nursery', or a place where people could refine navigational skills and sailing technology sheltered from the unpredictable cyclone belts to the north and south that make open-ocean sailing a very risky business (Irwin, 1989, 1992).

It wasn't until much later, around 31,00 BP, that people breached this apparent psychological and/or voyaging barrier and moved out into Remote Oceania, the region identified as the islands from the Reef/Santa Cruz group east to Vanuatu, New Caledonia, Fiji and the western edge of the Polynesian triangle and north into central and eastern Micronesia. The initial settlement of Remote Oceania was clearly associated with the Lapita cultural complex, though the origin of Lapita, which appears archaeologically in the Bismarck Archipelago around 3,500 BP, is hotly debated amongst Pacific prehistorians.

The Lapita Cultural Complex is most commonly recognized by its distinctive dentate stamped and red-slip decorated pottery. More precisely, it is identifiable by a suite of artefacts, unique settlement patterns and Neolithic subsistence economies. It is with Lapita that we first see the appearance of village settlements in the Pacific. These sites were generally located on small, off-shore islets or as stilt complexes over reef flats. Lapita subsistence was based on the cultivation of a range of plants, some of which were indigenous to Near Oceania while others were introductions from Island Southeast Asia. Lapita sites also are the first to conclusively and consistently contain skeletal remains of dogs, pigs and chickens, but also show a significant exploitation of fish and shellfish (Kirch, 1997).

In a recent review, Green (2003) outlined current models for Lapita origins in which he identified four sets of models. The first set, the 'Express Train to Polynesia' (ETP), focuses on a rapid dispersal from Southeast Asia (Taiwan) ultimately to Polynesia. These models involve people moving through Island Southeast Asia and Near Oceania with little or no contact with indigenous populations. The second set, referred to as the 'Bismarck Archipelago Indigenous Inhabitants' (BAII), represent the other extreme perspective: that there is no need to consider any major migration into Near Oceania to account for the appearance of the Lapita cultural complex. Instead, Lapita can be explained as an indigenous development within the Bismarck Archipelago. The 'Slow Boat to the Bismarcks' (SBB) is the third set of models. These focus on interactions occurring within Irwin's (1992) voyaging nursery region during the period from 6,000 to 3,500 BP, followed by rapid expansion out into Remote Oceania around 3,100 BP. The final set of models identified by Green is the 'Voyaging Corridor Triple I' (VC Triple-I), which is seen by some (Terrell and Welch, 1997) as a 'compromise solution' and not testable or predictive. This view builds on the Slow Boat model but allows for various components of the Lapita cultural complex to be the result of the *Intrusion* of new components along with the *Integration* of materials from indigenous inhabitants in Near Oceania and the *Innovation* or development of new, unique components.

While the Lapita cultural complex extended as far east as Fiji, Samoa and Tonga, the settlement of the rest of the Polynesian Triangle did not happen until later. After a hiatus of some 500 to 1,500 years, during which time Polynesian languages, culture and society developed, the Polynesians then moved out into central and east Polynesia. While there are several debates regarding the specific timing and sequence of settlement (Anderson and Sinoto, 2002; Kirch, 1986, 2000; Spriggs and Anderson, 1993), Polynesians settled Hawaii by 1,200 BP, Easter Island by 1,000 BP and New Zealand by 800 BP. The colonization of the Chatham Islands, located some 500 miles off of the east coast of New Zealand, represents the last major island group settled, which occurred by 500 years BP.

At about the same time as populations moved out to Central and East Polynesia, it appears that there was also movement of Polynesian populations westward. Several small islands in Vanuatu, New Caledonia, the Solomon Islands and the Carolines (in Micronesia) were settled by Polynesians apparently from the Samoa/Tonga region. Although the Polynesian populations in many cases are thought to have been intrusive arrivals, it appears that the Polynesian languages replaced those of the original inhabitants and, thus, these islands are known as the Polynesian outliers (Kirch, 2000).

The archaeology of Micronesia is not as well understood as that of the rest of Remote Oceania. The earliest dates for the settlement of Micronesia occur in the Marianas at about the same time period as the appearance of Lapita further to the south. These early Micronesian sites also include pottery, but analyses suggest it is not closely related to Lapita. Instead, it may come from a common ancestral tradition in the Philippines or elsewhere in island Southeast Asia (Kirch, 1997, 2000). In Palau, to the southwest of the Marianas, the earliest dates for human occupation are around 2,000 BP, but environmental evidence of human presence suggests that earlier occupation, as early as 3,000–4,000 BP, is possible (Kirch, 2000).

The settlement of the high, volcanic islands of the Caroline group is linked with the Lapita expansion from the Bismarcks-southeast Solomons-Vanuatu region. The settlement of the atolls found throughout the Carolines, Kiribati and the Marshalls occurred later and was intimately linked to changes in sea level that occurred within the last few thousand years. Geomorphological evidence (Dickinson, 2003) suggests that these low islands and atolls were not permanently above sea level and, therefore, were not available for human colonization until the first millennium AD at the earliest. It is generally accepted that the colonization of central and eastern Micronesia was a northern arm of a late Lapita expansion.

Linguistic evidence

Linguistic analyses have played a key role in developing models for Pacific settlement, particularly with regard to the origins of the Lapita peoples, associated with the spread of the Austronesian languages. The languages of the indigenous inhabitants of Near Oceania represent a highly diverse group, often referred to as 'Papuan'. This extreme diversity is not surprising, given the great time depth of human occupation in the area. These 'Papuan' or 'non-Austronesian' languages are difficult to define as a group, but belong to at least 12 very distinct language families that are mutually unintelligible. The languages are spoken today primarily in non-coastal and highland regions of New Guinea and in the more isolated parts of the Bismarcks, Bougainville and other islands in Near Oceania (Ross, 2001; Dunn et al., 2002).

The appearance of the Lapita cultural complex in Near Oceania is closely associated with the arrival of the Austronesian languages in the region. The Austronesian language family, in particular the Malayo-Polynesian subfamily, includes more than 1,000 languages spoken from Madagascar to Easter Island and throughout Southeast Asia. The other nine subfamilies of Austronesian were spoken exclusively in Taiwan, amongst the Taiwanese aboriginal populations, identifying Taiwan as the Austronesian homeland (Pawley and Ross, 1993). This, together with the apparent step-wise and rapid spread of the Malayo-Polynesian languages from there ultimately out into Remote Oceania, is what has led to the so-called 'Fast Train' models for Lapita origins (though it should be noted that the linguists did not come up with this model).

The Lapita dispersal in Oceania is clearly associated with the Oceanic subgroup of Austronesian languages. While this holds true for most of Remote Oceania, the languages of Guam and Palau, in western Micronesia, belong to a higher order sub-group of Western-Malayo-Polynesian, most closely related to the languages of the Philippines and Indonesia (Pawley, 2002).

The Polynesian languages, as was noted and frequently commented upon by the early European explorers, are all closely related and in many cases mutually intelligible. It has been suggested that they form a monophyletic unit that can be broken into smaller sub-groupings, all of which can be traced back to the Proto-Central-Pacific sub-group of Oceanic, the language spoken by the original Lapita inhabitants in Fiji, Samoa and Tonga (Kirch, 2000; Kirch and Green, 2001).

Human genetic evidence

In terms of addressing Pacific prehistory, most genetic studies have focused on identifying the origins of the Polynesians and their ancestors, the Lapita people. Given the colonization history, it is generally assumed that Polynesians are biologically a relatively representative remnant population of the original Lapita colonizers, less affected by admixture than the populations in western Remote Oceania. An alternative perspective is that they represent a unique sub-set of the original Lapita colonists and are, therefore, not necessarily representative of the population as a whole (Terrell, 1988).

When addressing population origins it is, of course, important to recognize that no modern populations today are 'pure' representatives of ancient populations. Modern populations are the result of a range of evolutionary and historic processes including drift, selection, migration and admixture, both historic and prehistoric. When trying to trace prehistoric population movements and identify origins, we need to acknowledge that independent factors affect all populations from the source to the 'end of the line'.

All of the genetic evidence, along with the linguistic and archaeological data, clearly point to an eastern as opposed to a South American (Heyerdahl, 1952) origin for all Pacific populations. However, the timing, numbers and specific source populations of those migrations, as well as the degree of association between genetic, linguistic and/or material cultural evidence, are the topics of debate and continued research.

Mitochondrial DNA variation

Given the emphasis on identifying the origins of the Polynesians, several specific molecular markers have been the focus of genetic studies in Remote Oceania. The most widespread and studied marker is a deletion of one copy of a 9 base pair (bp) repeat sequence

(CCCCCTCTA) in the COII/tRNA[lys] intergenic region and a marker generally referred to as the 'Polynesian motif' (Melton *et al.*, 1998; Redd *et al.*, 1995). The 9 bp deletion has a distribution that ranges from Madagascar to Easter Island, but is not found, or is found only rarely, in the highlands of New Guinea and amongst Australian aborigines (Betty *et al.*, 1996). This marker is generally considered to have originated in East Asia and is a defining marker of mitochondrial haplogroup B, which is also found in Native Americans. In addition to the 9 bp deletion, Remote Oceanic populations also have high frequencies of the 'Polynesian motif', a combination of three point mutations in the control region. These three transition substitutions, at positions 16217 (T-C), 16261 (C-T) and 16247 (A-G), appear to have occurred successively as people moved eastward across the Pacific. The combination of the 9 bp deletion and the full Polynesian motif, defining mtDNA haplogroup B4a1a1 (Trejaut *et al.*, 2005), reaches its highest percentage (over 85–100%) in east Polynesian populations (Sykes, 1995; Murray-McIntosh *et al.*, 1998), and is found throughout Remote Oceania and amongst Austronesian-speaking populations in the Bismarck Archipelago and along coastal regions of New Guinea and the Solomon Islands and at a high frequency amongst central and eastern Micronesian populations. The broad distribution led Lum and Cann (2000) to suggest that it might be better referred to as the 'Austronesian motif'. However, haplogroup B4a1a1 has not yet been found in Taiwan, Borneo or the Philippines.

Recent and increasing numbers of analyses of complete mitochondrial genomes is providing a much more refined picture of the evolution of the Polynesian motif (Friedlaender *et al.*, 2005, in press; Ingman and Gyllensten, 2003; Trejaut *et al.*, 2005). The B4a1a lineage, immediately ancestral to B4a1a1, has a broad distribution that includes several Taiwanese aboriginal groups (overall 8.74% of 640 tested, but particularly high in the Ami at 44.90% of 98), the Philippines (15.25% of 59), Indonesia (9.26% of 54) and the Moluccas (24.9% of 62) as well as coastal New Guinea and Remote Oceanic populations (Trejaut *et al.*, 2005). It has not been identified on the Asian mainland. Coalescence estimates for this lineage have been determined to be around 13,000 years (SE 3,841). The full motif, B4a1a1, has a coalescence estimate of 9,300 years (SE 2,500) and to date has only been identified in Near and Remote Oceania (Trejaut *et al.*, 2005). Interestingly this early Holocene date suggests that the full 'Polynesian motif' was present in Near Oceania before the appearance of the Lapita cultural complex. Unfortunately, complete mtDNA sequences from the Philippines and Eastern Indonesia, including populations in Wallacea, are not yet available, so at this point, the specific origin of the motif cannot be determined with any certainty.

In addition to the Polynesian motif, a limited number of Polynesian lineages represent non-Asian-derived mitochondrial sequences, in particular lineages belonging to haplogroups P and Q (discussed below). It has been suggested that several of these originated in Near Oceania (Lum and Cann, 2000; Friedlaender *et al.*, 2005) and are clear evidence of admixture between Near and Remote Oceanic populations at some point in prehistory or of mixed origins.

Lum and Cann (2000) have conducted the most widespread study of mtDNA variation amongst Micronesian populations. While they found that, like those of Polynesians, the most common mtDNA lineages in Micronesia were those possessing the 9 bp deletion and at least two, if not all three, of the point mutations of the 'Polynesian motif'. However, further analyses identify a number of unique point mutations within these Micronesian populations that differentiate them from Polynesian populations and identify differences between the various island groups. In particular, the Marianas have only a very low frequency of the Polynesian motif and are, therefore, distinct from other Micronesian populations. This is consistent with their linguistic classification, not belonging to the Oceanic sub-group, and with archaeological evidence suggesting settlement from the Philippines or even Taiwan (Kirch, 2000). Similarly, the mitochondrial diversity found in Yap and on Palau suggest a more complex history of interaction and direct gene flow with Southeast Asia and, in the case of Palau, with Near Oceania, results again consistent with linguistic and archaeological evidence (Lum and Cann, 2000).

More recently, significant in-depth studies of the mtDNA variation in Near Oceania are allowing us to tease apart evidence of population interactions in this very biologically and historically complex region (Merriwether *et al.*, 1999, 2005; Friedlaender *et al.*, 2002, 2005). The earliest settlement of Near Oceania appears to be most closely associated with two ancient mtDNA haplogroups, P and Q, which belong to the deep non-African N and M clades, respectively. Various dates have been calculated for the origins of these lineages, with P in the range of 50,000–60,000 BP and Q being slightly younger at about 45,000 BP (Forster *et al.*, 2001; Ingman and Gyllensten, 2003; Friedlaender *et al.*, 2005). Friedlaender and colleagues (2005) report data that clarify and expand on previous definitions of P and Q (Forster *et al.*, 2001) and identify significant structure within the P and Q haplogroups, with seven branches now identified within P (with internal branching in P1) and three major branches within Q.

Haplogroups P and Q make up about 90% of mtDNA lineages in New Guinea. They are not found on the Asian mainland and are rare in Island Southeast Asia. The distribution of P includes Australia, but generally Australia and New Guinea have separate sub-groups of P, with only the P3 haplotype found in both regions. Haplotype P1 is the most common sub-group of P found in Near Oceania where it is found throughout the New Guinea mainland. It is rare in the other islands of Near Oceania, except for those off of the Papuan tip, where Q is not found. P1 is also found at very low frequencies as far east as New Caledonia (Friedlaender *et al.*, 2005).

Haplogroup Q is much more common in the islands of Near Oceania, particularly on New Britain and northern Bougainville. Q1 and Q2 are also found at low frequencies in Remote Oceania extending into Polynesia where Q (undefined) is found at a very low frequency, less than 0.05 (Sykes *et al.*, 1995; Lum and Eann, 1998; Hagelberg *et al.*, 1999; Friedlaender *et al.*, 2002, 2005). Haplogroup Q has not yet been found along the south coast of New Guinea. Both P1 and Q1 are present in Micronesia (Lum and Cann, 1998, 2000; Friedlaender *et al.*, 2005).

Friedlaender and colleagues (2005) have identified clear star-like patterns of diversity within both P and Q, particularly in the P1 and Q1 lineages, suggesting population expansions that they estimate occurred 36,700 (SD 8,000) and 28,700 (SD 7,500) years ago respectively, or soon after first settlement of New Guinea (Friedlaender *et al.*, 2005 – note these are recalculations of the original expansion dates reported originally). Q2 expansion within Island Melanesia was determined to be approximately 10,000 years after Q1. They further suggest that 'the extremely localized distributions of specific haplotypes within the branches of Q and P are consistent with the highly restricted female movement within the region over the following millennia' (Friedlaender *et al.*, 2005: 1514).

In addition to the P, Q and B haplogroups found in Near and Remote Oceania, complete mitochondrial genome sequencing in the region has revealed a number of new and deep lineages belonging to Macrohaplogroup M (Merriwether *et al.*, 2005). Many branches of M are found throughout mainland and Southeast Asia and it is generally accepted that M was brought to Asia and ultimately Australasia as part of the first modern human migrations out of Africa, which proceeded along the southern coastal route (Kivisild *et al.*, 1999; Quintana-Murci *et al.*, 1999; Macaulay *et al.*, 2005).

These new ancient M lineages identified in Near Oceania, M27, M28 and M29 are not closely related to one another or to M lineages identified in Australian aboriginal, Indian or other Southeast Asian populations. M27 was found at its highest frequencies in central Bougainville, but also less frequently through the Solomon Islands and as far east as Vanuatu. The M28 lineage was relatively common in Near Oceania and in some populations of East New Britain it was found at 100% frequency. Interestingly, it is absent or very rare in Bougainville but was found as far east as Fiji. The authors suggest the distribution of this haplotype may be under-represented as sequencing of the commonly studied HVS 1 region will not identify one of the key defining mutations at site 16468. The third newly defined M haplotype, M29, may be distantly connected to Q lineages as it shares a defining transition at position 13500. It is found most frequently among the Tolai of East New Britain who, prior to major volcanic eruptions in the region, resided in southern New Ireland, and has also been identified in single samples from the Solomon Islands and Vanuatu (Merriwether *et al.*, 2005).

The combination of both studies of complete mitochondrial genomes and the fine-grained studies as those carried out by Friedlaender and Merriwether and colleagues are providing actual data that support what most anthropologists working in the region have thought – but many geneticists working in the region have ignored – that the population history in Near Oceania is particularly complex and that sampling a few individuals, or even several individuals from a single location on these islands, will not provide a representative sample of the region as a whole. Most importantly, they have shown that language affiliations and mitochondrial DNA haplotypes do not necessarily go hand in hand. While language may provide an indicator as to likely genetic affinities, much more in-depth knowledge of historic, prehistoric, cultural and linguistic histories in such complex regions is needed in order to explain modern-day diversity. One might expect that further fine-grained studies of modern populations in Near Oceania and elsewhere, along with increasing numbers of complete mitochondrial genome sequences from the region, will provide more evidence of a complicated human history.

Ancient mtDNA

While we are making progress in trying to understand prehistoric population movements by untangling the various components that contribute to modern human genetic variation, the only way that we can reliably know about the genetic makeup of prehistoric populations is through the analysis of ancient DNA (aDNA) extracted from archeological remains. Some of the earliest analyses of aDNA from human skeletal remains have focused on Pacific populations. In 1993 Hagelberg and Clegg applied recently developed techniques (Hagelberg and Clegg, 1991) to study the mtDNA of skeletal remains associated with Lapita sites in both Near and Remote Oceania. Their analysis involved screening for the presence or absence of the 9 bp deletion in the material. When their results indicated that none of the Lapita skeletons, which included material from Tonga, possessed the deletion, they concluded that Lapita origins were in 'Melanesia' and that 'the Lapita culture was carried from its Melanesian homeland into the central Pacific by indigenous inhabitants of island Melanesia, rather than by Austronesian-speaking migrants from Southeast Asia who settled the region en route to the eastern Pacific' (Hagelberg and Clegg, 1993: 168).

Hagelberg and Clegg's study also included analyses of skeletal material from more recent sites in Polynesia, all of which possessed the 9 bp deletion and the Polynesian motif or the 16217 and 16261 mutations, and material from Micronesia. The three Micronesian samples, from Kosrae in the Carolines, Yap and Afetna in the Marianas, were all dated to between 200 and 400 BP; only the Kosrae sample possessed the 9 bp deletion, but it had only the 16217 transition. In a later study, analyses of ancient mtDNA from

Easter Island skeletal remains that clearly date to pre-European periods all showed both the presence of the 9 bp deletion and the full Polynesian motif, finally laying to rest the issue of Amerindian origins for prehistoric Easter Islanders (Hagelberg *et al.*, 1994).

Unfortunately, the analyses of the Lapita material have never been replicated and, as the researchers themselves point out (Hagelberg and Clegg, 1993: 168), their conclusions rest solely on negative evidence. They also describe the poor preservation of the Lapita material and the fact that analyses of faunal material identified modern human DNA contamination. Since the publication of these early results, ancient DNA research has developed substantially, particularly in regard to the controls that are now considered necessary to eliminate, or at least identify modern contamination and verify positive results. The results of the Lapita material, therefore, have to be considered unsupported until the results have been replicated using the currently accepted aDNA protocols (Cooper and Poinar, 2000; O'Rourke *et al.*, 2000; O'Rourke, this volume). The recent discovery of an early Lapita cemetery in Teouma, Vanuatu (Stuart Bedford, unpublished data 2004), containing several well preserved human burials as well as several other Lapita burials identified in recent years from Vanuatu, New Caledonia and Fiji, will provide an opportunity to identify the mitochondrial signature of some of the first arrivals in Near Oceania.

Y chromosome variation

While much emphasis has been placed on studying mtDNA for understanding population origins, more recently we have been able to address the counterpart or male side of the story with analyses of the paternally inherited Y chromosome. Several studies have focused on the Y chromosomes found in Pacific populations, again, generally with a primary focus on identifying the origins of the Y chromosomes found in Polynesians. Since the analyses of Y chromosomes is relatively recent and are slightly more complex than those of mtDNA, researchers have focused on a number of different markers − SNPs, STRs and different combinations of both, so comparisons of results of different researchers are not always straightforward. This situation should improve in the future with the adoption of the nomenclature system proposed by the Y chromosome consortium (YCC) in 2002.

While the mtDNA in Polynesians is dominated by the island Southeast Asian markers, recent analyses of the paternally inherited Y chromosome suggest a more complicated history involving significantly more admixture with or input from indigenous Near Oceanic or Wallacean populations (Su *et al.*, 2000; Kayser *et al.*, 2000; Underhill *et al.*, 2001; Hurles *et al.*, 2002). Comparisons between Y chromosome and mtDNA studies clearly show, not surprisingly, that there has been more admixture by European men compared to European women in the Pacific (Hurles *et al.*, 1998; Underhill *et al.*, 2001).

Su and colleagues (2000), in their paper on identifying the origins of Polynesian Y chromosomes, investigated the distribution of 19 biallelic markers in Southeast Asian and Pacific populations. Their Polynesian sample was made up of one Samoan and 35 Tongan Y chromosomes plus 10 samples from the Polynesian outlier of Kapingamarangi. These Polynesian samples were pooled with Micronesian samples from Chuuk, Majuro, Kiribati, Guam, Palau, Pohnpei and Nauru for comparisons to identify population origins. They identified 15 Haplotypes from a total of 551 Y chromosomes studied and found the most haplotypic diversity (0.88) in Southeast Asia. Nine of the identified 14 Southeast Asian haplotypes were also found in their pooled Micronesian/Polynesian sample, along with one non-East Asian haplotype, H 17 (belonging to YCC Haplogroup M or Hg24, Jobling and Tyler-Smith, 2003) identified in 2 of 17 Y chromosomes from Chuuk. With the exception of Chuuk, the H 17 lineage was found exclusively in 'Melanesia' (represented by a total of: 6 samples from the Banks and Torres Islands, 10 from Maewo and 4 from Santo all in Vanuatu; 3 samples from the Nasioi, a 'Papuan' speaking group from Bougainville; and 90 samples from New Guinea) its apparent location of origin. In assessing the possibility of a Taiwanese origin for Polynesian Y chromosomes, Su *et al.* found, with the exception of one haplotype, H 6 (YCC lineage O3 or Hg 26, Jobling and Tyler Smith), the haplotype distributions of Taiwan aboriginals and Micronesian/Polynesian populations did not overlap significantly. This led the researchers to suggest that '. . . both the Taiwanese and Polynesian populations derive their ancestry in Southeast Asia' (2000: 8227).

Kayser and colleagues (2000), also addressing the question of Polynesian origins, focused their study initially on 28 males from the Cook Islands. Looking at variation in 8 Y chromosome SNPs (M4, M5, M9, M16, M21, M119, M122 and RPS4Y711), they found that all of their Cook Island samples belonged to one of three haplotypes defined by mutations at RPS4Y711, M9 and M122 (falling into YCC haplogroups C, K and O3, or Haplogroups 10 and 26 Jobling and Tyler-Smith). In order to identify the origin of these lineages, they assayed an additional 583 individuals from Asia, Australia and 'Melanesia' for these three markers and seven STRs (DYS19, 389I, 389II, 390, 391, 392 and 393). Finding significant length variation at Y-STR DYS390, they sequenced this marker from 319 individuals and identified four repetitive segments, 390.1 through 390.4. The Cook Island population had a high frequency (82%) of a deletion in the 390.3 segment, and in these individuals this deletion was also combined with a C to T mutation at position 711 of the RPS4Y gene. This combination of markers (DYS390.3 del/RPS4Y711T) was also found at lower frequencies (9−26%) in coastal New Guinea, New Britain, the Trobriand Islands, the Moluccas and Nusa Tenggara in eastern Indonesia. It was found in one individual from the highlands of New Guinea and one individual from southern Borneo, but not in any other Southeast Asian or Australian populations (Kayser *et al.*, 2000, 2001). In addition to the DYS390.3del/RPS4Y711T, the two other haplotypes found in the Cook Islanders were the East and Southeast Asian M122C/M9G type (YCC O3 lineage), found in 7% of Cook Islanders and the ancient M9G type which was found in all populations studied but also in many other non-Asian populations and is therefore not informative for addressing Pacific population origins (Kayser *et al.*, 2000).

Underhill *et al.* (2001), in a study of Y chromosome variation in New Zealand Maori, looked at a further defining marker in the DYS390.3del/RPS4Y711T haplogroup, defined by the M38 mutation (delineating YCC lineage C2). They identified this lineage (that they called ht3) in 42.6% of Maori Y chromosomes studied and at similar levels in 17 other Polynesians in their sample. This haplotype was originally identified in New Guinea (Underhill *et al.*, 2000) and was also identified in 58.8% of the 17 Indonesian samples that had previously been identified as belonging to the C haplogroup. The remaining non-European Y chromosomes identified in Maori belonged to the M9G and M9G/M122C lineages, both found at similar low frequencies as those reported by Kayser *et al.* (2000) for Cook Islanders.

In a study incorporating a slightly larger Polynesian sample consisting of 20 Cook Islanders, 25 Western Samoans and 34 Tongans – once all European specific Y haplotypes were removed – Hurles *et al.* (2002) focused on the hypervariable MSY1 combined with a number of binary markers and the 7 STRs studied by Kayser *et al.* (2000) to address the origins of Oceanic speaking populations. In the 390 Y chromosomes studied, they found the most variation in the MSY1 region, and when combined with microsatellite data identified 323 compound haplotypes that separate with no overlap between haplogroups.

Two haplogroups, Hg 10 (following Jobling and Tyler-Smith, 2003, or YCC Haplogroup C) and Hg 26 (defined by the ancient M9 mutation of YCC Macrohaplogroup K), accounted for 82% of the Y chromosomes they found in the Pacific. The 224 chromosomes belonging to Hg 26 could be assigned to a further eight different monophyletic sub-lineages based on diagnostic MSY1 mutational patterns, designated 26.1 through 26.8. These distinct lineages all have 'coherent geographical distributions' (Hurles *et al.*, 2002: 293), with extreme diversity identified in Vanuatu. In Polynesia, the only variant that was found in significant number was lineage 26.4. Lineage 26.5 was only identified in Micronesia (Kapingamarangi and Majuro), and 26.1 only in North Borneo.

Similarly, Haplogroup 10 could be further defined with three additional sub-lineages, 10.1 through 10.3. Lineage 10.2 was the most frequent lineage found in Polynesia at frequencies as high as 70% in the Cook Islands. The remaining Polynesian Hg 10 chromosomes were identified as 10.1. Both 10.1 and 10.2 were also found in Vanuatu, and 10.2 in New Guinea. None of the Hg 10 lineages were found in either Taiwan or the Philippines. The 10.3 lineages were only identified in Borneo. The third major haplogroup identified in the Pacific,

Hg 24 (YCC haplogroup M), was found at its highest frequency in New Guinea (63%) and it also was present in Vanuatu (7.3%) and Tonga (14.7%), but in none of the other populations studied by Hurles *et al.* (2002).

As with most genetic studies conducted in the Pacific, limited data is available for Y chromosome variations in Micronesia. Hurles and colleagues (2002) included two Micronesian populations in their study — representing Majuro, in the Marshall Islands, and Kapingamarangi in the Caroline group. While geographically considered to be in Micronesia, Kapingamarangi is known as a Polynesian outlier, one of the several islands in the western Pacific settled as a result of back migration from the Polynesian Triangle. This is apparent in the high frequency of the Haplogroup C lineages which are the most common lineages found in Polynesia. There are no easily identifiable ancestral sequences for the rest of the Y chromosome lineages found in Kapingamarangi and in Majuro, but more distantly they belong to the O Haplogroup, and are most closely related to lineages found in Borneo (Hurles *et al.*, 2002).

With regards to dating the origins of the various Y chromosome lineages and therefore commenting on the Lapita origins debates, all of the major Pacific lineages appear to be relatively young compared to some of the ancient mitochondrial lineage age estimates (Hurles *et al.*, 2002). The oldest lineages appear to be those belonging to Hg 10 and date to around 6,000 years, with the Hg 26 and Hg 24 lineages in the Pacific being even younger. However, given the distributions of the lineages, a Near Oceanic and/or eastern Indonesian origin for haplogroup 10 seems most likely, while the origin of haplogroup 26 cannot be more precisely identified to a region beyond island Southeast Asia (Hurles *et al.*, 2002). However, future more fine-grained Y chromosome studies, like those appearing for the mitochondrial genomes, will undoubtedly provide us with additional data and allow us to ask more specific questions if not providing more specific answers.

Biparentally inherited markers

Genetic research in the Pacific during the 1970s and 80s focused on several biparentally inherited markers, in particular HLA markers and those globin gene variants associated with thalassaemia (the edited volume by Hill and Serjeantson (1989) provides an excellent review of these studies). The distribution of malaria in the Pacific has had a significant impact, not only on the incidence of thalassaemia but also no doubt on the demography of both prehistoric and historic populations. Malaria is not found east or south of Vanuatu in the Pacific, and the removal of this selective pressure would allow for much more rapid population growth in the non-malarial areas of Remote Oceania.

It was the discovery of high rates of one particular globin gene variant associated with thalassaemia, the alpha 3.7III deletion, in Polynesian populations that provided the first clear evidence of admixture between 'Melanesian' populations and ancestral Polynesians (Hill *et al.*, 1985). This particular deletion is not present in Asia, but is found throughout coastal populations in Near Oceania and in western Remote Oceania, and is believed to have originated in the region. The fact that it is found at levels as high as 15% in Polynesia (Hill *et al.*, 1989), a region that has never harboured malaria, provided a clear biological link between Polynesians and their Lapita ancestors to the west. Another characteristic of Polynesian populations is the high frequency (often around 50%) of the alpha globin halplotypes Ia and IIa, which are extremely rare in Oceanic populations to the west, but are found in Southeast Asia (Hertzberg *et al.*, 1989).

A second alpha globin gene marker, the alpha 4.2 deletion, is also present at reasonably high levels in the Pacific. This mutation is found at its highest frequency (more than 50%) on the north coast of New Guinea. It is also found in South and Southeast Asia with gene frequencies around 1–5% and in Remote Oceania, at levels as high as 16% in some populations in Vanuatu and at a very low frequency (0.5%) in New Caledonia (Flint *et al.*, 1986). The alpha 4.2 deletion is not present in Polynesia. Both the alpha 3.7III and the alpha 4.2 deletions are found at low frequencies in Micronesian populations studied (O'Shaugnessy *et al.*, 1989).

More recently, Lum *et al.* (2002) have focused on neutral short tandem repeat (STR) loci from populations in Southeast Asia and the Pacific to identify population origins and affinities. They report a significant loss of heterozygosity from Southeast Asia eastwardly across the Pacific, suggesting a series of sequential bottleneck events during Oceanic colonization. However, they did find evidence of significant post-settlement gene flow, both regionally and between Near Oceanic populations and those of Island Southeast Asia and Remote Oceania. In particular, gene flow appeared to be high within Vanuatu, Fiji and Samoa, and between Samoa and the central and eastern Micronesian populations. Lower but significant levels of gene flow were suggested between New Guinea and Vanuatu and between Borneo and Vanuatu — all suggesting that there were several spheres of interaction within the Pacific, crossing the traditional geographic and linguistic boundaries. Overall, gene flow appeared to correlate more closely with geographic distance and less so with linguistic affinity, a finding supported by studies of mtDNA variation (Merriwether *et al.*, 1999, 2005; Friedlaender *et al.*, 2002, 2005).

Commensal studies: an alternative approach

While studies of human populations are providing increasing evidence regarding the ultimate origins of the Polynesians and evidence of population interactions in both Near and Remote Oceania, the lack of genetic variation within Polynesia has made addressing the more immediate origins of specific Polynesian populations virtually impossible. Faced with this problem, and interested in such specific questions as identifying the origin of the New Zealand Maori, we have attempted to address the question of the settlement of the Pacific using an alternative approach (Matisoo-Smith, 1994; Matisoo-Smith et al., 1998; Allen et al., 1996; Allen et al., 2001).

When Pacific peoples dispersed across the great ocean, they did not travel alone in their canoes. They transported a whole range of plants and animals to introduce to the islands they were to colonize. The animals associated with Lapita colonists include the dog (*Canis familiaris*), pig (assumed to be *Sus scrofa*), jungle fowl (*Gallus gallus*) and the Pacific rat (*Rattus exulans*). Other animals such as geckos, skinks and land snails may have been accidental introductions. Not all animals made it to all locations. For example, only the rat and the dog were successfully introduced to New Zealand and only the rat and chicken made it to Easter Island. Sometimes, the appearance of dogs and pigs on an island occurs early; they then disappear from the archaeological record, only to be re-introduced later in time. *Rattus exulans* was the only commensal animal that was introduced to virtually all islands that show evidence of Polynesian presence.

The Pacific rat, *Rattus exulans*, is the third most widely distributed rat worldwide, with a distribution that ranges from the Andaman Islands, through mainland and island Southeast Asia, and through-out the Pacific as far as Easter Island (Tate, 1935). It is believed to have its origin somewhere in island or peninsular Southeast Asia and was not believed to be in Near Oceania prior to the Holocene (Taylor et al., 1985; Roberts, 1994). *Rattus exulans* skeletal remains first appear in Remote Oceania in Lapita settlements and are present in all archaeological sites associated with both Lapita and with the later Polynesian settlement, generally in the earliest layers. The ubiquitous distribution, and the fact that remains appear in large numbers in archaeological middens, suggests that *R. exulans* was more than just a stowaway, that it was likely introduced as a food item. These rats do not swim and, therefore, cannot self disperse. As a result, a phylogeographic analysis of *R. exulans* populations should provide evidence of the origins of the canoes that transported the animals and, thus, shed light on the origins of both the Polynesians and the Lapita peoples.

Analyses of mtDNA diversity between *R. exulans* populations (Matisoo-Smith et al., 1998) identified a local general homeland region, incorporating the Southern Cook and Society Islands, for all East Polynesian rats. In addition to providing evidence on the origins of the canoes that brought *R. exulans* to the extremes of the Polynesian Triangle, the degree of variation in island rat populations provides clues to the overall mobility of ancient Polynesian voyagers. The high levels of mtDNA variation found in island populations of *R. exulans* suggest multiple introductions to islands from a number of sources (Matisoo-Smith et al., 1998). These results are consistent with analyses of mtDNA variation in New Zealand Maori that estimate the founding female population size. Using computer simulations and random selection of mtDNA lineages found in an East Polynesian founding population, Murray-McIntosh and colleagues (1998) suggest that a minimum of between 50 and 100 founding women would be necessary to account for the frequencies of lineages seen in New Zealand Maori today – significantly more than were likely to have been transported in a single canoe.

Most recently, we have extended our studies to aDNA extracted from archaeological remains and museum samples of *R. exulans* (Matisoo-Smith and Robins, 2004). In doing so, we were able to address the issue of ultimate Polynesian and Lapita origins. We attempted to test Green's (2003) models for Lapita and, based on the distribution of ancient *exulans* remains and our analyses of their mtDNA variation, we were able to reject the two extreme views: the 'Express Train' type models, in particular the Taiwan origin, and the 'Indigenous Origins', models suggesting the Bismarck Archipelago origins for Lapita. Our results instead supported the 'VC Triple I' model. We found evidence suggesting that there were significant interaction spheres extending from the Philippines through the Solomon Islands; but that the *exulans* that were taken out into Remote Oceania were an intrusive component of Lapita and that, given the current sampling, the most likely origin for those Remote Oceanic rats was Halmahera in Wallacea.

Analyses of the so called 'domesticated' animals carried into the Pacific are less straightforward than those of *R. exulans*, in that pigs, dogs and fowl introduced by the Lapita and Polynesian colonists are the same species as those introduced later by Europeans and, therefore, populations are subject to more recent admixture. Unfortunately pig, dog and chicken bones are not as numerous in most archaeological sites as rat bones, but preliminary results from Pacific dog bones (Savolainen *et al*., 2004), pig remains (Allen *et al*., 2001; Larson *et al*., 2005) and chicken bones (Storey *et al*., 2003) are promising.

The commensal approach has also recently been applied to plants. Hinkle (2004), has been looking at variation in *Cordyline*, carried out to Polynesia from the west, and Clarke (2003) has been investigating sequence variation in Polynesian sweet potatoes and bottle gourds, which were brought back by Polynesian voyagers from South America. The combinations of these various commensal approaches with the more traditional studies of human genetic variation will, perhaps, allow us to tease apart several components of human origins and interactions throughout prehistory.

Conclusion

So, what can we say about the settlement of the Pacific? Do the various forms of genetic data provide an overall picture of the peopling of Oceania, and how do they relate to the archaeological and linguistic models? We do see some consistent patterns in the genetic data that appear to represent original founding populations. The first arrivals to Sahul most likely carried ancient haplotypes from mitochondrial haplogroups P and Q and Y chromosomes containing the ancient M9 mutation. The estimated ages of these haplotypes correlate well with the current archaeological estimates for the first settlement of Sahul. The two mitochondrial lineages may even correlate with the two routes of ancient migration, with the P representing the western route and Q the eastern. Internal differentiation clearly occurred over the next 30,000 years or so, resulting in unique lineages in various parts of Near Oceania, followed by at least intermittent contacts with Southeast Asia for several thousands of years prior to the appearance of Lapita.

At least some, if not a significant portion of the first Lapita colonists in Remote Oceania most likely carried mitochondrial lineages belonging to haplogroup B and Y chromosomes belonging to haplogroup O3, which both originated in Southeast Asia. However, many of the age estimates for both Y chromosome and mtDNA lineages in Near Oceania, including the mtDNA B4a1a1 'Polynesian motif', pre-date the appearance of Lapita in Near Oceania and so we cannot necessarily take the apparent Southeast Asian origin as support for any type of 'Fast Train' models for Lapita origins. The Lapita people most likely also carried the Near Oceanic P and Q mtDNA lineages and similar Near Oceanic and/or east Indonesian derived C and M Y chromosome lineages to Remote Oceania. We can clearly see in all of the data available, not surprisingly, that there has been significant admixture and interaction, certainly between populations in Near Oceania and, in particular, western Remote Oceania.

Whereas Central/Eastern Micronesia was most likely part of this Lapita expansion, the western regions of Micronesia, particularly the Marianas and Palau, had initial settlement followed by regular interactions with Southeast Asian populations and, thus, we see significant variation biologically, culturally and linguistically in Micronesia. Polynesia, and in particular central/eastern Polynesia is the only region that represents a relatively homogeneous, monophyletic group biologically, culturally and linguistically (Green, 1991; Kirch and Green, 2001).

As a result of research over the last few years, perhaps we now have enough descriptive data from Pacific populations that we can start addressing the questions of process. Comparisons between mtDNA and Y chromosome distributions can provide evidence of kinship, differential migration and settlement strategies between men and women (e.g. see the work of Hage and Marck, 2002, 2003). This type of work, however, will require significant collaborations between geneticists, social anthropologists and historical linguists to reach its full potential. Much more fine-grained analyses of genetic variation in areas of complexity are also providing particularly interesting evidence of interaction and its impact on genetic variation and, it is hoped, these types of studies will be extended into other similarly complex regions. We certainly need a better understanding of the history and biological variation amongst indigenous populations of Island Southeast Asia. Genetic analyses of the most recently discovered, early Lapita-associated skeletal remains and comparisons with aDNA from later populations mean that we can start to more fully and reliably understand the processes of change through time in the Pacific. Further developments in aDNA techniques suggest that we might also be able to study aspects of prehistoric health through amplification of ancient pathogens preserved in bone and other tissues (Sallares and Gomzi, 2001).

The overall picture of the human settlement of the Pacific is one of complexity. We can no longer think of it in terms of any specific number of waves of migration, but rather in terms of a dynamic, ongoing process. If we are to fully understand the process, it will require the independent analyses of data from a range of disciplines, but, more importantly, it requires discussion and real collaboration between the practitioners of those disciplines. Together those researchers can develop a better understanding of the various histories of this fascinating and complex region, and it is in this type of collaboration that anthropological geneticists will really play their part.

References

Allen, J. S., T. N. Ladefoged, E. Matisoo-Smith, R. M. Roberts, W. Norman, H. Parata, S. Clout and D. M. Lambert. (1996). Maori prehistory: the genetic trail of the Kiore and Kuri. *Archaeology in New Zealand*, **39**(4), 291–5.

Allen, J. (2000). From beach to beach: the development of maritime economies in prehistoric Melanesia. *Modern Quaternary Research in SE Asia*, **16**, 137–76.

Allen, M. S., E. Matisoo-Smith and K. A. Horsburgh. (2001). Pacific 'Babes': pig origins, dispersals and the potentials of mitochondrial DNA analyses. *International Journal of Osteoarchaeology*, **11**, 4–13.

Anderson, A. J. and Y. Sinoto. (2002). New radiocarbon ages of colonization sites in East Polynesia. *Asian Perspectives*, **41**(2), 242–57.

Bellwood, P., G. Nitihaminoto, G. Irwin, G. Gunadi, A. Waluyo and D. Tanudirjo. (1998). 35,000 years of prehistory in the northern Moluccas. *Modern Quaternary Research in SE Asia*, **15**, 233–75.

Betty, D. A., A. N. Chin-Atkins, L. Croft, M. Sraml and S. Easteal. (1996). Multiple independent origins of the COII/tRNA^{Lys} intergenic 9-bp mtDNA deletion in aboriginal Australians. *American Journal of Human Genetics*, **58**, 428–33.

Dickinson, W. R. (2003). Impact of mid-Holocene hydro-isostatic highstand in regional sea level on habitability of islands in the Pacific Ocean. *Journal of Coastal Research*, **19**, 489–502.

Chappell, J. (1993). Late Pleistocene coasts and human migrations in the Austral region. In M. Spriggs, D.E. Yen, W. Ambrose, R. Jones, A. Thorne and A. Andrews (eds.), *A Community of Culture — The People and Prehistory of the Pacific. Occasional Papers in Prehistory 21*. The Australian National University: Canberra, pp. 43–7.

Clarke, A. (2003). Using DNA sequence data to trace the origins and dispersal of sweet potato and bottle gourd in Polynesia: Implications for human mobility. Paper presented at the Australasian Society for Human Biology Annual Conference, Auckland, December 2003.

Cooper, A. and H.N. Poinar. (2000). Ancient DNA: do it right or not at all. *Science*, **289**, 1139.

Dunn, M., A. Terrill and G. Reesink. (2002). The East Papuan languages: a preliminary typological appraisal. *Oceanic Linguistics*, **41**, 28–62.

Flint, J., A.V.S. Hill, D.K. Bowden, S.J. Oppenheimer, P.R. Sill, S.W. Serjeantson, J. Bana-Koiri, K. Bhatia, M. Alpers, A.J. Boyce, D.J. Weatherall and J.B. Clegg. (1986). High frequencies of alpha thalassaemia are the result of natural selection by malaria. *Nature*, **321**, 744–9.

Forster, P., A. Torroni, C. Renfrew and A. Rohl. (2001). Phylogenetic star contraction applied to Asian and Papuan mtDNA evolution. *Molecular Biology and Evolution*, **18**, 1864–81.

Friedlaender, J.S., F. Gentz, K. Green and A. Merriwether. (2002). A cautionary tale on ancient migration detection: Mitochondrial DNA variation in Santa Cruz Islands, Solomons. *Human Biology*, **74**, 453–71.

Friedlaender, J., T. Schurr, F. Gentz, G. Koki, F. Friedlaender, G. Horvat, P. Babb, S. Cerchio, F. Kaestle, M. Schanfield, R. Deka, R. Yanagihara and D.A. Merriwether. (2005). Expanding Southwest Pacific mitochondrial haplogroups P and Q. *Molecular Biology and Evolution*, **22**(6), 1506–17.

Friedlaender, J.S., F. Genta, F. Thompson, F. Kaestle, T. Schurr, G. Koki, C. Mgone, J. McDonough, L. Smith and D.A. Merriwether. (2005). Mitochondrial genetic diversity and its determinants in island Melanesia. In A.K. Pawley, R. Attenborogh, J. Golson and R. Hide (eds.), *Papuan Pasts: Investigations into the cultural linguistic and biological history of the Papuan speaking peoples*. Adelaide: Crawford House Australia.

Green, R.C. (1991). Near and Remote Oceania — disestablishing 'Melanesia' in culture history. In A. Pawley (ed.), *Man and a Half: Essays in Pacific anthropology and ethnobiology in honour of Ralph Bulmer*. The Polynesian Society, Auckland, pp. 491–502.

Green, R.C. (2003). In *Pacific Archaeology: Assessments and Anniversary of the First Lapita Excavation (July 1952)*, Sand, C. (ed.) (Le cahiers de l'Archeologie en Nouvelle-Caledonie, Vol. **15**. Noumea, New Caledonia), pp. 95–120.

Hage, P. and J. Marck. (2002). Proto-Micronesian kin terms, descent groups, and interisland voyaging. *Oceanic Linguistics*, **41**, 159–70.

Hage, P. and J. Marck. (2003). Matrilineality and the Melanesian origin of Polynesian Y chromosomes. *Current Anthropology*, **44**, 121–7.

Hagelberg, E. and J.B. Clegg. (1991). Isolation and characterization of DNA from archaeological bone. *Proceedings of the Royal Society, London Series B*, **244**, 45–50.

Hagelberg, E. and J.B. Clegg. (1993). Genetic polymorphisms in prehistoric Pacific islanders determined by analysis of ancient bone DNA. *Proceedings of the Royal Society, Lond. B*, **252**, 163–70.

Hagelberg, E., S. Quevedo, D. Turbon and J.B. Clegg. (1994). DNA from prehistoric Easter Islanders. *Nature*, **369**, 25–26.

Hagelberg, E., M. Kayser, M. Nagy, L. Roewer, H. Zimbahl, M. Krawczak, P. Lio and W. Schiefenhovel. (1999). Molecular genetic evidence for the human settlement of the Pacific: analysis of mitochondrial DNA, Y chromosome and HLA markers. *Philosophical Transactions of the Royal Society of London, Series B*, **354**, 141−52.

Hertzberg, M., K. N. P. Mickleson, S. W. Serjeantson, J. F. Prior and R. J. Trent. (1989). An Asian specific 9-bp deletion of mitochondrial DNA is frequently found in Polynesians. *American Journal of Human Genetics*, **44**, 504−10.

Heyerdahl, T. (1952). *American Indians in the Pacific: The theory behind the Kon-Tiki Expedition*. London: Allen and Unwin.

Hill, A. V. S., D. K. Bowden, R. J. Trent, D. R. Higgs, S. J. Oppenheimer, S. L. Thein, K. N. P Mickleson, D. J. Weatherall and J. B. Clegg. (1985). Melanesians and Polynesians share a unique alpha-thalassemia mutation. *American Journal of Human Genetics*, **37**, 571−80.

Hill, A. V. S. and S. J. Serjeantson. (1989). *The Colonization of the Pacific: A Genetic Trail*. Oxford: Clarendon Press.

Hill, A. V. S., D. F. O'Shaughnessy and J. B. Clegg. (1989). Haemoglobin and globin gene variants in the Pacific. In A. V. S. Hill and S. J. Serjeantson (eds.), *The Colonization of the Pacific: A Genetic Trail*. Oxford: Clarendon Press, pp. 246−85.

Hinkle, A. E. (2004). Distribution of a male sterile form of 'Ti' (*Cordyline fruticosa*) in Polynesia and its ethnobotanical significance. *Journal of the Polynesian Society*.

Hurles, M. E., C. Irven, J. Nicholson, P. G. Taylor, F. R. Santos, J. Loughlin, M. A. Jobling and B. C. Sykes. (1998). *American Journal of Human Genetics*, **63**, 1793−1806.

Hurles, M. E., J. Nicholson, E. Bosch, C. Renfrew, B. C. Sykes and M. A. Jobling. (2002). Y chromosomal evidence for the origins of Oceanic-speaking peoples. *Genetics*, **160**, 289−303.

Ingman, M. and U. Gyllensten. (2003). Mitochondrial genome variation and evolutionary history of Australian and New Guinean aborigines. *Genome Research*, **13**, 1600−6.

Irwin, G. J. (1989). Against, across and down the wind: the first exploration of the Pacific Islands. *Journal of the Polynesian Society*, **98**, 167−206.

Irwin, G. J. (1992). *The Prehistoric Exploration and Colonisation of the Pacific*. Cambridge: Cambridge University Press.

Jobling, M. A. and C. Tyler-Smith. (2003). The human Y chromosome: an evolutionary marker comes of age. *Nature Reviews*, **4**, 598−612.

Kayser, M., S. Brauer, G. Weiss, P. A. Underhill, L. Roewer, W. Schiefenhövel and M. Stoneking. (2000). Melanesian origin of Polynesian Y chromosomes. *Current Biology*, **10**, 1237−46.

Kayser, M., S. Brauer, G. Weiss, W. Schiefenhövel, P. A. Underhill and M. Stoneking. (2001). Independent histories of human Y chromosomes from Melanesia and Australia. *American Journal of Human Genetics*, **68**, 173−90.

Kirch, P. V. (1986). Rethinking East Polynesian prehistory. *Journal of the Polynesian Society*, **95**, 9−40.

Kirch, P. V. (1997). *The Lapita Peoples: Ancestors of the Oceanic World*. Oxford: Blackwell.

Kirch, P. V. (2000). *On the Road of the Winds: An Archaeological History of the Pacific Islands Before European Contact*. Berkeley: University of California Press.

Kirch, P. V. and R. C. Green. (2001). *Hawaiki, Ancestral Polynesia: An Essay in Historical Anthropology*. Cambridge: Cambridge University Press.

Kivisild, T., M. J. Bamshad, K. Kaldma, *et al.* (15 coauthors), (1999). Deep common ancestry of Indian and western Eurasian mitochondrial DNA lineages. *Current Biology*, **9**, 1331–4.

Larson, G., K. Dobney, U. Albarella, M. Fang, E. Matisoo-Smith, J. Robins, S. Lowden, H. Finlayson, T. Brand, E. Willerslev, P. Rowley-Conwy, L. Andersson and A. Cooper. (2005). Worldwide phylogeography of wild boar reveals multiple centres of pig domestication. *Science*, **307**, 1618–21.

Leavesley, M. G., M. I. Bird, L. K. Fifield, P. A. Hausladen, G. M. Santos and M. L. diTada. (2002). Buang Merabak: Early evidence for human occupation in the Bismarck Archipelago, Papua New Guinea. *Australian Archaeology*, **54**, 55–7.

Lum, J. K. and R. L. Cann. (1998). mtDNA and language support a common origin of Micronesians and Polynesians in Island Southeast Asia. *American Journal of Physical Anthropology*, **105**, 109–19.

Lum, J. K. and R. L. Cann. (2000). mtDNA lineage analyses: origins and migrations of Micronesians and Polynesians. *American Journal of Physical Anthropology*, **113**, 151–68.

Lum, J. K., L. B. Jorde and W. Schiefenhovel. (2002). Affinities among Melanesians, Micronesians and Polynesians: a neutral biparental genetic perspective. *Human Biology*, **74**, 413–30.

Macaulay, V., C. Hill, A. Achilli, C. Rengo, D. Clarke, W. Meehan, J. Blackburn, O. Semino, R. Scozzari, F. Cruciani, A. Taha, N. K. Shaari, J. M. Raja, P. Ismail, Z. Zainuddin, W. Goodwin, D. Bulbeck, H.-J. Bandelt, S. Oppenheimer, A. Torroni and M. Richards. (2005). Single, rapid coastal settlement of Asia revealed by analysis of complete mitochondrial genomes. *Science*, **308**, 1034–6.

Matisoo-Smith, E. (1994). The human colonisation of polynesia. A novel approach: genetic analyses of the polynesian rat (*Rattus exulans*). *Journal of the Polynesian Society*, **103**, 75–87.

Matisoo-Smith, E., R. M. Roberts, G. J. Irwin, J. S. Allen, D. Penny and D. M. Lambert. (1998). Patterns of prehistoric mobility indicated by mtDNA from the Pacific rat. *Proceedings of the National Academy of Science, USA*, **95**, 15145–50.

Matisoo-Smith, E. and J. Robins. (2004). Origins and dispersals of Pacific peoples: evidence from mtDNA phylogenies of the Pacific rat. *Proceedings of the National Academy of Sciences, USA*, **101**(24), 9167–72.

Melton, T., R. Peterson, A. J. Redd, N. Saha, A. S. M. Sofro, J. Martinson and M. Stoneking. (1998). Polynesian genetic affinities with Southeast Asian populations identified by mtDNA anlaysis. *American Journal of Human Genetics*, **57**, 403–14.

Merriwether, D. A., J. S. Friedlaender, J. Mediavilla, C. Mgone, F. Gentz and R. Ferrell. (1999). Mitochondrial DNA variation is an indicator of Austronesian influence in Island Melanesia. *American Journal of Physical Anthropology*, **110**, 243–70.

Merriwether, D. A, J. A. Hodgson, F. R. Friedlaender, R. Allaby, S. Cerchio, G. Koki and J. S. Friedlaender. (2005). Ancient mitochondrial M halpogroups identified in the Southwest Pacific. *Proceedings of the National Academy of Science, USA*, **102**(37), 13034–39.

Murray-McIntosh, R. P., B. J. Scrimshaw, P. J. Hatfield and D. Penny. (1998). Testing migration patterns and estimating founding population size in Polynesia using mtDNA sequences. *Proceedings of the National Academy of Science, USA*, **95**, 9047–52.

O'Connell, J.F. and J. Allen. (2004). Dating the colonization of Sahul (Pleistocene Australia-New Guinea): a review of recent research. *Journal of Archaeological Science*, **31**, 835–53.

O'Connor, S. and Veth, P. (2000). The world's first mariners: Savannah dwellers in an island environment. *Modern Quaternary Research in SE Asia*, **16**, 99–137.

O'Rourke, D.H., M.G. Hayes and S.W. Carlyle. (2000). Ancient DNA studies in physical anthropology. *Annual Reviews in Anthropology*, **29**, 217–42.

O'Shaughnessy, D.F., A.V.S. Hill, D.K. Bowden, D.J. Weatherall and J.B. Clegg. (1989). Globin genes in Micronesia: origins and affinities of Pacific island people. *American Journal of Human Genetics*, **46**, 144–55.

Pawley, A. and R.C. Green. (1973). Dating the dispersal of the Oceanic languages. *Oceanic Linguistics*, **12**, 1–67.

Pawley, A. and M. Ross. (1993). Austronesian historical linguistic and culture history. *Annual Reviews in Anthropology*, **22**, 425–59.

Pawley, A. (2002). The Austronesian dispersal: languages, technologies and people. In P. Bellwood and C. Renfrew (eds.), *Examining the Farming/Language Dispersal Hypothesis*. Cambridge: McDonald Institute for Archaeological Research, pp. 251–73.

Quintana-Murci, L., O. Semino, H.-J. Bandelt, G. Passarino, K. McElreavey and A.S. Santachiara-Berecetti. (1999). Genetic evidence of an early exit of Homo sapiens sapiens from Africa through eastern Africa. *Nature Genetics*, **23**, 437–41.

Redd, A.J., N. Takezaki, S.T. Sherry, S.T. McGarvey, A.S.M. Sofro and M. Stoneking. (1995). Evolutionary history of the COII/tRNA^{Lys} intergenic 9 base pair deletion in human mitochondrial DNAs from the Pacific. *Molecular Biology and Evolution*, **12**, 604–15.

Roberts, R.M. (1994). Origin, dispersal routes and geographic distribution of *Rattus exlans* (Peale), with special reference to New Zealand. *Pacific Science*, **45**, 123–30.

Ross, M. (2001). Is there an East Papuan phylum? Evidence from pronouns. In A. Pawley, M. Ross and D. Tryon (eds.), *The Boy from Bundaberg: Studies in Melanesian linguistics in honour of Tom Dutton*. Canberra: Pacific Linguistics, pp. 310–21.

Sallares, R. and S. Gomzi. (2001). Biomolecular archaeology of malaria. *Ancient Biomolecules*, **3**, 195–213.

Savolainen, P., T. Leitner, A.N. Wilton, E. Matisoo-Smith and J. Lundeberg. (2004). A detailed picture of the origin of the Australian Dingo, obtained from the study of mitochondrial DNA. *Proceedings of the National Academy of Sciences, USA*, **101**(33), 12387–90.

Spriggs, M.J.T. and A.J. Anderson. (1993). Late colonisation of East Polynesia. *Antiquity*, **67**, 200–17.

Storey, A., D. Burley and D. Yang. (2003). Differential preservation of avian remains: New hope for ancient DNA in open air site samples. Paper presented at the Australasian Society for Human Biology Annual Conference, Auckland, December 2003.

Su, B., L. Jin, P. Underhill, J. Martinson, N. Saha, S.T. McGarvey, M.D. Shriver, J. Chu, P. Oefner, R. Chakraborty and R. Deka. (2000). Polynesian origins: insights from the Y-chromosome. *Proceedings of the National Academy of Science, USA*, **97**, 8225–8.

Sykes, B., A. Lieboff, J. Low-Beer, S. Tetzner and M. Richards. (1995). The origins of the Polynesians: an interpretation from mitochondrial lineage analysis. *American Journal of Human Genetics*, **57**, 1463–75.

Tate, G. H. H. (1935). Rodents of the genera *Rattus* and *Mus* from the Pacific Islands. *Bulletin of the American Museum of Natural History*, **68**, 145–78.

Taylor, J. M., J. H. Calaby and H. M. Van Deusen. (1985). A revision of the genus *Rattus* (Rodentia, Muridae) in the New Guinea Region. *Bulletin of the American Museum of Natural History*, **173**, 177–336.

Terrell, J. E. (1988). History as a family tree, history as an entangled bank: constructing images and interpretations of prehistory in the South Pacific. *Antiquity*, **62**, 642–57.

Terrell, J. E and R. L. Welsch. (1997). Lapita and the temporal geography of prehistory. *Antiquity*, **71**, 548–72.

Trejaut, J. A., T. Kivisild, J. H. Loo, C. L. Lee, C. L. He, C. J. Hsu, Z. Y. Li and M. Lin. (2005). Traces of archaic mitochondrial lineages persist in Austronesian-speaking Formosan populations. *PLoS Biology*, 3(8), e247.

Underhill, P. A., P. Shen, A. A. Lin, L. Jin, G. Passarino, W. H. Yang, E. Kauffman, B. Bonné-Tamir, J. Bertranpetit, P. Francalacci, M. Ibrahim, T. Jenkins, J. R. Kidd, S. Q. Mehdi, M. T. Seielstad, R. S. Wells, A. Piazza, R. Davis, M. W. Feldman, L. L. Cavalli-Sforza and P. J. Oefner. (2000). Y chromosome sequence variation and the history of human populations. *Nature Genetics*, **26**, 358–61.

Underhill, P. A., G. Passarino, A. A. Lin, S. Marzuki, P. J. Oefner, L. L. Cavalli-Sforza and G. K. Chambers. (2001). Maori origins, Y-chromosome haploypes and implications for human history in the Pacific. *Human Mutation*, **17**, 271–80.

Vayda, A. P. and R. A. Rappaport. (1963). Island cultures. In F. R. Fosberg (ed.), *Man's Place in the Island Ecosystem*. Honolulu: Bishop Museum Press, pp. 133–44.

White, J. P. and J. O'Connell. (1982). *A Prehistory of Australia, New Guinea and Sahul*. Sydney: Academic Press.

Chapter 15

The Prehistoric Colonization of the Americas

Francisco M. Salzano

Departamento de Genética, Instituto de Biociências, Universidade Federal do Rio Grande do Sul, Brazil

An old problem

America's 'discovery' by the Europeans of the fifteenth century posed a question: who were the strange people who inhabited the land? A papal bull from Paul III (1468–1549) solemnly recognized in 1537 their human status. Since they were humans they could only be descendants of Adam and Eve, more specifically of Noah's grand-children. Arias Montanus was able to find a resemblance between the words Peru and Ophir, one of Noah's descendants, therefore setting the question of their origin, at least in biblical terms (Pericot y García, 1962; Lavallée, 2000)

Equally bizarre was Florentino Ameghino's contention that all humanity had originated from the Argentinian Pampean region. Most of these suggestions or hypotheses that were put forward before the twentieth century or early in the 1900s have now only historical (or mythological) interest. Some of them have been summarized in Salzano and Callegari-Jacques (1988) and Crawford (1998). Since no paleoanthropological findings of high antiquity were found in America the questions were centred on three basic issues: (a) from where did the Amerindians come? (b) how many waves of immigration occurred?, and (c) when did they arrive? Researchers from different disciplines considered these questions and diverse, some-times contradictory answers had been made. They will be considered in the following sections.

The colonization process: non-genetic evidences

Geology and archaeology

The present consensus is that the prehistoric colonization of the New World should have occurred via the Bering Land Bridge, formed as a result of lower sea levels which were present there at the time. Human occupation of Beringia (defined by an eastern boundary located at the Mackenzie River in Canada and a western boundary at the Lena River in Siberia) dates back to an interval between the end of the Last Glacial Maximum cold peak (around 16,500 years before present or BP) and the beginning of the Holocene (9,100 years BP). Firmly documented archaeological sites are not present, however, until the middle of the Late Glacial (11,500 years BP to 10,500 years BP) (Hoffecker and Elias, 2003; Fagan, 2004).

Another question important for the early colonization of America concerns the availability of routes that would be open for the southern migration. Two such routes have been considered, the first giving an interior access to North America (the so-called 'ice-free corridor'), and the other an access through the coast. In relation to the first, the corridor probably remained closed by the Laurentide and Cordilleran ice sheets until about 11,000 years BP; while the northwest coast route could have been opened earlier, by 14,500 years BP (Hoffecker and Elias, 2003; Fagan, 2004).

Dixon (2001) devised a model for the colonization of the Americas that would have occurred along the coast. According to him, there is no need to think of human migration as a specific event, the metaphor of a 'wave' of migration being misleading. But he would identify, from a technological perspective, two major colonizing events. The first would be by the ancestors of the Clovis/Nenana complexes between 11,500 BP to 13,500 BP. These people would use atlatl darts tipped with bifacially flaked stone end blades lashed to harpoon-like heads seated on bone foreshafts. The second would be by those who carried the American Paleoarctic tradition, who introduced the bow and arrow. They would have entered the continent about 10,500 BP, and they would manufacture composite projectile points characterized by the insertion of razor sharp stone microblades along the sides of bone and antler projectile points. Later migrations would occur through the Cordillerian East Coast, Interior, and High Arctic regions. Ames (2003) provided archaeological details about the Northwest Coast sequences (from Yakutat Bay, Alaska to Cape Mendocino, California), while Hall et al. (2004) presented several arguments in favour of a coastal-entry model.

The Coastal-Migration hypothesis was tested by Surovell (2003) through a simulation model which included as parameters the width of the coastal corridor, maximum population growth rate, distances that could be travelled between occupied and unoccupied territories, and return rates for coastal and inland ecosystems. Three events were singled out: first inland migration, arrival in South America, and arrival on the coast at the latitude of Monte Verde, in Chile. The model was trying to answer the paradox that the Monte Verde site in Chile has an estimated age of 12,500 BP, while the Clovis site in New Mexico, USA has an estimated age one thousand years younger (11,500 BP). If the Amerindians could reach a coastal distance of 13,400 km southward 1,000 years prior to inland migration into

North America, the hypothesis would be met. The conclusion was that the Coastal-Migration hypothesis alone could not explain the dates discrepancy.

Until recently the picture for the colonization of the Americas was fairly simple: the first Americans were identified with the Clovis people, spear-hunters who had crossed the land bridge from Siberia through the ice-free corridor. Decimating the large game with their bifacial fluted projectile points, they would have moved south through upland grasslands until they reached the tip of South America just one thousand years after their entrance into the continent.

The findings today, as indicated in the previous paragraphs, point to alternative ways of thinking. First, the colonization seems to have occurred by opportunistic foragers, who collected a wide range of foods with simple stone tools flaked on only one side, and who occupied the most diverse habitats. Most well-documented sites contain no megafaunal bones contemporary with the human occupations, but instead fruits, nuts, vegetables, fish, shellfish, and small game animals. Secondly, a curious pattern emerges when the most well-established dates of these Paleoindian sites are considered (Table 15.1). No clear north-south cline is observed; instead most of the dates occur in the 10,000–11,000 BP range, independently of geographical location. It is as if a near instantaneous radiation of people occurred at about this time.

The geologic and archaeological findings, therefore, are not able to give a clear, simplified picture of America's colonization. Part of the problem, of course, may be due to diverse availability, related to differential site preservation in the north, centre, and south of the continent. As also stressed by Surovell (2003), site localization is a function not simply of the length of occupation of a region, but of the cumulative occupation, expressed as the number of persons occupying a region multiplied by time; regions with long cumulative occupation spans would be more likely to produce early dates than those with short cumulative occupation spans, even if people first arrived in the latter. A final, simpler explanation would be that the researchers would hesitate to accept or tend to reject earlier dates that would contradict their hypotheses!

Perhaps more important than simple dates are the ways of living of these peoples during these early days. Nuñez et al. (2002) provided an elegant analysis for the type of adaptation that humans utilized in the occupation of the Atacama Desert in Chile between 10,500 BP to 7,000 BP and again after 4,000 BP. As was emphasized by Dillehay (2002) the study in question is an example of how the collation of data from different sources can help to understand past events.

Microscopic plant crystals (phytoliths), collected from the Las Vegas site in Ecuador and directly dated from 9,320 to 10,130 years BP, provided evidence for the human cultivation of Cucurbita (squash and gourd) at that time, suggesting an independent emergence of plant food production in South America, contrary to the view that

Table 15.1. Selected information on archaeological sites which provided data on Amerindian early settlements (10,000 years BP or earlier).

Site and location	Date (years before present)	References
Mesa, Alaska	10,240	1
Old Crow, Canada	27,000	1,2
Bluefish Caves, Canada	12,290–12,900	2,3
Nenana, Alaska	10,700–11,800	1,4
Colby, USA	10,864	1
Union Pacific Mammoth	11,280	1
Vail, USA	10,500	1
Debert, USA	10,500	1
Meadowcroft, USA	14,000–16,000	2,4
Dent, USA	10,750	1
Daisy Cave, USA	10,650	1
Arlington Springs, USA	10,960	1
Shawnee Minnisink, USA	10,560	1
Domebo, USA	10,900	1
Clovis, USA	10,900–11,500	1,2
Cactus Hill, USA	10,920	1
Lehner, USA	10,950	1
Murray Springs	10,880	1
Taimar Taima, Venezuela	11,800–12,600	2,4,5
El Abra, Colombia	11,210–12,400	1,2
Tibitó, Colombia	11,700	1,2
Las Vegas, Ecuador	10,840	1
Pedra Pintada, Brazil	11,075	1,5
Pedra Furada, Brazil	12,000	2,4,5
Serra da Capivara, Brazil	11,060	6
Baixão do Perna, Brazil	10,530	1,2
Chão do Caboclo, Brazil	11,000	1
Brejo da Madre de Deus, Brazil	11,060	1
Guitarrero Cave, Peru	10,535	1
Quebrada Jaguay, Peru	11,105	1
Quebrada Tacahuay, Peru	10,770	1
Goiás, Brazil	10,750	1
Santa Elina, Brazil	10,000	5
Lapa do Boquête, Brazil	11,000–12,000	1,5
Lapa do Dragão, Brazil	11,000	1
Lapa Vermelha, Brazil	11,000	5
Rio Uruguai, Brazil	12,700	1,5
Quereo, Chile	11,000–11,400	5
Tagua-Tagua, Chile	11,000–11,400	5
Cerro La China, Uruguay	11,150	1
Monte Verde, Chile	12,500	2,4,5
Los Toldos, Argentina	12,600	2
Cueva del Medio, Chile	10,550	1
Fell's Cave, Chile	10,300–11,000	1
Tres Arroyos, Chile	10,600	1

References: 1. Roosevelt *et al.* (2002); 2. Lavallée (2000); 3. Hoffecker and Elias (2003); 4. Dixon (2001); 5. Prous and Neves (2000); 6. Lessa and Guidon (2002).

this event originated in Mesoamerica (Piperno and Stothert, 2003; Bryant, 2003). It also appears that contrary to early beliefs, complex regional settlement patterns and large-scale transformation of local landscapes were present in Amazonia at the time of the European discovery and even previously (Roosevelt *et al.*, 1996, 2002; Heckenberger *et al.*, 2003).

Paleoanthropology and morphology

What was the morphology of these First Americans? Measurements in Paleoindian crania have been performed at an increasing pace in recent times, appropriately analysed with modern statistical tools, and a comparison with more recent Amerindian remnants showed a clear difference, Paleoindians, as characterized by the famous Kennewick and Luzia skulls, are distinguished by long and narrow neurocrania, low, narrow and projecting faces, while historical specimens, on the contrary, showed short and wide neurocrania, high, wide and retracted faces. Most recent studies showing these and other differences between the two sets are those of Brace *et al.* (2001), Jantz and Owsley (2001), González-José *et al.* (2001a), and Neves *et al.* (2003, 2004, 2005), Hubbe *et al.* (2004). The latter authors also provided a comprehensive list of previous investigations which led to this result.

A Two Main Biological Components Model was then presented by Neves *et al.* (2003), which took into consideration also the suggestion of Dixon (2001) described in the previous section. An 'Australo-Melanesian-like' population, relying on atlatl darts as the primary weapon would have entered the Americas around 14,000 years BP. They would expand southward in the New World along the Pacific Coast, but at the Panama Isthmus they would separate in three directions to the south. The first group would migrate along the Pacific coast, the other along the Atlantic coast, and the third inward into the Amazon basin. This would explain the several early dates of occupancy (10,000–12,000 years BP) throughout the South American territory.

When climatic conditions ameliorated in North America, the other migration would take place around 11,000 years BP, their members using bow and arrow. These people would have a cranial morphology similar to that seen in present-day Northern Asians and Native Americans.

In Neves *et al.*'s (2003) view, there was isolation between these two stocks and replacement of the first by the later migrants. However, González-José *et al.* (2003), studying 33 skulls of Baja California dated at early historical times for 24 variables, stored at the Regional Museum of La Paz and National Museum of Anthropology and History of Mexico, showed that they presented clearer affinities with Paleoamerican remains than with modern Amerindians. They concluded that the Baja California peninsula may have provided a refuge for these early people, who would have been kept isolated from the most recent migrants, with Mongoloid characteristics, by the Sonoran desert. Dillehay (2003), however, doubts that the isolation could be so effective as to prevent a merging of characteristics.

What clues do we have concerning the health of these individuals? Steckel *et al.* (2002) considered this question, devising a Health Index, a multiple-attribute measure of this state as expressed in bones, that would consider the following characteristics: stature, signs of iron-deficiency anaemia, enamel hypolasias, dental health, degenerative joint disease, trauma, and skeletal infections. A total of 12,520 skeletons were considered, of which 9,826 were from Native Americans, covering a time span from about 4,000 BP to the early twentieth century. The index average values among the Amerindians ranged from 64.0 among Central Americans to 78.1 for Eastern North Americans, the value for South Americans being 75.1 (perfect health would display an index of 100). Very good health was observed among coastal Brazilians, coastal Ecuadorians and coastal Georgians, while the worst value was found at Hawikku, in New Mexico. A temporal trend for declining health was observed, reflecting acculturation problems.

The prevalences of parasites in prehistoric Amerindians can in principle also furnish data about past migrations. In relation specifically to intestinal parasitism Confalonieri *et al.* (1991) reviewed the previous results. The presence of *Enterobius vermicularis* in coprolites dated to 10,000 years BP, and the data suggested a route of infected people along the Pacific coast up to Argentina. The infection was not observed in Brazilian remains. More difficult to interpret are the findings related to *Ancylostoma duodenale* and *Trichuris trichiura*. Since these parasites need a passage through the soil for their reproduction, the Beringian route for the prehistoric migrations was considered unlikely due to low soil temperatures. On the other hand, *Ascaris lumbricoides* seems to be a North American Indian infection that did not reach South America in prehistoric times.

Regional analyses are useful, since they can provide details that continental evaluations would miss. Two examples of such endeavours will be summarized here. The first was the work by Varela and Cocilovo (2002) on 237 crania from the Azapa Valley and Coast, Chile (18°S–30°S). The individuals, studied for six variables, lived between 3,500 BP to >1,470 AD. The evidence supported the ethnohistorical hypothesis that suggested a clear socioeconomic differentiation between the coastal and valley populations during the Late Intermediate periods. As a matter of fact, strong biological isolation was evident after the Early Intermediate, the coastal groups specializing in sea resources while the valley populations relied on agricultural or pastoral sources for survival.

The second example involves studies in a region located further south than that of the first investigation (37°S–55°S). First González-José *et al.* (2001b), based on 253 skulls subdivided in nine subsamples and studied for 20 nonmetric traits, tested four models for the settlement of Patagonia. The one that best fitted the data was M4, which implied two different migratory waves, one originating the southern and the other the northern groups. In another investigation González-José *et al.* (2002) studied 441 skulls divided in nine samples from six Patagonian and three Fueguian populations tested for 13 measurements. A long history of isolation between Fueguians and Patagonians was inferred.

A word of caution should be given in relation to these studies. Despite claims to the contrary (Jantz and Owsley, 2001), cranial morphology is quite plastic, as is documented by the effects of skull deformation habits that were widespread in some Amerindian groups. Selective factors may also be present, obscuring historical ancestral-descendant relationships.

Linguistics

Locke (2001) asserted that 'One can think of few issues within the field of biological anthropology that are more important than the evolution of language'. This view, however, was not accepted in the relatively recent past, and as a matter of fact the *Societé de Linguistique de Paris* approved a motion, in 1866, prohibiting any mention to the origin of language in its texts (Franchetto and Leite, 2004). The fact is that after boundless speculation this study received the support of a series of hard data, both from linguistics itself as well as from related disciplines, turning, therefore, to a respectable area of science. Phylogenetic approaches have been recently summarized by Locke (2001), Hauser *et al.* (2002), Bever and Montalbetti (2002), and Culotta and Hanson (2004). Diamond and Bellwood (2003), on the other hand, tried to relate the expansion of agriculture with language development on a worldwide scale.

Amerindian languages have been the subject of comparative studies from as long ago as the nineteenth century. Reviews of those previous works can be found in Wilbert (1968), Rodrigues (1986), Greenberg (1987) and Cavalli-Sforza *et al.* (1994); and our group (Salzano *et al.*, 2005) has recently investigated how far the genetic data would favour one of three alternatives proposed by eminent linguists for the relationships among the four most important lowland South American Native language families (Arawak, Carib, Ge, Tupi). Greenberg's (1987) classification did not fit either Y-chromosome (Bolnick *et al.*, 2004) or mtDNA (Hunley and Long, 2005) data.

A pioneer in the investigation of the rate of linguistic change with time was Morris Swadesh. In recent times, however, Johanna Nichols can be singled out as the most important linguist in this area of endeavour, especially in relation to its application to Amerindian origins (Adler, 2000).

Nichols (2002) ably summarized her views concerning the prehistoric colonization of the New World. She stressed that while there are about 6,000 languages on earth, which can be grouped into around 300 genealogical families and many structural types,

150 of these families and the full range of structural types are native to the Americas. This diversity seems far too high for the Clovis-type colonization model. A series of analyses and simulations indicated that the human settlement of the Americas should have begun over 20,000 years ago, and that the process involved three main trajectories. Immigrants from Siberia spread southward chiefly along the coast with, from time to time, spread eastward in the interior; while after the end of glaciation northward and eastward movements occurred in North America. Five large areal populations were distinguished: (a) western North America, extending to the Rocky Mountains; (b) eastern North America, from this chain of mountains to the Atlantic; (c) Mesoamerica, that is, southern Mexico and Central America; (d) western South America; and (e) eastern South America. They form a network with a hub centred in the vicinity of the eastern Gulf of Mexico and/or the eastern Caribbean, from which there were spreads in several directions. Mesoamerica, eastern North America and eastern South America form a closely linked network, with western North America and western South America more distant. The point to be emphasized is that the main factor in these relationships was not outside migration, but autoctonous diversification.

Some of Johanna Nichols' views were criticized by her colleagues (Adler, 2000), and specifically Nettle (1999) argued that her assumption that the diversity of linguistic stocks increases linearly with time is not valid, and that on the contrary the high Amerindian linguistic diversity points to a recent colonization of the continent. These diverging views rely on basically different models of linguistic fissioning and extinction; therefore, no decision can be reached at the present time. But Nettle (1999) did not present any alternative model for the prehistoric American colonization.

The colonization process: genetic evidences

Blood groups and proteins

During all the twentieth century a wealth of information on the so-called 'classical' polymorphisms has accumulated in relation to Amerindian populations. Reviews can be found in Salzano and Callegari-Jacques (1988) and Crawford (1998). Cavalli-Sforza et al. (1994) have specifically considered in what way this information could be used to evaluate the question of Amerindian origins. They compiled data from 115 populations, which were grouped into 23 linguistic units, investigated in relation to 73 genetic markers. The study involved the calculation of genetic distances, phylogenetic trees, and principal component analysis. Additionally, dendrograms of individual tribes, as well as geographic maps of single genes, were used.

What were the main conclusions of this study? They observed two major clusters in the genetic tree, one involving the Arctic and the other the Amerind populations; and the Arctic cluster showed a secondary split into Na-Dene and Eskimo. Therefore, they concluded that the data were indicating the occurrence of three main waves of migration for the colonization of the New World, as had been postulated by Greenberg *et al.* (1986) on the basis of linguistic, dental, and blood group plus protein gene frequency results. But they also asserted that the Na-Dené and Eskimo could have separated in Beringia, which constitutes the scenario proposed, on the basis of mtDNA data, by those who suggested just one migrational event for the early colonization of the Americas (Bonatto and Salzano, 1997a,b).

As for the question of dating, comparison between all Amerinds and all northern Mongoloids led to an estimated time of divergence of 31,000 years BP. Cavalli-Sforza *et al.* (1994) discussed why they believed that this could be an overestimate. They also stressed that this was the date of separation presumably still on the Asian mainland, and that the date of passage of the Bering strait could have been later. From the tree that they had obtained they estimated that the Na-Dené/Eskimo split should have occurred (probably still in Asia) about 18,000 years BP.

DNA markers

Mitochondrial DNA

Mitochondria are organelles found in the cell's cytoplasm, where they occur in large (1.0–1.5 thousand) copies. Their inheritance is strictly maternal (except in few, anomalous cases), and their genome in humans has 16,568 base pairs arranged in a circular fashion which does not recombine. With the progress in molecular testing techniques, several studies surveyed this whole genome in different populations (for instance, Ingman and Gyllesten, 2001) and the abundance of copies per cell favours its investigation in ancient, prehistoric samples.

These characteristics make mtDNA an excellent material for genetic and evolutionary studies, and indeed Amerindians have been extensively investigated in relation to this organelle. By the use of distinctive markers it is possible to distinguish at least five (A, B, C, D, X) haplogroups in samples from different individuals, and material of diverse antiquity. For instance, Salzano (2002) lists 13 samples in a total of 338 individuals who lived from 150 years to 8,000 years ago in the region. Marked differences in haplogroup prevalences were found, with absence of A and B among 17 ancient Aleuts, and of A and B in 58 Fueguian-Patagonian remains. Studies in extant groups were of course much more extensive, and included 90 samples and a total of 3,829 persons. Haplogroup D was the most frequent (67%) in a sample of 57 Aleuts, while A was the most common (65%–76%) in North American Eskimo and Na-Dené, as well as in Mexico and Central America (76%). In North and

South American Amerinds, the haplogroup prevalences were more uniform, with a small excess (respectively 36% and 32%) of B.

How were these different data interpreted in relation to the peopling of the Americas? Table 15.2 reviews the interpretations of 11 studies. The techniques employed in these studies varied from high density restriction fragment length polymorphism (RFLP) analyses to sequencing either of the high variable segment I (HVS-I), of both HVS-I and HVS-II, extensive portions outside these regions, or of the whole genome.

Before we examine the figures in Table 15.2 it is important that we remember the limitations of the methods employed. The first problem is the calibration of the mtDNA substitution rate, since the derived time estimations are based on the relationship between molecular diversity and elapsed time to generate it. Coalescence times of a given molecule do not necessarily exactly match population histories. Most of these analyses ignore population structure, and in the case of highly variable sites back mutations may occur, blurring the relationships which are being investigated. The concept

Table 15.2. | Peopling of the Americas as viewed from mtDNA data.

No. of waves	Haplogroups	Type of study	Arrival time (years before present)	References
4	A–D	HVS-I+HVS-II sequences	14,000–21,000	1
3	A,C,D	RFLP	26,000–34,000	2
	B	RFLP	12,000–15,000	2
	A	RFLP	7,000–9,000[1]	2
2	A–D,X	HVS-I sequences	20,000–25,000	3
	A	HVS-I sequences	11,000[1]	3
1	A–D	HVS-I+HVS-II sequences	30,000–40,000	4
–	X	RFLP+HVS-I sequences	12,000–36,000	5
1	A–D	HVS-I sequences	23,000–37,000	6
3	A,C,D	RFLP	24,000–48,000	7
	B	RFLP	13,000–18,000	7
	X	RFLP	13,000–35,000	7
1	A–D	8.8 kb outside the D-loop	12,000–19,000	8
2	A,C,D	Complete sequences	26,000–32,000	9
	D	Complete sequences	10,000	9
–	A–D	HVS-I sequences	13,000–19,000[2]	10
–	X	Complete sequences	11,000–18,000	11

[1]Beringian expansion, leading to Eskimo and Na-Dené peoples.
[2]Divergence time between Central and South Amerindians.

References: 1. Horai *et al.* (1993); 2. Torroni *et al.* (1994); Wallace (1995); 3. Forster *et al.* (1996); 4. Bonatto and Salzano (1997b); 5. Brown *et al.* (1998); 6. Stone and Stoneking (1998); 7. Schurr (2000); 8. Silva *et al.* (2003); 9. Bandelt *et al.* (2003); 10. Fuselli *et al.* (2003); 11. Reidla *et al.* (2003).

of 'waves' of migration was already criticized at the beginning of this chapter; and the search for a potential source of migrants in the Old World assumes that the source population remained stable and unchanged over many thousands of years (Bandelt *et al.*, 2003; Eshleman *et al.*, 2003).

With these reservations in mind, let us look at the figures in Table 15.2. The number of waves of migration assumed varied from one to four, and the arrival times of the main bulk of migrants (carrying haplogroups A-D or A, C, D) from 12,000 to 48,000 years BP. These wide differences are due to different assumptions which have to be made, as was mentioned in the previous paragraph. The late arrival of persons carrying exclusively the A, B or D haplogroups was not confirmed in most recent analyses, and in relation to haplogroup X, confined to North American Amerindians (Dornelles *et al.*, 2005), there is also no clear indication that their carriers arrived independently, since the values obtained (11,000 to 36,000 years BP) present about the same range as that obtained for the other haplogroups. Different test procedures also do not explain earlier or later dates.

From where did the Paleoindians emerge and expand into the Americas? Candidates are northern China, southeastern Siberia or Mongolia, and in terms of mtDNA, since the Altaians exhibit all four Asian and American Indian-specific haplogroups (A-D) in higher frequencies than those observed in Mongolians or Chinese, and are the only ones who carry the X haplogroup in southern Siberia (Derenko *et al.*, 2001), they are the best candidates for the hypothetical source population.

Mitochondrial DNA can also be important for the unravelling of regional patterns of migration. Thus, complete sequencing of the DNA of 30 Aleuts of the Commander Islands and seven Sireniki Eskimo from Chukotka showed that they are closely related; and the geographic specificity and remarkable intrinsic diversity of one of the lineages (D2) supported the refugial hypothesis. The latter assumed that the Eskimo-Aleut founding population originated in Beringian/southwestern Alaskan refugia during the early postglacial period rather than from a more recent migration from the interior of Siberia (Derbeneva *et al.*, 2002).

Major prehistoric movements in North America have been evaluated by Malhi *et al.* (2002) and Eshleman *et al.* (2003). Based on extensive surveys using RFLPs and HVS-I sequences described in these and previous publications of the group, they were able to detect patterns of population expansions in the region and to test seven specific hypothesis proposed by archaeologists and linguists, with variable results.

The mtDNA of the Taino, extinct inhabitants of the Caribbean, showed only two (C and D) haplogroups and markedly reduced diversity. These findings led Laloueza-Fox *et al.* (2001) to suggest that South America (where C and D are more frequent) should have been the homeland of the Taino ancestors, confirming linguistic and archaeological results, and that they had lost diversity in their migratory movement, which followed the chain configuration of the Antillean islands.

Spatial structure involving especially haplogroups A and D in Colombia suggested to Keyeux *et al.* (2002) two distinct migrational waves of ancestors who expanded through South America, one along the Pacific Coast, and the other along a lowland Amazonian trail. Fuselli *et al.* (2003) also found evidence for differences in western and eastern South American Indian populations; they dated the divergence time between Central and South Amerindians between 13,000 and 19,000 years BP.

As the number of complete mtDNA sequences available for analyses increases, it is becoming clear that natural selection has to be considered in the interpretation of its American and worldwide variability. Mishmar *et al.* (2003) evaluated the substitution rates in several regions of this organelle and identified at least three genes (*ABP6*, *cytochrome b*, and *cytochrome oxidase I*) whose variability may be functionally significant. They suggested that at least one of the selective factors may be climate.

Y chromosome

Genes in the non-recombining portion of the Y chromosome (NRY) can be considered as a counterpart of those present in the mtDNA; as the latter, they do not recombine across generations and are uniparentally inherited, in this case along the paternal line. The earlier studies suggested almost no or very restricted variability in this portion of DNA, but later findings documented more variation, although it is generally less marked than those found in the mtDNA or autosome regions.

The number of American Native populations tested for NRY markers is higher than 50 (Salzano, 2002), but the techniques and systems considered [pulsed-field electrophoresis; an alphoid highly variable system; RFLPs; short tandem repeat (STR); single nucleotide (SNP), or insertion/deletion polymorphisms; sequence; high-performance liquid chromatography (DHPLC)] were highly variable, as was the number of markers studied. To further complicate matters, different nomenclatures had been used until a consensus had been achieved (Y Chromosome Consortium, 2002). This nomenclature is based in a binary, hierarchical system. The lineages (haplogroups) are defined on the basis of the most derived mutation at a given biallelic locus, and an asterisk for the exclusion of other markers that define other lineages within the clade. Additional characterization can be made by the use of highly variable STRs, which would then define specific haplotypes.

Earlier investigations suggested a single founder Native American Y lineage, and therefore was consistent with a unique migratory 'wave' into the continent. This lineage, using the current nomenclature, is characterized by a C→T mutation at marker M3 within the P-M45Y lineage, and occurs only in Amerindians in high frequency

(77%; Bortolini *et al.*, 2003). A second lineage, marked by a nucleotide substitution in the RPS4Y gene, was however subsequently discovered at lower frequencies (Bergen *et al.*, 1999), which was restricted to eastern Asia and Native Na-Dené, North and Central Americans. It was then postulated that it had been introduced at a later date in the Americas (Karafet *et al.*, 1999; Lell *et al.*, 2002).

From where would these migratory events originate? Lell *et al.* (2002) suggested that the first migration departed from southern Middle Siberia, and the second from the Lower Amur/Sea of the Okhkotsk region. However, as noted by Bortolini *et al.* (2003), this proposal is inconsistent with the generally high frequency of haplogroup K-M9 in eastern Siberia and its absence in the Americas. The principal-component analysis performed by the latter authors suggested a close genetic relatedness between the Chipewa and Cheyenne (both North American Indians) and certain populations of central/southern Siberia (notably the Kets, Yakut, Selkup, and Altai). This pattern agrees with the distribution of mtDNA haplogroup X, absent from eastern Siberia but present in the Altaians of southern Central Siberia. The relationship between the Ket and Na-Dené languages (Greenberg, 1996) also favours this region as the original home of the First Americans.

As for the age of these migrations, Bianchi *et al.* (1998) derived an ancestral founder haplotype, OA, of the DYS199T (M3) lineage, calculated as the Y microsatellite mutation rate the figure of 0.0012, and from data obtained from seven such loci, assuming a generation time of 27 years, derived the date of entry in the Americas as 22,770 years BP.

Other estimates showed widely different numbers: (a) 2,100–30,000 years BP (Underhill *et al.*, 1996); (b) 9,000–11,000 years BP (Ruiz-Linares *et al.*, 1999); (c) 7,650 years BP (Karafet *et al.*, 1999); (d) 17,700 (for the entire Q lineage) and 27,500 (for the entire C lineage) (Hammer and Zegura, 2002).

More recent values were obtained by Seielstad *et al.* (2003) and Bortolini *et al.* (2003). The first used two methods of estimation, both based on the accumulation of variation at 15 microsatellite loci in the newly observed M242-T lineage, found in low frequencies in Eurasia and in 100% of Amerindians who do not carry the RPS4Y mutation. It was assumed, based on extensive typing, that the M242 T mutation occurred before the M3 change and therefore very close in time to the entry of the First Americans in the continent. One method relies on assumptions about effective population sizes and population growth rates (Su *et al.*, 1999). Using several estimates about these variables, the microsatellite mutation rate, and generation times, Seielstad *et al.* (2003) arrived to what they considered the best estimate of 14,000–15,000 years BP. The second method they used, developed by Stumpf and Goldstein (2001) relies on an inference of the ancestral microsatellite haplotype, mutation rates and generation times. Again, varying these parameters, they arrived to a best estimate of 15,000–18,000 years BP.

Bortolini *et al.* (2003), used a third method, based on the mean average square distance between the inferred ancestral haplotype of a haplogroup and all its observed descendants. With a generation time of 25 years and a Y microsatellite mutation rate of 0.0018 they calculated the age for the Q-M242 lineage in Amerinds as 13,611 years BP, and in Mongolians as 15,416 years BP. The similarity between the estimates suggested to them an entry in the Americas soon after the occurrence of the mutation. Another point of interest is related to lineage Q-M19, which occurs only in the Ticuna (67%) and Wayuu (10%), and not in 24 other Amerindian populations (including some occupying areas intermediate between the places where these two South American Indian tribes live). The estimated age for Q-M19 is similar to that obtained for its parental haplogroup Q-M3 ($\approx$7,000–8,000 years BP), suggesting that population isolation and possibly the process of tribalization of Native Americans started soon after the initial dispersal in the region.

Finally, Zegura *et al.* (2004) studied not less than 63 binary polymorphisms and 10 STRs on a sample of 2,344 chromosomes from 18 Native American populations (with an exception all from North or Central America, while 23 of the 25 populations studied by Bortolini *et al.*, 2003, were from South America) plus 28 Asian and five European groups. Differently from the conclusions of Bortolini *et al.* (2003), favouring two major migrations, they found that their data pointed to a single entry of the ancestors of Native Americans in the New World. Their divergence time estimates ranged from 10,100 to 17,200 years BP irrespective of statistical method, population comparison, or haplogroup employed. In accordance with the opinion of Tarazona-Santos and Santos (2002) they believe that the R lineage, considered by Karafet *et al.* (1999) and Lell *et al.* (2002) to be indicative of a second wave of migration, was actually introduced in Amerindians by non-Native admixture.

X and autosomes

Much fewer studies were performed in the X, as compared to the Y chromosome investigations (Salzano, 2002). The reason for this is simple. While the non-recombining region provides an opportunity for the evaluation of a series of historical events (as outlined in the previous section), the X chromosome regions present a dynamics of change that is not amenable to easy analyses (either exclusive inheritance, which however is not free from recombination; or pseudoautosomal patterns in the homologous X/Y portions). The six studies which included X chromosome variability in American Natives listed by Salzano (2002) did not address the problem of the colonization of the New World.

Hey (2005) devised a new method of analysis that allows estimation of founding population sizes, changes in them, time of population formation, and gene flow. He then applied it to the question of the colonization of the New World using data from seven autosome, one X-chromosome, and mtDNA loci. According to his results the founding population of the New World was very small (less than 80 individuals) and arrived at a relatively late (maybe around 14,000 years ago) time.

The human leucocyte antigen (HLA) system variability has been used, however, for some inferences related to the problem under consideration. Tokunaga *et al.* (2001) using serology-level HLA-A and HLA-B, as well as HLA-DRB1 sequence-level data, stressed the strong genetic affinities between Native Americans and Northeast Asians, especially the Ainu, considered to be descendants of the Upper Paleolithic peoples of the region. Tsuneto *et al.* (2003), based on HLA-DRB1 allele frequencies, verified a clear separation between a large number of Amerindian populations and representatives of the Na-Dené and Eskimo. Arnaiz-Villena *et al.* (2000) and Gómez-Casado *et al.* (2003), also considering the HLA-DRB1 variability, stressed the uniqueness of Amerindians and their separate clustering from Siberian/Na-Dené/Eskimo groups, but from this simple finding elaborated a series of strange hypotheses (repeated in exactly the same words in the two papers) concerning the peopling of the Americas.

Data from other autosome systems are sparse and generally did not consider the problem examined here. Studies of our group, including beta-globin cluster haplotypes (Bevilaqua *et al.*, 1995), dopamine receptor DRD2 and DRD4 genes (Hutz *et al.*, 2000), 15 short tandem repeat polymorphisms (STRPs) (Hutz *et al.*, 2002), the highly variable *Alu* insertion of the 3'UTR of the *LDLR* gene (Heller *et al.*, 2004) and polymorphic L1 and *Alu* insertions (Mateus Pereira *et al.*, 2005) generally suggest the absence of major multiple migration events and a single origin for the first colonizers of the American continent.

Viruses, bacteria, and fungi

Since the pioneering studies of Neel *et al.* (1994), a series of studies have been developed trying to relate microorganism parasite variability with past Amerindian migrations. Several investigated the T-cell lymphotropic virus types I and II (review in Salzano, 2002). Since HTLV-II was present in high frequencies in American Natives but not in Siberian ethnic populations, it was suggested that the first migrants to America should have derived mainly from Mongolia and Manchuria.

The human polyomavirus JC (JCV) was also studied in several investigations involving Amerindians (Agostini *et al.*, 1997; Sugimoto *et al.*, 2002; Zheng *et al.*, 2003; Pavesi, 2004). The distinctiveness of Asian/Amerindian strains from those derived from other human populations was stressed by all of them, with no distinction between the Amerind and Na-Dené peoples. On the other hand, the Eskimo carried a unique genotype belonging to type A, which led to a separation from the Amerind/Na-Dené cluster.

Three loci of *Helicobacter pylori*, a Gram-negative bacterium that colonizes the human stomach and that may lead to increased risk for gastric cancer and peptic ulcer disease, were studied in Amazonian

(Venezuelan) Amerindians by Ghose *et al.* (2002). They found among them East Asian genotypes not present in the Mestizo population, suggesting that *H. pylori* had been introduced in the Americas with the first prehistoric migrants.

Coccidioides immitis, the etiologic agent of coccidioidomycosis, was also studied using nine microsatellite and three other genes in North and South American Natives by Fisher *et al.* (2001). They estimated that the introduction of this parasite into South America should have occurred within the last 9,000–140,000 years.

Synthesis

Table 15.3 tries to collate the diverse pieces of evidence previously discussed. As can be seen, conclusions using different kinds of data, and sometimes even the same genetic system, are many times different. But at present the most likely picture that emerges is one of a single major migration without significant discontinuities in time. The First Americans should have entered the continent at least 15,000

Table 15.3. Main recent hypotheses for the prehistoric colonization of the Americas.

Type of evidence	No. of waves	Regions of origin	Time of entrance before present
Geology and archaeology	2	–	10,500
Paleoanthropology and morphology	2	–	(a) 14,000 (b) 11,000
Linguistics	1	Siberia	> 20,000
Blood groups and proteins	3	–	(a) 31,000 (b) 18,000
Mitochondrial DNA	4	–	14,000–21,000
	3	–	(a) 26,000–34,000/24,000–48,000 (b) 12,000–15,000/13,000–18,000 (c) 7,000–9,000/13,000–35,000
	2	–	(a) 20,000–25,000/26,000–32,000 (b) 11,000–10,000
	1	Altai Mountains, Southern Siberia	(a) 30,000–40,000 (b) 23,000–37,000 (c) 12,000–19,000
Y chromosome	2	(a) Southern Middle Siberia (b) Lower Amur/Sea of the Okhkotsk region	(a) 22,770 (b) 14,000–25,000 (c) 15,000–18,000 (d) 13,611
	1	–	10,100–17,200
Autosomes	1	–	–
Parasites	1	Mongolia and Manchuria	9,000–140,000

years BP, probably using the Pacific coast route, and the main source of these migrants should have been the Altai Mountains of southern Siberia. These are basically the same conclusions arrived at by Mulligan *et al.* (2004) and Schurr and Sherry (2004) after other extensive reviews of the genetic information.

References

Adler, R. (2000). Voices from the past. *New Scientist*, **165**(2227), 36–40.

Agostini, H. T., Yanagihara, R., Davis, V., Ryschkewitsch, C. F. and Stoner, G. L. (1997). Asian genotypes of JC virus in Native Americans and in a Pacific Island population: markers of viral evolution and human migration. *Proceedings of the National Academy of Sciences, USA*, **94**, 14542–6.

Ames, K. M. (2003). The Northwest Coast. *Evolutionary Anthropology*, **12**, 19–33.

Arnaiz-Villena, A., Vargas-Alarcón, G., Granados, J., Gómez-Casado, E., Longas, J., Gonzáles-Hevilla, M., Zúñiga, J., Salgado, N., Hernández-Pacheco, G., Guillén, J. and Martinez-Lazo, J. (2000). HLA genes in Mexican Mazatecans, the peopling of the Americas and the uniqueness of Amerindians. *Tissue Antigens*, **56**, 405–16.

Bandelt, H-J., Hermstadt, C., Yao, Y-G., Kong, Q-P., Kivisild, T., Rengo, C., Scozzari, R., Richards, M., Villems, R., Macaulay, V., Howell, N., Torroni, A. and Zhang, Y-P. (2003). Identification of Native American founder mtDNAs through the analyses of complete mtDNA sequences: some caveats. *Annals of Human Genetics*, **67**, 512–24.

Bergen, A. W., Wang, C. Y., Tsai, J., Jefferson, K., Dey, C., Smith, K. D., Park, S. C., Tsai, S. J. and Goldmann, D. (1999). An Asian-Native American paternal lineage identified by RPS4Y resequencing by microsatellite haplotyping. *Annals of Human Genetics*, **63**, 63–80.

Bever, T. and Montalbetti, M. (2002). Noam's ark. *Science*, **298**, 1565–6.

Bevilaqua, L. R. M., Mattevi, V. S., Ewald, G. M., Salzano, F. M., Coimbra, C. E. A. Jr., Santos, R. V. and Hutz, M. H. (1995). Beta-globin gene cluster haplotype distribution in five Brazilian Indian tribes. *American Journal of Physical Anthropology*, **98**, 395–401.

Bianchi, N. O., Catanesi, C. I., Bailliet, G., Martinez-Marignac, V. L., Bravi, C. M., Vidal-Rioja, L. B., Herrera, R. J. and López-Camelo, J. S. (1998). Characterization of ancestral and derived Y-chromosome haplotypes of New World Native populations. *American Journal of Human Genetics*, **63**, 1862–71.

Bolnick, D. A., Shook, B. A., Campbell, L. and Goddard, I. (2004). Problematic use of Greenberg's linguistic classification of the Americas in studies of Native American genetic variation. *American Journal of Human Genetics*, **75**, 519–23.

Bonatto, S. L. and Salzano, F. M. (1997a). A single and early migration for the peopling of the Americas supported by mitochondrial DNA sequence data. *Proceedings of the National Academy of Sciences, USA*, **94**, 1866–71.

Bonatto, S. L. and Salzano, F. M. (1997b). Diversity and age of the four major mtDNA haplogroups, and their implications for the peopling of the New World. *American Journal of Human Genetics*, **61**, 1413–23.

Bortolini, M-C., Salzano, F. M., Thomas, M. G., Stuart, S., Nasanen, S. P. K., Bau, C. H. D., Hutz, M. H., Layrisse, Z., Petzl-Erler, M. L., Tsuneto, L. T., Hill, K., Hurtado, A. M., Castro-de-Guerra, D., Torres, M. M., Groot, H., Michalski, R.,

Nymadawa, P., Bedoya, G., Bradman, N., Labuda, D. and Ruiz-Linares, A. (2003). Y-chromosome evidence for differing ancient demographic histories in the Americas. *American Journal of Human Genetics*, **73**, 524–39.

Brace, C. L., Nelson, A. R., Seguchi, N., Oe, H., Sering, L., Qifeng, P., Yongyi, L. and Tumen, D. (2001). Old World sources of the first New World human inhabitants: a comparative craniofacial view. *Proceedings of the National Academy of Sciences*, **98**, 10017–22.

Brown, M. D., Hosseini, S. H., Torroni, A., Bandelt, H-J., Allen, J. C., Schurr, T. G., Scozzari, R., Cruciani, F. and Wallace, D. C. (1998). mtDNA haplogroup X: an ancient link between Europe/Western Asia and North America? *American Journal of Human Genetics*, **63**, 1852–61.

Bryant, V. M. (2003). Invisible clues to New World plant domestication. *Science*, **299**, 1029–30.

Cavalli-Sforza, L. L., Menozzi, P. and Piazza, A. (1994). *The History and Geography of Human Genes*. Princeton, NJ: Princeton University Press.

Confalonieri, U., Ferreira, L. F. and Araújo, A. (1991). Intestinal helminths in Lowland South American Indians: some evolutionary interpretations. *Human Biology*, **63**, 863–73.

Crawford, M. H. (1998). *The Origins of Native Americans. Evidence from Anthropological Genetics*. Cambridge, UK: Cambridge University Press.

Culotta, E. and Hanson, B. (eds.) (2004). Evolution of language. *Science*, **303**, 1315–35.

Derbeneva, O. A., Sukernik, R. I., Volodko, N. V., Hosseini, S. H., Lott, M. T. and Wallace, D. C. (2002). Analysis of mitochondrial DNA diversity in the Aleuts of the Commander Islands and its implications for the genetic history of Beringia. *American Journal of Human Genetics*, **71**, 415–21.

Derenko, M., Grzybowski, T., Malyarchuk, B. A., Czarny, J., Miscicka-Sliwka, D. and Zakharov, I. A. (2001). The presence of mitochondrial haplogroup X in Altaians from South Siberia. *American Journal of Human Genetics*, **69**, 237–41.

Diamond, J. and Bellwood, P. (2003). Farmers and their languages: the first expansions. *Science*, **300**, 597–603.

Dillehay, T. D. (2002). Climate and human migrations. *Science*, **298**, 764–5.

Dillehay, T. D. (2003). Tracking the First Americans. *Nature*, **425**, 23–4.

Dixon, E. J. (2001). Human colonization of the Americas: timing, technology and process. *Quaternary Science Reviews*, **20**, 277–99.

Dornelles, C. L., Bonatto, S. L., Freitas, L. B. and Salzano, F. M. (2005). Is haplogroup X present in extant South American Indians? *American Journal of Physical Anthropology*, **127**, 439–48.

Eshleman, J. A., Malhi, R. S. and Smith, D. G. (2003). Mitochondrial DNA studies of Native Americans: conceptions and misconceptions of the population prehistory of the Americas. *Evolutionary Anthropology*, **12**, 7–18.

Fagan, B. M. (2004). *The Great Journey. The Peopling of Ancient America*. Gainesville: University Press of Florida.

Fisher, M. C., Koenig, G. L., White, T. J., San-Blas, G., Negroni, R., Alvarez, I. G., Wanke, B. and Taylor, J. W. (2001). Biogeographic range expansion into South America by *Coccidioides immitis* mirrors New World patterns of human migration. *Proceedings of the National Academy of Sciences, USA*, **98**, 4558–62.

Forster, P., Harding, R., Torroni, A. and Bandelt, H-J. (1996). Origin and evolution of Native American mtDNA variation: a reappraisal. *American Journal of Human Genetics*, **59**, 935–45.

Franchetto, B. and Leite, Y. (2004). *Origens da Linguagem*. Rio de Janeiro: Jorge Zahar Editor.

Fuselli, S., Tarazona-Santos, E., Dupanloup, I., Soto, A., Luiselli, D. and Pettener, D. (2003). Mitochondrial DNA diversity in South America and the genetic history of Andean highlanders. *Molecular Biology and Evolution*, **20**, 1682–91.

Ghose, C., Perez-Perez, G. I., Dominguez-Bello, M-G., Pride, D. T., Bravi, C. M. and Blazer, M. J. (2002). East Asian genotypes of *Helicobacter pylori* strains in Amerindians provide evidence for its ancient human carriage. *Proceedings of the National Academy of Sciences, USA*, **99**, 15107–11.

Gómez-Casado, E., Martínez-Lazo, J., Moscoso, J., Zamora, J., Martín-Villa, M., Perez-Blas, M., Lopez-Santalla, M., Gramajo, P. L., Silvera, C., Lowy, E. and Arnaiz-Villena, A. (2003). Origin of Mayans according to HLA genes and the uniqueness of Amerindians. *Tissue Antigens*, **61**, 425–36.

González-José, R., Dahinten, S. L., Luis, M. A., Hernández, M. and Pucciarelli, H. M. (2001a). Craniometric variation and the settlement of the Americas: testing hypotheses by means of R-matrix and matrix correlation analyses. *American Journal of Physical Anthropology*, **116**, 154–65.

González-José, R., Dahinten, S. and Hernández, M. (2001b). The settlement of Patagonia: a matrix correlation study. *Human Biology*, **73**, 233–48.

González-José, R., García-Moro, C., Dahinten, S. and Hernández, M. (2002). Origin of Fuegian-Patagonians: an approach to population history and structure using R matrix and matrix permutation methods. *American Journal of Human Biology*, **14**, 308–20.

González-José, R., González-Martín, A., Hernández, M., Pucciarelli, H. M., Sardi, M., Rosales, A. and van der Molen, S. (2003). Craniometric evidence for Paleoamerican survival in Baja California. *Nature*, **425**, 62–5.

Greenberg, J. H. (1986). The 'Greenberg' hypothesis. *Science*, **274**, 1447.

Greenberg, J. H. (1987). *Language in the Americas*. Stanford, CA: Stanford University Press.

Greenberg, J. H., Turner, C. G. II and Zegura, S. L. (1986). The settlement of the Americas: a comparison of the linguistic, dental, and genetic evidence. *Current Anthropology*, **27**, 477–97.

Hall, R., Roy, D. and Boling, D. (2004). Pleistocene migration routes into the Americas: human biological adaptations and environmental constraints. *Evolutionary Anthropology*, **13**, 132–44.

Hammer, M. F. and Zegura, S. L. (2002). The human Y chromosome haplogroup tree: nomenclature and phylogeography of its major divisions. *Annual Review of Anthropology*, **31**, 303–21.

Hauser, M. D., Chomsky, N. and Fitch, W. T. (2002). The faculty of language: what is it, who has it, and how did it evolve? *Science*, **298**, 1569–79.

Heckenberger, M. J., Kuikuro, A., Kuikuro, U. T., Russell, J. C., Schmidt, M., Fausto, C. and Franchetto, B. (2003). Amazonia 1492: pristine forest or cultural parkland? *Science*, **301**, 1710–14.

Heller, A. H., Salzano, F. M., Barrantes, R., Krylov, M., Benevolenskaya, L., Arnett, F. C., Munkhbat, B., Munkhtuvshin, N., Tsuji, K., Hutz, M. H., Carnese, F. R., Goicoechea, A. S., Freitas, L. B. and Bonatto, S. L. (2004). Intra and intercontinental molecular variability of an *Alu* insertion in the 3'UTR of the *LDLR* gene. *Human Biology*, **76**, 591–604.

Hey, J. (2005). Of the number of New World founders: a population genetic portrait of the peopling of the Americas. *Plos Biology*, **3**(6), 1–11.

Hoffecker, J. F. and Elias, S. A. (2003). Environment and archaeology of Beringia. *Evolutionary Anthropology*, **12**, 34–49.

Horai, S., Kondo, R., Nakagawa-Hattori, Y., Hayashi, S., Sonoda, S. and Tajima, K. (1993). Peopling of the Americas, founded by four major lineages of mitochondrial DNA. *Molecular Biology and Evolution*, **10**, 23–47.

Hubbe, M., Neves, W. A., Atui, J. P. V., Cartelle, C. and Pereira da Silva, M. A. (2004). A new early human skeleton from Brazil: support for the 'Two Main Biological Components Model' for the settlement of the Americas. *Current Research in the Pleistocene*, **21**, 77–81.

Hunley, K. and Long, J. C. (2005). Gene flow across linguistics boundaries in Native North American populations. *Proceedings of the National Academy of Sciences, USA*, **102**, 1312–17.

Hutz, M. H., Almeida, S., Coimbra, C. E. A. Jr, Santos, R. V. and Salzano, F. M. (2000). Haplotype and allele frequencies for three genes of the dopaminergic system in South American Indians. *American Journal of Human Biology*, **12**, 638–45.

Hutz, M. H., Callegari-Jacques, S. M., Almeida, S. E. M., Armborst, T. and Salzano, F. M. (2002). Low levels of STRP variability are not universal in American Indians. *Human Biology*, **74**, 791–806.

Ingman, M. and Gyllensten, U. (2001). Analysis of the complete human mtDNA genome: methodology and inferences for human evolution. *Journal of Heredity*, **92**, 454–61.

Jantz, R. L. and Owsley, D. W. (2001). Variation among early North American crania. *American Journal of Physical Anthropology*, **114**, 146–55.

Karafet, T. M., Zegura, S. L., Posukh, O., Osipova, L., Bergen, A., Long, J., Goldman, D., Klitz, W., Harihara, S., de Knijff, P., Wiebe, V., Griffiths, R. C., Templeton, A. R. and Hammer, M. F. (1999). Ancestral Asian source(s) of New World Y-chromosome founder haplotypes. *American Journal of Human Genetics*, **64**, 817–31.

Keyeux, G., Rodas, C., Gelvez, N. and Carter, D. (2002). Possible migration routes into South America deduced from mitochondrial DNA studies in Colombian Amerindian populations. *Human Biology*, **74**, 211–33.

Lalueza-Fox, C., Calderón, F. L., Calafell, F., Morera, B. and Bertranpetit, J. (2001). mtDNA from extinct Tainos and the peopling of the Caribbean. *Annals of Human Genetics*, **65**, 137–51.

Lavallée, D. (2000). *The First South Americans. The Peopling of a Continent from the Earliest Evidence to High Culture*. Salt Lake City, UT: University of Utah Press.

Lell, J. T., Sukernik, R. I., Starikovskaya, Y. B., Su, B., Jin, L., Schurr, T. G., Underhill, P. A. and Wallace, D. C. (2002). The dual origin and Siberian affinities of Native American Y chromosomes. *American Journal of Human Genetics*, **70**, 192–206.

Lessa, A. and Guidon, N. (2002). Osteobiographic analysis of Skeleton I, sítio Toca dos Coqueiros, Serra da Capivara National Park, Brazil, 11,060 BP: first results. *American Journal of Physical Anthropology*, **118**, 99–110.

Locke, J. L. (2001). Rank and relationships in the evolution of spoken language. *Journal of the Royal Anthropological Institute*, **7**, 37–50.

Malhi, R. S., Eshleman, J. A., Greenberg, J. A., Weiss, D. A., Shook, B. A. S., Kaestle, F. A., Lorenz, J. G., Kemp, B. M., Johnson, J. R. and Smith, D. G. (2002). The structure of diversity within New World mitochondrial DNA haplogroups: implications for the prehistory of North America. *American Journal of Human Genetics*, **70**, 905–19.

Mateus Pereira, L. H., Socorro, A., Fernandez, I., Masleh, M., Vidal, D., Bianchi, N. O., Bonatto, S. L., Salzano, F. M. and Herrera, R. J. (2005). Phylogenetic information in polymorphic L1 and *Alu* insertions from East Asians and Native American populations. *American Journal of Physical Anthropology*, **128**, 171–84.

Mishmar, D., Ruiz-Pesini, E., Golik, P., Macaulay, V., Clark, A. G., Hosseini, S., Brandon, M., Easley, K., Chen, E., Brown, M. D., Sukernik, R. I., Olckers, A. and Wallace, D. C. (2003). Natural selection shaped regional mtDNA variation in humans. *Proceedings of the National Academy of Sciences, USA*, **100**, 171–6.

Mulligan, C. J., Hunley, K., Cole, S. and Long, J. C. (2004). Population genetics, history, and health patterns in Native Americans. *Annual Review of Genomics and Human Genetics*, **5**, 295–315.

Neel, J. V., Biggar, R. J. and Sukernik, R. I. (1994). Virologic and genetic studies relate Amerind origins to the indigenous people of the Mongolia/ Manchuria/southeastern Siberia region. *Proceedings of the National Academy of Sciences, USA*, **91**, 10737–41.

Nettle, D. (1999). Linguistic diversity of the Americas can be reconciled with a recent colonization. *Proceedings of the National Academy of Sciences, USA*, **96**, 3325–9.

Neves, W. A., Prous, A., González-José, R., Kipnis, R. and Powell, J. (2003). Early Holocene human skeletal remains from Santana do Riacho, Brazil: implications for the settlement of the New World. *Journal of Human Evolution*, **45**, 19–42.

Neves, W. A., González-José, R., Hubbe, M., Kipnis, R., Araujo, A. G. M. and Blasi, O. (2004). Early Holocene human skeletal remains from Cerca Grande, Lagoa Santa, Central Brazil, and the origins of the first Americans. *World Archaeology*, **36**, 479–501.

Neves, W. A., Hubbe, M., Okumura, M. M., González-José, R., Figuti, L., Eggers, S. and de Blasis, P. A. D. (2005). A new early Holocene human skeleton from Brazil: implications for the settlement of the New World. *Journal of Human Evolution*, **48**, 403–14.

Nichols, J. (2002). The First American languages. In Jablonski, N. G. (ed.), *The First Americans*. San Francisco, CA: California Academy of Sciences, pp. 273–93.

Núñez, L., Grosjean, M. and Cartajena, I. (2002). Human occupations and climate change in the Puna de Atacama, Chile. *Science*, **298**, 821–4.

Pavesi, A. (2004). Detecting traces of prehistoric human migrations by geographic synthetic maps of polyomavirus JC. *Journal of Molecular Evolution*, **58**, 304–13.

Pericot y Garcia, L. (1962). *América Indígena*. Barcelona: Salvat.

Piperno, D. R. and Stothert, K. E. (2003). Phytolith evidence for early Holocene *Cucurbita* domestication in southwest Ecuador. *Science*, **299**, 1054–7.

Prous, A. and Neves, W. A. (2000). A primeira descoberta da América. In Aguilar, N. (ed.), *Catálogo da Mostra do Redescobrimento*. São Paulo: Associação Brasil 500 Anos, pp. 72–97.

Reidla, M., Kivisild, T., Metspalu, E., Kaldma, K., Tambets, K., Tolk, H-V., Parik, J., Loogväli, E-L., Derenko, M., Malvarchuk, B., Bermisheva, M., Zhadanov, S., Pennarun, E., Gubina, M., Golubenko, M., Damba, L., Fedorova, S., Gusar, V., Grechanina, E., Mikerezi, I., Moisan, J-P., Chaventré, A., Khusnutdinova, E., Osipova, L., Stepanov, V., Voevoda, M., Achilli, A., Rengo, C., Rickards, O., De Stefano, G. F., Papiha, S., Beckman, L.,

Janicijevic, B., Rudan, P., Anagnou, N., Michalodimitrakis, E., Koziel, S., Usanga, E., Geberhiwot, T., Hernstadt, C., Howell, N., Torroni, A. and Villems, R. (2003). Origin and diffusion of mtDNA haplogroup X. *American Journal of Human Genetics*, **73**, 1178–90.

Rodrigues, A. D. (1986). *Línguas Brasileiras. Para o Conhecimento das Línguas Indígenas*. São Paulo: Edições Loyola.

Roosevelt, A. C., Costa, M. L., Machado, C. L., Michab, M., Mercier, N., Valladas, H., Feathers, J., Barnett, W., Silveira, M. I., Henderson, A., Sliva, J., Chernoff, B., Reese, D. S., Holman, J. A., Toth, N. and Schick, K. (1996). Paleoindian cave dwellers in the Amazon: the peopling of the Americas. *Science*, **272**, 373–84.

Roosevelt, A. C., Douglas, J. and Brown, L. (2002). The migrations and adaptations of the First Americans Clovis and pre-Clovis viewed from South America. In Jablonski, N. G. (ed.), *The First Americans*. San Francisco, CA: California Academy of Sciences, pp. 159–235.

Ruiz-Linares, A., Ortiz-Barrientos, D., Figueroa, M., Mesa, N., Munera, J. G., Bedoya, G., Velez, I. D., Garcia, L. F., Perez-Lezaun, A., Bertranpetit, J., Feldman, M. W. and Goldstein, D. B. (1999). Microsatellites provide evidence for Y chromosome diversity among the founders of the New World. *Proceedings of the National Academy of Sciences, USA*, **96**, 6312–17.

Salzano, F. M. (2002). Molecular variability in Amerindians: widespread but uneven information. *Anais da Academia Brasileira de Ciências*, **74**, 223–63.

Salzano, F. M. and Callegari-Jacques, S. M. (1988). *South American Indians. A Case Study in Evolution*. Oxford: Clarendon Press.

Salzano, F. M., Hutz, M. H., Salamoni, S. P., Rohr, P. and Callegari-Jacques, S. M. (2005). Genetic support to proposed patterns of relationship among Lowland South American languages. *Current Anthropology*, **4b** (supplement), S121–9.

Schurr, T. G. (2000). Mitochondrial DNA and the peopling of the New World. *American Scientist*, **88**, 246–53.

Schurr, T. G. and Sherry, S. T. (2004). Mitochondrial DNA and Y chromosome diversity and the peopling of the Americas: evolutionary and demographic evidence. *American Journal of Human Biology*, **16**, 420–39.

Seielstad, M., Yuldasheva, N., Singh, N., Underhill, P., Oefner, P., Shen, P. and Wells, R. S. (2003). A novel Y-chromosome variant puts an upper limit on the timing of first entry into the Americas. *American Journal of Human Genetics*, **73**, 700–5.

Silva, W. A. Jr, Bonatto, S. L., Holanda, A. J., Ribeiro-dos-Santos, A. K., Paixão, B. M., Goldman, G. H., Abe-Sandes, K., Rodriguez-Delfin, L., Barbosa, M., Paçó-Larson, M. L., Petzl-Erler, M. L., Valente, V., Santos, S. E. B. and Zago, M. A. (2003). Correction: mitochondrial DNA variation in Amerindians. *American Journal of Human Genetics*, **72**, 1346–8.

Steckel, R. H., Rose, J. C., Larsen, C. S. and Walker, P. L. (2002). Skeletal health in the Western Hemisphere from 4000 B.C. to the present. *Evolutionary Anthropology*, **11**, 142–55.

Stone, A. C. and Stoneking, M. (1998). mtDNA analysis of a prehistoric Oneota population: implications for the peopling of the New World. *American Journal of Human Genetics*, **62**, 1153–70.

Su, B., Xiao, J., Underhill, P., Deka, R., Zhang, W., Akey, J., Huang, W., Shen, D., Luo, J., Chu, J., Tan, J., Shen, P., Davis, R., Cavalli-Sforza, L. L., Chakraborty, R., Xiong, M., Du, R., Oefner, P., Chen, Z. and Jin, L. (1999). Y-chromosome evidence for a northward migration of modern humans into Eastern Asia during the last Ice Age. *American Journal of Human Genetics*, **65**, 1718–24.

Sugimoto, C., Hasegawa, M., Zheng, H-Y., Demenev, V., Sekino, Y., Kojima, K., Honjo, T., Kida, H., Hovi, T., Vesikari, T., Schalken, J. A., Tomita, K., Mitsunobu, Y., Ikegaya, H., Kobayashi, N., Kitamura, T. and Yogo, Y. (2002). JC virus strains indigenous to Northeastern Siberians and Canadian Inuits are unique but evolutionarily related to those distributed throughout Europe and Mediterranean areas. *Journal of Molecular Evolution*, **55**, 322–35.

Stumpf, M. P. and Goldstein, D. B. (2001). Genealogical and evolutionary inference with the human Y chromosome. *Science*, **291**, 1738–42.

Surovell, T. A. (2003). Simulating coastal migration in New World colonization. *Current Anthropology*, **44**, 580–91.

Tarazona-Santos, E. and Santos, F. R. (2002). The peopling of the Americas: a second major migration? *American Journal of Human Genetics*, **70**, 1377–80.

Tokunaga, K., Ohashi, J., Bannai, M. and Juji, T. (2001). Genetic link between Asians and Native Americans: evidence from HLA genes and haplotypes. *Human Immunology*, **62**, 1001–8.

Torroni, A., Neel, J. V., Barrantes, R., Schurr, T. G. and Wallace, D. C. (1994). Mitochondrial DNA 'clock' for the Amerinds and its implications for timing their entry into North America. *Proceedings of the National Academy of Sciences, USA*, **91**, 1158–62.

Tsuneto, L. T., Probst, C. M., Hutz, M. H., Salzano, F. M., Rodriguez-Delfin, L., Zago, M. A., Hill, K., Hurtado, A. M., Ribeiro-dos-Santos, A. K. C. and Petzl-Erler, M. L. (2003). HLA Class II diversity in seven Amerindian populations. Clues about the origins of the Aché. *Tissue Antigens*, **62**, 512–26.

Underhill, P. A., Jin, L., Zemans, R., Oefner, P. J. and Cavalli-Sforza, L. L. (1996). A pre-Colombian Y chromosome-specific transition and its implications for human evolutionary history. *Proceedings of the National Academy of Sciences, USA*, **93**, 196–200.

Varela, H. H. and Cocilovo, J. A. (2002). Genetic drift and gene flow in a prehistoric population of the Azapa Valley and Coast, Chile. *American Journal of Physical Anthropology*, **118**, 259–67.

Wallace, D. C. (1995). Mitochondrial DNA variation in human evolution, degenerative disease, and aging. *American Journal of Human Genetics*, **57**, 201–23.

Wilbert, J. (1968). Loukotka's classification of South American Indian languages. In Loukotka, C., *Classification of South American Indian Languages.* Los Angeles: Latin American Center, University of California, pp. 7–23.

Y Chromosome Consortium (2002). A nomenclature system for the tree of human Y-chromosomal binary haplogroups. *Genome Research*, **12**, 339–48.

Zegura, S. L., Karafet, T. M., Zhivotovsky, L. A. and Hammer, M. F. (2004). High-resolution SNPs and microsatellite haplotypes point to a single, recent entry of Native American Y chromosomes into the Americas. *Molecular Biology and Evolution*, **21**, 164–75.

Zheng, H-Y., Sugimoto, C., Hasegawa, M., Kobayashi, N., Kanayama, A., Rodas, A., Mejia, M., Nakamichi, J., Guo, J., Kitamura, T. and Yogo, Y. (2003). Phylogenetic relationships among JC virus strains in Japanese/Koreans and Native Americans speaking Amerind or Na-Dené. *Journal of Molecular Evolution*, **56**, 18–27.

Chapter 16

Anthropological Genetics: Present and Future

Henry C. Harpending

Department of Anthropology, University of Utah, Salt Lake City, UT, USA

Introduction

When Michael Crawford asked me to write a final chapter for this volume he suggested that I write some 'science fiction' and that I should 'let myself go'. I will take him at his word in what follows, assessing several areas of anthropological genetics in terms of their likely short-term directions and consequences. I discuss anthropological genetics specifically rather than any of the rest of human genetics since there is no shortage of science fiction available about the latter.

If there is a coherent theme in my story it is that some areas of our discipline will soon find themselves at the centre of public debates about important political, social justice and educational issues. This will be caused by market forces in some cases and by new data about human genetic diversity that will force a drastic reassessment of our views about the possibility of group equalities in contemporary industrial societies. As a discipline there is a sense in which we have had our heads hidden in the sand, earnestly teaching our students about lactase and HbS and about G6PD while maintaining the fiction, or failing to deny the fiction, that humans don't really differ at all in traits that have large visible social consequences. Other areas of our discipline will in the near future enjoy less visibility than they do now, for example the effort to read global human history from patterns in the neutral genome. Instead, admixture with archaic populations and assimilation of archaic genes, together with pervasive natural selection, have obscured much of the history we have been trying to see in our genome.

Anthropological genetics theory

Anthropological genetics came into its own in the 1970s when typing technologies became easily available and several groups undertook field studies examining local patterns of genetic variation, the most important of which were those of lowland South American Indians by James Neel and his associates. In parallel Newton Morton and others (Morton, 1973) were doing innovative studies on the genetic structure of human populations, the relationship over large areas between dispersal, population density, and the decline of genetic similarity with distance. Soon afterward there appeared the important review volume of Cavalli-Sforza and Bodmer (1971) and the first volume of this series (Crawford and Workman, 1973), which brought together in one easily accessible place the relevant demographic and population genetics theory and important population data then available.

Most of the theory of that tradition was neutral theory, theory about how genetic variation should behave in populations if it was not affected by natural selection.

There is an unbroken line from then to today in the study of neutral genetic diversity in our species. Early on many of us were essentially trying to use genetic data to do sociology. Genetic markers could provide biological indicators of mating structure and relations among groups, and there was the prospect of a social science that would encompass both culture and biology.

The technologies that we developed worked very well (see Crawford, this volume, Chapter 1). We could measure aspects of local population history and of mating structure with our biological indicators. Unfortunately at about the same time cultural anthropology seemed to lose interest in human social behaviour, and biology became a topic to be feared and avoided in many departments. For a vision of where anthropology should have gone I recommend Alan Fix's (1999) monograph.

The debate about modern human origins reinvigorated the study of human neutral diversity and changed the time and area focus from that of small regions over centuries to the whole species over tens of millennia. We are now not so concerned with marker frequency differences among valleys; we are concerned with marker frequency differences over the globe. We have gone from using a handful of markers to hundreds, even thousands, in these comparisons but nothing has really changed. Among major human groups the fraction of diversity, computed as some variant of Wright's F_{st} statistic, that is among (rather than within) populations is approximately 1/8. This value has been known for 35 years or so and it is the answer to a vast corpus of pointless unnecessary literature about whether or not there are human races. It is easy to attribute meanings to this simple number: for example Lewontin (1972) persuasively argued that 1/8 was a small number and that human group differences were therefore trivial. On the other hand, one could argue as well that a kinship coefficient of 1/8 corresponds to differences among sets of half siblings, and few consider kinship between half siblings to be trivial.

The amount of ongoing discussion and debate about human race is difficult to understand since the data are clear and no one disagrees about them. Instead the issues under debate are peripheral, often merely semantic, and they have easy answers. 'Should' race be used in epidemiology? Of course it should if it provides useful prior information about patients. On the other hand, information about the race of patients will be useless as soon as we discover and can type cheaply the underlying genes that are responsible for the associations. Can races be enumerated in any unambiguous way? Of course not, and this is well known not only to scientists but also to anyone on the street. A recent report on this issue from a federally sponsored group (Race, Ethnicity, and Genetics Working Group, 2005) was so devoid of substance that it was discussed on the popular weblog of John Hawks (2005) under the topic of 'Your Tax Dollars at Work'.

The volume of new DNA data that has appeared over the last several decades has provided one important new insight about global human diversity: African populations contain absolutely more diversity than populations outside Africa, and there is a diversity cline away from Africa. This strong pattern was never apparent from classical markers nor from electrophoretic data. The greater African diversity was hidden because the systems that were typed were developed in European populations and the most diverse systems were chosen from them. The important review of human genetic diversity by Cavalli, Menozzi and Piazza (1964), for example, did not show the elevated diversity of Africa.

The high African diversity is the most important genetic support for the Recent African Origin model of anatomically modern humans, according to which we appeared as essentially a new species somewhere in Africa and colonized the earth, displacing archaic human populations in the process. The early support from genetics for this model was furnished by mitochondrial DNA. The important patterns in human mtDNA were the relatively recent coalescence of human mtDNA, the topology of the gene tree in which populations outside Africa were represented by a subtree of the global tree, and later the apparent star-like structure of the tree, indicating a major population expansion 40 to 80 thousand years ago, just when the earliest modern-looking fossils appear in the record. There was a problem with the timing of the origin of modern humans from the mismatch distributions (Harpending et al., 1993): if there was a single expansion event then all humans, or perhaps all humans outside Africa, should carry the signature of the same event in their mitochondrial genotypes. Instead there was great heterogeneity in mismatch peaks, indicating that ancestors of African populations expanded first, perhaps 80 kya, then Asians at 60 kya or so, then Europeans at 40 kya (Harpending et al., 1993). This heterogeneity is almost certainly a signature of a cascade of expansion events, essentially the kind of wave-of-advance of a new phenotype described by Eswaran *et al.* (2005). A wave of advance destroys diversity as it progresses since the small wavefront population accounts for most of the growth. Such a process accounts handily for the overall low diversity and small effective size of our species. We do not need to postulate any bottlenecks or episodes of size reduction in our species at all, such dramatic events probably never happened.

It is clear that mitochondrial DNA was replaced by some kind of out-of-Africa event but, unfortunately, the nuclear genome shows little or no evidence of such a replacement event (Harpending and Eswaran, 2005; Harpending and Rogers, 2000; Eswaran *et al.*, 2005; Pearson *et al.*, 2003). If the history of mtDNA were in fact the demographic history of our species then it would unambiguously show up in a neutral nuclear genome and it does not. One possible explanation is that there was a near replacement out of Africa but also a lot of assimilation of archaic nuclear genes into the modern human populations. A second possibility is that there is a lot of ongoing selection in contemporary human populations, so much that important episodes in our species' history are obscured. My own prediction is that both of these explanations will turn out to be correct.

Almost all of the theory about race differences is theory about neutral loci and almost all of the data have been from neutral markers. There seems little prospect of any fundamental new development in this tradition and I predict that it will wither as we turn our attention to loci that have been or are under selection. For decades we have tacitly regarded natural selection as a nuisance that might occasionally obscure the history we have been seeking in the passive neutral genome. There are exceptions of course: we all teach about the malaria defence polymorphisms and about lactase deficiency and its association with dairying. But these have been visible exceptions to our preoccupation with neutral markers. For years the malaria polymorphisms stood alone as clear plausible examples of selection in humans; they seemed to be very unusual cases, and I even stopped teaching about them in introductory human evolution courses. Now some us think there may be pervasive strong ongoing natural selection in our species, strong enough to obscure the signatures of ancient demography that we have been trying to read for a decade or so.

Direct indications of this pervasive selection are just starting to appear as population samples are available to be scanned for millions of SNPs (Single Nucleotide Polymorphisms). Several kinds of patterns may be signatures of recent selection. First, regions of the genome where the number of singleton or rare SNPs is greater than neutral theory suggests could be consequences of a recent selective sweep. Like population growth, positive selection causes a star-like gene genealogy. In such a gene tree many mutations will have only one or a few descendants in a sample. Second, a newly successful gene will have spread rapidly so that there has not been time for recombination to occur in the chromosomal neighbourhood, leading to large regions of linkage disequilibrium around the selected gene. Third, a human region can be compared with that of an outgroup species such as the chimp: selection is suspected if the ratio of coding to non-coding substitutions is especially high. Several such scans have appeared (Carlson *et al.*, 2005; Nielsen *et al.*, 2005). A pattern here and elsewhere is that there are more detectable cases of ongoing selection outside Africa than there are in Africa. This is not surprising given that we are mostly an African species adapting to new environments outside Africa.

Forensics

DNA technology has made spectacular contributions to forensics. The daily papers are full of stories of old crimes solved, of innocents released from death row, of identification of human remains. Like desktop computers 30 years ago, this technology is too good to remain in the hands of experts and professionals.

Market forces will soon bring us kits so that everyone can do forensics at home. There will be rules and laws passed to try to keep the genie in the bottle, and there may even be an arms race between institutional users like law enforcement and private individuals. It will be fun to watch, like the present arms race between the music industry and illegal music file sharing over the internet. But the personal computer revolution and the internet have developed the public taste for technology and for information, and our own personal DNA databases will be impossible to pass up. Whose lipstick is on your collar? Did the party guests snoop in the bathroom cabinets? Which dog left the mess in the front yard? We will enjoy personal technology to answer questions like these as easily as the internet lets us read weblogs.

Meanwhile the market is likely to come up with countermeasures — think of bleach dispensers inside office paper shredders. The first generation of countermeasures will probably be bogus, pushed by the same folks who fill our email boxes with spam advertising phony drugs. If real countermeasures appear, law enforcement will certainly try to suppress them with the same zeal that the music industry tries to suppress file sharing. Another amusing arms race may follow.

Privacy will be something old people remember. This is not as distressing as it might be: after all it is only within the last few millennia, when many people lived in large anonymous urban centres, that humans have had privacy as we enjoy it. In small-scale face-to-face societies, even in small towns and villages, there is little or no privacy.

Medical genetics

While we ordinarily consider medical genetics a separate discipline, the ranks of medical geneticists are full of anthropologists and scientists with training and interests in anthropology. The small-scale isolated communities that we typically study have important advantages for medical studies (Terwilliger and Lee, Chapter 3, this volume). While a spectacular amount of human genetic information has been produced by the Human Genome project and its offshoots, my impression is that the medical benefits have been far fewer so far than the those in the promises that motivated the data collection. There are a lot of weak correlations in the literature and numerous local findings that do not generalize. For example a chromosomal region that is involved in risk for schizophrenia in Japan, say, is likely not to replicate in Rome or Nairobi.

Cochran *et al.* (2000) present an argument from evolutionary theory that many diseases with large fitness effects cannot be caused by deleterious mutations. Their basic argument is simple: common genetic variants with large effects on fitness could not persist since they would be eliminated by selection. Since the largest deleterious mutation rates known are on the order of 10^{-4} per generation, any disease with a total effect on fitness greater than this is probably not genetic. Instead, they propose, such diseases are caused by parasites and infectious agents. According to their perspective apparent genetic causes may just be genetic variations in responses to pathogens. They point out, for example, that leprosy is highly heritable but it is caused (in the most direct sense of 'cause') by a pathogen. The recent Nobel prize in medicine for the year 2005 to Marshall and Warren for elucidating the role of *Helicobacter pylori* in gastritis and peptic ulcer shows that their model is entirely plausible. Notice that humans may vary genetically in the susceptibility to *H. pylori*, leading to familiality of peptic ulcer.

My prediction is that genetics and genomics will lose much of its central place in medical research over the next several decades and infectious diseases will enjoy a resurgence in funding. Right now there appears to be an unfortunate imbalance in research priority.

Human history

Anthropological genetics has not lived up to its promise to clarify human demographic history and unravel the origins of human diversity. A decade or so ago anthropological geneticists brought great promise to the table: it seemed for example that pattern in human mtDNA variation had ended the battle between the multiregional and single origin theories of the origin of modern humans. Since those heady days data from the nuclear genome has brought little to the table save for confusion and ambiguity, and most of us are far less certain of a single origin scenario than we were a decade ago. In a review of a recent large conference at Cambridge on human origins, Barnham (2005) remarks on the relative lack of presence of genetics in debates about current issues in the field.

The literature is still dominated by anecdotes about haplotypes, usually either of mtDNA or nrY. It is difficult to know what to make of this literature since the stories range from the interesting and plausible (Haak *et al.*, 2005) to the absurd (Macauley *et al.*, 2005). The problem with the literature is that there is no way to replicate any of it and there is no good reason that the history of a single locus should have a whole lot to do with the histories of the populations in which it occurs. It will take a lot of simulations, together with successful reconstructions of the simulated histories by practitioners of the haplotype arts, before most of us pay a lot of attention to this literature when it describes events older than the last few millennia. This leads to my prediction that we will focus much more on local and regional history in the next few decades and less on trying to trace the global human diaspora with haplotypes.

My predictions presuppose the absence of dramatic new technologies. If, for example, someone succeeds in obtaining reliable long nuclear sequences from Neanderthals then all bets are off. Such sequences could tell us whether substantial admixture occurred between archaics and modern humans and whether, as it appears, any of the new successful alleles that are undergoing sweeps in our species are derived from archaics.

Ongoing selection

The first of the current generation of socially interesting (read 'controversial') polymorphisms undergoing strong selection was described by Ding *et al.* (2002) at the D4 dopamine receptor locus. There is a repeat polymorphism within the coding region: the common human haplotype has four 48bp repeats (4R) at the locus, while the variant with seven repeats (7R) is spreading rapidly. There are known consequences for behaviour of the 7R variant: it predisposes to childhood ADHD and novelty seeking, and individuals with 7R are less altruistic than others according to self-reports (Bachner-Mellman *et al.*, 2005). The mechanism of selection is not known. There are two hypotheses in the literature, neither of which is very convincing.

Chen *et al.* (1999) suggested that the allele might predispose its bearers to disperse, so that 7R would be concentrated in populations who had recently moved long distances. Their hypothesis accounts for the high frequency of 7R in American Indian populations with the world's highest frequencies in South American groups. Harpending and Cochran (2002) proposed the alternative hypothesis that 7R was favoured under conditions of chronic local warfare and raiding and that it was selected against in groups where raiding was rare. The latter hypothesis accounts for the absence of 7R among Kalahari Bushmen and among populations in East Asia with a long history of intense peasant agriculture.

More recently Bruce Lahn's group at the University of Chicago has described two haplotypes, one at the ASPM (abnormal spindle-like microcephaly associated) locus and one at the microcephalin locus, that are undergoing strong positive selection in humans (Evans *et al.*, 2005; Mekel-Bobrov *et al.*, 2005). These loci, which regulate brain size in mammals, can cause microcephaly when they are badly damaged. In both cases a single haplotype has been under strong positive selection: in microcephalin the favoured haplotype apparently started its sweep nearly 40 thousand years ago while at the ASPM locus the favoured haplotype started spreading only six thousand years ago. The microcephalin D haplotype (D for derived) is common in Europe and Asia and is nearly fixed in New World populations, while the much newer ASPM D is spreading rapidly in Europe and Asia but has not yet reached New World populations. Puzzlingly, neither new haplotype is at very high frequency in Africa.

There are some other evolving variants in our species that are concerned with development and/or function of the central nervous system. Rockman *et al.* (2005) discuss ongoing evolution at the pro-dynorphin promoter: prodynorphin is a precursor of endogenous opioids and peptides in the brain. Hardy *et al.* (2005) argue that a long inversion associated with reproductive success in Europeans was assimilated from Neanderthals. The locus, MAPT, is involved with neurodegenerative disease.

While we have clear textbook examples of ongoing or very recent selection in humans in response to disease and to diet, these newly described loci have direct consequences for behaviour (DRD4) or are likely to. It is probably significant that the first few such loci that are well characterized all have to do with the brain. Most anthropologists have been trained in the Boasian social science tradition that firmly separates mind and body: disease defences or adaptations to diet are not threats to that tradition while genetic diversity that affects cognition and behaviour is a direct threat. Other loci are known that are now or will soon become part of the threat, for example the androgen receptor (Comings *et al.*, 2002) or the serotonin transporter (Noblett and Coccaro, 2005). Both of these loci vary among populations and both of them are associated with aspects of social behaviour.

My prediction is that the whole Boasian edifice will soon come crashing down, accompanied by an unpleasant public uproar. Anthropological geneticists will be in the middle of it all and we are not very well prepared for it. Our textbooks evade the issue of human group differences in cognitive and behavioural propensities and, often with great piety, repeat the mantra that there are no such differences except to the extent that they are 'social constructs' having nothing to do with gene differences.

Our heads have not always been in the sand. The classic text by Cavalli-Sforza and Bodmer (1971), a foundation document for our discipline, discussed human quantitative traits using blood pressure, stature, and intelligence test scores as primary examples. Today a text might discuss blood pressure, or stature, but never intelligence. One of the clearest and best studied human traits, one of great social and economic importance, is effectively hidden from our students and, by extension, journalists and public intellectuals who deserve better from us. Groups differ in intelligence, as do individuals. In spite of 35 years of vigorous denial that these differences are associated with gene differences, I can think of no convincing or even plausible evidence that environmental causes, as we usually think of environment, have much of anything to do with group differences (Rushton and Jensen, 2005). The underlying biology of intelligence is being revealed at a rapid rate (Gray and Thompson, 2004) yet most of us and our students remain blind to all of it.

My prediction, in my capacity here as an author of science fiction, is that we are in for a turbulent decade. Groups, whether they be by sex or race or class, certainly do not have equal potentials in all things, yet in the United States our educational system is facing a law (the *No Child Left Behind* act) that mandates that they all must achieve equal performance. Something has to break soon, and with the breakage will come drastic redefinitions of our shared public notions of fairness and social justice.

References

Bachner-Melman, R., Gritsenko, I., Nemanov, L., Zohar, A.H., Dina, C. and Ebstein, R.P. (2005). Dopaminergic polymorphisms associated with self-report measures of human altruism: a fresh phenotype for the dopamine D4 receptor. *Mol Psychiatry*, **10**(4), 333–35.

Barnham, L. (2005). On 'Rethinking the human revolution' or 'That dog won't hunt'. Review of colloquium: Rethinking the human revolution: new behavioural and biological perspectives on the origins and dispersal of modern humans, Cambridge, England, 7–11 September 2005. Before Farming [online version], 3 article 5.

Carlson, C.S., Thomas, D.J., Eberle, M.A., Swanson, J.E., Livingston, R.J., Rieder, M.J. *et al.* (2005). Genomic regions exhibiting positive selection identified from dense genotype data. *Genome Res*, **15**(11), 1553–65.

Cavalli-Sforza, L.L. and Bodmer, W.F. (1971). *The Genetics of Human Populations*. San Francisco: Freeman.

Cavalli-Sforza, L.L., Menozzi, P. and Piazza, A. (1994). *The History and Geography of Human Genes*. Princeton, NJ: Princeton University Press.

Chen, C., Burton, M., Greenberger, E. and Dmitrieva, J. (1999). Population migration and the variation of Dopamine D4 receptor (DRD4) frequencies around the globe. *Evolution and Human Behavior*, **20**, 309–24.

Cochran, G.M., Ewald, P.W. and Cochran, K.D. (2000). Infectious causation of disease: an evolutionary perspective. *Perspect Biol Med*, **43**(3), 406–48.

Comings, D.E., Muhleman, D., Johnson, J.P. and MacMurray, J.P. (2002). Parent-daughter transmission of the androgen receptor gene as an explanation of the effect of father absence on age of menarche. *Child Dev*, **73**(4), 1046–51.

Crawford, M.H. and Workman, P.L. (1973). *Methods and theories of anthropological genetics (School of American Research. Advanced seminar series)*. University of New Mexico Press.

Ding, Y.C., Chi, H.C., Grady, D.L., Morishima, A., Kidd, J.R., Kidd, K.K., Flodman, P., Spence, M.A., Schuck, S., Swanson, J.M., Zhang, Y.P. and Moyzis, R.K. (2002). Evidence of positive selection acting at the human dopamine receptor D4 gene locus. *Proc Natl Acad Sci USA*, **99**(1), 309–14.

Eswaran, V., Harpending, H. and Rogers, A.R. (2005). Genomics refutes an exclusively African origin of humans. *Journal of Human Evolution*, **49**(1), 1–18.

Evans, P.D., Gilbert, S.L., Mekel-Bobrov, N., Vallender, E.J., Anderson, J.R., Vaez-Azizi, L.M., Tishkoff, S.A., Hudson, R.R. and Lahn, B.T. (2005). Microcephalin, a gene regulating brain size, continues to evolve adaptively in humans. *Science*, **309**(5741), 1717–20.

Fix, A.G. (1999). *Migration and Colonization in Human Microevolution*. New York: Cambridge University Press.

Gray, J.R. and Thompson, P.M. (2004). Neurobiology of intelligence: science and ethics. *Nat Rev Neurosci*, **5**(6), 471–82.

Haak, W., Forster, P., Bramanti, B., Matsumura, S., Brandt, G., Tanzer, M., Villems, R., Renfrew, C., Gronenborn, D., Alt, K.W. and Burger, J. (2005). Ancient DNA from the First European Farmers in 7500-Year-Old Neolithic Sites. *Science*, **310**, 1016–18.

Hardy, J., Pittman, A., Myers, A., Gwinn-Hardy, K., Fung, H.C., de Silva, R., Hutton, M. and Duckworth, J. (2005). Evidence suggesting that Homo neanderthalensis contributed the H2 MAPT haplotype to Homo sapiens. *Biochem Soc Trans*, **33**(Pt 4), 582–85.

Harpending, H. and Cochran, G. (2002). In our genes. *Proc Natl Acad Sci USA*, **99**(1), 10–12.

Harpending, H. and Eswaran, V. (2005). Tracing modern human origins. *Science*, **309**(5743), 1995–7; author reply 1995–7.

Harpending, H.C., Sherry, S.T., Rogers, A.R. and Stoneking, M. (1993). The genetic structure of ancient human populations. *Curr. Anthrop*, **34**, 483–96.

Harpending, H. and Rogers, A. (2000). Genetic perspectives on human origins and differentiation. *Annu Rev Genomics Hum Genet*, **1**, 361–85.

Hawks, J. (2005). johnhawks.net::paleoanthropology, genetics, and evolution. from http://johnhawks.net/weblog

Lewontin, R.C. (1972). The apportionment of human diversity. In Dobzhansky, T.H., Hecht, M.K. and Steere, W.C. (eds.), *Evolutionary Biology: Volume 6*. New York: Appleton-Century-Crofts, pp. 381–98.

Macaulay, V., Hill, C., Achilli, A., Rengo, C., Clarke, D., Meehan, W. et al. (2005). Single, rapid coastal settlement of Asia revealed by analysis of complete mitochondrial genomes. *Science*, **308**(5724), 1034–36.

Mekel-Bobrov, N., Gilbert, S.L., Evans, P.D., Vallender, E.J., Anderson, J.R., Hudson, R.R., Tishkoff, S.A. and Lahn, B.T. (2005). Ongoing adaptive evolution of ASPM, a brain size determinant in Homo sapiens. *Science*, **309**, 1720–22.

Morton, N.E. (1973). Genetic Structure of Populations: Proceedings of a Conference sponsored by the University of Hawaii and dedicated to Sewall Wright. Honolulu: distributed by University Press of Hawaii.

Nielsen, R., Bustamante, C., Clark, A.G., Glanowski, S., Sackton, T.B., Hubisz, M.J. *et al.* (2005). A scan for positively selected genes in the genomes of humans and chimpanzees. *PloS Biol*, 3(6), e170.

Noblett, K.L. and Coccaro, E.F. (2005). Molecular genetics of personality. *Curr Psychiatry Rep*, **7**(1), 73–80.

Pearson, O.M., Stone, A.C. and Eswaran, V. (2003). On the diffusion-wave model for the spread of modern humans. *Current Anthropology*, **44**(4), 559–61.

Race, Ethnicity, and Genetics Working Group. (2005). The use of racial, ethnic, and ancestral categories in human genetics research. *American Journal of Human Genetics*, **77**, 519–32.

Rockman, M.V., Hahn, M.W., Soranzo, N., Zimprich, F., Goldstein, D.B. and Wray, G.A. (2005). Ancient and Recent Positive Selection Transformed Opioid cis-Regulation in Humans. *PLoS Biol*, 3(12), e387.

Rushton, J.P. and Jensen, A.R. (2005). Thirty years of research on race differences in cognitive ability. *Psychology Public Policy and Law*, **11**(2), 235–94.

Index